Advanced Science Research International Symposium 2009

(ASR 2009)

AA008648

Positron, Muon and Other Exotic Particle Beams for Materials and Atomic/Molecular Sciences
Journal of Physics: Conference Series Volume 225

Tokai, Japan
10-12 November 2009

ISBN: 978-1-61738-656-5
ISSN: 1742-6588

Printed from e-media with permission by:

Curran Associates, Inc.
57 Morehouse Lane
Red Hook, NY 12571

Some format issues inherent in the e-media version may also appear in this print version.

Printed by Curran Associates, Inc. (2010)

Institute of Physics
Dirac House, Temple Back
Bristol BS1 6BE UK

Phone: 44 1 17 929 7481
Fax: 44 1 17 920 0979

techtracking@iop.org

Additional copies of this publication are available from:

Curran Associates, Inc.
57 Morehouse Lane
Red Hook, NY 12571 USA
Phone: 845-758-0400
Fax: 845-758-2634
Email: curran@proceedings.com
Web: www.proceedings.com

TABLE OF CONTENTS

PREFACE

It is our great pleasure to deliver the proceedings of ASR2009, the Advanced Science Research International Symposium 2009. ASR2009 is part of a series of symposia which is hosted by the Japan Atomic Energy Agency, Advanced Science Research Center (JAEA-ASRC), and held every year with different scientific topics. ASR2009 was held at Tokai in Japan from 10–12 November 2009. In total, 102 participants, including 29 overseas scientists, made 44 oral presentations and 64 poster presentations.

In ASR2009 we have focused on material and atomic/molecular science research using positrons, muons and other exotic particle beams. The symposium covered all the fields of materials science which use such exotic particle beams. Positrons, muons and other beams have similar and different features. For example, although positrons and muons are both leptons having charge and spin, they give quite different information about materials. A muon mainly detects the local magnetic state of the solid, while a positron detects crystal imperfections and electron momenta in solids. Other exotic particle beams also provide useful information about materials which is not able to be obtained with muons or positrons. Therefore, the complementary use of particle beams, coupled with an understanding of their relative advantages, leads to greater excellence in materials research. This symposium crossed the fields of muon science, positron science, unstable-nuclei science, and other exotic particle-beam science. We therefore believe that ASR2009 became an especially important meeting for finding new science with exotic particle beams.

Finally, we would like to extend our appreciation to all the participants, committee members, and support staff for their great efforts to make ASR2009 a fruitful symposium.

ASR2009 Chairs
Wataru Higemoto and Atsuo Kawasuso
Advanced Science Research Center, Japan Atomic Energy Agency

Y Shirai, Kyoto Univ.
R Suzuki, AIST
A Uedono, Univ. Tsukuba

Local organizing committee (JAEA)
M Maekawa
Y Fukaya
T U Ito
A Yabuuchi
K Ninomiya
T Hirade
W Higemoto
A Kawasuso
S Sakurai

Secretariat (JAEA)
H Sekino

Cooperation
The Physical Society of Japan
Positron Science Society Society of Muon and Meson Science of Japan
International Society for μSR Spectroscopy

All papers published in this volume of *Journal of Physics: Conference Series* have been peer reviewed through processes administered by the proceedings Editors. Reviews were conducted by expert referees to the professional and scientific standards expected of a proceedings journal published by IOP Publishing.

Microscopic indicator for thermodynamic stability of hydrogen storage materials provided by muon-spin spectroscopy

Jun Sugiyama[1], Yutaka Ikedo[2], Tatsuo Noritake[1], Kazutoshi Miwa[1], Shinichi Towata[1], Tatsuo Goko[3], Oren Ofer[3], Martin Månsson[4], Eduardo J. Ansaldo[3], Jess H. Brewer[3,5], and Kim H. Chow[6]

[1] Toyota Central Research & Development Laboratories Inc., Nagakute, Aichi 480-1192, Japan
[2] Muon Science Laboratory, KEK, Tsukuba, Ibaraki 305-0801, Japan
[3] TRIUMF, 4004 Wesbrook Mall, Vancouver, BC, V6T 2A3 Canada
[4] LNS, ETH Zürich and Paul Scherrer Institut, CH-5232 Villigen PSI, Switzerland
[5] CIfAR and Department of Physics and Astronomy, University of British Columbia, Vancouver, BC, V6T 1Z1 Canada
[6] Department of Physics, University of Alberta, Edmonton, AB, T6G 2G7 Canada

E-mail: e0589@mosk.tytlabs.co.jp

Abstract. In search of a high-capacity hydrogen storage system, we have investigated the thermodynamic properties of borohydrides [$M(BH_4)_n$]. Using positive muon-spin rotation and relaxation (μ^+SR), we have acquired experimental data for a wide range of different powder samples below ambient temperature. Zero-field μ^+SR measurements indicate the formation of the H-μ^+-H system in $LiBH_4$, $NaBH_4$, KBH_4 and $Ca(BH_4)_2$, but not in $Mg(BH_4)_2$. It is also found that the amplitude of the HμH signal ($A_{H\mu H}$) varies with the electronegativity (χ_P) of M^{n+}. This is because, when χ_P of M^{n+} is small, $[BH_4]^-$ should be more negative, resulting in an increase in the electron-density of H^- ions. Therefore, $A_{H\mu H}$ increases with decreasing χ_P. Since the thermodynamic stability of $M(BH_4)_n$ also depends on χ_P, $A_{H\mu H}$ is thought to be a microscopic indicator for the stability of borohydrides.

1. Introduction

Vehicular applications of fuel cells face several huge barriers, among which the key component is still a safe, light, high-capacity hydrogen storage system, despite a long history of research into hydrogen storage materials [1]. All the alloys that are currently used, including the Ti-based compounds, have a disadvantage for practical applications due to their small gravimetric H_2 density, which makes them too heavy to carry on automobiles. It is well recognized that complex borohydrides are possible candidates for future onboard hydrogen storage materials [2, 3]; however, their slow dehydrogenation kinetics and high operating temperature (for releasing H)— i.e., hydrogen desorption temperature (T_d) need to be drastically improved if they are to enable a practical hydrogen storage system. Furthermore, although the dehydrogenation and rehydrogenation are known to involve multi-process reactions, their mechanisms are still not fully understood [4, 5]. For instance, $LiBH_4$, which exhibits the highest gravimetric H_2 density (18.5 mass%), liberates three of four hydrogen atoms and decomposes into LiH and boron through four intermediate states [6]. However, only one intermediate state $Li_2(B_{12}H_{12})$ has been characterized [4, 7] so far.

The nature of the "hydrogen bond" between positive muons (μ^+) and anions has been extensively investigated by a positive muon-spin rotation and relaxation (μ^+SR) technique, since the discovery of the $F^-\mu^+F^-$ (or "FμF") ion in metal fluoride crystals such as LiF, NaF, CaF_2 and BaF_2 [8]. The FμF system is easily identified *via* its characteristic muon spin oscillation signal in a zero magnetic field (ZF-) μ^+SR spectrum, due to a dipole-dipole coupling in the collinear ^{19}F-μ^+-^{19}F spin system. The FμF system has also been found in organic polymers, *e.g.* Teflon [9] and Nafion [10], despite a lack of crystal symmetry in these polymers. The formation of the FμF system is thus a rather common phenomenon in compounds containing F^- ions [11, 12]. Similar hydrogen bond systems with other anions *e.g.* "Hμ^+H" and/or "Hμ^+" have been reported in a Kagomé compound $ZnCu_3(OH)_6Cl_3$ [13], in a layered cobalt dioxide $[Ca_{0.85}OH]_{1.16}[CoO_2]$ [14] and in $NaAlH_4$ [15], the last of which was studied as a potential hydrogen storage material for fuel cells.

In $NaAlH_4$, the volume fraction of HμH ($V_{H\mu H}$) was proposed to be an indicator for the dehydrogenation kinetics *via* comparison of $V_{H\mu H}$ between pure and Ti-doped $NaAlH_4$ [15, 16]. However, although the discovery of HμH demonstrates that μ^+SR can contribute significantly to research in hydrogen storage materials, $NaAlH_4$ is unlikely to be a good solution for future onboard systems, due to its relatively small gravimetric H_2 density (compared to $LiBH_4$) and its insufficient rehydrogenation capacity. We have, therefore, performed a systematic μ^+SR investigation of all the five $M(BH_4)_n$ systems, for which pure powder samples are available. We have found that the HμH system is a unique probe of both structural and dynamic/kinematic properties of H in these compounds.

2. Experimental

Powder samples of $LiBH_4$, $NaBH_4$, KBH_4, and $Ca(BH_4)_2$ were purchased from Aldrich Co. LTD, and $Mg(BH_4)_2$ from FZK (Forschungszentrum Karlsruhe). Since the samples are unstable in air, they were packed in an In- or Au-sealed aluminum powder cell in an Ar-filled glove box. The μ^+SR experiments were performed on the **M20** surface muon beam line at TRIUMF, using an experimental setup and techniques described elsewhere [17].

3. Results and Discussion

Figure 1(a) shows the temperature (T) variation of the ZF-μ^+SR time spectrum for $LiBH_4$ in the T range between 5 and 250 K. One can clearly see a damped oscillation, particularly below 150 K. Since there are no magnetic moments in $LiBH_4$, such oscillation indicates the formation of the HμH system in $LiBH_4$. In fact, the ZF- spectrum is well fitted by a combination of the HμH signal, a static Gaussian Kubo-Toyabe (G_{KT}) signal (from other muon sites experiencing disordered magnetic fields due to nuclear moments), and a background (BG) signal from the powder cell (which is made of high-purity aluminum);

$$A_0\,P_{ZF}(t) = A_{H\mu H}P_{H\mu H}(t)\exp(-\lambda_{H\mu H}t) + A_{KT}G_{KT} + A_{BG}\,. \qquad (1)$$

Here A_0 is the initial ($t = 0$) asymmetry, $P_{ZF}(t)$ is the muon spin polarization function in ZF, $A_{H\mu H}$, A_{KT} and A_{BG} are the asymmetries of the three signals, $\lambda_{H\mu H}$ is the exponential relaxation rate of the precessing signal, and σ_G is the Gaussian relaxation rate. Here $P_{H\mu H}(t)$ is the muon spin relaxation function caused by a colinear three spin 1/2 system, given in [8]:

$$P_{H\mu H}(t) = \frac{1}{2} + \frac{1}{6}\cos(\sqrt{3}\omega_d t) + \frac{1+1/\sqrt{3}}{6}\cos(\frac{3+\sqrt{3}}{2}\omega_d t) + \frac{1-1/\sqrt{3}}{6}\cos(\frac{3-\sqrt{3}}{2}\omega_d t)\,, \qquad (2)$$

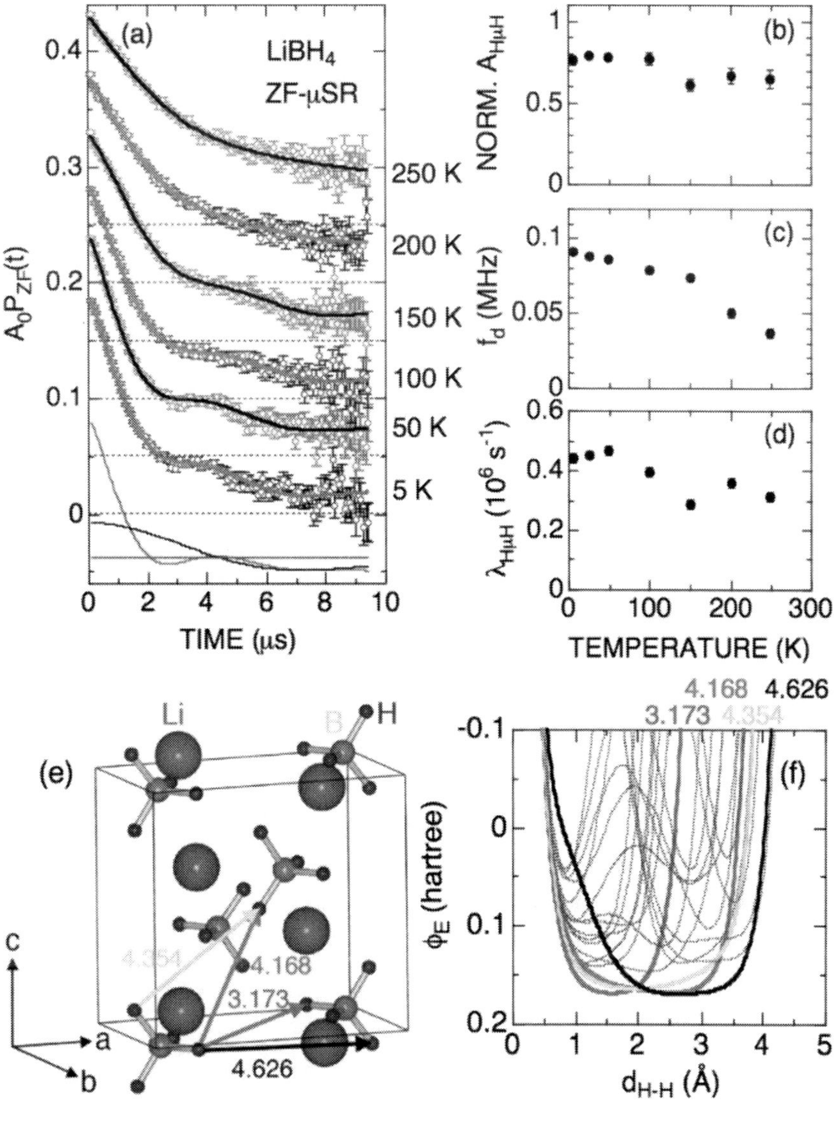

Figure 1. Temperature dependence of (a) the ZF-μ^+SR time spectrum for $LiBH_4$; (b) the normalized $A_{H\mu H}$; (c) f_d; and (d) $\lambda_{H\mu H}$, crystal structure of (e) orthorhombic $LiBH_4$ below ~ 381 K, and (f) the calculated electrostatic potential (ϕ_E) along the H-H bond in the distance range between 2.45 and 4.69 Å, in which 21 combinations are included. Bold solid lines in (a) represent the fitting result using Eq. (1) and the bottom three solid lines show the three components in Eq. (1). Each spectrum is offset by 0.05 for clarity of display. The μ^+SR parameters in (b)-(d) were obtained by fitting the ZF-μ^+SR time spectra using Eq. (1). Error bars in (c) are comparable to or below the size of solid circles. Arrows and numbers in (e) represent the distance to the neighboring hydrogens, at the center of which μ^+ locates, based on the prediction from the ϕ_E calculations. In (f), the lowest 4 ϕ_Es are represented by a bold line, while the rest 17 by a narrow (blue) dotted line.

where ω_d is the muon precession frequency due to the nuclear dipole field of the proton in the H-μ^+ bond. From ω_d, it is possible to calculate the distance r between the μ^+ and the nucleus according to [8]:

$$\omega_{\rm d} \equiv 2\pi \cdot f_{\rm d} = \mu_0 \hbar \frac{\gamma_\mu \gamma_N}{4\pi r^3}, \tag{3}$$

Here γ_μ is the gyromagnetic ratio of μ^+ and γ_N is the nuclear gyromagnetic ratio of ^{1}H.

Figures 1(b)-1(d) show the T dependences of the μ^+SR parameters for $LiBH_4$. The normalized $A_{\rm H\mu H}$ [$N_{A_{\rm H\mu H}} \equiv A_{\rm H\mu H}/(A_{\rm H\mu H} + A_{\rm KT})$] reaches $\sim$ 0.8 at 5 K and decreases monotonically with increasing T. However, even at 250 K, $N_{A_{\rm H\mu H}} \sim 0.6$. As T increases from 5 K, $f_{\rm d}$ also decreases monotonically up to 280 K, which means that the increase in r with T is probably connected to thermal fluctuations/vibrations of the BH_4 anions. Using Eq (3), $2r$ is estimated to be 3.16-4.42 Å. According to electrostatic potential ($\phi_{\rm E}$) calculations using a first-principles DFT+GGA method [18], there are four stable μ^+ sites for the HμH system; namely, the center between the neighboring hydrogens with the distance $d_{\rm H-H}$ = 3.173, 4.168, 4.354, and 4.626 Å [Figs. 1(e) and 1(f)]. These values are in good agreement with $2r$ estimated from $f_{\rm d}$, indicating the change in occupancy of μ^+ at the center of the four $d_{\rm H-H}$ with T. Particularly, since $2r$ = 3.16 Å at 5 K, μ^+ locates at the center of the shortest $d_{\rm H-H}$ among four stable sites in the ground state. The monotonic decrease in $\lambda_{\rm H\mu H}$ with T implies an increase in homogeneity of the local (nuclear magnetic) environment caused by thermal fluctuation/vibration. That is, the local variation of "distance" for the H-μ bond in the HμH system may be averaged out by local motion, while the atomic membership of HμH itself is maintained. This is consistent with the T dependences of $N_{A_{\rm H\mu H}}$ and $f_{\rm d}$. It should be noted that μ^+ starts to diffuse at high T, at which its thermal energy is comparable to the barrier energy between the potential minima. In order to clarify such diffusive behavior, we need to measure ZF-spectra above 250 K together with longitudinal field spectra to distinguish the μ^+ motion from the HμH signal.

In order to learn the nature of the Hμ^+H system in other borohydrides, Fig. 2 shows the ZF-spectrum for $LiBH_4$, $NaBH_4$, $MgBH_4$, KBH_4 and $CaBH_4$ at 5 K. Except for $MgBH_4$, all the ZF-spectra exhibit oscillations, indicating formation of an HμH unit. The results of fits to the ZF-spectra obtained at 5 K are summarized in Table 1. In contrast to $LiBH_4$, there are mainly two muon sites for the HμH system in the tetrahedral $NaBH_4$ lattice. The estimated r suggests that the μ^+ is located at the center of the second nearest neighboring hydrogens [3.427 Å away, Fig. 3(d)].

Table 1. HμH parameters obtained by fitting the ZF-spectra at 5 K using Eqs. (2)-(3).

M for $M(BH_4)_2$	$f_{\rm d}$ (MHz)	$2r$ (Å)	$N_{A_{\rm H\mu H}}$
Li	0.0974(2)	3.16(2)	0.82(4)
Na	0.0854(1)	3.30(1)	0.99(1)
Mg	0	∞	-
K	0.106(3)	3.07(3)	0.7399(1)
Ca	0.119(2)	2.96(2)	0.34(8)

Furthermore, we found a dynamic change in the HμH signal accompanying a structural phase transition of $NaBH_4$ at T_c =189.9 K from a low-T tetragonal phase to a high-T cubic phase due to an order-disorder transition of the BH_4 tetrahedra [Figs. 3(d) and 3(e)] [19–21]. The HμH system is stable in the ordered state, but would be unstable in the disordered state due to the

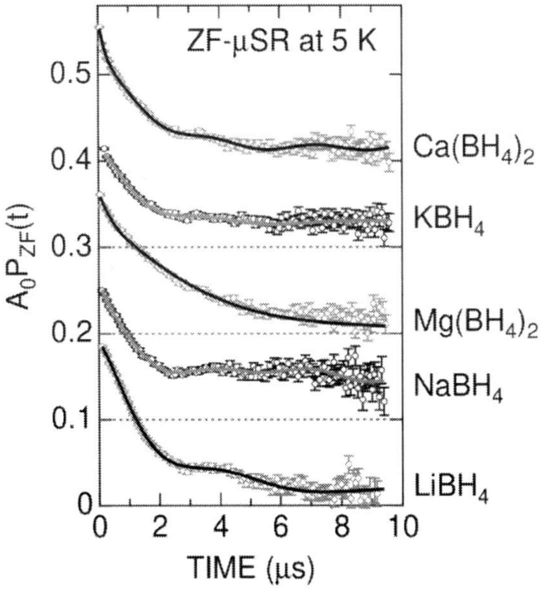

Figure 2. ZF-μ^+SR time spectra for $LiBH_4$, $NaBH_4$, $Mg(BH_4)_2$, KBH_4 and $Ca(BH_4)_2$ at 5 K. Solid lines represent the fitting result using Eq. (1). Each spectrum is offset by 0.1 for clarity of display.

rotation of the $[BH_4]^-$ group [Fig. 3]. We, therefore, fitted the ZF-spectrum using Eq. (1) plus an additional Hμ signal $[A_{H\mu}P_{H\mu}(t)\exp(-\lambda_{H\mu}t)]$, where $P_{H\mu}(t)$ is given by [9].

$$P_{H\mu}(t) = \frac{1}{6} + \frac{1}{6}\cos(\omega_d t) + \frac{1}{3}\cos(\frac{1}{2}\omega_d t) + \frac{1}{3}\cos(\frac{3}{2}\omega_d t) . \tag{4}$$

This means that μ^+ forms a "$H^+[BH_4]^-$"-like μBH_4 molecule at high T instead. In addition, despite the fact that the transition is thought to be second-order, the present μ^+SR result suggests the presence of the $\mu[BH_4]$ state even below T_c, probably due to microscopic vibration/rotation of the BH_4 tetrahedra.

Although structural considerations are likely to explain the magnitude of $N_{A_{H\mu H}}$ for $LiBH_4$, $NaBH_4$, and KBH_4, they offer no acceptable reasons for the absence of the HμH signal in $Mg(BH_4)_2$ and the rather small $N_{A_{H\mu H}}$ in $Ca(BH_4)_2$. We, therefore, wish to discuss the stability of $N_{A_{H\mu H}}$ from another viewpoint. According to theoretical and experimental work on the thermodynamic stability of $M(BH_4)_n$ (M = Li, Na, K, Ca, Mg, Zn and Sc), the stability of $M(BH_4)_n$ is likely to show a good correlation with the electronegativity (χ_P) of cations M^{n+} [Fig. 4(a)] [22]; that is, larger χ_P, implies less thermodynamic stability. This indicates that, although the bonding between M^{n+} and $n[BH_4]^-$ is basically ionic, the charge transfer from $[BH_4]^-$ to M^{n+} is a key feature for the stability of $M(BH_4)_n$. This situation is thus expected to be drastically changed if the liberated H ions are coupled with the neighboring $[BH_4]^-$ anions. In other words, the stability of the Hμ^+H system in $M(BH_4)_n$ should also be a good indicator for their performance of dehydrogenation. In fact, one can clearly see a good correlation between T_d and $N_{A_{H\mu H}}$ [Fig. 4(b)], where T_d was determined by thermal desorption spectroscopy measurements using gas chromatography [22].

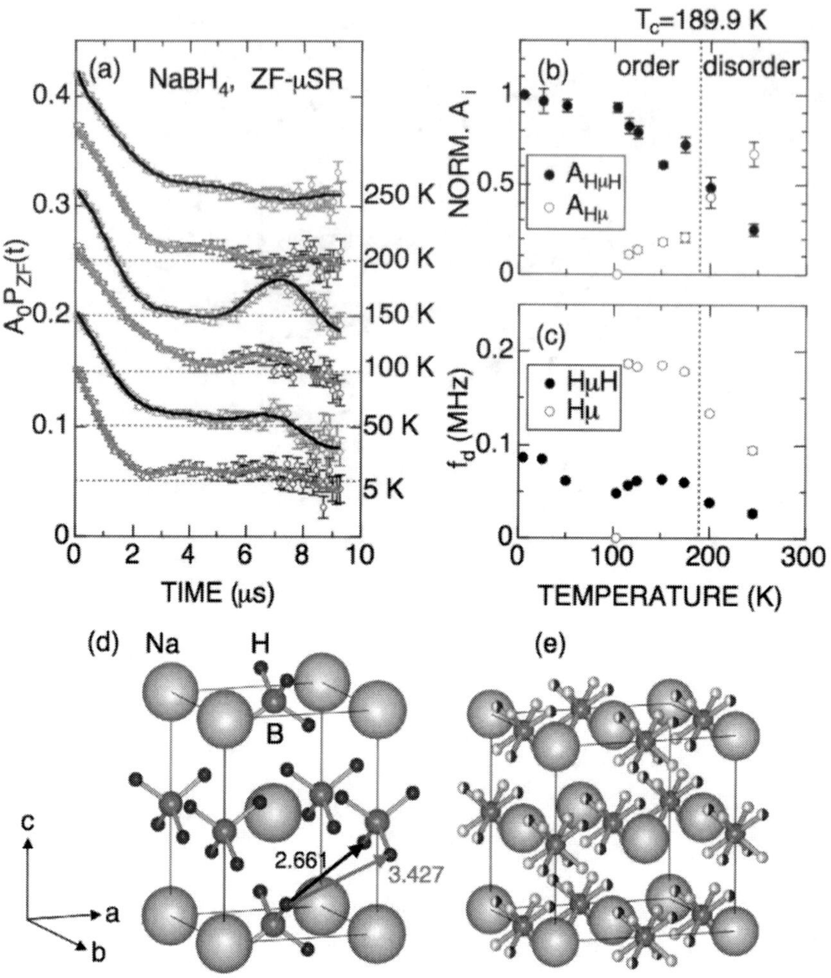

Figure 3. Temperature dependences of (a) the ZF-μ^+SR time spectrum for $NaBH_4$; (b) $N_{A_{\mathrm{H}\mu\mathrm{H}}}$ and $N_{A_{\mathrm{H}\mu}}$; and (c) f_d, and the crystal structure (d) below and (e) above $T_c = 189.9$ K. In (a), the solid lines represent the fitting result using Eq. (1) plus an additional Hμ signal [9]. Each spectrum in (a) is offset by 0.05 for clarity of display.

As already proposed by Kadono and coworkers [15], proton-like hydrogens would be generated during the desorption reaction of $M(BH_4)_2$. In order to accelerate this reaction, the generated hydrogens should be removed from the materials through a solid state diffusion process. The presence of a stable HμH (and/or Hμ) system indicates the formation of a $[BH_4]^-$-H^+-$[BH_4]^-$ (and/or H^+-$[BH_4]^-$) bond in $M(BH_4)_2$ during the desorption reaction. Since such a bond clearly reduces the reaction rate, it is expected that additional thermal energy is necessary to decompose the $M(BH_4)_2$ with high $N_{A_{\mathrm{H}\mu\mathrm{H}}}$ compared with that for the materials with low $N_{A_{\mathrm{H}\mu\mathrm{H}}}$. This is a reasonable explanation for the linear relationship in Fig. 4(b). This also demonstrates that μ^+SR is a unique tool for investigating the nature of $LiBH_4$ and related hydrogen storage materials *via* the HμH signal. Finally, we wish to emphasize that, although the relationship between T_d and χ_P was found empirically for $M(BH_4)_n$ [Fig. 4(a)], the underlying mechanism has been successfully clarified by the present μ^+SR experiment. In fact, there was no experimental evidence regarding which parameters were predominant for determining T_d. μ^+SR is, therefore, a unique tool to

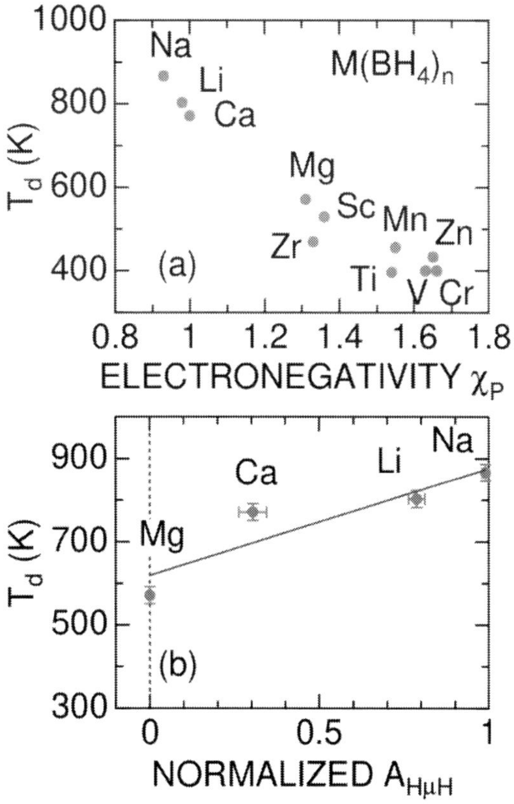

Figure 4. The relationship between (a) hydrogen desorption temperature (T_d) and electronegativity (χ_P) [22]; and (b) T_d and $N_{A_{H\mu H}}$. Note that T_d for KBH_4 was not reported in Ref. [22].

investigate the thermal stability of $M(BH_4)_n$ and related compounds.

4. Acknowledgments

This work was performed at TRIUMF, Vancouver, Canada and we are thankful to the staff of TRIUMF for help with the μ^+SR experiments. We also appreciate Prof. S. Orimo and Dr. H-W. Li of Tohoku Univ. for valuable discussions. JHB is supported at UBC by CIfAR, by NSERC of Canada and (through TRIUMF) by NRC of Canada, KHC by NSERC of Canada and (through TRIUMF) by NRC of Canada. This work is partially supported by the NEDO "Development for Hydrogen Production, Transportation and Storage System" Project. All images involving crystal structure were made with a computer program VESTA.

References

[1] Schlapbach L and Zütteltel A 2001 *Nature* **414** 353
[2] Orimo S, Nakamori Y, Eliseo J R, Züttel A and Jensen C M 2007 *Chem. Rev.* **107** 4111
[3] Züttel A, Borgschulte A and Orimo S 2007 *Scrip. Mater.* **56** 823
[4] Orimo S, Nakamori Y, Ohba N, Miwa K, Aoki M, Towata S and Züttel A 2006 *Appl. Phys. Lett.* **89** 021920
[5] Hwang S J, Bowman R C, Reiter J J W, Rijssenbeek J, Soloveichik G L, Zhao J C, Kabbour H and Ahn C C 2008 *J. Phys. Chem. Lett.* C **112** 3164

Journal of Physics: Conference Series **225** (2010) 012051 doi:10.1088/1742-6596/225/1/012051

[6] Züttel A, Wenger P, Rentsch S, Sudan P, Mauron P and Emmenegger C 2003 *J. Power Sources* **118** 1
[7] Ohba N, Miwa K, Aoki M, Noritake T, Towata S, Nakamori Y, Orimo S and Züttel A 2006 *Phys. Rev.* B **74** 075110
[8] Brewer J H, Kreitzman S R, Noakes D R, Ansaldo E J, Harshman D R and Keitel R 1986 *Phys. Rev.* B **33** 7813(R)
[9] Nishiyama K, Nishiyama S W and Higemoto W 2003 *Physica* B **326** 41
[10] Sugiyama J 2004-2006 *RIKEN-RAL Muon Facility Report 2004-2006* **6** 62
[11] Lancaster T, Blundell S J, Baker P J, Hayes W, Giblin S R, McLain S E, Pratt F L, Salman Z, Jacobs E A, Turner J F C and Barnes T 2007 *Phys. Rev.* B **75** 220408(R)
[12] MacFarlane W A, Schick-Martin D, Egilmez M, Fan I, Song Q, Chow K H, Cordier S, Perrin C and Goren S D 2009 *Physica* B **404** 622
[13] Mendels P, Bert F, de Vries M A, Olariu A, Harrison A, Duc F, Trombe J C, Lord J S, Amato A and Baines C 2007 *Phys. Rev. Lett.* **98** 077204
[14] Ikedo Y, Sugiyama J, Nozaki H, Russo P L, Andreica D, Amato A, Månsson M, Shizuya M, Isobe M and Takayama-Muromachi E 2009 *Physica* B **404** 607
[15] Kadono R, Shimomura K, Satoh K H, Takeshita S, Koda A, Nishiyama K, Akiba E, Ayabe R M, Kuba M and Jensen C M 2008 *Phys. Rev. Lett.* **100** 026401
[16] Bogdanovic B and Schwickardi M 1997 *J. Alloys Comp.* **253** 1
[17] Kalvius G M, Noakes D R and Hartmann O 2001 *Handbook on the Physics and Chemistry of Rare Earths* vol 32 (Amsterdam: Elsevier Science) pp 55–451
[18] Miwa K, Ohba N, Towata S, Nakamori Y and Orimo S 2004 *Phys. Rev.* B **69** 245120
[19] Johnston H L and Hallett N C 1953 *J. Am. Chem. Soc.* **75** 1467
[20] Fischer P and Züttel A 2001 *PSI Annual Report*
[21] Allis D G and Hudson B S 2004 *Chem. Phys. Lett.* **385** 166
[22] Nakamori Y, Miwa K, Ninomiya A, Li H, Ohba N, Towata S, Züttel A and Orimo S 2006 *Phys. Rev.* B **74** 045126

Muons and Frustrated Magnetism in $NiGa_2S_4$ and $Pr_2Ir_2O_7$

D E MacLaughlin[1], Y Nambu[2,4], Y Ohta[2], Y Machida[2], S Nakatsuji[2] and O O Bernal[3]

[1]Dept. of Physics & Astronomy, Univ. of California, Riverside, California 92521, USA
[2]Institute for Solid State Physics, Univ. of Tokyo, Kashiwa 277-8581, Chiba, Japan
[3]Dept. of Physics & Astronomy, California State Univ., Los Angeles, California 90032-4221, USA

E-mail: macl@physics.ucr.edu

Abstract. Geometrically frustrated magnets are of interest because of the novel phenomena that arise from their exotic ground states and low-lying excitations. Muon spin rotation and relaxation (μSR) is a sensitive probe of magnetism on the local (atomic) distance scale, and is an attractive tool for the study of frustrated magnets. The muon carries a unit electric charge, however, which can have an appreciable effect on local properties. We discuss two cases where such an effect might be involved. In the 2D triangular antiferromagnet $NiGa_2S_4$ the 'magnetic' specific heat is field-independent up to 7 T, suggesting nonmagnetic excitations, but μSR experiments reveal Ni spin freezing below ~9 K and strong magnetic fluctuations down to 25 mK. Comparison with Ga nuclear quadrupole resonance data suggests, however, that the muon charge does not cause this discrepancy. In the pyrochlore iridate $Pr_2Ir_2O_7$ muon spin relaxation due to a distribution of quasistatic fields is observed over a wide temperature range. The data strongly suggest hyperfine-enhanced ^{141}Pr nuclear magnetism, which requires a nonmagnetic Pr^{3+} ground state. This may be due to lifting of the Pr^{3+} non-Kramers degeneracy by the muon electric field or, at least in part, a property of a spin-liquid-like many-body ground state.

1. Introduction

Geometrical frustration in solids refers to situations where the geometry of a crystal lattice prohibits simultaneous satisfaction of all interactions between atoms [1]. In general frustration can lead to macroscopic degeneracy of the (classical or quantum) ground state and a very large density of low-lying excited states, due to the combinatorics involved in distributing the frustrated interactions. This in turn leads to novel phenomena not found in unfrustrated systems such as suppression of the ordering temperature, sometimes to $T = 0$. Magnets with geometrically frustrated interactions provide a testing ground for understanding the interplay between frustration, magnetic order, and a wide variety of exotic states.

Magnetic resonance techniques use *spin probes* (nuclei, electrons, muons) to determine magnetic properties on the atomic scale [2]. Electronic and/or nuclear spins in the environment of a spin probe produce a local field $\boldsymbol{B}^{\mathrm{loc}}$ at the spin-probe site. This local field fluctuates randomly

[4] Present address: Dept. of Physics & Astronomy, Johns Hopkins Univ., Baltimore, Maryland 21218, USA.

in time due to thermal excitations, and may also have a quasistatic component $\langle \boldsymbol{B}^{\rm loc} \rangle$.[5] The local nature of the spin probe renders magnetic resonance studies complementary to experiments that perform bulk averages, such as measurements of thermodynamic and transport quantities, and to diffraction and scattering techniques (neutron, x-ray, ...), which are reciprocal-space probes.

In time-differential muon spin rotation (td-μSR) [3, 4] the spin probes are spin-polarized muons implanted in the sample. Usually positive muons (μ^+) are used, which can be considered short-lived impurities after they come to rest at interstitial sites in the sample. The μ^+ spins precess in the resultant of $\boldsymbol{B}^{\rm loc}(t)$ and any applied field, and eventually decay: $\mu^+ \rightarrow e^+ + \nu_e + \overline{\nu}_\mu$ (mean lifetime 2.2 μs). The direction of the decay positron emission is correlated with the μ^+ spin orientation at the time of decay. Thus the μ^+ spin-polarization component in a given direction is proportional to the difference between parallel and antiparallel count rates for that direction. The time evolution of this *asymmetry* is the primary measurement in td-μSR.

In transverse-field μSR (TF-μSR) muon spin precession in a magnetic field perpendicular to the μ^+ polarization results in oscillations of the asymmetry at the μ^+ precession frequency. In zero applied field (ZF-μSR) or longitudinal field (LF-μSR) the only precession arises from components $\boldsymbol{B}_\perp^{\rm loc}$ of the local field perpendicular to the μ^+ spin direction. Thermal spin excitations give rise to fluctuations of $\boldsymbol{B}_\perp^{\rm loc}$, which dephase or 'relax' the μ^+ spin polarization.

μSR experiments are useful for studies of geometrically frustrated magnets (see *e.g.* [5]) for several reasons. Muon spins are local probes in real space; unlike neutron diffraction, they do not depend on spatial correlation or long-range order to detect quasistatic magnetism or spin fluctuations. They are therefore ideally suited to studies of disordered spin freezing, which is often found in geometrically frustrated magnets. ZF-μSR is an extremely sensitive probe of weak quasistatic magnetism, being able to detect quasistatic moments as weak as 10^{-3} $\mu_{\rm B}$. The dynamic relaxation is strong and easy to measure.

The electric field from the μ^+ charge can, however, perturb the phenomenon under study. In metals such effects are minimized by screening of the charge; superconducting and magnetic transition temperatures observed in μSR are typically the same as the bulk values, indicating that the muon charge does not have a major effect on the ordered state. μ^+-induced effects have been observed [6, 7], however, and should as a rule be searched for carefully in all μSR data.

In the present paper we discuss μSR experiments in two prototypical geometrically frustrated antiferromagnets, the two-dimensional triangular insulator $NiGa_2S_4$ [8, 9] and the 3D pyrochlore Kondo lattice $Pr_2Ir_2O_7$ [10, 11], where there is reason to suspect μ^+-induced effects. We shall argue that in $NiGa_2S_4$ such effects probably do not affect the results qualitatively, whereas in $Pr_2Ir_2O_7$ the μ^+ charge is more important. Even so, μSR experiments seem to yield reliable information for both materials.

2. $NiGa_2S_4$

2.1. Structure, specific heat

In the quasi-2D antiferromagnet $NiGa_2S_4$ the Ni^{2+} ions ($S = 1$) form a geometrically frustrated two-dimensional triangular sublattice [8]. Neutron scattering experiments reveal the short-range spin-correlated state at low temperatures shown in figure 1. The correlation length is short (~25 Å), and there is no long-range order. The 'magnetic' specific heat $C_{\rm M}$ (*i.e.* after correction for the phonon contribution) is shown in figure 2. It is unusual in a number of respects. There are two maxima of $C_{\rm M}(T)$, one centered around the Curie-Weiss temperature $|\theta_{\rm W}| \approx 80$ K and the other at $T_{\rm peak} \approx 10$ K, where the magnetic susceptibility also exhibits a weak maximum. This

[5] We note that $\langle \boldsymbol{B}^{\rm loc} \rangle$ is sampled only for times shorter than the lifetime of the probe spin state, and thus appears stationary if its fluctuations are sufficiently slow, or 'quasistatic'. Quasistatic and strictly static fields cannot be distinguished by magnetic resonance measurements, which merely put a lower limit (*viz.* the probe-spin lifetime) on the correlation time for fluctuations of $\langle \boldsymbol{B}^{\rm loc} \rangle$.

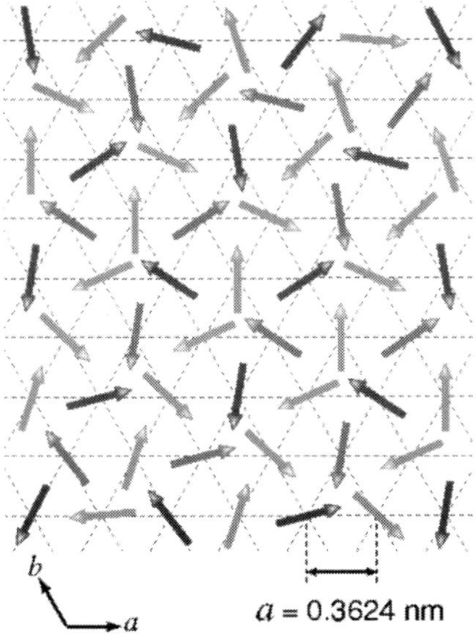

Figure 1. low-temperature Ni-spin correlations in $NiGa_2S_4$. From [9].

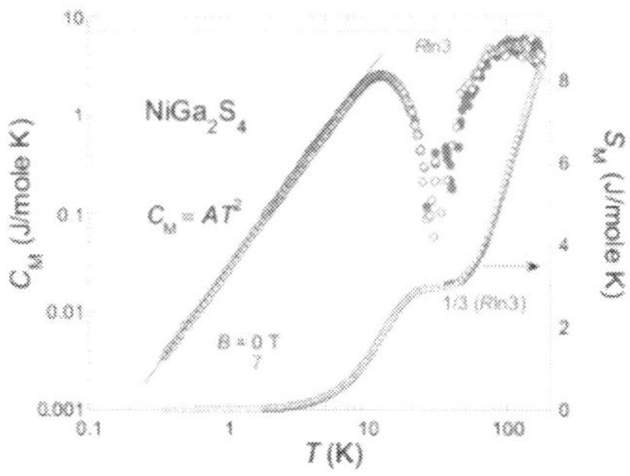

Figure 2. Dependence of magnetic specific heat C_M and entropy S_M on temperature and magnetic field in $NiGa_2S_4$. From [9].

seems to indicate freezing of single-spin excitations below $\sim|\theta_W|$, whereas more complicated highly degenerate low-energy states due to magnetic frustration in the triangular lattice are involved below 10 K. At low temperatures $C_M \propto T^2$, and at all temperatures C_M is independent of applied field up to 7 T to within the experimental resolution of a few percent. There was no evidence for bulk spin freezing in the early data, and it was suggested that at low-temperatures $NiGa_2S_4$ is a 'viscous' (*i.e.* slowly-fluctuating) spin liquid.

2.2. μSR; comparison with NQR

In contrast, μSR experiments [12, 13] reveal the onset of a quasistatic μ^+ field $\langle \boldsymbol{B}_{\rm loc} \rangle$ in $NiGa_2S_4$ below a freezing temperature $T_{\rm f} \approx 9$ K (figure 3). The oscillations are strongly damped, indicative of a broad distribution of $\langle \boldsymbol{B}_{\rm loc} \rangle$ consistent with a short correlation length; the average field is $\langle B_{\rm loc} \rangle_{\rm av} \approx 2.5$ kG. The oscillation frequency ω_μ, which in conventional magnets is proportional to the order parameter in the magnetically ordered phase, is well described by the mean-field Brillouin function for $S = 1$ [figure 3(a)].

The dynamic relaxation in $NiGa_2S_4$ is also widely distributed, as evidenced by the sub-exponential polarization decay (upward curvature in a semilog plot of decay curve versus time; data not shown) observed in both μSR and ^{69}Ga nuclear quadrupole resonance (NQR) [14] experiments. A relaxation function of the so-called 'stretched exponential' form $\exp[-(\Lambda t)^\beta]$ can be fit to the data; here Λ is a typical rate associated with the distribution and $\beta < 1$ qualitatively parametrizes the distribution width. The temperature dependence of Λ in $NiGa_2S_4$ is given in figure 3(b). Data were taken in a dilution refrigerator for temperatures below ~3 K, and in a gas-flow cryostat above ~2 K. The low-temperature data are less accurate because in the dilution refrigerator a significant fraction of muons stop in the silver cold finger.

In figure 3(b) it can be seen that Λ increases strongly as $T_{\rm f}$ is approached from above, as expected from critical slowing down at a phase transition. Below $T_{\rm f}$ Λ decreases to a minimum at ~2 K, and remains $\approx 5\ \mu{\rm s}^{-1}$ down to the lowest temperatures of measurement. A lower bound for the Ni-spin correlation time at 25 mK is $\Lambda^{-1} \approx 2 \times 10^{-7}$ s $\gg \hbar/k_{\rm B}T_{\rm f} \approx 10^{-12}$ s. Thus in $NiGa_2S_4$ the Ni spins are either completely frozen or else form a very viscous spin liquid.

The behaviour of Λ Below ~1 K is in contrast to the zero-field ^{69}Ga spin-lattice relaxation rate $1/T_1$ measured using NQR [14], shown in figure 4.[6] NQR and μSR are both local spin-

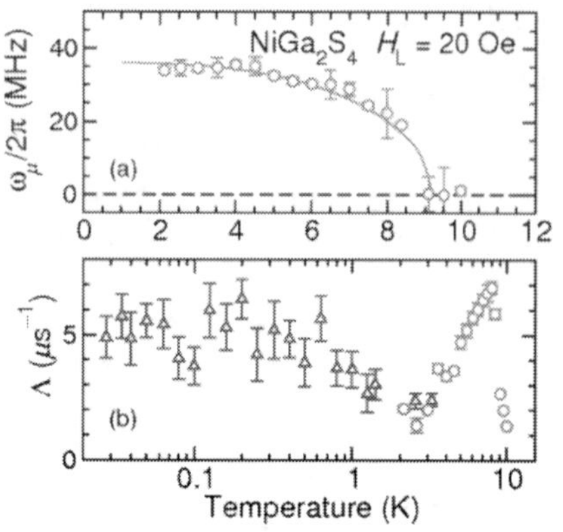

Figure 3. (a) Spontaneous μ^+ precession frequency ω_μ and (b) dynamic μ^+ spin relaxation rate Λ in $NiGa_2S_4$. Circles: data above ~2 K from gas-flow cryostat (from [13]). Triangles: data below ~3 K from dilution refrigerator.

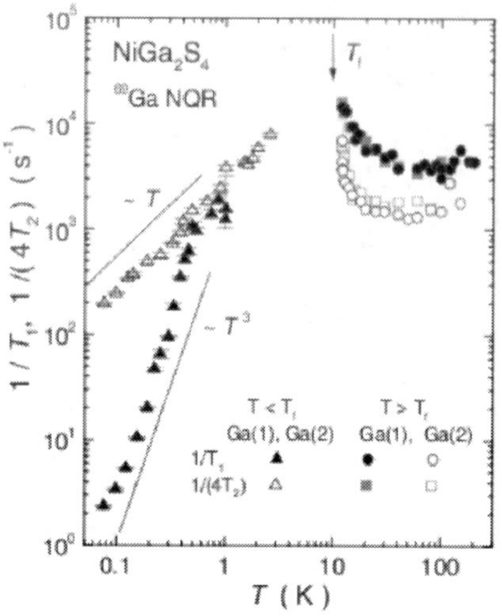

Figure 4. ^{69}Ga NQR relaxation rates $1/T_1$ and $1/T_2$ in $NiGa_2S_4$. From [14].

probe techniques, and in general would be expected to yield similar results. But below ~1 K $1/T_1$(NQR) follows a T^3 law, compared to the temperature-independent μ^+ relaxation rate Λ.

2.3. Effect of muon charge?

The field independence of C_M in $NiGa_2S_4$ (figure 2) suggests a low-temperature state without spin freezing. This seems to conflict with the μSR evidence for at least quasistatic spin freezing and the strong magnetic field dependence of the μSR relaxation at low temperatures [13, 15]. It has been suggested [16] that the low-temperature state is in fact nonmagnetic, but the muon itself "... uncovers a frozen magnetic moment in its vicinity." However, there is also evidence for spin freezing from ^{69}Ga NQR spectra below ~2 K [14], which are broadened by about the same field distribution width as obtained from the μSR quasistatic relaxation rate.[7] Unlike muons, ^{69}Ga nuclei are native to the material and do not perturb the Ni^{2+} magnetism.

The difference between μSR and NQR relaxation rates at low temperatures (figures 3 and 4) also suggests an effect of the muon on its magnetic environment. A similar discrepancy is found in the kagomé lattice-like compound volborthite [$Cu_3V_2O_7(OH)_2{\cdot}2H_2O$], where it has been argued [5, 17] that, on the contrary, the effect is intrinsic: the symmetric position of ^{51}V NMR nuclei with respect to the kagomé lattice filters out dominant spin fluctuations with a particular spatial configuration. Similarly, in $NiGa_2S_4$ the ^{69}Ga nuclei are centered on triangles in the Ni planes, raising the possibility of a similar picture in which no effect of the μ^+ charge is involved. This is of course speculation, and more work is needed to identify the fluctuation modes in $NiGa_2S_4$. But the case for μ^+-induced magnetism is at least open to question.

[6] Along with the spin-echo relaxation rate $1/T_2$, which we do not consider here.

[7] Between ~2 K and T_f the ^{69}Ga NQR spin-echo signal is 'wiped out', *i.e.* decays too quickly to be observable [14].

It should be noted that $T_N = 0$ for the ideal 2D Heisenberg antiferromagnet [18], with or without frustration, but perturbations (anisotropy, interlayer interactions, defects, ...) can produce a nonzero transition temperature. In this light the NQR and μSR evidence for a transition in $NiGa_2S_4$ is perhaps not surprising, although the transition may be more exotic [19]. But the extreme field independence of the specific heat remains difficult to understand.

3. $Pr_2Ir_2O_7$

$Pr_2Ir_2O_7$ is a metallic compound in which inelastic neutron scattering experiments [10] indicate a well-isolated Pr^{3+} Γ_3 non-Kramers doublet crystalline electric field (CEF) ground state. Transport properties suggest a Kondo effect even below a RKKY interaction temperature ~20 K [11]; geometrical frustration has been suggested as a mechanism for suppression of magnetic order and a spin-liquid state.

LF-μSR experiments in weak applied fields have been performed to characterize local quasistatic and dynamic magnetism in this compound [20]. Kubo-Toyabe (K-T) relaxation functions [21] are observed between 25 mK and 200 K, indicating a Gaussian distribution of quasistatic $\langle \boldsymbol{B}_{\rm loc} \rangle$ over this temperature range. The K-T functional form is expected from nuclear dipolar fields, but the $T \to 0$ quasistatic μ^+ spin relaxation rate $\Delta(0) \approx 8\ \mu s^{-1}$ is enhanced by a factor of ~50 over the expected value for ^{141}Pr nuclei[8] [20] and indicates a randomly-oriented quasistatic moment of order $0.1\mu_B$. The temperature dependence of Δ, shown in figure 5, shows no evidence of a magnetic phase transition. Instead, $\Delta(T)$ decreases smoothly with increasing temperature but remains nonzero up to 125 K. The data rule out ordering of the full Pr^{3+} CEF ground-state moment (3.0 μ_B) down to 25 mK.

3.1. Hyperfine-enhanced nuclear magnetism

We argue [20] that this quasistatic field distribution is due to *hyperfine enhancement* of the ^{141}Pr nuclear magnetism [22, 23], in which ^{141}Pr nuclear moments are enhanced by a Van Vleck-like mechanism when the Pr^{3+} ions are in a nonmagnetic (*e.g.* singlet) ground state. Evidence that the origin of this relaxation is nuclear rather than electronic magnetism comes from a plot of Δ versus the molar bulk susceptibility $\chi_{\rm mol}$, shown in figure 6. For small $\chi_{\rm mol}$ (high temperatures)

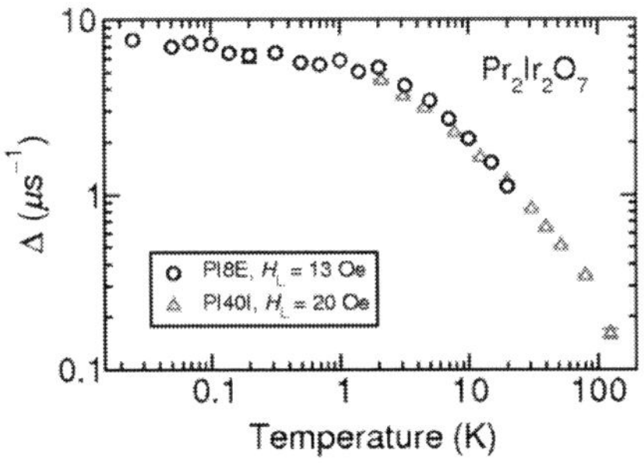

Figure 5. Temperature dependence of weak LF-μSR quasistatic relaxation rate Δ in $Pr_2Ir_2O_7$, 25 mK $\leq T \leq$ 125 K.

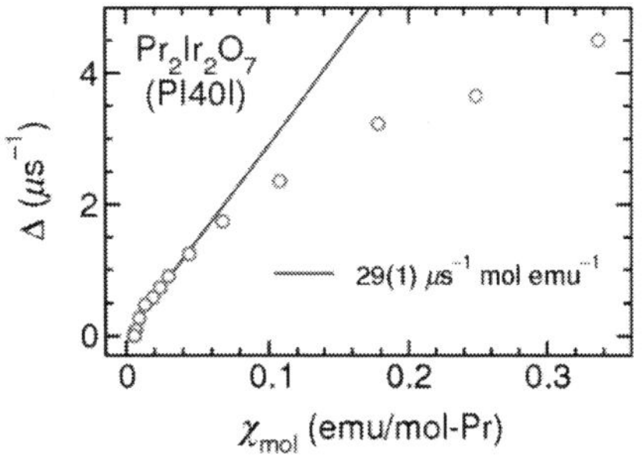

Figure 6. Relaxation rate Δ versus molar bulk susceptibility $\chi_{\rm mol}$, with temperature an implicit parameter, in $Pr_2Ir_2O_7$, 2 K $\leq T \leq$ 125 K.

$\Delta \propto \chi_{\rm mol}$, as expected from the hyperfine enhancement mechanism: the enhancement factor is $a_{4f}\chi$, where a_{4f} is the ^{141}Pr atomic hyperfine coupling constant [22, 23]. The slope in figure 6 yields a reasonable estimate of the unenhanced ^{141}Pr dipolar field [20].

[8] Ir and O nuclear magnetic moments are too small to contribute appreciably.

3.2. *Effect of μ^+ charge*

Assuming this picture, the downward curvature in Δ versus $\chi_{\rm mol}$ at low temperatures (figure 6) suggests that the local susceptibility $\chi_{\rm loc}$ of Pr^{3+} ions is reduced in the neighborhood of μ^+ sites. This would happen if the Pr^{3+} non-Kramers degeneracy were lifted by the μ^+ electric field. Such an effect would also contribute to, and could conceivably explain completely, the nonmagnetic (singlet) Pr^{3+} ground state required for hyperfine-enhanced nuclear magnetism [20].

To test for muon-induced reduction of $\chi_{\rm loc}$ we have measured the μ^+ frequency shift $K = \Delta\omega_\mu/\omega_\mu$ using TF-μSR [3]. In paramagnets K is a measure of the local susceptibility: $K = a_\mu \chi_{\rm loc}$, where a_μ describes the coupling between electronic spins and the μ^+ spin [2]. In figure 7 K is plotted against $\chi_{\rm mol}$, with temperature an implicit parameter, before and after correction for the Lorentz and demagnetization fields. These are quite significant and the correction is not accurate, due to uncertainties in sample dimensions and the powder packing fraction. The error bars on the corrected shift $K_{\rm corr}$ are mainly due to these uncertainties in the correction, and are not random between points. The curvature in $K_{\rm corr}$ versus $\chi_{\rm mol}$ (figure 7), like that in Δ versus $\chi_{\rm mol}$ (figure 6), is evidence for a deviation of $\chi_{\rm loc}$ from the bulk $\chi_{\rm mol}$, *i.e.* for muon-induced modification of $\chi_{\rm loc}$ [6, 7]. Other mechanisms, such as contributions from CEF excited states or the Kondo effect [24], would not lead to the observed hyperfine enhancement.

Figure 8 gives Δ versus $K_{\rm corr}$. It can be seen that the relation is much more linear than

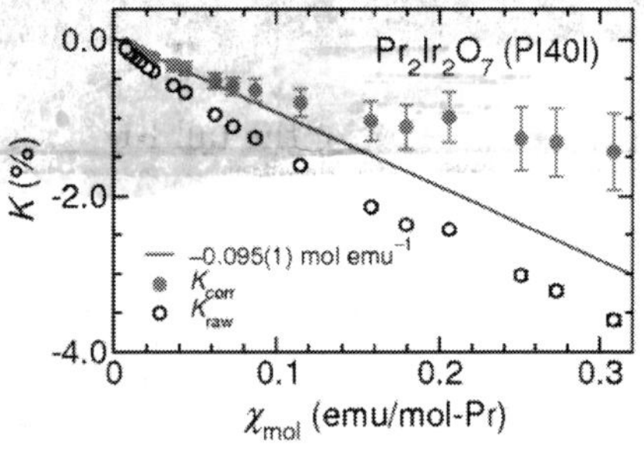

Figure 7. Raw and corrected μ^+ frequency shift versus molar bulk susceptibility $\chi_{\rm mol}$, with temperature an implicit parameter, in $Pr_2Ir_2O_7$.

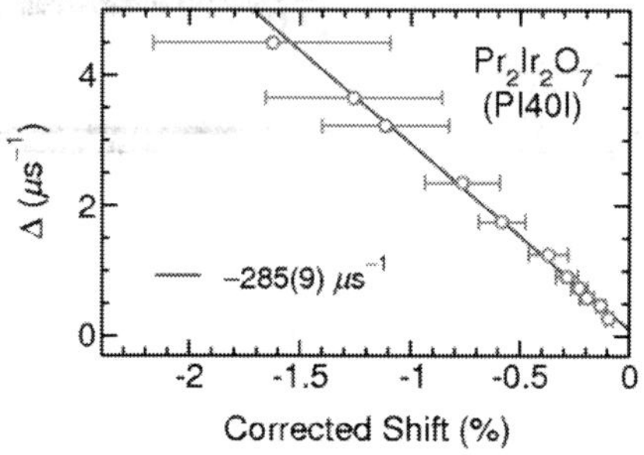

Figure 8. Relaxation rate Δ versus corrected μ^+ frequency shift $K_{\rm corr}$, with temperature an implicit parameter, in $Pr_2Ir_2O_7$.

in figure 6 over the entire temperature range, even if the correction is not accurate. We conclude that Δ tracks the local susceptibility accurately, yielding further evidence that the K-T relaxation in $Pr_2Ir_2O_7$ is due to hyperfine-induced ^{141}Pr nuclear magnetism.

3.3. *Phase transition*

Recent specific heat measurements indicate a phase transition in some $Pr_2Ir_2O_7$ samples at $T_c \sim 0.8$ K, apparently correlated with Pr-Ir stoichiometry [25]. Examples are shown in figure 9. As shown in figure 10(a), however, ZF-μSR experiments on similar samples do not exhibit the large increase of quasistatic μ^+ relaxation rate below T_c expected for a magnetic transition. For the full non-Kramers doublet ground-state moment of 3.0 μ_B the quasistatic relaxation would not be directly observable, but would cause a decrease of the initial muon asymmetry (signal amplitude) to $\sim$1/3 of its value above T_c. No such decrease is observed [figure 10(b)]. There is also no feature in the dynamic μ^+ relaxation rate at T_c [figure 10(a)].

The splitting of the ground-state Pr^{3+} doublets noted above [20] could suppress the quasistatic magnetism locally. In metals this effect should be quite local because of screening of the μ^+

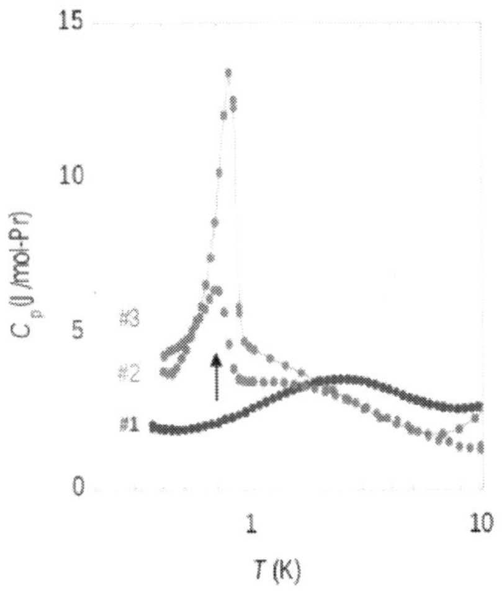

Figure 9. Low-temperature specific heat of three samples of $Pr_2Ir_2O_7$. Data from [25].

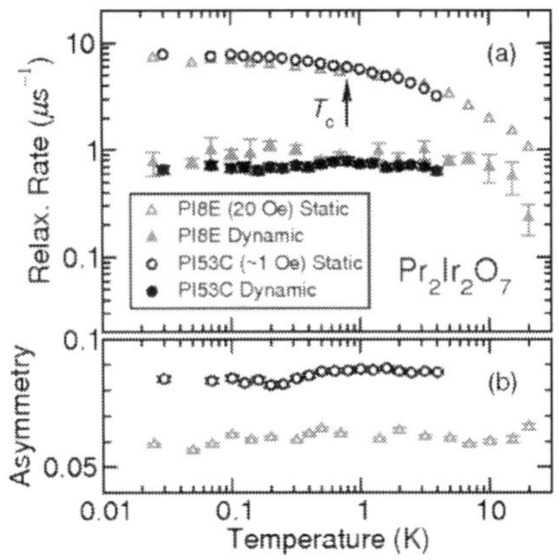

Figure 10. (a) Quasistatic (open symbols) and dynamic (filled symbols) μ^+ spin relaxation rates, and (b) μSR asymmetry (signal amplitude) in $Pr_2Ir_2O_7$, for samples with (PI53C) and without (PI8E) a specific-heat anomaly at T_c.

charge; in the intermetallics $PrNi_5$ [6] and $PrIn_3$ [7] significant CEF modification is confined to nearest-neighbour Pr^{3+} ions. Lattice-sum calculations of the dipolar field at representative μ^+ sites in $Pr_2Ir_2O_7$ show, however, that even for complete suppression out to 25 Å, including ~900 Pr ions, the expected contribution to $\langle B_{\rm loc}\rangle$ from more distant Pr ions would be ~40 G $\approx 5\ \mu s^{-1}/\gamma_\mu$. Onset of such a contribution at T_c would be a ~20% effect, *i.e.* easily visible; its absence [figure 10(a)] is evidence that the phase transition in $Pr_2Ir_2O_7$ is nonmagnetic.

4. Conclusions

4.1. $NiGa_2S_4$

In this compound NQR and μSR experiments appear at first sight to both confirm and contradict each other. Quasistatic spin freezing is seen by both techniques, but it is not clear whether or not the muon charge has a significant influence on the low-temperature spin dynamics since the temperature dependence of dynamic spin relaxation is different for the two techniques (*cf.* section 2.2). This may be due to differences in local surroundings and consequent filtering of spin fluctuations as in volborthite [5], but the question needs to be studied in more detail. The most important issue in $NiGa_2S_4$ remains how to reconcile the field-independent specific heat with the evidence from NQR and μSR for disordered quasistatic magnetism.

4.2. $Pr_2Ir_2O_7$

Here the evidence for μ^+-induced modification of the μSR data (nonlinear K versus $\chi_{\rm mol}$) is fairly clear. It remains uncertain, however, whether μ^+-induced CEF level splitting is the only mechanism for a Pr^{3+} singlet ground state or, alternatively, whether a many-body singlet ground state associated with the frustrated Pr^{3+} spin system also contributes.

Even though the μ^+ electric field lifts the non-Kramers degeneracies of nearest-neighbor and possibly several near-neighbor Pr^{3+} ions, a significant local field from more distant Pr^{3+} spins would remain if these were frozen, with or without long-range order (section 3.3). The absence of μSR evidence for Pr^{3+} magnetism (other than the $4f$ polarization involved in the nuclear hyperfine enhancement mechanism) strongly suggests that in $Pr_2Ir_2O_7$ the ground state, low-lying excitations, and phase transition at 0.8 K are all nonmagnetic in character.

Acknowledgments
We are grateful for technical assistance from the TRIUMF Centre for Molecular and Materials Science, where these experiments were carried out. We wish to thank E J Ansaldo and K Onuma for assistance with the experiments, and P Dalmas de Réotier, H Kawamura, R F Kiefl, R R P Singh, A Yaouanc, and S Zhao for useful discussions. This work was partially supported by U.S. NSF Grant nos. 0422671 and 0801407 (Riverside) and 0604015 (Los Angeles), by a Grant-in-Aid (No. 21684019) from the Japanese Society for the Promotion of Science (JSPS), and also by Grants-in-Aid for Scientific Research on Priority Areas (Nos. 17071003 and 19052003) from the Ministry of Education, Culture, Sports, Science and Technology (MEXT), Japan.

References
[1] Ramirez A P 1994 *Ann. Rev. Mater. Sci* **24** 453–480
[2] Slichter C P 1990 *Principles of Magnetic Resonance* 3rd ed (New York: Springer-Verlag)
[3] Schenck A 1985 *Muon Spin Rotation Spectroscopy: Principles and Applications in Solid State Physics* (Bristol & Boston: A. Hilger)
[4] Blundell S J 1999 *Contemp. Phys.* **40** 175–192
[5] Mendels P, Olariu A, Bert F, Bono D, Limot L, Collin G, Ueland B, Schiffer P, Cava R J, Blanchard N, Duc F and Trombe J C 2007 *J. Phys.: Condens. Matter* **19** 145224–1–9
[6] Feyerherm R, Amato A, Grayevsky A, Gygax F N, Kaplan N and Schenck A 1995 *Z. Phys. B* **99** 3–13
[7] Tashma T, Amato A, Grayevsky A, Gygax F N, Pinkpank M, Schenck A and Kaplan N 1997 *Phys. Rev. B* **56** 9397–9405
[8] Nakatsuji S, Nambu Y, Tonomura H, Sakai O, Jonas S, Broholm C, Tsunetsugu H, Qiu Y and Maeno Y 2005 *Science* **309** 1697–1700
[9] Nakatsuji S, Nambu Y, Onuma K, Jonas S, Broholm C and Maeno Y 2007 *J. Phys.: Condens. Matter* **19** 145232–1–7
[10] Machida Y, Nakatsuji S, Tonomura H, Tayama T, Sakakibara T, van Duijn J, Broholm C and Maeno Y 2005 *J. Phys. Chem. Solids* **66** 1435–1457
[11] Nakatsuji S, Machida Y, Maeno Y, Tayama T, Sakakibara T, van Duijn J, Balicas L, Millican J N, Macaluso R T and Chan J Y 2006 *Phys. Rev. Lett.* **95** 087204–1–4
[12] Yaouanc A, Dalmas de Réotier P, Chapuis Y, Marin C, Lapertot G, Cervellino A and Amato A 2008 *Phys. Rev. B* **77** 092403–1–4
[13] MacLaughlin D E, Nambu Y, Nakatsuji S, Heffner R H, Shu L, Bernal O O and Ishida K 2008 *Phys. Rev. B* **78** 220403(R)–1–4
[14] Takeya H, Ishida K, Kitagawa K, Ihara Y, Onuma K, Maeno Y, Nambu Y, Nakatsuji S, MacLaughlin D E, Koda A and Kadono R 2008 *Phys. Rev. B* **77** 054429–1–13
[15] MacLaughlin D E, Nambu Y, Nakatsuji S, Ishida K, Heffner R H, Shu L and Bernal O O 2009 *J. Phys.: Conf. Ser.* **145** 012040–1–5
[16] Varma C 2008 Mysterious order for spins on a triangular lattice *J. Club Condens. Matter Phys.* November 2008
[17] Bert F, Mendels P, Bono D, Olariu A, Ladieu F, Trombe J C, Duc F, Baines C, Amato A and Hillier A 2006 *Physica B* **374-375** 134–137
[18] Chakravarty S, Halperin B I and Nelson D R 1989 *Phys. Rev. B* **39** 2344–2371
[19] Kawamura H, Yamamoto A and Okubo T 2009 Z_2-vortex ordering of the triangular-lattice Heisenberg antiferomagnet arXiv:0909.0121
[20] MacLaughlin D E, Ohta Y, Machida Y, Nakatsuji S, Luke G M, Ishida K, Shu L and Bernal O O 2009 *Physica B* **404** 667–670
[21] Hayano R S, Uemura Y J, Imazato J, Nishida N, Yamazaki T and Kubo R 1979 *Phys. Rev. B* **20** 850–859
[22] Murao T 1971 *J. Phys. Soc. Jpn.* **31** 683–690
[23] Bleaney B 1973 *Physica (Utrecht)* **69** 317–329
[24] Ohishi K, Heffner R H, Ito T U, Higemoto w, Morris G D, Hur N, Bauer E D, Sarrao J L, Thompson J D, MacLaughlin D E and Shu L 2009 *Phys. Rev. B* **80** 125104–1–7
[25] Ohta Y, Machida Y and Nakatsuji S Low-temperature phase transition in $Pr_2Ir_2O_7$ unpublished

μSR study of $CeRhIn_5$ under applied pressure

R H Heffner[1,2], T Goko[3,4], D Andreica[5], K Ohishi[7], W Higemoto[2], T U Ito[2], A Amato[6], J Spehling[8], H -H Klauss[8], E D Bauer[1], J D Thompson[1] and Y J Uemura[3]

[1] Los Alamos National Laboratory, Los Alamos, NM, 87545 USA
[2] ASRC, Japan Atomic Energy Agency, Tokai-mura, Ibaraki 319-1195, Japan
[3] Dept. of Physics, Columbia University, New York, NY 10027 USA
[4] TRIUMF, Vancouver, British Columbia, V6T 2A3 Canada
[5] Faculty of Physics, Babes-Boyai University, 400084 Cluj-Napoca, Romania
[6] Laboratory for Muon Spectroscopy, PSI, CH-5232 Villigen PSI, Switzerland
[7] Advanced Meson Science Laboratory, Riken, Wako, 351-0198 Japan
[8] Institut für Festkörperphysik, TU Dresden, D-01069 Dresden, Germany

E-mail: muonphysics@hotmail.com

Abstract. At ambient pressure $CeRhIn_5$ is an incommensurate antiferromagnet with Néel temperature T_N = 3.8 K. The application of pressure reduces T_N and yields (1) a state in which AFM coexists with superconductivity, and then (2) a superconducting state in the absence of AFM. We report transverse-field muon spin relaxation (μSR) measurements at pressures 2.07 and 2.26 GPa in the pure superconducting state of single-crystalline $CeRhIn_5$, which has a tetragonal structure. μSR is sensitive to the local field distribution produced by the vortex lattice in a mixed superconducting state, allowing a measurement of the magnetic field penetration depth λ(T). We measured $\lambda_{ab}(T)$ for applied field along the c-axis as a function of temperature T, and find $\lambda_{ab}(0)$ = 372(5) nm and 338(6) nm for P = 2.07 GPa and 2.26 GPa, respectively. The temperature dependence of the superfluid density $\rho_s(T) \propto \lambda^{-2}(T)$ was found to be $\rho_s(T) \propto T^n$, with n = 1.9(2).

1. Introduction

$CeRhIn_5$ is an example of a system exhibiting competing broken symmetry ground states near a quantum critical point (QCP) [1]. At ambient-pressure $CeRhIn_5$ is an incommensurate, local-moment ($\mu = 0.8\ \mu_B$) antiferromagnet with Néel temperature T_N = 3.8 K [2]. As pressure is applied T_N increases slightly and then decreases, extrapolating towards zero in the region 2.3-2.4 GPa. For P ≅ 0.8 – 1.8 GPa, bulk superconductivity coexists with magnetic order, and for P > 1.8 GPa superconductivity is found in the apparent absence of magnetic order. $CeRhIn_5$ is thus an example of a quantum-critical system where the magnetic order parameter is driven to zero by applied pressure and where magnetism and superconductivity interact.

Several important magnetic and superconducting properties of this system have been determined. At ambient pressure the magnetic order has been found to be an incommensurate helix along the c-axis with moments in the a-b plane and propagation vector **k** = (0.5,0.5,δ=0.298), with δ changing as pressure exceeds about 1 GPa [2]. Nuclear quadrupole resonance (NQR) experiments [3] exhibit an additional commensurate phase above 1.7 GPa, though this has not been verified by neutron scattering [2]. An unconventional superconducting order parameter (having a symmetry lower than that of the tetragonal lattice) has been found from NQR [4] and field-oriented specific heat [5] measurements.

Muon spin relaxation (μSR) experiments can yield important additional information. Beams of polarized positive muons can be produced with sufficient momentum to penetrate high pressure cells enabling muon experiments on materials under applied pressure. The μ^+ occupies one or more interstitial lattice sites and serves as a probe of the local magnetic field, allowing measurements of the magnetic and superconducting properties of materials. The volume fraction of ordered magnetic states can also be easily determined.

In a type II superconductor such as $CeRhIn_5$ the applied magnetic field penetrates in discreet, quantized vortices. The field in the sample falls off between the vortices with a characteristic length λ, the magnetic field penetration depth. This yields an inhomogeneous internal field distribution which is sensed as a relaxation of the muon's spin polarization in a μSR experiment [6].The penetration depth is related to the superfluid density ρ_s by the relation $\lambda^{-2} \propto \rho_s/m^*$, where m^* is the effective mass of the paired electrons. Here we report μSR studies of the penetration depth λ(T) in a single crystal of $CeRhIn_5$ under small applied magnetic field H = 0.05 T at pressures P = 2.07 and 2.26 GPa.

2. Experimental Procedures

2.1 High pressure cell and $CeRhIn_5$ sample

The high-pressure experiments were performed on the GPD spectrometer at the Paul Scherer Institute (PSI) in Villigen, Switzerland, using backward-decay positive muons. In addition, some measurements were taken at the GPS spectrometer at PSI to obtain spectra below T_N in a nearly background free environment to establish the form of the fitting functions for the data taken in the pressure cell.

A pressure cell made from MP35N alloy was used, with about a 5 mm diameter cylindrical sample space for pressures P < 2.5 GPa. The cell was thermally anchored to a helium-3 cryostat, allowing measurements between about T = 0.25 – 300 K. The pressure medium inside the cell was Daphne oil. The applied pressure was measured in-situ at low temperatures by ac susceptibility using the pressure dependence of the superconducting transition temperature of a small piece of In wire located near the sample inside the pressure cell [7]. The pressure determinations were performed at temperatures close to those of the μSR measurements themselves.

The sample was a cylindrically-shaped $CeRhIn_5$ single crystal about 5 mm in diameter and about 17 mm long with its b-axis along the cylinder axis. The incoming muon beam momentum (and, hence, spin polarization S_μ) was aligned in the horizontal plane perpendicular to the sample b-axis (the vertical direction of the GPD spectrometer). The applied field H could therefore be rotated in the a-c plane of the sample by rotating the cell through an angle θ about its vertical axis. In this way H could be oriented along the c-axis of the sample, as described below.

2.2 Transverse field μSR signals

In a static magnetic field $|\mathbf{B}|$ oriented transversely to the muon spin ($\mathbf{B} \perp \mathbf{S}_\mu$, denoted TF below) the muon relaxation function is given by $G(t)\cos(2\pi\nu t + \phi)$, where $2\pi\nu = \gamma_\mu|\mathbf{B}|$ and the muon's gyromagnetic ratio $\gamma_\mu = 2\pi \times 135.54$ MHz/T. The function G(t) describes the relaxation of the spin polarization, in our case by an inhomogeneous field distribution [6]. Figure 1a shows the Fourier transformed μSR signals for data taken with an empty cell in H = 0.05 T. Good fits to the muon relaxation function for these data were obtained with a stretched exponential function $G_c(t) = A_c \exp(-(\lambda_c t)^\beta)$.

Inside the sample, and above the superconducting transition temperature T_c, the muons predominantly experience randomly-oriented, quasi-static nuclear dipole fields. Below T_c muons stop randomly with respect to the flux-line-lattice (FFL), and the resultant inhomogeneous field distribution causes a dephasing of the ensemble of muon spins which can be reasonably approximated by a Gaussian relaxation function (figure 1b below). (More complicated expressions which take into account the asymmetric field distribution produced by the FLL [9] are not justified in this experiment due to the large background produced by the pressure cell.) Hence, the transverse muon relaxation function G(t) was fit by a sum of two terms corresponding to muons stopping in the sample and the cell, respectively:

$$G(t) = A_l \exp(-\sigma^2 t^2/2) \cos(2\pi\nu_l t + \phi) + A_c \exp(-(\lambda_c t)^\beta) \cos(2\pi\nu_c t + \phi), \quad (1)$$

with fixed values $A_l = 0.25$ and $A_c = 0.75$. The values for A_l and A_c were determined by fits to the amplitudes of the oscillating signal below T_N in zero applied field (ZF). Under these conditions the ZF relaxation from the empty cell was found to be well described by $G(t) = A_1 \exp(-\lambda_0 t)\, G_{KT}(t) + A_2$, where $G_{KT}(t)$ is the well-known Kubo-Toyabe relaxation function [10] and λ_0 is a fit parameter. Thus, because the cell produced no oscillation signal, the fraction of muons stopping in the sample could be determined by the relative amplitude of the oscillating signal in ZF below T_N. Note that the sample filling factor of ≅ 25% measured in this way agrees well with a geometrical calculation of the area subtended by the sample in the beam.

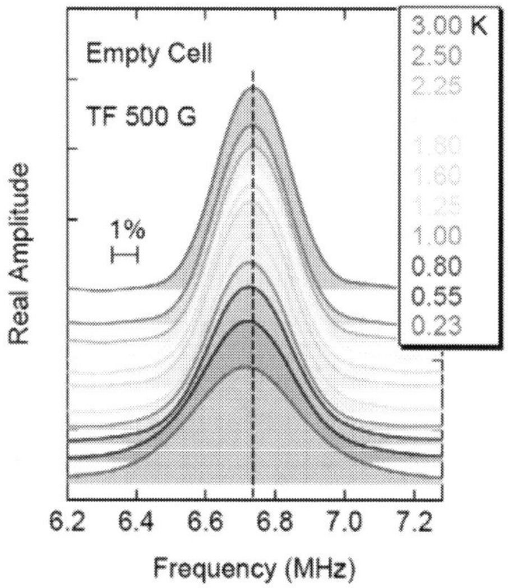

Figure 1a. Fourier transformed frequency spectra for cell alone in H = 0.05 T. The curves (color on line) range from 3.00 K to 0.23 K (top to bottom).

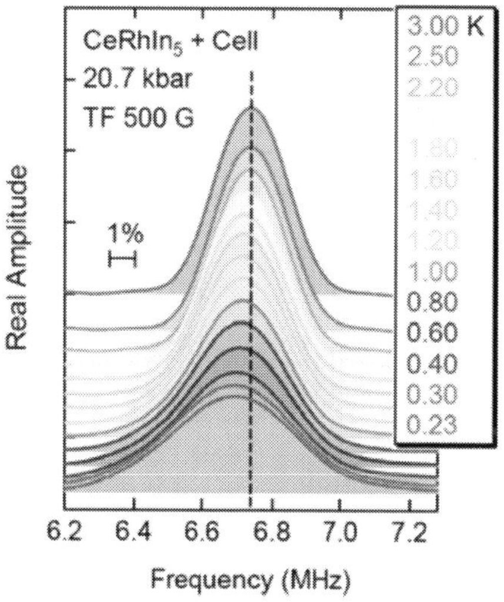

Figure 1b. Fourier transformed frequency spectra for sample plus cell in H = 0.05 T. The curves (color on line) range from 3.00 K to 0.23 K (top to bottom).

2.3 Sample alignment

The measured value of λ depends on the orientation of the applied magnetic field with respect to the crystalline axes in a non-cubic material. When the field is oriented along the tetragonal c-axis, superconducting screening currents flow in the a-b plane and one measures λ_{ab}. In $CeRhIn_5$ the internal fields **B** below T_N lie predominantly along the c-axis for the measured μ^+ interstitial sites [8]. Therefore, the sample could be aligned so that S_μ was perpendicular to the c-axis by maximizing the amplitudes of the muons' precession signals as a function of the angle θ in zero applied field. In this way, the applied field could be oriented along the sample's c-axis.

3. Determination of $\lambda_{ab}(T)$

The TF data were fit to equation (1) with the cell parameters λ_c and β held fixed at their values determined from the cell-only data. All data were taken in a field-cooled manner to establish an equilibrium field distribution in the superconducting state. Figure 2 shows the temperature dependence of the parameters $\sigma(T)$ and $\nu(T)$ in 0.05 T transverse field taken at P = 2.07 and 2.26 GPa. At these pressures the sample is in the pure superconducting state. Several features are evident. Above $T_c \cong 2.2$ K $\sigma(T)$ is roughly independent of temperature and pressure, corresponding to relaxation in the paramagnetic state of $CeRhIn_5$. Below $T_c \cong 2.2$ K $\sigma(T)$ increases in value due to the inhomogeneous field distribution produced by the FLL, and $\nu_l(T)$ decreases in value due to the Meissner screening by the supercurrents.

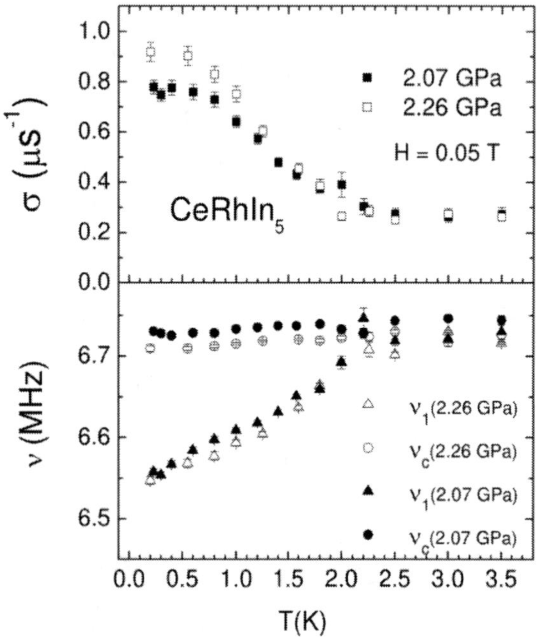

Figure 2. Temperature dependence of the Gaussian relaxation rate σ and frequencies ν_l and ν_c at pressures 2.07 and 2.26 GPa from fits to equation (1).

To obtain the signal $\sigma_v(T)$ from the FLL one unfolds the Gaussian widths according to $\sigma_v^2 = \sigma^2 - \sigma_b^2$, where σ_b is the average temperature-independent background rate above T_c. These unfolded data are shown in figure 3, using $\sigma_b = 0.271\ \mu s^{-1}$ for P = 2.07 GPa and $\sigma_b = 0.269\ \mu s^{-1}$ for P = 2.26 GPa. Also shown in figure 3 is a fit to an empirical formula for the superfluid density $\rho_s(T)$ appropriate for unconventional superconductors [11]:

$$\rho_s(0) - \rho_s(T) \propto T^n . \tag{2}$$

Here we note that $\sigma_v \propto \rho_s/m^*$, and we assume that the effective mass m^* is independent of temperature. We obtain $\sigma_v(0) = 0.91(3)\ \mu s^{-1}$ and $n = 1.83(27)$ for P = 2.26 GPa and $\sigma_v(0) = 0.75(2)\ \mu s^{-1}$ and $n = 1.94(27)$ for 2.07 GPa. T_c was fixed in the fits at 2.2 K.

To estimate $\lambda_{ab}(0)$ from $\sigma_v(0)$ we use a well-known relation [12] valid for $H \ll H_{c2}$ and given in equation (3). $H_{c2}(0)$ is greater than 10 T for the pressures of interest [13].

$$\sigma_v^2/\gamma_\mu^2 = \langle \Delta B^2 \rangle = 0.00371\ \phi_0^2/\lambda_{ab}^4, \tag{3}$$

where $\langle \Delta B^2 \rangle$ is the mean square field distribution caused by the FLL and $\phi_0 = hc/2e = 2.07 \times 10^{-15}$ Tm^2 is the magnetic flux quantum. This yields $\lambda_{ab}(0) = 372(5)$ nm and 338(6) nm for P = 2.07 GPa and 2.26 GPa, respectively. The errors are statistical. Note that no mean-free-path corrections have yet been made to these values.

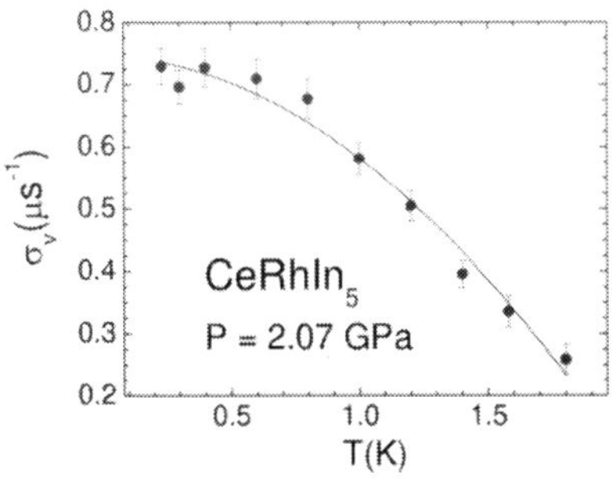

Figure 3a. Temperature dependence of $\sigma_v(T)$ at P = 2.07 GPa with fit (solid curve) to equation (2). $\sigma_v(0) = 0.75(2)\ \mu s^{-1}$, n = 1.94(27). T_c fixed at 2.2 K.

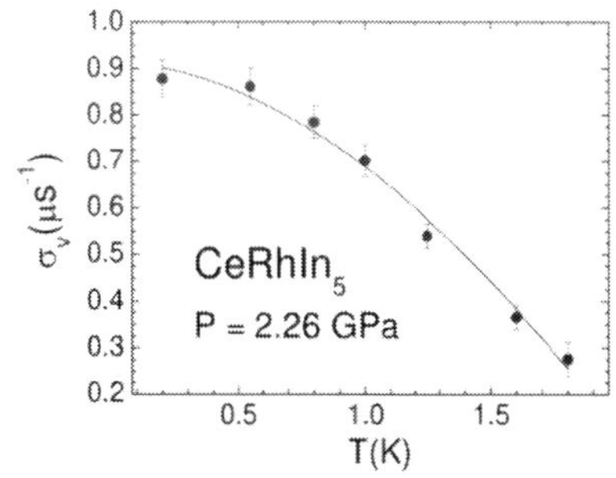

Figure 3b. Temperature dependence of $\sigma_v(T)$ at P = 2.26 GPa with fit (solid curve) to equation (2). $\sigma_v(0) = 0.91(3)\ \mu s^{-1}$, n = 1.83(27). T_c fixed at 2.2 K.

4. Conclusions

We have measured the TF-μSR relaxation rate at P = 2.07 GPa and 2.26 GPa in the pure superconducting state of $CeRhIn_5$ and extracted the temperature dependence and magnitude of the magnetic field penetration depth (or, alternatively, the superfluid density). We find that the density $\rho_s(0)$ is slightly higher at the higher pressure, which may result from the increased conduction electron itinerancy with pressure in this system. The temperature dependence of $\rho_s(T)$ at the two pressures is essentially the same, and can be parameterized by equation (2) with $\langle n \rangle \cong 1.9(2)$. We note that a simple (BCS) two-fluid model of superconductivity gives n = 4. Thus, our results show a power-law temperature dependence, indicating nodes in the superconducting order parameter, and a significantly weaker temperature dependence than for a BCS superconductor, both consistent with an unconventional order parameter. Finally, we estimate the Ginsburg-Landau coherence length ξ_{GL} from the upper critical field as $\xi_{GL}^2 = \phi_0/2\pi H_{c2}(0)$, yielding $\xi_{GL} < 5.6$ nm for $H_{c2}(0) > 10$ T [13]. This yields $\kappa = \lambda_{ab}/\xi > 63$, taking the average of the measured values of λ_{ab} quoted above. This value can be compared to $CeCoIn_5$ ($\kappa \cong 69$) and $CeIrIn_5$ ($\kappa \cong 26$), both ambient-pressure superconductors.

References

[1] Park T, Ronning F, Yuan H Q, Salamon M B, Movshovich R, Sarrao J L and Thompson J D 2006 *Nature* **440** 65; Knebel G, Aoki D, Braithwaite D, Salce B and Flouquet J 2006 *Phys. Rev.* B **74**, 020501(R)

[2] Aso N, Kiwamu I, Yoshizawa H, Fujiwara T, Uwatoko Y, Chen Gen-Fu, Sato N K and Miyake K 2009 *J. Phys. Soc. Jpn.* **78** 073703 and references therein

[3] Yashima M, Mukuda2009 *Phys. Rev.* B **79** 214528

[4] Kohori Y, Yamato Y, Imamoto Y and Kohara T 2000 Eur. Phys. J. B **18** 1601

[5] Park T, Bauer E D and Thompson J D 2008 *Phys Rev. Lett.* **101** 177002; Park T and Thompson J D 2009 *New J. Phys.* **11** 055062

[6] Schenck A 1985 *Muon Spin Rotation Spectroscopy: Principles and Applications to Solid State Physics* (Hilger, Bristol)
[7] Eiling A and Schilling J S 1981 *J. Phys. F: Metal Phys.* **11** 623
[8] Schenck A, Andreica D, Gygax F N, Aoki D and Onuki Y 2002 *Phys. Rev.* B **66** 144404
[9] Sonier J E, Brewer J and Kiefl R F 2000 *Rev. Mod. Phys.* **72** 769
[10] Kubo R and Toyabe T 1967 in *Magnetic Resonance and Relaxation* (ed. R. Blinc, North Holland, Amsterdam) 810
[11] Sigrist M 2005 *Lectures on the Physics of Highly Correlated System IX: Ninth Training Course* (ed. Avella A and Mancini F) *American Institute of Physics 0-7354-0279-5/05* p. 165
[12] Brandt E H 1988 *Phys. Rev.* B **37** 2349
[13] Ida Y, Settai R, Ota Y, Honda F and Onuki Y 2008 *J. Phys. Soc. Jpn.* **77** 084708; Knebel G, Aoki D, Brison J-P and Flouquet J 2008 *J. Phys. Soc. Jpn.* **77** 114704

Acknowlegements
DA acknowledges partial financial support from the Romanian CNCSIS project 444/2009.

International Symposium on Advanced Science Research 2009 IOP Publishing
Journal of Physics: Conference Series **225** (2010) 012013 doi:10.1088/1742-6596/225/1/012013

Muon Knight shift measurements in $CeCoIn_5$ below 17 kOe

W Higemoto[1], A Koda[2], R Kadono[2], K Ohishi[3], Y Haga[1], H Shishido[4,5], R Settai[4] and Y Ōnuki[4]

[1]Advanced Science Research Center, Japan Atomic Energy Agency, Tokai, Ibaraki 319-1195, Japan
[2]Muon Science Laboratory, Institute of Materials Structure Science, High Energy Accelerator Research Organization, Tsukuba, Ibaraki 305-0801, Japan
[3]Advanced Meson Science Laboratory, RIKEN, Wako, Saitama 351-0198, Japan
[4]Graduate School of Engineering Science, Osaka University, Toyonaka, Osaka 560-8531, Japan

E-mail: `higemoto.wataru@jaea.go.jp`

Abstract. We report muon Knight shift measurements in single crystals of $CeCoIn_5$ below 17 kOe. The muon Knight shift decreases with decreasing temperature below T_c and the reduction of the Knight shift is roughly explained by the Fermi liquid relation. The magnetic field dependence of the reduction of the muon Knight shift is not in proportion to $\sqrt{H}$. This fact is inconsistent with simple expectation for a d-wave superconductor.

1. Introduction

Unconventional superconductivity in heavy fermion systems has attracted interest and inspired extensive experimental and theoretical research over the last several years. The compound $CeCoIn_5$ has a $HoCoGa_5$-type tetragonal crystal structure and has been reported to be superconducting at ambient pressure with a transition temperature of 2.3 K[1]. Unconventional superconductivity has been reported in several studies and the symmetry of the Cooper pairing in $CeCoIn_5$ most likely belongs to $d_{x^2-y^2}$[2]. Knight shift K measurements by nuclear magnetic resonance (NMR) spectroscopy have shown that K decreases at temperatures below T_c, indicating spin singlet superconductivity[3, 4]. The muon spin rotation (μSR) above 20 kOe has also revealed the reduction of K at temperatures below T_c[5]. Recently, it was noted that a paramagnetic effect plays an important role in superconductivity near H_{c2} in $CeCoIn_5$, and the Fulde-Ferrell-Larkin-Ovchinnikov (FFLO) phase, or another exotic phase, has been proposed. To understand the unique features of the superconductivity near H_{c2}, the fundamental nature of superconductivity, namely, its properties at a low temperature and in a low field, are especially important. In this paper, we report a muon Knight shift in the normal and superconducting phases of $CeCoIn_5$ in the weak magnetic region below 17 kOe to elucidate the basic character of superconductivity in the low field region.

[5] Present address: Department of Physics, Kyoto University, Kyoto 606-8502, Japan

2. Experimental

Conventional μSR measurements under transverse fields (TF) were carried out at the M15 beamlines of TRIUMF in Vancouver, Canada. A positive muon beam with a momentum of 29 MeV/c was implanted into the specimen, and the time evolution of the muon spin polarization was recorded. For measurements in a dilution refrigerator below 3 K, the specimen was glued by Apiezon-N grease to a silver sample holder. In a helium flow cryostat, a background-free signal was obtained. Magnetic fields up to 17 kOe were applied parallel to the c-axis of the specimen and perpendicular to the initial muon spin direction.

3. Results and Discussions

First, we describe muon Knight shift measurements in the normal conducting phase. Figure 1 shows the fast Fourier transform (FFT) of the μSR spectra obtained for $CeCoIn_5$ under a field of 17 kOe at 6.4 K and 100 K. Here, two peaks can be seen in each of the FFT spectra, denoted as f_1 and f_2 in Fig. 1. We analyzed the μSR spectra by fitting the data to a sum of two components with Gaussian damping:

$$P(t) = \Sigma_{i=1,2} A_i \exp(-\sigma_i^2 t^2) \cos(2\pi f_i t + \phi). \tag{1}$$

The muon Knight shift of the i-th component, K_i, is then defined as follows:

$$K_i = \frac{f_i - f_{ext}}{f_{ext}}.$$

Here, f_{ext} is the external field frequency, defined as $f_{ext} = \gamma_\mu B_{ext}/2\pi$, and B_{ext} is the external field that is determined by the frequency of muons stopped in the sample holder [made of $CaCO_3$ (^{4}He cryostat) or silver (in the dilution refrigerator)]. In Fig. 2, the muon Knight shift is plotted against susceptibility (Clogston-Jaccarino plot, or K-χ plot); the susceptibility data are measured at 17 kOe for the same specimen. Here, we note that the frequency shift of f_2 is too small to be analyzed precisely. Consequently, K_2 is difficult to obtain precisely and hereafter only K_1 is discussed. Above about 40 K, a linear relation in the K-χ plot can be seen, as is expected for the normal state. The muon Knight shift is expressed as $K(T) = K_s(T) + K_{orb}$, where $K_s(T)$ and K_{orb} are the spin and orbital Van-Vleck part of the Knight shift, respectively, and K_{orb} is independent of temperature. The spin part of the muon Knight shift is expressed as $K_s = (1/N_A\mu_B)A_{hf}\chi$. Here, A_{hf} is the hyperfine coupling constant and $A_{hf} = A_{dip} + A_c$, where A_{dip} and A_c correspond to the contribution from the dipolar coupling and contact hyperfine coupling, respectively. The estimated value of A_{hf} above 40 K is -1.84(7) kOe/μ_B. Deviation from linearity in the K-χ plot appears below ~40 K. We note that similar results have been obtained in NMR experiments[3, 4].

Next, we describe the muon Knight shift measurement in the superconducting phase. The frequency shift Δf in the superconducting phase is expressed by $2\pi\Delta f/\gamma_\mu = KH_{ext} + H_{FLL} + H_{dia}$, where H_{FLL} is the difference between the external field and the field at the van Hove singularity (corresponding to the field near the saddle points) of the field profile in the flux line lattice (FLL) state,[7] and H_{dia} is the magnetic field shift due to Meissner diamagnetism. At 17 kOe and 10 kOe, we find no broadening of the FFT spectra below T_c due to FLL formation, indicating that the inhomogeneity of the magnetic field in the specimen due to FLL formation is quite small, except for the vortex core, and H_{FLL} is negligibly small. The Meissner diamagnetic component is estimated to be about -0.4 Oe (-1.8 Oe) at 17 kOe (10 kOe) by using the relation $H_{dia} = -H_{c1}\ln(\beta e^{-1/2}d/\xi)/\ln\kappa$[8], which leads to a negligibly small value of 0.002 % (0.018 %) for the Knight shift at 17 kOe (10 kOe). Here, we use the penetration depth λ = 550 nm[6], the lower critical field $H_{c1} \sim \frac{\Phi_0}{4\pi\lambda^2}$=23 Oe, the Ginzburg-Landau (GL) parameter κ=67, the GL coherence length ξ ~8.2 nm[9] , the numerical constant β=0.381 for a triangular lattice[8], and

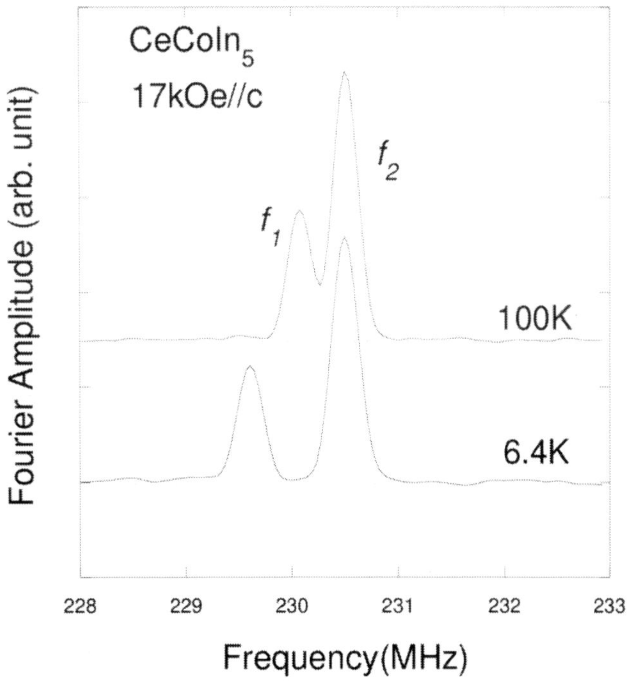

Figure 1. FFT spectra of $CeCoIn_5$ at 6.4 K and 100 K.

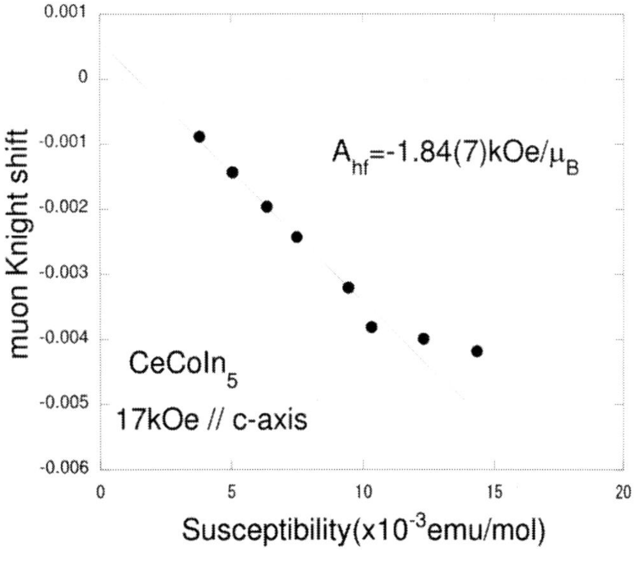

Figure 2. Clogston-Jaccarino plot (K-χ plot) of $CeCoIn_5$ above 2 K.

the vortex lattice spacing d=38 nm and 49 nm at 17 kOe and 10 kOe, respectively. Thus, we conclude that the effect of FLL formation on the Knight shift is negligible.

Figure 3 shows the temperature dependence of the muon Knight shift below 2.5 K. The muon

Knight shift exhibits significant dependence on temperature below T_c, indicating spin-singlet superconductivity. The expected reduction of the Knight shift following the superconducting transition can be estimated by using the Fermi liquid relation. In an f-electron system, the local magnetic susceptibility at low temperatures is possibly dominated by the Van-Vleck contribution. Thus, the contribution of quasiparticles to the Knight shift is not easily estimated only from the K-χ plot. Here, we apply the following Fermi liquid relation to the estimation of the quasiparticle contribution to the Knight shift in the present specimen[10],

$$\Delta K_s = A_{hf}\chi^{qp} = A_{hf}\frac{\gamma g_J^2 \mu_B^2 J(J+1)}{\pi^2 k_B^2} \times R, \quad (2)$$

where χ^{qp} represents the contribution of quasiparticles to spin susceptibility. We use the electronic specific heat coefficient γ = 350 mJ/molK^2[1], g_J=6/7, J = 5/2 for Ce^{3+} ($4f^1$ configuration). Since the Wilson ratio R (2> R >1) in the heavy fermion state is of the order of unity, we set R=1. Using the obtained hyperfine coupling constant A_{hf}, we estimate ΔK_s ~0.06 % for K_1, which is indicated by the arrow in Fig. 3. Observed reductions($|K_1(T_c) - K_1(0)|$) of the muon Knight shifts are roughly consistent with the value estimated by using Eq. (2). This indicates that the spin part of the muon Knight shift is explained by the Fermi liquid relation and consistent with the results of NMR experiments[3, 4]. In Fig. 4,

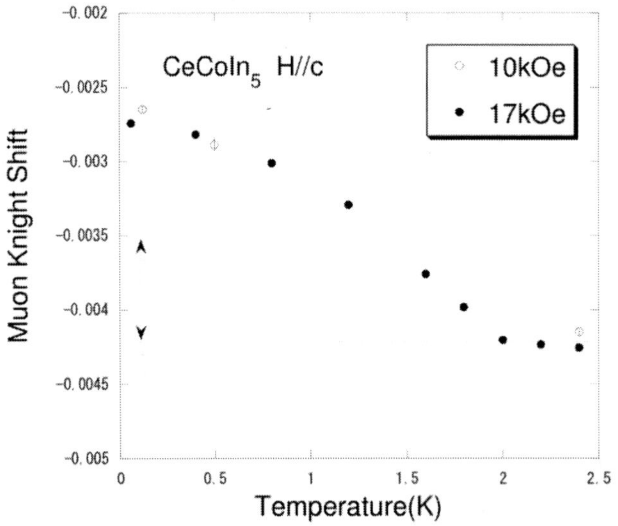

Figure 3. Temperature dependence of muon Knight shift K_1. Arrow indicates estimated reduction of the Knight shift in the superconducting phase according to Eq. (2)

the magnetic field dependence of the reduction of the muon Knight shift, $|K_1(T_c) - K_1(0)|$, is plotted. Above 20 kOe, the data in Ref. 5 is used. In general, K is proportion to the density of state $N(0)$. For a d-wave superconductor, it has been suggested that the density of state $N(0) \propto \sqrt{H}$, primarily originates from the continuous spectrum that is concentrated in the vicinity of the gap nodes and outside the vortex core[11]. Indeed, the magnetic field dependence of the specific heat of $CeCoIn_5$ follows the above relation[12]. However, the magnetic field dependence of the muon Knight shift is not proportional to $\sqrt{H}$[13]. Although we have no clear explanation for this inconsistency at present, this fact might indicate that the contribution from the vortex core is not reflected in the current muon Knight shift. Therefore, further investigation is required.

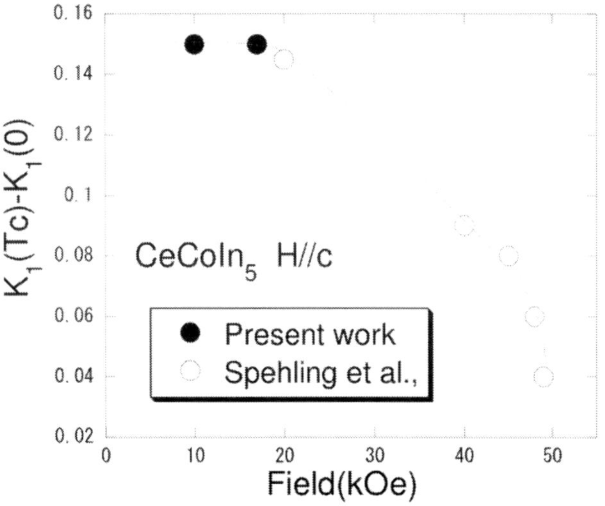

Figure 4. Magnetic field dependence of reduction of muon Knight shift, $|K_1(T_c) - K_1(0)|$ in $CeCoIn_5$. Open circles denote data taken from Ref. 5.Solid line is guide for eyes.

Acknowledgments

The authors acknowledge the staff of TRIUMF. This work was partially supported by a Grant-in-Aid for Scientific Research from the Ministry of Education, Culture, Sports, Science and Technology, Japan.

References

[1] Petrovic C, Pagliuso P G, Hundley M F, Movshovich R, Sarrao J L, Thompson JD, Fisk Z and Monthoux P 2001 *J. Phys.: Condens. Matter* **13** L337
[2] An K, Sakakibara T, Settai R, Ōnuki Y, Hiragi M, Ichioka M and Machida K 2010 *Phys. Rev. Lett.* **104** 037002
[3] Kohori Y, Yamato Y, Iwamoto Y, Kohara T, Bauer E D, Maple M B and Sarrao J L 2001 *Phys. Rev.* B **64** 134526
[4] Curro N J, Simovic B, Hammel P C, Pagliuso P G, Sarrao J L, Thompson J D and Martins G B 2001 *Phys. Rev.* B **64**, 180514
[5] Spehling J, Heffner R H, Sonier J E, Curro N, Wang C H, Hitti B, Morris G, Bauer E D, Sarrao J L, Litterst F J and Klauss H H 2009 *Phys. Rev. Lett.* **102** 207004
[6] Higemoto W, Koda A, Kadono R, Kawasaki Y, Haga Y, Aoki D, Settai R, Shishido H and Ōnuki Y 2002 *J. Phys. Soc. Jpn.* **71** 1023
[7] For example, Sonier J E, Brewer J H and Kiefl R F 2000 *Rev. Mod. Phys.* **72** 769
[8] de Gennes P G in *Superconductivity of Metals and Alloys*, 1966 W.A. Benjamin, Inc., New York, Amsterdam
[9] Settai R, Shishido H, Ikeda S, Murakawa Y, Nakashima M, Aoki D, Haga Y, Harima H and Ōnuki Y 2001 *J. Phys.: Condens. Matter* **13** L627
[10] Tou H, Ishida K and Kitaoka Y 2005 *J. Phys. Soc. Jap.* **74** 1245
[11] Volovik G E 1993 *JETP Lett.* **58** 469
[12] Aoki H, Sakakibara T, Shishido H, Settai R, Ōnuki Y, Miranovic P and Machida K 2004 *J. Phys.: Condens. Matter* **16** L13
[13] By formation of a flux line lattice, the peak position of the frequency distribution (*e.g.* Fig. 1) shifts to lower frequency. Therefore, the shift in flux lattice formation increases the Knight shift in this case.

International Symposium on Advanced Science Research 2009 IOP Publishing
Journal of Physics: Conference Series **225** (2010) 012042 doi:10.1088/1742-6596/225/1/012042

Magnetism in $Ce_2Rh(In,Sn)_8$ heavy-fermion compound

Kazuki Ohishi[1], Takao Suzuki[1], Robert H Heffner[2,3], Takashi U Ito[2], Wataru Higemoto[2], and Eric D Bauer[3]

[1]Advanced Meson Science Laboratory, Nishina Center for Accelerator-Based Science, RIKEN, Wako 351-0198, Japan
[2]Advanced Science Research Center, Japan Atomic Energy Agency, Tokai, Ibaraki 319-1195, Japan
[3]Los Alamos National Laboratory, Los Alamos, NM 87545, USA

E-mail: kazuki.ohishi@riken.jp

Abstract. Measurements of muon spin relaxation have been performed on $Ce_2RhIn_{8-x}Sn_x$ ($x = 0.0, 0.1, 0.5$ and 0.7) single crystals down to 0.3 K. A clear muon spin precession signal was observed in $x = 0.0$ and 0.1 samples due to antiferromagnetic ordering with magnetic volume fraction of ~75% below T_N. While an increase of relaxation rate was observed in $x = 0.5$ below 0.5 K, no precession signal was observed, suggesting the suppression of antiferromagnetism with increasing Sn concentration.

1. Introduction

Heavy-fermion compounds possess a characteristic temperature T^* signifying a crossover from localized behavior at high temperatures to coherent behavior at low temperatures, causing a reduced low-temperature resistivity and high Sommerfeld constant. This behavior is explained as the onset of coherent conduction electron scattering by the Kondo lattice of f spins (associated with low-temperature hybridization between f electrons and conduction electrons); at high temperatures the f electrons scatter the conduction electrons independently as local impurities. Recently, however, a semi-quantitative model has been proposed which significantly improves our understanding of the relation between the single-site Kondo effect and the formation of a Kondo liquid in the presence of intersite coupling [1].

It has been observed that heavy-fermion Knight shift anomalies, which are a deviation from a linear relation between the shift K and the magnetic susceptibility χ, occur at low temperatures. Generally, K is expressed as $K = A\chi/(N_A\mu_B)$, where A is the hyperfine coupling constant, N_A is Avogadro's number and μ_B is the Bohr magneton. Possible reasons for a non-linear K-χ relation can involve a change of the hyperfine coupling constant A with temperature or the onset of an additional component of χ, or both concurrently. An example of the former involves the depopulation of low-lying f electron crystalline-electric-field levels [2, 3], which can change A. A recently-introduced example of the latter is the so-called two-fluid model of heavy fermion formation [1, 4, 5, 6, 7, 8]. Here, in addition to the high-temperature χ representing the local f moments, a low-temperature susceptibility component χ_{KL}, characteristic of the coherent Kondo-liquid state below temperature T^*, is introduced. In our recent study of

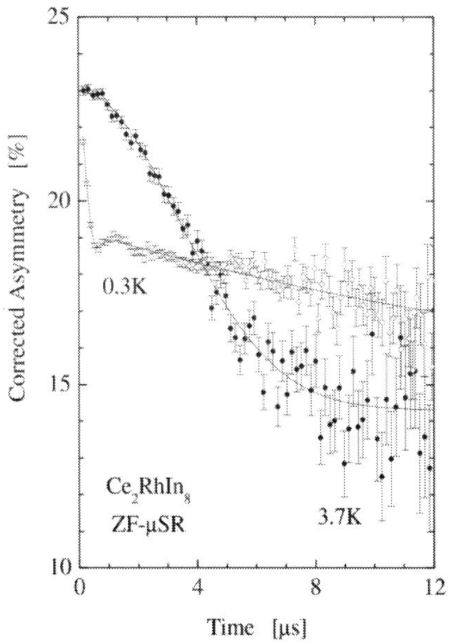

Figure 1. ZF-μSR time spectra in Ce_2RhIn_8 above and below T_N.

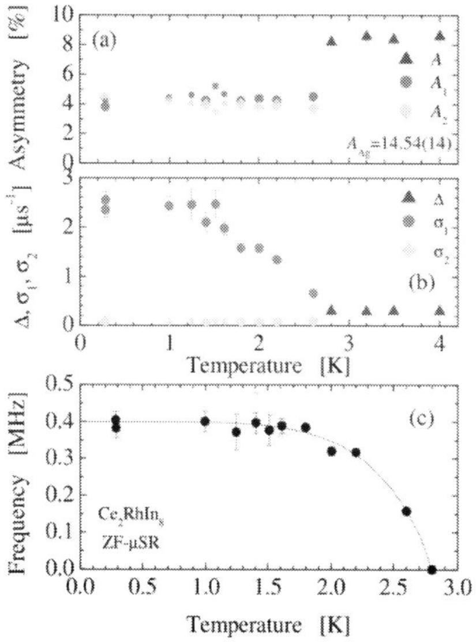

Figure 2. Temperature dependence of the (a) asymmetries, (b) relaxation rates and (c) frequency in Ce_2RhIn_8.

muon Knight shifts to investigate the formation of the heavy-fermion state in single crystals of $(Ce_{1-x}La_x)_2IrIn_8$ [8], we found that $T_c^* = 20.0(6)$ K and $T_a^* = 15.2(1.2)$ K for applied field $H \parallel c$-axis and $H \parallel a$-axis in Ce_2IrIn_8, respectively. For the Ce diluted systems $(Ce_{1-x}La_x)_2IrIn_8$, $x = 0.1, 0.25, 0.50, 0.70$ and 0.90, T^* decreases linearly for $x \leq 0.5$, reaching zero near $x = 0.7$, indicating the reduction of intersite f-spin correlations with Ce dilution.

Ce_2RhIn_8 is a known antiferromagnet with $T_N = 2.8$ K [9] and a crystal structure which is isostructural with that of Ce_2IrIn_8. Recently, Bauer *et al.* [10] have succeeded in substituting Sn for In in $Ce_2RhIn_{8-x}Sn_x$, and their specific heat studies suggest that the antiferromagnetic order is suppressed by Sn substitution for a critical concentration of about $x \sim 0.5$, suggesting a quantum critical point. Compared to the La-doped systems $[(Ce_{1-x}La_x)_2IrIn_8]$ the number of f electrons is kept constant in the Sn-doped systems, and the magnetic interaction changes.

We have performed muon Knight shift measurement and muon spin rotation and relaxation (μSR) measurements in order to investigate the development of the heavy-fermion and magnetic ground states in $Ce_2RhIn_{8-x}Sn_x$ ($x = 0, 0.1, 0.5, 0.7$). In this paper, we present preliminary results regarding the magnetic ground states in these systems.

2. Experimental Details

Single crystalline samples were grown by the In-flux method at Los Alamos National Laboratory [11]. Conventional time-differential μSR experiments were carried out over a temperature between 0.3 K and 10 K under zero applied field (ZF) with the ARGUS spectrometer at the RIKEN-RAL Muon Facility in the U.K. which provides pulsed beams of nearly 100% spin-polarized muons. Samples were mounted on high-purity silver plates in a ^{3}He cryostat with their c-axis parallel to the initial muon spin polarization ($c \parallel P_\mu$).

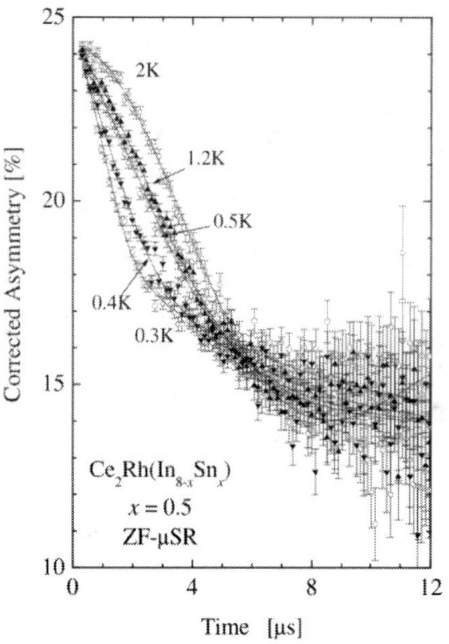

Figure 3. ZF-μSR time spectra in $Ce_2RhIn_{8-x}Sn_x$ ($x = 0.5$) at various temperatures.

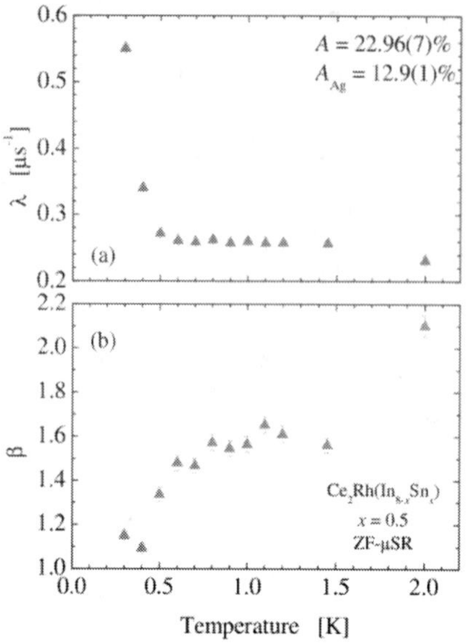

Figure 4. Temperature dependence of the (a) relaxation rate λ and the exponent β in $Ce_2RhIn_{8-x}Sn_x$ ($x = 0.5$).

3. Results and Discussion

Figure 1 shows ZF-μSR time spectra at 3.7 K and 0.3 K in Ce_2RhIn_8. At temperatures above T_N, the time spectra consist of two components, one ($P_z(t)$) showing slow relaxation due to random local fields from nuclear magnetic moments, and the other nearly independent of time. More specifically, we have

$$A(t) = AP_z(t) + A_{Ag}, \tag{1}$$

where A and A_{Ag} are the initial μ-e decay asymmetries from the sample and silver sample holder, respectively, and $P_z(t) = \exp(-\Delta^2 t^2/2)$, with Δ/γ_μ being the rms value of the random local field. (Here $\gamma_\mu = 2\pi \times 135.54$ MHz/T is the muon gyromagnetic ratio.) The time spectra were well reproduced by the above formula with $\Delta = 0.303(7)\ \mu s^{-1}$. The relaxation rate Δ is mostly due to ^{113,115}In nuclear moments, and is slightly larger than found previously [12]. For $T < T_N$, we found that the spectra are well described by

$$A(t) = A_1 \exp(-\sigma_1^2 t^2) \cos(2\pi f t + \phi) + A_2 \exp(-\sigma_2^2 t^2) + A_{Ag}, \tag{2}$$

where σ_1 and σ_2 are the relaxation rates, f is the muon spin precession frequency under a local field H_{loc} and ϕ is the initial phase. As shown in Fig. 1, a clear precession signal was observed below T_N. Figure 2 shows the temperature dependence of the fractions A, A_1 and A_2, the rates σ_1 and σ_2, and the frequency f. The temperature dependence of f is well reproduced by the empirical expression $f = f_0[1-(T/T_N)^\alpha]^\gamma$ with $T_N = 2.79(1)$ K, $f_0 = 0.40(1)$ MHz, $\alpha = 5.5(1.0)$ and $\gamma = 0.8(1)$ [solid line in Fig. 2(c)]. The volume fraction of the antiferromagnetic component at 0.3 K was estimated to be $\sim 75\%$. These results are quantitatively consistent with earlier literature [12, 13], except that the observed frequency is a little smaller.

A fitting procedure similar to that for the parent compound was applied to the $Ce_2RhIn_{8-x}Sn_x$ ($x = 0.1$) data. A clear muon spin precession signal was observed in $x = 0.1$, but T_N is sharply reduced compared to $x = 0$. T_N was estimated to be 1.8 K. The temperature dependence of the frequency is similar in behavior to Fig. 2(c).

Figure 3 shows the ZF-μSR time spectra in a sample with $x = 0.5$ at various temperatures. While the relaxation rate above 2 K is almost independent of temperature, there is a gradual increase with decreasing temperature. As shown in figure 3, no muon spin precession signal was observed at the lowest temperature ($T = 0.3$ K), suggesting the absence of magnetic ordering at this temperature. The time spectra were described by Eq. (1) with stretched exponential function $P_z(t) = \exp[-(\lambda t)^{\beta}]$, where λ is the relaxation rate and β is the power of the exponent. The temperature dependence of the fitted parameters is shown in Fig. 4. The lineshape shows a change from Gaussian ($\beta = 2$) to single exponential decay ($\beta = 1$) with decreasing temperature. The rate λ is almost independent of temperature above 0.5 K, and starts to increase rapidly with decreasing temperature below 0.5 K. Such fast relaxation without any detectable oscillation is a common signature of disordered magnetism that is either static or dynamically fluctuating. Application of longitudinal fields showed that these magnetic signals completely decoupled between 100–200 mT at 0.3 K, consistent with a static distribution of local fields. It suggests a spin-glass like phase which was observed in isostructural material Ce_2IrIn_8 [12]. However, further experiments below 0.3 K are necessary to clarify the magnetic ground state.

No muon spin precession was observed in $x = 0.7$ sample (not shown here). The data for $x = 0.7$ were fitted with a stretched exponential function, and show similar behavior to that observed in the $x = 0.5$ material, except that λ for $x = 0.7$ does not show the increase below 0.5 K seen in Fig. 4. Instead $\lambda(T)$ is approximately constant at $\sim 0.29\ \mu s^{-1}$ between 2 K and 0.3 K.

4. Conclusions

We have performed ZF- and LF-μSR measurements in Sn-doped single crystals of $Ce_2RhIn_{8-x}Sn_x$ ($x = 0.0, 0.1, 0.5$ and 0.7). We find a signature of magnetic order corresponding to a static local field in the $x = 0.0$ and 0.1 samples. For the $x \geq 0.5$ samples, we find no coherent precession of the muon spin, and, hence, no magnetic order. Instead, we observe an increase of relaxation rate below 0.5 K in $x = 0.5$ and a temperature independent relaxation rate at all measured temperatures in $x = 0.7$. While a more detailed analysis is still in progress, our microscopic μSR measurements find that a suppression of magnetic ordering occurs with increasing Sn concentration in $Ce_2RhIn_{8-x}Sn_x$, consistent with specific heat measurements [10].

References

[1] Yang Yi-Feng, Fisk Zachary, Lee Han-Oh, Thompson J D, and Pines D 2008 *Nature* **454** 611
[2] MacLaughlin D E, Peña O, and Lysak M J 1981 *Phys. Rev. B* **23** 1039
[3] Ohama T, Yasuoka H, Mandrus D, Fisk Z, and Smith J L 1995 *J. Phys. Soc. Jpn.* **64** 2628
[4] Nakatsuji S, Yeo S, Balicas L, Fisk Z, Schlottmann P, Pagliuso P G, Moreno N O, Sarrao J L, and Thompson J D 2002 *Phys. Rev. Lett.* **89** 106402
[5] Nakatsuji S, Pines D, and Fisk Z 2004 *Phys. Rev. Lett.* **92** 016401
[6] Curro N J, Young B -L, Schmalian J, and Pines D 2004 *Phys. Rev. B* **70** 235117
[7] Yang Y -F, and Pines D 2008 *Phys. Rev. Lett* **100** 096404
[8] Ohishi K, Heffner R H, Ito T U, Higemoto W, Morris G D, Hur N, Bauer E D, Sarrao J L, Thompson J D, MacLaughlin D E, and Shu L 2009 *Phys. Rev. B* **80** 125104
[9] Thompson J D, Movshovich R, Fisk Z, Bouquet F, Curro N J, Fisher R A, Hammel P C, Hegger H, Hundley M F, Jaime M, Pagliuso P G, Petrovic C, Phillips N E and Sarrao J L 2001 *J. Magn. Magn. Mater.* **266-230** 5
[10] Bauer E D, unpublished result
[11] Macaluso R T, Sarrao J L, Moreno N O, Pagliuso P G, Thompson J D, Fronczek F R, Hundley M F, Malinowski A and Chan J Y 2003 *Chan. Chem. Mater.* **15** 1394
[12] Morris G D, Heffner R H, Moreno N O, Pagliuso P G, Sarrao J L, Dunsiger S R, Nieuwenhuys G J, MacLaughlin D E, and Bernal O O 2004 *Phys. Rev. B* **69** 214415
[13] Schenck A, Gygax F N, Ueda T, and Ōnuki Y 2004 *Phys. Rev. B* **70** 054415

International Symposium on Advanced Science Research 2009 IOP Publishing
Journal of Physics: Conference Series **225** (2010) 012021 doi:10.1088/1742-6596/225/1/012021

μ^+ diffusion in cubic f-electron compounds observed by high transverse field μ^+SR

T. U. Ito[1], W. Higemoto[1], K. Ohishi[2], N. Nishida[3], R. H. He ner[1,4], Y. Aoki[5], H. S. Suzuki[6], T. Onimaru[7], H. Tanida[7], and S. Takagi[8]

[1] Advanced Science Research Center, Japan Atomic Energy Agency, Tokai, Ibaraki 319-1195, Japan
[2] Advanced Meson Science Laboratory, RIKEN, Wako, Saitama 351-0198, Japan
[3] Department of Physics, Tokyo Institute of Technology, Meguro, Tokyo 152-8551, Japan
[4] Los Alamos National Laboratory, Los Alamos, New Mexico 87545, USA
[5] Department of Physics, Tokyo Metropolitan University, Hachioji, Tokyo 192-0397, Japan
[6] National Institute for Materials Science, Tsukuba, Ibaraki 305-0047, Japan
[7] Department of Quantum Matter, ADSM, Hiroshima University, Higashi-Hiroshima, Hiroshima, 739-8530, Japan
[8] Department of Physics, Tohoku University, Sendai, Miyagi 980-8578, Japan

E-mail: `ito.takashi15@jaea.go.jp`

Abstract. Positive muon ($^+$) di usion was observed in cubic f-electron compounds $PrPb_3$ and $SmAg_2In$ by means of the $^+$ spin rotation ($^+$SR) method in high transverse eld applied along the [001] direction. The applied eld splits one crystallographic site into two magnetically inequivalent sites. The $^+$ di usion between these magnetically di erent sites drastically in uences the $^+$SR frequencies, so that two lines are gradually merged into one line at increasing temperature due to the motional narrowing e ect. The $^+$ hopping rate Ω derived from the $^+$SR line shape follows the Arrhenius law, namely, $\Omega = \Omega_0 \exp(-U/k_BT)$ in both cases. The preexponential factor Ω_0 suggests that the dominant mechanism of $^+$ di usion is classical over-the-barrier hopping in $PrPb_3$ for $140<T<200$ K and phonon-assisted quantum tunneling in $SmAg_2In$ for $40<T<70$ K.

1. Introduction

The di usion of positive muons ($^+$) in condensed matter has been investigated for many years using the $^+$ spin rotation/relaxation ($^+$SR) technique. These $^+$SR studies have provided deep insight into quantum di usion of a light charged particle at low temperatures, as well as a hopping process at high temperatures [1, 2]. These processes are closely related to hydrogen dynamics in materials since the $^+$ can be regarded as a light isotope of ^{1}H. Knowledge of $^+$ di usion is therefore important, not only for fundamental science, but also for industrial applications.

Information about motion of the $^+$ can be deduced from the dynamical properties of the hyper ne eld that the $^+$ feels in materials. The dynamical nature of the hyper ne eld is often investigated by monitoring narrowing of a single $^+$SR line caused by the $^+$ motion between equivalent interstitial sites. For precise determination of the $^+$ hopping rate, the local eld at an interstitial site must be constant in magnitude and static within the $^+$SR time-window over the temperature range of interest. However, this condition frequently breaks down in f-electron compounds, in which magnetism of lanthanide or actinide ions causes a temperature-dependent $^+$ spin relaxation. In such cases, the high transverse eld (TF) $^+$SR method is useful for investigation of the $^+$ dynamics.

The high TF- $^+$SR study of the $^+$ di usion is based on the observation of nuclear spin di usion by NMR [3]. A high TF applied along a certain crystal axis induces a magnetic moment at the ion which leads to symmetry lowering at the probe site. The magnetic environment of the probe site can split into more than two di erent environments even if the probe site is crystallographically unique. Di usion of the probe between the magnetically inequivalent sites drastically in uences the NMR or $^+$SR lines in frequency space, where multiple lines are gradually merged into one line with increasing temperature due to the motional narrowing e ect.

The rst high TF- $^+$SR investigation of the $^+$ di usion in *f*-electron compounds was performed by Alexandrowicz *et al.* in $PrIn_3$ which has a cubic $AuCu_3$-type structure; high sensitivity of this technique to the $^+$ di usion was demonstrated [4]. In this pioneering paper it was suggested that isomorphic $PrPb_3$ is a candidate system potentially amenable for such experiments. In this paper, we report experimental results using high TF- $^+$SR in $PrPb_3$, as well as successful observation of $^+$ motion in $SmAg_2In$, which has a cubic Heusler-type crystal structure.

2. Experiment

Single-crystalline samples of $PrPb_3$ and $SmAg_2In$ were grown by the Bridgeman method. The $^+$SR measurements were carried out using the HiTime spectrometer at the M15 beam line, TRIUMF, Canada. A single-crystalline sample with a (001) plane surface was mounted on a sample holder. A magnetic eld was applied along the [001] direction and the positive muons were implanted into the sample with an initial $^+$ spin polarization perpendicular to the [001] direction. The magnitude of the applied eld was 2 T for $PrPb_3$ and 6 T for $SmAg_2In$.

3. $^+$SR line shape in $PrPb_3$ and analysis

The $^+$SR line shape in $PrPb_3$ in the paramagnetic state is shown in Fig. 1 (left). Two lines with peaks at ω_1 and ω_2 with the intensity ratio of 2:1 were observed in the temperature range below 160 K. Here the implanted $^+$ is almost stationary at an interstitial site, within the $^+$SR time-window. This low temperature behavior has been previously reported in Ref. [5, 6]. Among possible interstitial sites, only the *d* site at (0.5, 0, 0) is consistent with this splitting. This site is located at the midpoint of two Pr ions, as depicted in the inset of Fig. 1 (left). The two lines are gradually merged into one line with increasing temperature as a result of the motional narrowing.

The $^+$ hopping rate Ω between the two magnetic sites can be extracted from the $^+$SR line shape with the Anderson-Weiss approach, assuming a Markov-type hopping process [7]. The $^+$SR line shape is expressed using the notation of Ref. [3] as follows,

$$I(\omega') = \frac{L^3 + 3L^2\Omega + \frac{4}{3}\delta^2\Omega + L[\omega' \quad \frac{2}{3}\delta\omega' + \delta^2 + \frac{9}{4}\Omega^2]}{L^4 + (\omega'^2 \quad \delta^2)^2 + 3L^3\Omega + \Omega L(3\omega'^2 + 2\delta\omega' + 3\delta^2) + \frac{1}{4}\Omega^2(3\omega' + \delta)^2 + L^2[2(\omega'^2 + \delta^2) + \frac{9}{4}\Omega^2]}, \tag{1}$$

where $\omega_0 = (\omega_1 + \omega_2)/2$, $\delta = (\omega_2 \quad \omega_1)/2$, $\omega' = \omega \quad \omega_0$, L is the intrinsic Lorentzian linewidth of the signals at $\omega_{1,2}$ in the absence of hopping. It should be noted that this theory requires that the $^+$SR frequencies ω_1 and ω_2 expected for *static* $^+$ are to be used in the motional narrowing regime at high temperatures. We estimated these values from low temperature data, assuming a linear relation between the peak frequencies and the susceptibility . The result of linear ts to the peak frequencies between 130 K and 150 K is shown in Fig. 1 (right) with solid lines. Using values for ω_1 and ω_2 derived from these linear functions, we tted the $^+$SR line shape data with Eq. 1 and obtained satisfactory results, as shown in Fig. 1 (left) by the solid curves.

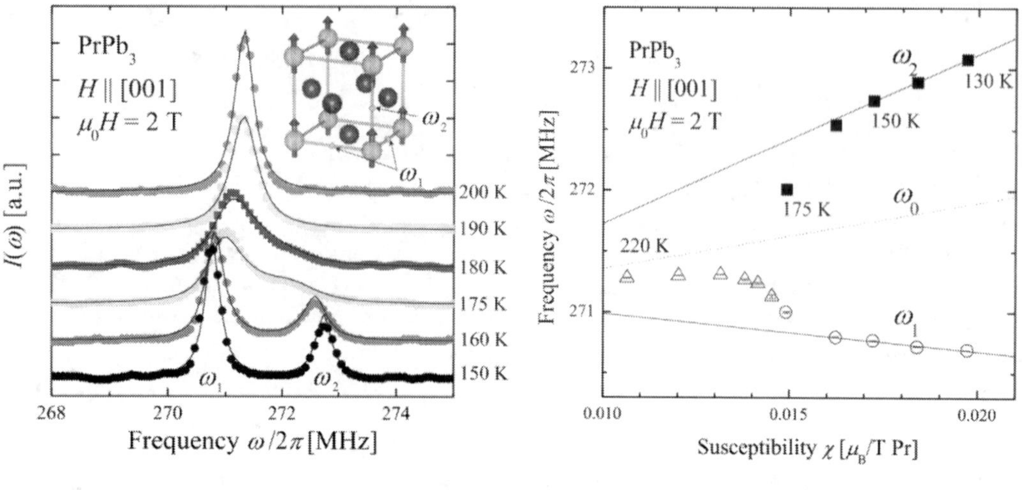

Figure 1. Left: the $^+$SR line shape in $PrPb_3$ under a eld of 2 T applied along the [001] direction. The solid curves are the best ts with Eq. 1. The inset shows a schematic view of a unit cell and representative $^+$ sites. The thick arrows represent the magnetic moments induced by the external eld at the Pr site along the eld direction. Right: the $^+$SR frequencies as functions of the susceptibility . The solid lines $\omega_1(T)$ and $\omega_2(T)$ represent the linear ts to the data between 130 K and 150 K. The dotted line corresponds to $\omega_0(T) = (\omega_1 + \omega_2)/2$.

In all of the ts L was xed to be 0.142 s ^{1}, which was determined from the data at 130 K. The only free parameter Ω was obtained as a function of temperature, and is discussed in Sec. 5.

4. $^+$SR line shape in $SmAg_2In$ and analysis

The $^+$SR line shape in $SmAg_2In$ in the paramagnetic state is shown in Fig. 2 (left). Two lines with peaks at ω_1 and ω_2 with the intensity ratio of 1:2 were observed in the $^+$ localization regime below 42.8 K. Note that the intensity ratio is opposite to that of the signal in $PrPb_3$. Among possible interstitial sites, only the d site at (0.25, 0.25, 0) is consistent with this splitting. This site is located at the face center of the cube constructed by four Sm and four In atoms at the corners, and one Ag atom at the center, as depicted in the inset of Fig. 2 (left). The same site was suggested in isomorphic $PrAg_2In$ from the $^+$ spin relaxation rate in zero applied eld [8]. The two lines are gradually merged into one line at increasing temperature, as was observed in $PrPb_3$.

In this case the $^+$SR line shape is expressed as

$$I(\omega') = \frac{L^3 + 3L^2\Omega + \frac{4}{3}\delta^2\Omega + L[\omega' + \frac{2}{3}\delta\omega' + \delta^2 + \frac{9}{4}\Omega^2]}{L^4 + (\omega'^2 \quad \delta^2)^2 + 3L^3\Omega + \Omega L(3\omega'^2 \quad 2\delta\omega' + 3\delta^2) + \frac{1}{4}\Omega^2(3\omega' \quad \delta)^2 + L^2[2(\omega'^2 + \delta^2) + \frac{9}{4}\Omega^2]}. \tag{2}$$

As before, the $^+$SR frequencies ω_1 and ω_2 used in the motional narrowing regime were estimated from the linear relation between the peak frequencies and in the localization regime. The result of linear ts to the peak frequencies between 20 K and 42.8 K is shown in Fig. 2 (right) by the solid lines. Using values for ω_1 and ω_2 derived from these linear functions, we tted the $^+$SR line shape data with Eq. 2, and obtained the satisfactory results as shown in Fig. 2 (left) by the solid curves. In these ts L was xed to be 0.133 s ^{1}, which was determined from the data at 42.8 K. The only free parameter Ω was obtained as a function of temperature, and is discussed in Sec. 5.

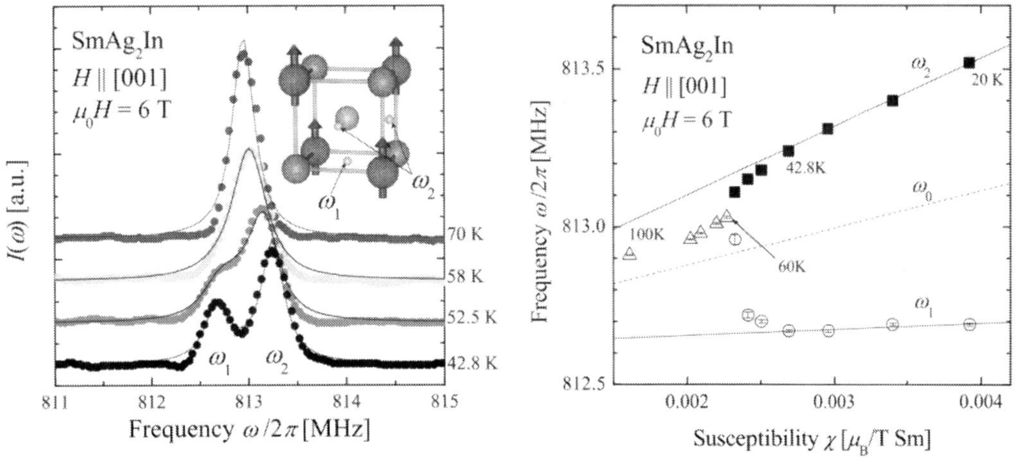

Figure 2. Left: the +SR line shape in $SmAg_2In$ under a eld of 6 T applied along the [001] direction. The inset shows a schematic view of a part of a unit cell and representative + sites. The solid curves are the best ts with Eq. 2. The thick arrows represent the magnetic moments induced by the external eld at the Sm site along the eld direction. Right: the +SR frequencies as functions of the susceptibility . The solid lines $\omega_1(T)$ and $\omega_2(T)$ represent linear ts to the data between 20 K and 42.8 K. The dotted line corresponds to $\omega_0(T) = (\omega_1 + \omega_2)/2$.

5. Results and Discussion

The + hopping rate Ω in $PrPb_3$ and $SmAg_2In$ derived from the analytic treatment of the +SR line shape is shown by the Arrhenius plot in Fig. 3. The apparent linear behavior suggests that the Ω follows a thermal activation-type temperature dependence, namely,

$$\Omega(T) = \Omega_0 \exp\left(\frac{U}{k_B T} \right). \tag{3}$$

Satisfactory ts were obtained using this equation as shown in Fig. 3 with solid lines. This indicates that classical over-the-barrier hopping or phonon-assisted quantum tunneling should be dominant in the present cases; even in the latter case the $\Omega(T)$ approximately follows Eq. 3 when $T < U/k_B$. The preexponential factor Ω_0 and the *potential barrier* U are shown in Table 1, together with a parameter U for $PrIn_3$ obtained in Ref. [4].

The value of Ω_0 is useful to discriminate between the classical over-the-barrier hopping and the phonon-assisted quantum tunneling by comparing the value expected for a zero-point vibrational frequency of the + in a potential well, which is typically in the order of 10^{13} s 1 [9]. In the case of $PrPb_3$, the Ω_0 value is comparable to this value, indicating that the classical over-the-barrier hopping is dominant above 140 K. On the other hand, the Ω_0 value for $SmAg_2In$ is considerably smaller than this characteristic value. This suggests that the the phonon-assisted quantum tunneling comes into play in $SmAg_2In$ in the temperature range between 40 K and 70 K, although the zero-point vibrational frequency of the + in this compound has not been accurately determined. The low value of U/k_B also supports this interpretation.

6. Summary

The + di usion was observed in cubic f-electron compounds $PrPb_3$ and $SmAg_2In$ by means of the high TF- +SR technique in magnetic elds applied along the [001] direction. The + di usion between magnetically inequivalent sites drastically in uences the +SR frequencies,

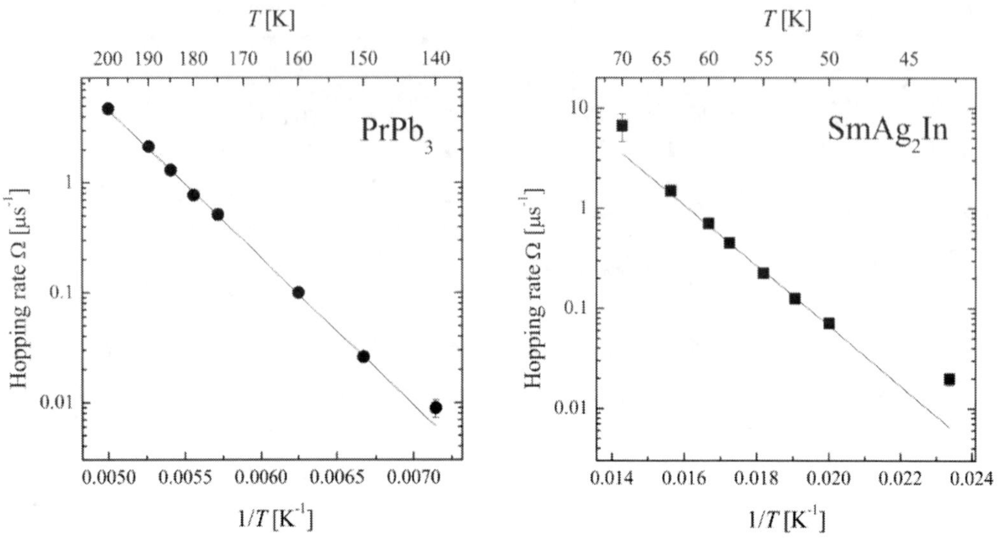

Figure 3. The Arrhenius plot of the $^{+}$ hopping rate Ω in $PrPb_3$ (left) and $SmAg_2In$ (right). The solid lines are the best ts with Eq. 3.

Table 1. The $^{+}$ hopping parameters in $PrPb_3$, $PrIn_3$, and $SmAg_2In$.

	Ω_0 [s ^{1}]	U/k_B [K]
$PrPb_3$	2.3±0.8 10^{13}	3093±60
$PrIn_3$ [4]	—	3260±45
$SmAg_2In$	7.1±1.5 10^{10}	694±11

where two lines are gradually merged into one line at increasing temperature due to the motional narrowing e ect. The $^{+}$ hopping rate Ω derived from the $^{+}$SR line shape follows the Arrhenius law, namely, $\Omega = \Omega_0 \exp(U/k_BT)$ in both cases. The preexponential factor Ω_0 suggests that the dominant mechanism of $^{+}$ di usion is the classical over-the-barrier hopping in $PrPb_3$ for $140<T<200$ K and the phonon-assisted quantum tunneling in $SmAg_2In$ for $40<T<70$ K.

Acknowledgement
We thank sta of TRIUMF for experimental assistance. This work was supported by a Grant-in-Aid for Scienti c Research on Priority Area "Skutterudite", Ministry of Education, Culture, Sports, Science and Technology, Japan, KEK-MSL Inter-University Program for Oversea Muon Facilities, and 21st Century COE Program at Tokyo Tech "Nanometer-Scale Quantum Physics".

References
[1] Kadono R *et al.* 1989 *Phys. Rev.* B **39** 23
[2] Luke G M *et al.* 1991 *Phys. Rev.* B **43** 3284
[3] Abragam A 1961 *Principles of Nuclear Magnetism* (Clarendon Oxford) 447-451
[4] Alexandrowicz G *et al.* 1999 *Phys. Rev. Lett.* **82** 1028
[5] Ito T U *et al.* 2007 *J. Mag. Mag. Mat.* **310** 743
[6] Ito T U *et al.* 2009 *Phys. Rev. Lett.* **102** 096403

[7] Anderson P W *et al.* 1954 *J. Phys. Soc. Jpn.* **9** 316
[8] MacLaughlin D E *et al.* 2000 *Phys. Rev.* B **61** 555
[9] Storchak V G and Prokof'ev N V 1998 *Rev. Mod. Phys.* **70** 929

Journal of Physics: Conference Series **225** (2010) 012055 doi:10.1088/1742-6596/225/1/012055

SR study on the spin singlet state in the $S = 1/2$ cluster magnet $GaNb_4S_8$

Y Tabata[1], Y Kajinami[1], T Waki[1], I Watanabe[2] and H Nakamura[1]
[1]Department of Materials Science and Engineering, Kyoto University, Kyoto 606-8501, Japan
[2]Advanced Meson Science Laboratory, RIKEN Nishina Center, Wako 351-0198, Japan

E-mail: y.tabata@ht4.ecs.kyoto-u.ac.jp

Abstract. $GaNb_4S_8$ is a $S = 1/2$ Nb_4-cluster magnet exhibiting a structural phase transition at $T_S = 32$ K. We have investigated the magnetic ground state of the compound by means of the zero-field (ZF) and the longitudinal-field (LF) μSR techniques. In the ZF-μSR spectra, no muon spin precession was observed below T_S, revealing an absence of magnetic long-range orders. Instead, a stretched exponential relaxation was observed, which remains under the LF up to 3950 G. The field dependence of the relaxation rate obeys the Redfield's equation, indicating that a dynamic component of electronic spin fluctuations is dominant below T_S. These μSR results can be understood as sporadic dynamics of spin fluctuations in a spin singlet sea and support the spin singlet state of $GaNb_4S_8$ due to the formation of Nb_8 octamer from two Nb_4 tetramers.

A mixed-valent magnetic cluster system, in which more than one atoms share unpaired electrons within the cluster unit, is very attractive as an intermediate system between the localized spin system and the itinerant electron magnet. As a category of such type materials, ternary chalcogenides AB_4X_8 (A = Ga, Al, Ge; B = V, Mo, Nb, Ta; X = S, Se) with the cubic $GaMo_4S_8$ structure (space group $F\bar{4}3m$) are of particular interest [1, 2, 3, 4]. In these compounds, cubic $(B_4X_4)^{n+}$ and tetrahedral $(AX_4)^{n-}$ ions are weakly coupled in the NaCl manner, resulting in hopping conduction between the clusters. They are in general semiconducting with a narrow gap. In the $(B_4X_4)^{n+}$ ion, one can find a tetrahedral cluster (tetramer) of the magnetic B-atoms, where d-electrons form cluster orbitals. The combination of intra- and inter-cluster interactions leads a rich variety in their magnetic properties.

$GaNb_4S_8$ has attracted attention as a sort of Mott insulator, where 4d-electrons are localized at Nb_4 clusters due to electron correlations, and also as one of pressure-induced superconductors [1]. In the molecular orbital schemes, the tetrahedral Nb_4 cluster in $(Nb_4S_4)^{5+}$ shares seven 4d-electrons, resulting in one unpaired electron at the highest orbital (HOMO). Thus, the Nb_4 tetramer has a local moment with $S = 1/2$. The experiments support the proposed molecular orbital schemes. The magnetic susceptibility obeys the Curie-Weiss law with the effective moment $\mu_{eff} = 1.73\ \mu_B$/f.u. [1]. The negative large Weiss constant $\theta = -298$K suggests antiferromagnetic (AF) coupling among clusters. The compound exhibits a structural phase transition at $T_s = 32$ K from the high temperature cubic state to the tetragonal state ($P\bar{4}2_1m$) [2]. It is possibly due to the Jahn-Teller type instability of the triply degenerate HOMO of Nb_4 tetramers. A detailed study [5] indicates the magnetic ground state in $GaNb_4S_8$ is a non-magnetic spin singlet state. Refering lattice symmetry [2], it is suggested the spin singlet state is realized by forming Nb_8 octamer from two Nb_4 tetramers

In this article, we report results of μSR experiments in zero-field (ZF) and longitudinal-field (LF). The results clearly support an absence of static internal field and support the spin singlet ground state in $GaNb_4S_8$.

All polycrystalline samples were prepared by solid state reaction from pure elements sealed in evacuated quartz tubes. Final heat treatment was performed at 900 °C for 1 days after several heat treatments and intermediate regrinds. Sample quality was checked by powder X-ray diffraction at room temperature and the temperature (T) dependence of the magnetic susceptibility measured by commercial type SQUID magnetometer (MPMS). ZF and LF μSR experiments were made at the RIKEN-RAL Muon Facility at the Rutherford-Appleton Laboratory in the UK using a pulsed positive surface muon beam in the T-range between 4.5 and 100 K under LF of 0 – 3950 G.

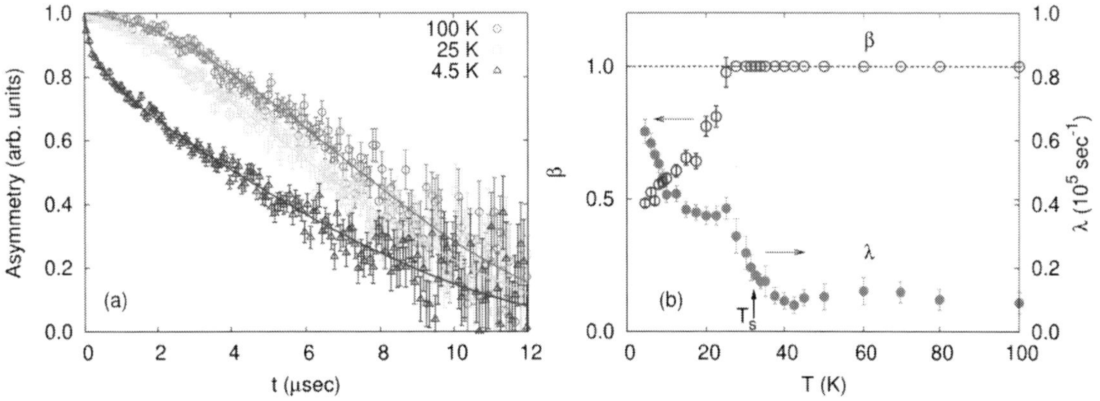

Figure 1. (color online)(a) Typical Examples of ZF-μSR spectra at $T = 4.5$, 25 and 100 K. Solid curves represent the fit by the damped KT function described in the text. Below T_s , the spectra are fit with the phenomenological stretched exponential function. (b) The T-dependences of λ and β. β is fixed to be 1 above 30 K.

Figure 1 shows the result of the ZF μSR experiment. The ZF μSR spectra, shown in Fig. 1 (a), exhibit no oscillation due to the muon precession caused by static internal field even at the lowest temperature. It clearly indicates an absence of magnetic long-range order below T_s. Above T_s, the relaxation is dominated by Gaussian-like depolarization and can be fitted by the common damped Kubo-Toyabe (KT) function $P_\mu(t) = \exp(-\lambda t)G_z^{\rm KT}(\Delta, t)$ with $G_z^{\rm KT}(\Delta, t) = \frac{1}{3} + \frac{2}{3}(1 - \Delta^2 t^2)\exp(-\frac{1}{2}\Delta^2 t^2)$. Δ/γ_μ (γ_μ: muon gyromagnetic ratio) is the width of the static Gaussian field distribution and λ is the damping rate associated with an additional relaxation process mostly due to electron spin fluctuations. Above T_s, we obtained nearly T-independent $\Delta \simeq 1 \times 10^5$ sec^{-1} and $\lambda \simeq 1 \times 10^4$ sec^{-1} by treating Δ and λ as free parameters. The value of Δ corresponds to a tiny static field of $\Delta/\gamma_\mu \sim 1$ G, which is ascribed to dipolar fields coming from randomly oriented nuclear spins. With decreasing T, the damping becomes faster as shown in Fig. 1 (a), and the above function turns to be inappropriate. Thus, we phenomenologically applied a stretched exponential function of $P_\mu(t) = \exp[-(\lambda t)^\beta]G_z^{\rm KT}(\Delta, t)$ to fit the low-T data, where β is a free fitting parameter. In addition, the fast damping part tends to mask the Gaussian deporlarization due to small static nuclear dipolar field. Hence, we fixed Δ to be 1×10^5 sec^{-1} to analyze the low T data with a consideration of the T-independence of Δ at high T. We show T-dependences of λ and β in Fig. 1 (b). One can see a gradual increase of λ as decreasing T passing through T_s. Furthermore, β gradually deviate from 1 around 30 K and reaches to 0.5 at the lowest T, suggesting gradual development of distribution of relaxation time. The enhancement of λ suggests a development of dynamic fluctuating field. Another possible origin of the enhancement of λ is the development of static field having a Lorentzian-type distribution, as observed in spin glass materials [6].

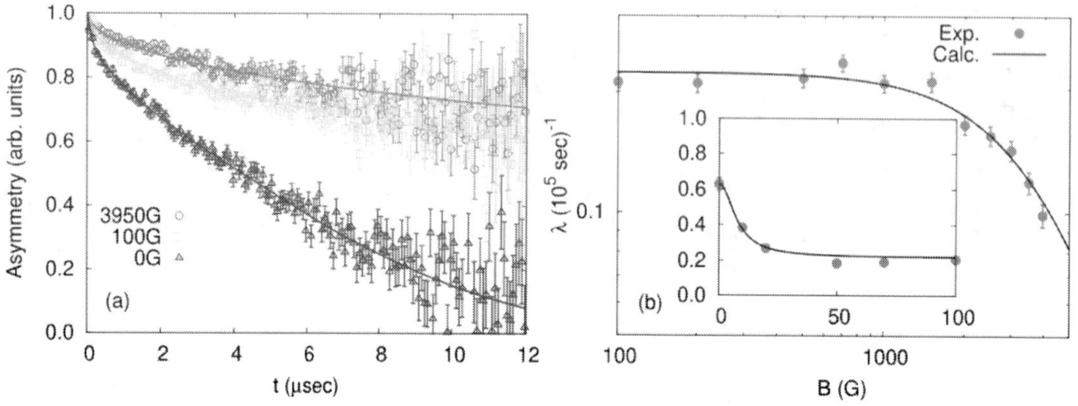

Figure 2. (a) Typical examples of LF-μSR spectra at $T = 4.5$ K. Solid curves represent the fit by the product of a stretched exponential function and a LF KT function described in the text. β is fixed to be 0.5. (b) The field dependence of λ. In the main figure, the data above 100 G are shown. Inset shows the data below 100 G. Solid curve represents the fit by the sum of two Redfield's equations, described in Eq. 1

To clarify the origin of the enhancement of λ in low T, we show the results of the LF μSR experiment at the lowest $T = 4.5$ K in Fig. 2. First, we should note that the relaxation still maintain under the longitudinal field of 3950 G and it comes from dynamic fluctuating field. Figure 2 (b) shows a field dependence of λ. In the analysis of the data under LF, we used the product of a stretched exponential function and a LF KT function $P_\mu(t) = \exp[-(\lambda t)^\beta] G_z^{\rm KT}(\Delta, t; B)$. We assumed the field-indepndent value of $\beta = 0.5$ for the simplicity. The assumed function describes the experimental data well, as shown in Fig. 2 (a). The field dependence of λ suggests two components of λ; namely, one showing a rapid decoupling by applied LF and the other being more insensitive to LF. The field dependence of λ obeys a sum of two Redfield's equations,

$$\lambda(B) = \lambda_{\rm a}(B) + \lambda_{\rm b}(B) = \frac{2\Lambda_{\rm a}^2 \tau_{\rm ca}}{1 + (\gamma_\mu B \tau_{\rm ca})^2} + \frac{2\Lambda_{\rm b}^2 \tau_{\rm cb}}{1 + (\gamma_\mu B \tau_{\rm cb})^2}, \tag{1}$$

where $\lambda_{\rm a}(B)$ and $\lambda_{\rm b}(B)$ correspond to the field-sensitive and -insensitive components of λ, respectively. Equation 1 describes the field dependence of λ well, as shown in Fig. 2 (b). $\Lambda_{\rm a}/\gamma_\mu$ and $\Lambda_{\rm b}/\gamma_\mu$ are magnitudes of fluctuating fields and $\tau_{\rm ca}$ and $\tau_{\rm cb}$ are characteristic correlation times of the fluctuations. By fitting the experimental data, we obtained $\Lambda_{\rm a}/\gamma_\mu = 1.4$ G, $\Lambda_{\rm b}/\gamma_\mu = 22$ G $\tau_{\rm ca} = 1.5$ μsec and $\tau_{\rm cb} = 3.0$ nsec. It clearly indicates a presence of dynamic fluctuating field and the faster fluctuating component is dominant. The relatively large $\tau_{\rm ca}$ may suggests that the $\lambda_{\rm a}$ comes from "quasi" static field due to random spin-freezing, and the much smaller value of $\Lambda_{\rm a}/\gamma_\mu$ than $\Lambda_{\rm b}/\gamma_\mu$ indicates its fraction is very small (< 10 %). Hence we can conclude that the enhancement of λ in low T originates from dynamic fluctuating spins of electrons mostly.

The enhancement of λ is often seen in the spin singlet systems [7, 8, 9], although the origin is still under debate. In $SrCr_8Ga_4O_{19}$ [7], which was proposed to have a spin liquid type ground state, the sporadic dynamics of unpaired spins migrating in a spin singlet sea. Its dynamic feature is similar to $GaNb_4S_8$. However, the value of β in $SrCr_8Ga_4O_{19}$ is ~ 2 at $T \to 0$, being quite different from the result in $GaNb_4S_8$ in which β reaches to 0.5 as approaching to $T = 0$. The value of $\beta \simeq 0.5$ suggests that the unpaired spins are fixed spatially and are fluctuating dynamically. The small magnitude of fluctuating field $\Lambda_b/\gamma_\mu = 22$ G indicates a small number of unpaired spins, which may be created by the defect in the sample or by the muon polaronic effect. Destruction of the spin singlet pairs due to the muon polaronic effect was proposed in the organic spin-Peierls compound $MEM(TCNQ)_2$ [8]. It is noted that an enhancement of λ was not observed in the metallic spin singlet material Mo_3Sb_7 [10]. In the metallic system, the charge of muon can be screened by conduction electrons and the muon polaronic effect should be weaker. Thus, we speculate that the sporadic unpaired spins in the insulating spin singlet systems such as $GaNb_4S_8$ are mostly created by the muon polaronic effect.

This study was supported by Grant-in-Aid for Scientific Research on Priority Areas "Novel States of Matter Induced by Frustration", and Grant-in-Aid for Young Scientists (Start-up) from MEXT, Japan.

[1] Pocha R, Johrendt D, Ni B and Abd-Elmeguid M M 2005 *J. Am. Chem. Soc.* **127** 8732
[2] Jakob S, Muller H, Johrendt D, Altmannshofer S, Scherer W, Rayaprol S and Pöttgen R 2007 *J. Mater. Chem.* **17** 3833
[3] Pocha R, Johrendt D and Pöttgen R 2000 *Chem. Mater.* **12** 2882
[4] Nakamura H, Ikeno R, Motoyama G, Kohara T, Kajinami Y and Tabata Y 2009 *J. Phys.: Conf. Series* **145** 012077
[5] Waki T, Kajinami Y, Tabata Y, Yoshida M, Takigawa M, Watanabe I and Nakamura H arXiv:0906.5116
[6] Uemura Y J, Yamazaki T, Harshman D R, Senba M and Ansaldo E J 1985 *Phys. Rev. B* **31** 546
[7] Uemura Y J, Keren A, Kojima K, Le L P, Luke G M, Wu W D, Ajiro Y, Asano T, Kuriyama Y, Mekata M, Kikuchi H and Kakurai K 1994 *Phys. Rev. Lett.* **73** 3306
[8] Lovett B W, Blundell S J, Pratt F L, Jestädt Th, Hayes W, Tagaki S and Kurmoo M 2000 *Phys. Rev. B* **61** 12241
[9] Baker P J,Blundell S J, Pratt F L, Lancaster T, Brooks M L, Hayes W, Isobe M, Ueda Y, Hoinkis M, Sing M, Klemm M, Horn S and Classen R 2007 *Phys. Rev. B* **75** 094404
[10] Koyama T, Yamashita H, Takahashi Y, Kohara T, Watanabe I, Tabata Y and Nakamura H 2008 *Phys. Rev. Lett.* **101** 126404

Journal of Physics: Conference Series **225** (2010) 012017 doi:10.1088/1742-6596/225/1/012017

Comparative μ^+SR study of the zigzag chain compounds $NaMn_2O_4$ & $LiMn_2O_4$

Y Ikedo[1], J Sugiyama[2], O Ofer[3], M Månsson[4], H Sakurai[5], E Takayama-Muromachi[5], E J Ansaldo[3], J H Brewer[3,6] and K H Chow[7]

[1] Muon Science Laboratory, KEK, Ibaraki, Japan
[2] Toyota Central R&D Laboratories Inc., Aichi, Japan
[3] TRIUMF, Vancouver, Canada
[4] LNS, ETH Zurich and Paul Scherrer Institut, Villigen PSI, Switzerland
[5] National Institute for Materials Science, Ibaraki, Japan
[6] CIfAR and Department of Physics and Astronomy, University of British Columbia, Vancouver, BC, Canada
[7] Department of Physics, University of Alberta, Edmonton, Canada

E-mail: `e0589@mosk.tytlabs.co.jp`

Abstract. The magnetic nature of the $CaFe_2O_4$-type $NaMn_2O_4$ and $Li_{0.92}Mn_2O_4$, in which Mn ions form a zigzag chain along the b-axis, have been investigated by muon-spin rotation and relaxation (μ^+SR) and susceptibility (χ) measurements in the temperature range between 1.9 and 250 K using polycrystalline samples. Weak transverse-field μ^+SR measurements revealed the existence of a bulk antiferromagnetic (AF) transition for NMO at $T_N = 39$ K and for $Li_{0.92}Mn_2O_4$ at $T_N = 44$ K, although the $\chi(T)$ curve did not exhibit a clear anomaly around T_N for either compound. Below T_N, however, the zero-field μ^+SR spectra in both samples did not show the oscillatory signal characteristic of long-range magnetic order. Instead, a slowly relaxing tail is observed. Comparisons with the calculated internal magnetic field for the AF ordered phase suggests that the AF spin structure is disordered and/or rapidly fluctuating. In the paramagnetic region above 100 K, a nuclear magnetic field in $Li_{0.92}Mn_2O_4$ was found to be dynamic, whereas that in $NaMn_2O_4$ was almost static. This implies the diffusive motion of Li^+ ions in $Li_{0.92}Mn_2O_4$ even below 250 K.

1. Introduction

The nature of the ground state of the quasi-1D (Q1D) spin system, having characteristic 1D *zigzag* chains in the structure, is attracting considerable theoretical and experimental interests [1, 2]. In these systems, there exists a competition between the first nearest neighbor (FNN) and the second nearest neighbor (SNN) interactions. Furthermore, the interaction between the *zigzag* chains, i.e., the inter-chain interaction, plays an essential role in the formation of magnetic order/disorder in the real systems, since a *zigzag* chain is never isolated. The spin density of the *zigzag* chain is the yet other significant factor that determines the ground state of the compounds.

We have initiated experimental studies of *zigzag* chain compounds to understand how the competition among FNN, SNN, and inter-chain interactions affects the magnetic ground state by means of positive muon-spin rotation/relaxation (μ^+SR). Recently, two new *zigzag* chain

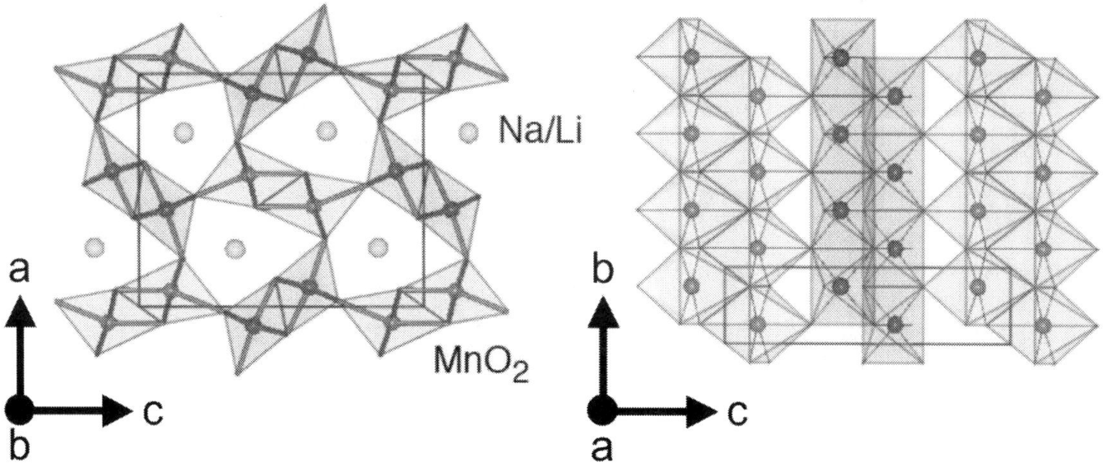

Figure 1. Crystal structure of NMO. The Mn_2O_4 double-chains, i.e., *zigzag* chains are formed by a network of edge-sharing MnO_6 octahedra align along the *b*-axis so as to make an irregular hexagonal 1D channel. The Na (or Li) ions are located at the center of the 1D channel.

compounds AMn_2O_4 (A = Na and Li) have been synthesized using a high-pressure technique . The valence state of the Mn ions is naturally expected to be +3.5, although both compounds are insulating even at ambient T. Their crystal structure is described as a $CaFe_2O_4$(CFO)-type orthorhombic system with space group *Pnma* (see Fig. 1). In $NaMn_2O_4$ (NMO) a cooperative Jahn- Teller distortion suggests charge ordering of the Mn^{3+} and Mn^{4+} ions, while for $Li_{0.92}Mn_2O_4$ (LMO) there is no indication of charge-order.

2. Experimental

A polycrystalline sample of NMO was prepared by a solid-state reaction technique under a pressure of 6 GPa using Na_2O_2 and Mn_2O_3 powders as starting materials [3]. A mixture of the two powders was packed in an Au capsule in an Ar-filled glove-box, then heated at 1300°C for 1 hour under P=6 GPa, and finally quenched to ambient T. For the synthesis of the CFO-type LMO, the spinel-type LMO powder was, at first, prepared by a conventional solid-state reaction technique at ambient pressure using Li_2CO_3 and Mn_2O_3 powders as starting materials [4]. A mixture of the two powders was heated at 850°C. Then, the obtained spinel-type LMO powder was packed in an Au capsule, heated at 1300°C for 1 hour under P=6 GPa, and finally quenched to ambient T. A powder x-ray diffraction (XRD) analysis showed that the NMO sample was almost single phase with the CFO-type orthorhombic system of space group *Pnma* at ambient T, while the LMO sample consisted of the $CaFe_2O_4$-type orthorhombic phase and Li_2MnO_3, as in the previous report. The volume fraction of the Li_2MnO_3 impurity phase was estimated to range around 10%. DC susceptibility (χ) measurements show the presence of an AF transition at T_N = 15 K for NMO, but the absence of a clear AF transition for LMO.

Seven polycrystalline discs with 6 mm diameter and 5 mm thickness were set on the muon-veto sample holder. Conventional time-differential μ^+SR spectra were collected at the **M20** surface muon channel at TRIUMF, Vancouver, Canada, using an experimental setup and techniques described elsewhere [5].

3. Results

As we reported previously [6], the weak transverse field (wTF-) measurements were carried out with applying field of 50 Oe below 80 K. As decreasing T, the wTF asymmetries, which corresponds to the volume fraction V_F of paramagnetic (PM) phases in a sample, showed a clear drop at 39 K, indicating occurrence of a magnetic transition at $T_N = 39$ K. The discrepancy of T_Ns between that obtained by susceptibility measurement and wTF-μSR spectra is probably due to the difference of time windows between these methods. In order to investigate the magnetic properties below T_N, we measured the μ^+SR spectrum with zero external field (ZF) down to 1.9 K. However, the ZF-spectrum exhibits no oscillation but a fast relaxation in an early time domain even at 1.9 K. Moreover, since the relaxation rate is too fast for μ^+SR, we do not observe the full asymmetry (~ 0.24) at $T \leq 30$ K. This implies the existence of a large fluctuating internal magnetic field below T_N. The ZF-spectrum is phenomenologically fitted by a combination of three exponential relaxation functions;

$$\begin{aligned} A_0 P_{ZF}(t) &= A_1 \exp(-\lambda_1 t) \\ &+ A_2 \exp(-\lambda_2 t) \\ &+ A_3 \exp(-\lambda_3 t), \end{aligned} \qquad (1)$$

where A_i (i=1, 2, and 3) is the the asymmetries of the three components of the ZF-μ^+SR signal, and λ_i is their relaxation rate. In addition, in order to obtain reasonable values for the three signals even under the absence of the full asymmetry at $t = 0$, the total asymmetry is fixed as 0.24, i.e., $\Sigma A_i = A_0 = 0.24$.

Figure 2 shows the T dependences of (a) the normalized asymmetries ($N_{A_i} \equiv A_i/A_0$, with i =1, 2, and 3), (b) λ_1, and (c) λ_2 and λ_3 for NMO. As T decreases from T_N, N_{A_1} increases rapidly down to ~ 30 K, and then levels off to a constant value (~ 0.7) with further lowering T. In contrast to N_{A_1}, both N_{A_2} and N_{A_3} decrease rapidly with decreasing T below T_N, and then they are almost T-independent below ~ 30 K. The $\lambda_1(T)$ curve exhibits roughly a similar T dependence to the $N_{A_1}(T)$ curve, while both the $\lambda_2(T)$ and $\lambda_3(T)$ curves are less T-dependent even below the vicinity of T_N. Considering the magnitude of $N_{A_1}(\sim 2/3)$ at low T and a very small magnitude of λ_2 and λ_3 compared with λ_1, both A_2 and A_3 signals are most likely to be the "1/3 tail" signal.

In order to further elucidate the magnetic behavior around T_N, we also carried out ZF-μ^+SR measurements in the T range between 35 and 100 K. As a result, the relaxation rate was found to increase rapidly with decreasing T from 100 K. In fact, it is difficult to fit the spectrum using a simple exponential relaxation function [$\exp(-\lambda t)$] in the whole T range between 35 and 100 K, but the ZF-μ^+SR spectrum is well fitted by a power exponentially relaxing signal;

$$A_0 P_{ZF}(t) = A_{ZF} \exp(-(\lambda_{ZF} t)^{\beta}), \qquad (2)$$

Note that "a power exponentially relaxing signal" has been observed for many dense-moment disordered magnetic systems [7, 8] in a paramagnetic state. The fit results are shown in Fig. 3. As T decreases from 100 K, $\beta \sim 1.4$ [between Gaussian relaxation ($\beta = 2$) and simple exponential ($\beta = 1$)] down to 60 K. Then, β decreases rapidly as T is lowered further, dropping below β=0.5 (root exponential for a dilute disordered magnet) [9]. Ultimately, β approaches a value of 1/3, indicating the existence of a dense disordered phase [10]. This means a gradual increase in the number density of localized magnetic moments, which are detectable with the muon-time scale, with decreasing T below ~ 60 K. Accompanying the change in β, λ_{ZF} also increases very rapidly below 70 K with decreasing T, suggesting a broadening of the distribution of H_{int}. This result is, therefore, consistent with the scenario that disordered moments appear below ~ 60 K and develops with decreasing T, and finally magnetic order or disorder completes below T_N.

As we described in previous paper [6], the wTF-μ^+SR measurements on LMO also demonstrate the existence of a bulk magnetic transition located at T_N=44 K, although there

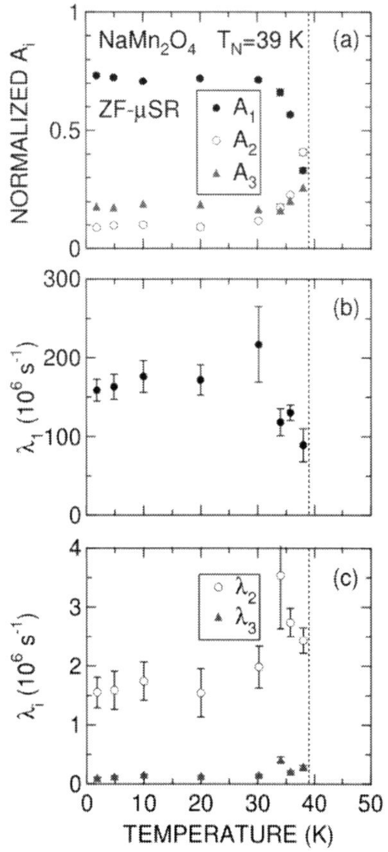

Figure 2. T dependences of (a) $N_{A_i}(= A_i/A_0$, with i =1, 2, and 3), (b) λ_1, and (c) λ_2 and λ_3 for NMO. The data was obtained by the fitting of the ZF-spectrum using Eq. (1).

are no clear anomalies in the $\chi(T)$ curve around T_N, as in the case for NMO. the ZF-spectrum for LMO is very similar to that for NMO. However, the absence of the fast relaxing signal in the early time domain of the ZF-spectrum for LMO suggests the existence of a larger internal magnetic field in LMO at low T than that in NMO.

4. Discussion

Unfortunately, there are no oscillations in the ZF-spectrum for both NMO and LMO even at 1.9 K. It is therefore impossible to determine the muon sites from the ZF-spectrum at T below T_N. However, since the ZF-spectrum exhibits a Kubo-Toyabe behavior above 100 K for both compounds, the implanted muons are most likely to *see* the internal magnetic field due to nuclear magnetic moments of ^{23}Na, ^{7}Li, ^{6}Li, and ^{55}Mn at high T. More correctly, the ZF-spectrum is well fitted by the following function;

$$A_0\,P_{ZF}(t) \quad = \quad A_{KT}\exp(-\lambda_{KT}t)G^{DGKT}(\Delta,\nu,t), \qquad (3)$$

where $G^{DGKT}(\Delta,\nu,t)$ is a dynamic Gaussian Kubo-Toyabe function, λ_{KT} is its relaxation rate, Δ is the static width of the local frequency at the disordered sites, and ν is the field fluctuation rate. The exponential relaxation is likely to come from the spin fluctuation of the d electrons

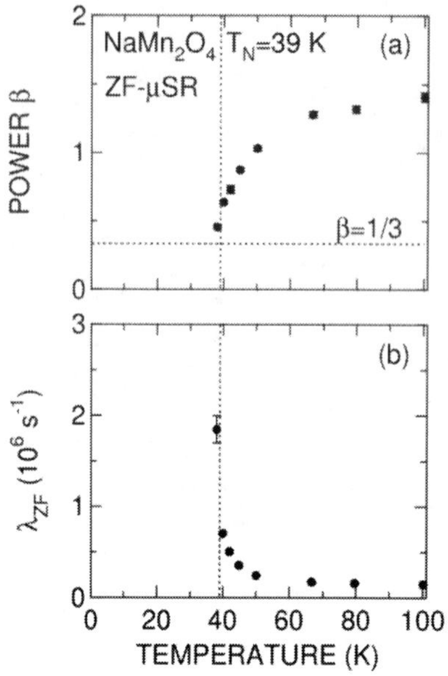

Figure 3. Temperature dependences of (a) β, and (b) $\lambda_{\rm ZF}$ for the NMO powder, particularly above $T_{\rm N}$. The ZF data were obtained by fits using Eq. (2).

Table 1. Δ, $\lambda_{\rm KT}$, and $A_{\rm KT}$ for NMO and LMO. The data were obtained by a global fitting of the ZF- and LF-spectra at T between 100 and 250 K using Eq. (3).

compound	Δ (10^6 s^{-1})	$\lambda_{\rm KT}$ (10^6 s^{-1})	$A_{\rm KT}$
NMO	0.116(2)	0.052(1)	0.232(4)
LMO	0.204(4)	0.033(2)	0.220(3)

probably due to a local Jahn-Teller instability of the Mn^{3+} ions with $t_{2g}^3e_g^1$ configuration, as in the case of the spinel LMO, while $G^{\rm DGKT}(\Delta,\nu,t)$ corresponds to the nuclear magnetic field. A preliminary fit using Eq. (3) for the ZF-spectrum at each T between 100 and 250 K indicated that the three parameters, i.e., Δ, $\lambda_{\rm KT}$, and $A_{\rm KT}$ are approximately independent of T. All of the ZF and LF-spectra were, therefore, fitted by Eq. (3) using common Δ, $\lambda_{\rm KT}$, and $A_{\rm KT}$— i.e., a global fitting technique (see Table 1).

On the other hand, electrostatic potential calculations suggest that there are four stable muon sites in the crystallographic lattice, as in the case of NaV_2O_4 [11]. That is, muons locate at the vicinity of the O^{2-} ions but ~1Å away [see Fig. 4(a)]. Then, dipole field calculations at the four sites provide that $\Delta = 0.32(1)\times10^6$ s^{-1} for NMO and $\Delta = 0.41(4)\times10^6$ s^{-1} for LMO. Although the calculations reproduce the relationship $\Delta_{\rm NMO} < \Delta_{\rm LMO}$, the calculated values are larger by 150-200% than the measured ones. Making comparison with the sites predicted by the potential calculations, the real muon sites are likely to deviate far away from the O^{2-} ions, probably due to a variation of the atomic positions with T. Furthermore, if we fix $\lambda_{\rm KT} = 0$ in Eq. (3), the global fit yields that $\Delta = 0.324(1)\times10^6$ s^{-1} for NMO and $\Delta = 0.307(3)\times10^6$ s^{-1} for LMO, although the fitting accuracy is rather poorer than that using a non-zero $\lambda_{\rm KT}$. This would imply a possible interaction between the spin fluctuation of Mn ions and nuclear magnetism, for the details are unknown.

Figures 4(b) and 4(c) show the T dependences of Δ and ν for NMO and LMO. It is found

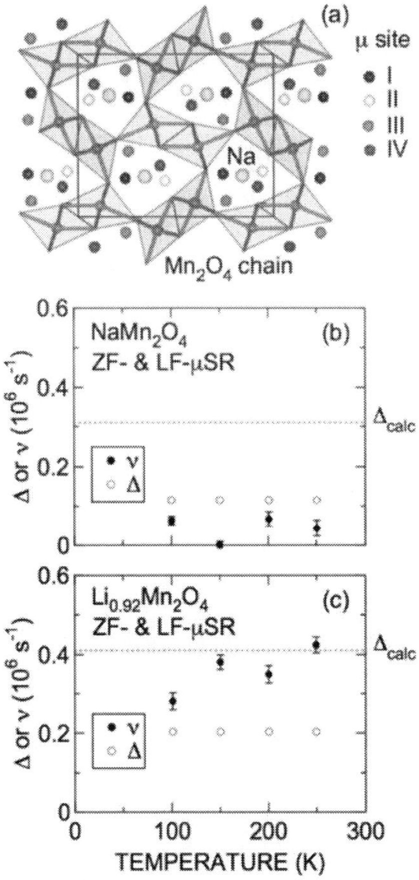

Figure 4. (a) possible four muon sites (I - IV) predicted by electrostatic potential calculations and T dependence of the field distribution width (Δ) and the field fluctuation rate (ν) above 100 K for (b) NMO and (c) LMO. $\Delta_{\rm calc}$ means the calculated Δ for the muon sites in (a).

that $\nu_{\rm NMO} \leq 0.06 \times 10^6$ s^{-1} in the whole T range measured, while $\nu_{\rm LMO}$ increases with increasing T. Furthermore, $\nu_{\rm LMO}$ is about 4 times larger than $\nu_{\rm NMO}$ even at 100 K. This indicates the diffusive motion of the Li ions even at 100 K in the LMO lattice. Indeed, the electrochemical property of LMO was reported in Ref. [4], in order to investigate its performance as a cathode material for reversible Li-ion batteries. Assuming that $\nu_{\rm LMO}$ corresponds to the jump rate of the Li ions between the neighboring sites, the diffusion coefficient of the Li ions ($D_{\rm Li}$) would be given by [12];

$$D_{\rm Li} = \frac{1}{2} Z_v s^2 \nu, \tag{4}$$

where Z_v is the fraction of the vacancy of the Li site and s is the jump distance. Here, we also assume the diffusion along the b-axis, i.e., along the 1D channel, because it is most unlikely that the Li^+ ions jump accros the Mn_2O_4 *zigzag* chain. Therefore, s is equivalent to the b-axis length and Z_v is 0.08 from the neutron study [4]; that is, the $D_{\rm Li}$ is thus estimated as $0.136(1) \times 10^{-10}$ cm^{2}s^{-1} at 250 K, although $D_{\rm Li}$ evaluated by the other techniques for LMO is currently unavailable. We should here note that the diffusive motion in LMO is most likely enhanced by the Li vacancies ($\sim$ 8%), and would not be an intrinsic feature of the stoichiometric

Table 2. The internal magnetic field ($H_{\rm int}$) for NMO and LMO estimated by dipole field calculations.

compound	AF structure		Site I	Site II	Site III	Site IV	average
NMO	CaV_2O_4-type	$H_{\rm int}/\mu_{\rm ord}$ (Oe/μ_B)	1397	1276	2317	1367	1600 ± 200
LMO	CaV_2O_4-type	$H_{\rm int}/\mu_{\rm ord}$ (Oe/μ_B)	1567	1296	1465	1775	1520 ± 90
NMO	random	$H_{\rm int}/\mu_{\rm ord}$ (Oe/μ_B)	1066	972	1365	948	1090 ± 90
LMO	random	$H_{\rm int}/\mu_{\rm ord}$ (Oe/μ_B)	1173	1004	1386	1103	1170 ± 70

LMO. In order to further investigate the diffusive motion of the Li ions, we need to measure the ZF- and LF-spectra at high T using a pulse-muon facility, such as, ISIS in the UK or J-PARC in Japan.

Here, we discuss the magnitude of the internal magnetic field ($H_{\rm int}$) in the AF phase below T_N. The AF spin structure of the related *zigzag* chain compound CaV_2O_4 was determined by neutron experiments [13, 14]; that is, AF order along the two 1D legs in the *zigzag* chain, and AF order between the *zigzag* chains. This AF structure is also supported by NMR and μ^+SR experiments [15, 11]. Hence, in order to roughly estimate $H_{\rm int}$ below T_N, it would be reasonable to assume the same AF structure for NMO and LMO. We also estimated $H_{\rm int}$ for the fully disordered phase, i.e., a spin-glass-like phase, although the $\chi(T)$ curve for NMO is likely to suggest the AF order. The result of dipole field calculations for the AF phase is summarized in Table 2.

Assuming that $\mu_{\rm ord}$ is comparable to $\mu_{\rm eff}$ for the both compounds, $H_{\rm int} = 5360$ Oe for NMO and 6080 Oe for LMO at low T. Note that this estimation provides the upper limit of $H_{\rm int}$. Then, $H_{\rm int}$ is directly converted into the muon-spin precession frequency, using the muonic gyromagnetic ratio $[\gamma_\mu/(2\pi) = 13.553$ kHz/Oe]. We should, therefore, expect an oscillatory signal with 73 MHz for NMO and 82 MHz for LMO. Since both frequencies range in the μ^+SR timescale, the absence of well-defined oscillatory signal in the ZF-spectrum indicates either a broad distribution of $H_{\rm int}$ or a rapidly fluctuating $H_{\rm int}$. Both are consistent with the result of the heat capacity measurements on LMO [4] and neutron and μ^+SR results on the spinel LMO [16, 17]. In order to determine whether the magnetic nature is static or dynamic, particularly for a disordered state, an appropriate longitudinal field (LF) is usually applied to decouple the effect of the disordered moments. However, because of the lack of the meaningful signal in an early time-domain of the ZF-apectrum for the both samples, it is very difficult to deduce the nature of the magnetic phase by LF measurements.

As described in previous paper, the χ measurements imply that the magnitudes of the effective magnetic moment of Mn ions and the paramagnetic Curie temperature for LMO are larger than those for NMO [6]. This would suggest that $H_{\rm int}$ for LMO is larger than that for NMO. Indeed, if we ignore the oscillatory signal from the Li_2MnO_3 phase in the ZF-spectrum for LMO, the spectrum eventually consists of only the tail component even at 1.9 K. This suggests that in LMO, a rapidly relaxing signal is hidden below 0.01 μs, in contrast to the case for NMO. Thus, it is reasonable that $H_{\rm int}$(LMO) > $H_{\rm int}$(NMO). This could also indicate a different spin structure between NMO and LMO, but the details are unknown based only on the μ^+SR experiment. In order to further clarify the AF nature for both compounds, we unambiguously need to perform a neutron experiment.

5. Summary

By means of positive muon-spin rotation and relaxation (μ^+SR), we clarified the magnetic nature of NMO and LMO in which the Mn ions form a *zigzag* Mn_2O_4 chain. For NMO (LMO), weak transverse field (wTF-) and zero field (ZF-) μ^+SR measurements suggest the appearance of a random internal magnetic field at T below 60 K (80 K) and the existence of a bulk magnetic transition at $T_N = 39$ K (44 K). Although the spin structure below T_N is still unknown, the AF order is thought to be highly disordered, as in the case of a spinel LMO.

Acknowledgments

We thank the staff of TRIUMF for help with the μ^+SR experiments. YI and JS are partially supported by the KEK-MSL Inter-University Program for Oversea Muon Facilities, and JHB is supported at UBC by CIfAR, NSERC of Canada, and (through TRIUMF) by NRC of Canada, KHC by NSERC of Canada and (through TRIUMF) by NRC of Canada, and ETM by WPI Initiative on Materials, Nanoarchitronics, MEXT, Japan. This work is also supported by Grant-in-Aid for Scientific Research (B), 19340107, MEXT, Japan. All images involving crystal structure were made with VESTA.

References

[1] Masuda T, Zheludev A, Manaka H, Regnault L-P, Chung J-H and Qiu Y: 2006 *Phys. Rev. Lett.* **96** 047210
[2] Sugiyama J, Nozaki H, Ikedo Y, Mukai K, Andreica D, Amato A, Brewer J H, Ansaldo E J: Morris G D, Takami T and Ikuta H, 2006 *Phys. Rev. Lett.* **96** 197206
[3] Akimoto J, Awaka J, Kijima N, Takahashi Y, Maruta Y, Tokiwa K and Watanabe T: 2006 *J. Solid State Chem.* **179** 169-174
[4] Yamaura K, Huang Q, Zhang L, Takada K, Baba Y, Nagai T, Matsui Y, Kosuda K and Takayama-Muromachi E: 2006 *J. Ame. Chem. Soc.* **128** 9448-9456
[5] Kalvius G M, Noakes D R and Hartmann O: 2001 *Handbook on the Physics and Chemistry of Rare Earths*, ed Gschneidner Jr J A, Eyring and Lander G H, **32**, Ch. 206 (Amsterdam Elsevier Science), and references cited therein
[6] Sugiyama J, Ikedo Y, Ofer O, Månsson M, Ansaldo E J, Brewer J H, Chow K H, Sakurai H and Takayama-Muromachi E 2009 *J. Phys. Soc. Jpn.* **78** 084715
[7] Noakes D R, Kalvius G M, Wappling R, Stronach C E, White Jr M F, Saito H and Fukamichi K: 1998 *Phys. Lett. A* **238** 197-202
[8] Noakes D R, Kalvius G M and Hartmann O: 2002 *Phys. Rev. B* **65** 132413
[9] Uemura Y J, Yamazaki T, Hayano R S, Nakai R and Huang C Y: 1980 *Phys. Rev. Lett.* **45** 583-587
[10] Ogielski A T: 1985 *Phys. Rev. B* **32** 7384-7398
[11] Sugiyama J, Ikedo Y, Goko T, Ansaldo E J, Brewer J H, Russo P L, Chow K H and Sakurai H: 2008 *Phys. Rev. B* **78** 224406
[12] Borg R J and Dienes G J: 1988 *An Introduction to Solid State Diffusion* (San Diego, Academic Press) pp. 57-60
[13] Hastings J M, Corliss L M, Kunnmann W and La Placa S: 1967 *J. Phys. Chem. Solids* **28** 1089-1092
[14] Pieper O, Corliss L M, Kunnmann W and La Placa S: 2008 *Experimental report of BENSC*, PHY-01-2088-EF
[15] Zong X, Suh B J, Niazi A, Yan J Q, Schlagel D L, Lograsso T A and Johnston D C: 2008 *Phys. Rev. B* **77** 014412
[16] Greedan J E, Wiebe C R, Wills A S and Stewart J R: 2002 *Phys. Rev. B* **65** 184424
[17] Sugiyama J, Mukai K, Ikedo Y, Russo P L, Suzuki T, Watanabe I, Brewer J H, Ansaldo E J, Chow K H, Ariyoshi K and Ohzuku T: 2007 *Phys. Rev. B* **75** 174424

μ^+SR study on triangular antiferromagnet $LiCrO_2$

Y Ikedo[1], J Sugiyama[2], K Mukai[2], M Månsson[3], T Goko[4], D Andreica[5] and A Amato[6]

[1] Muon Science Laboratory, KEK, Ibaraki, Japan
[2] Toyota Central R&D Laboratories Inc., Aichi, Japan
[3] LNS, ETH Zurich and Paul Scherrer Institut, Villigen PSI, Switzerland
[4] TRIUMF, Vancouver, Canada
[5] Faculty of Physics, Babes-Bolyai University, Cluj-Napoca, Romaina
[6] Laboratory for Muon Spin Specroscopy, Paul Scherrer Institut, Villigen PSI, Switzerland

E-mail: `e0589@mosk.tytlabs.co.jp`

Abstract. In order to elucidate the magnetic nature of the lithium chromium dioxide $LiCrO_2$, in which the Cr^{3+} ions form a two-dimensional triangular lattice in the CrO_2 plane, we have performed a positive muon-spin rotation and relaxation (μ^+SR) experiment using a powder sample at the temperatures below 155 K. Weak-transverse-field measurements indicated the existence of a bulk antiferromagnetic (AF) transition at T_N=61.2 K. The ZF-μ^+SR measurements indicated that $LiCrO_2$ is a pure paramagnet for $T \geq 62.5$ K, since its internal magnetic field (H_{int}) can be explained by solely nuclear magnetic moments. This means that, contrary to previous suggestions by susceptibility and heat capacity measurements, no short-range order exists for $T \geq 62.5$ K. However, ZF-μ^+SR detected the change in H_{int} from a low-T static state to a high-T dynamic state at 115 K, most likely connected to a change in the position/motion of the Li ions.

1. Introduction

Historically the two-dimensional triangular lattice (2DTL) system has served as a playground for novel ideas about unconventional phases of frustrated anti-ferromagnets. The layered cobalt/nickel oxides has been heavily investigated e.g. $LiCoO_2$ (electrode in Li-ion batteries) and Na_xCoO_2 (thermoelectric properties and superconductor) [1, 2, 3]. However, experimental studies on the chromium dioxides have been very limited. Recently a newfound interest has started as shown by both experimental [4] and theoretical work [5].

For $LiNiO_2$ [2] and $Na_{0.75}CoO_2$ [6], an A-type antiferromagnetic (AF) order is present at low temperature, with ferromagnetic (FM) order within the NiO_2 planes and AF order between the planes. However, for $LiCrO_2$, experimental work [7], as well as theoretical studies [5], show that the intraplane magnetic interaction is strongly AF, while the interplane is very weak and also AF. Susceptibility (χ) data [7] show clear AF order below $T_N = 62$ K. Further, $\chi(T)$ curve does not follow Curie-Weiss law below 300 K [8]. This Indicates that a possible short-range order sets in well above T_N. However, no experimental study is to be found in the published literature.

As we reported in previous paper [10], wTF measurements indicated the existence of a bulk AF transition at T_N=61.2 K (see Fig 2). Below T_N, ZF-μ^+SR spectra showed a clear spontaneous oscillation, indicating the formation of static long-range order. The observed ZF-spectrum was well reproduced by the AF spin structure proposed by neutron measurements [11]. In this

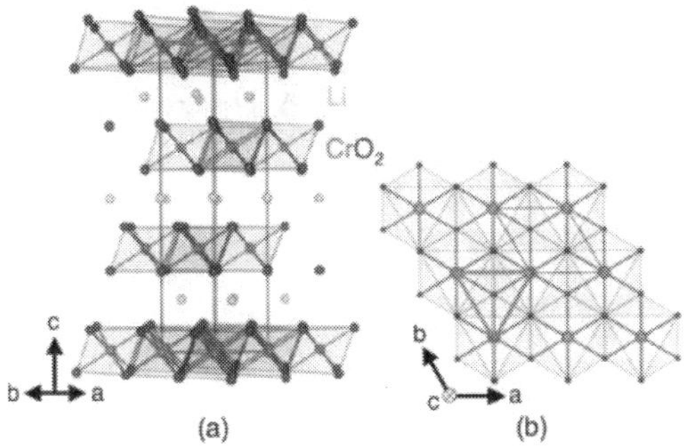

Figure 1. Crystal structure of $LiCrO_2$ showing (a) alternating stacks of Li and CrO_2 layers along the c-axis and (b) top-view along the c-axis demonstrating the triangular configuration of the Cr-ions.

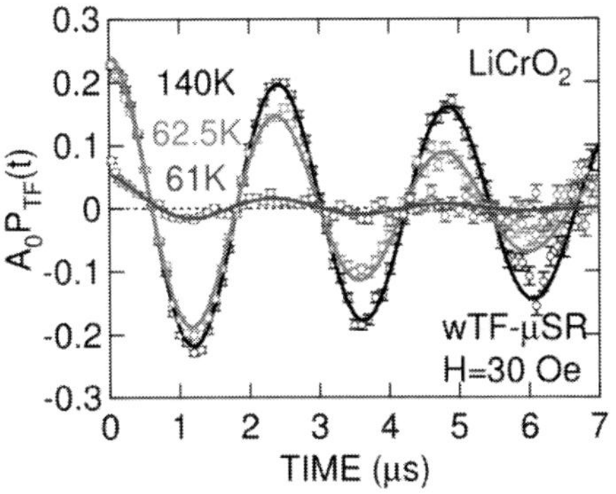

Figure 2. Temperature dependence of the wTF (= 30 Oe) μ^+SR time spectra. A_0 is the initial (t=0) asymmetry and $P_{TF}(t)$ is the muon spin polarization function. These spectra show a clear decrease of the A_0 between 61 K and 62.5 K, indicating the existence of AF transition.

paper, we mainly report the μ^+SR experiment on $LiCrO_2$ above T_N, in order to clarify the non Curie-Weiss behavior in $\chi(T)$ curve for $LiCrO_2$.

2. Experimental Details

A powder sample of $LiCrO_2$ was prepared at Osaka City University by a solid-state reaction technique using reagent grade $LiOH{\cdot}H_2O$ and Cr_2O_3 powders as starting materials. Powder x-ray diffraction (XRD) analysis showed that the sample was single phase with a rhombohedral system of space group $R\bar{3}m$; the a_H-axis length was 2.899 Å and the c_H-axis length was 14.418 Å. An inductively coupled plasma (ICP) analysis indicated that the Li/Cr ratio was 1.03.

μ^+SR spectra were measured at the Swiss Muon Source (SμS), Paul Scherrer Institut, Villigen, Switzerland. By using the surface muon beamline μE1 and the Dolly spectrometer, zero-field (ZF), weak transverse-field (wTF), and weak longitudinal-field (LF) spectra were collected for 1.7 K $\leq T \leq$ 155 K. The experimental setup and techniques were described in detail elsewhere [9].

3. Experimental Results

The ZF-spectrum exhibits a Kubo-Toyabe behavior just above T_N [see Fig. 3(a)]. This means that the implanted muons *see* the internal magnetic field (H_{int}) due to nuclear magnetic moments of ^{7}Li and ^{6}Li, together with a minor contribution from ^{53}Cr. Contrary to previous suggestions [8], this also excludes the possibility of the formation of any short-range order above T_N. However, as T increases from 80 K, the minimum in the ZF spectrum ($\sim$ 6 μs at 80 K) shifts towards a later time domain and the relaxation rate decreases with T. This implies that H_{int} starts to fluctuate above 80 K. Indeed, the ZF-spectrum are well fitted by a dynamic Gaussian Kubo-Toyabe function [$G^{DGKT}(\Delta, \nu, t)$] for H_{int} due to the fluctuating nuclear magnetic moments;

$$A_0 P_{ZF,LF}(t) = A_{KT} G^{DGKT}(\Delta, \nu, t), \tag{1}$$

where Δ is the static width of the local frequency at the disordered sites, and ν is the field fluctuation rate. A preliminary fit using Eq. (1) for the ZF-spectrum at each T between 62.5 and 155 K indicated that the two parameters, A_{KT} and Δ, are approximately independent of T. All of the ZF-spectra were, therefore, fitted by Eq. (1) using common A_{KT} and Δ. The global fit provides that $A_{KT} = 0.2309 \pm 0.0002$ and $\Delta = (0.3124 \pm 0.0001) \times 10^6$ s^{-1}.

Figure 3(b) shows the T dependence of ν for $LiCrO_2$ together with the T-independent $\Delta^{PM}_{measure}$. The $\nu(T)$ curve shows a *dull* but step-like change from almost 0 to $\sim 0.96 \times 10^6$ s^{-1} in the T range between 80 and 140 K. This demonstrates the appearance of dynamical fluctuation of H_{int} above $\sim$80 K. In addition, weak longitudinal field (LF = 20 Oe) measurements at 65 K and 140 K supports the $\nu(T)$ and $\Delta(T)$ curve obtained by the global fit for the ZF-spectra. Moreover, the $\chi_{FC}(T)$ curve starts to deviate from the $\chi_{ZFC}(T)$ curve below approximately 130 K [10]. This evidences the intrinsic change in a magnetic environment of $LiCrO_2$ around 130 K. Therefore, it is very reasonable to consider the change in Li motion/position as an origin of the increase in ν above 80 K.

4. Discussion

According to electrostatic potential calculations, positive muons are most likely to locate in the vicinity of the O^{2-} ions; that is, $\sim$ 1Å away from the O^{2-} ions along the c-axis. Taking into account the nuclear magnetic moments of ^{7}Li, ^{6}Li, and ^{53}Cr, the dipole field calculations at muon site in the paramagnetic state provided that $\Delta^{PM}_{calc} = (0.485 \pm 0.002) \times 10^6$ s^{-1}. This is roughly in agreement with the experimental result, although the calculated Δ^{PM}_{calc} is 55% larger [see Fig. 3(b)] than the measured $\Delta^{PM}_{measure}$[$= (0.3124 \pm 0.0001) \times 10^6$ s^{-1}]. Consequently, the muon sites, which are predicted by the electrostatic potential calculations, are found to give reasonable H_{int} both for the AF ordered state and the paramagnetic state.

The $\nu(T)$ curve demonstrates a clear change of H_{int} from a low-T static state to a high-T dynamic state at $T_c \sim 115$ K, at which $\nu = \frac{1}{2}[\nu(65\text{ K}) + \nu(155\text{ K})]$. Since ν appears to level off to a constant value above 140 K [see dotted curve in Fig. 3(b)], the change at T_c is most unlikely due to a diffusive motion of the Li^+ ions. This is because the relationship between ν and T is expected to be given by $\nu \propto \exp(-1/T)$ for Li diffusion [12]. Therefore, T_c is more likely to be associate with a structural transition caused by the change in the position of Li^+ ions. In other words, T_c could be explained by an order-disorder transition of the Li^+ ions; i.e., the Li ions could slightly shift their position from the regular site at low T. Moreover, Δ is expected to be insensitive to such small displacement of the Li^+ ions, when the Li^+ ions locate in the vicinity of the regular Li site.

Such transition would induce a charge disproportionation of the Cr ions; i.e., $Cr^{3+} \rightarrow Cr^{3+\delta} + Cr^{3-\delta}$, as in the case for $Na_{0.5}CoO_2$ [13, 14]. However, even for the mixed valence state of the Co ions in $Na_{0.5}CoO_2$, δ was reported below 0.2 [13, 14]. Hence, it would be very difficult to detect δ for $LiCrO_2$, even if such charge disproportionation occurs.

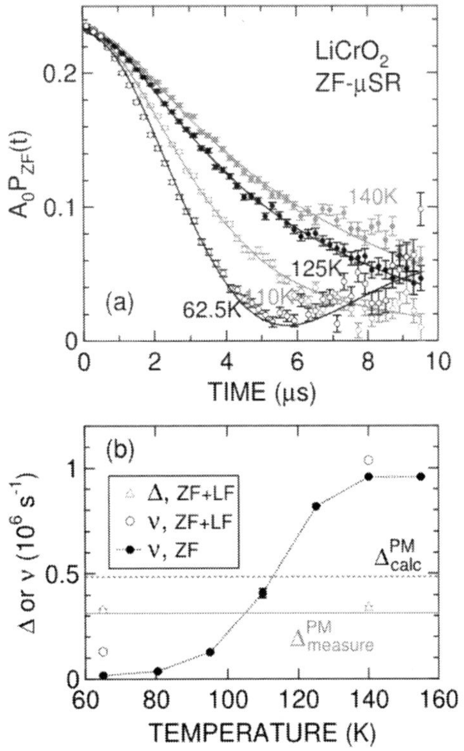

Figure 3. Temperature dependences of (a) ZF-μ^+SR spectrum above T_N and (b) the field distribution width (Δ) shown as triangles and field fluctuation rate (ν) as circles. The data in (b) were obtained by fitting the μ^+SR time spectra to Eq. (1). Closed symbols are obtained from a *global* fit to ZF-spectra only, while open symbols are obtained from a *global* fit to both ZF and longitudinal field (LF = 20 Oe) spectra at 65 and 140 K. The solid horizontal line in (b) represents a common $\Delta_{\rm measure}^{\rm PM}(= 0.3124 \times 10^6~{\rm s}^{-1})$ for a global fit, while the dashed horizontal line shows a prediction from dipole field calculations $\Delta_{\rm calc}^{\rm PM}$. Dotted curve serves as a guide to the eye.

Actually, although we attempted to fit the ZF-spectrum by $G^{\rm DGKT}(\Delta, \nu, t)\exp(-\lambda_{\rm Cr}t)$, where the exponential relaxation should correspond to the fluctuation of the Cr spins, the $\lambda_{\rm Cr}(T)$ curve exhibits no crucial anomaly around T_c. It should also be noted that a Kubo-Toyabe function is only an *approximation of the nuclear dipole contribution* to the ZF-μ^+SR spectrum. Hence, although the spectrum is well fitted by a Kubo-Toyabe function, there is still a possibility that the dynamics observed above T_N originate from electronic moments. In order to further investigate the details of this transition, we need to measure the ZF- and LF-spectra at high T using a pulse-muon facility, such as, ISIS in the UK or J-PARC in Japan. Furthermore, it is more preferable to perform a precise structural analysis of $LiCrO_2$ as a function of T by both neutron and x-ray (from a synchrotron radiation source) diffraction.

Finally, we wish to address the possible role of the order-disorder transition of the Li ions for the magnetic nature of $LiCrO_2$. In contrast to $NaCrO_2$, for which no long-range AF order is observed by neutron measurements down to the lowest T measured [15, 16], a long-range AF order is clearly detected in $LiCrO_2$ by neutron measurements [11], for reasons currently unknown. As in the case for $Na_{0.5}CoO_2$ [17, 18] and $K_{0.5}CoO_2$ [19, 20], the Li ordering in $LiCrO_2$ below 115 K is expected to alter the charge distribution within the CrO_2 planes, and could hereby stabilize the long-range order below T_N. Also, since the Kubo-Toyabe behavior is observed for $T \geq 62.5$ K, it is most unlikely that (previously suggested) short-range order is present above T_N in $LiCrO_2$. In order to explain non Curie-Weiss behavior of the $\chi(T)$ curve at high T, we thus need another mechanisms; such as, a Li disordering above 115 K inducing slight delocalization of the charge/spin state of the Cr ions, resulting in the difference between the $\chi_{FC}(T)$ and $\chi_{ZFC}(T)$ curve below 130 K. Also, such Li disordering would provide an additional contribution to C_p, which is probably difficult to distinguish from a magnetic contribution by only subtracting C_p for $LiCoO_2$ from C_p for $LiCrO_2$ [8], as in the case for $Na_{0.62}CoO_2$ [21].

5. Conclusions

By means of a positive muon-spin rotation and relaxation (μ^+SR) technique, we have clarified the magnetic nature of $LiCrO_2$, in which the Cr^{3+} ions form a two-dimensional triangular lattice in the CrO_2 plane. Below $T_N = 61.2$ K, zero-field (ZF-) μ^+SR measurements demonstrate the formation of static AF order. The observed ZF-spectrum was well explained by the AF spin structure proposed by neutron measurements. Furthermore, the ZF-μ^+SR measurements clearly indicated the absence of magnetic-order for $T \geq 62.5$ K, whether it is long-ranged or short-ranged. Instead, a dynamic change in the Li motion was detected at approximately $T = 115$ K, probably due to an order-disorder transition within the Li planes.

Acknowledgments

This work was performed at the Swiss Muon Source, Paul Scherrer Institut, Villigen, Switzerland and we are thankful to Robert Scheuermann for assistance with the μ^+SR experiments. YI and JS are partially supported by the KEK-MSL Inter-University Program for Overseas Muon Facilities. This work is also supported by Grant-in-Aid for Scientific Research (B), 19340107, MEXT, Japan. All images involving crystal structure were made with VESTA.

References

[1] Sugiyama J, Brewer J H, Ansaldo E J, Itahara H, Tani T, Mikami M, Mori Y, Sasaki T, Hébert S and Maignan A 2004 *Phys. Rev. Lett.* **92** 017602
[2] Sugiyama J, Mukai K, Ikedo Y, Russo P L, Nozaki H, Andreica D, Amato A, Ariyoshi K and Ohzuku T 2008 *Phys. Rev. B* **78** 144412
[3] Takada K, Sakurai H, Takayama-Muromachi E, Izumi F, Dilanian R A and Sasaki T 2003 *Nature* **422** 53
[4] Olariu Mendels A P, Bert F, Ueland B G, Schiffer P, Berger R F and Cava R J 2006 *Phys. Rev. Lett.* **97** 167203
[5] Mazin I I 2007 *Phys. Rev. B* **75** 094407
[6] Bayrakci S P, Mirebeau I, Bourges P, Sidis Y, Enderle M, Mesot M, Chen D P, Lin C T and Keimer B 2008 *Phys. Rev. Lett.* **94** 157205
[7] Delmas C, Le Flem G, Fouassier C and Hagenmuller P 1978 *J. Phys. Chem. Solids* **39** 55
[8] Alexander L K, Büttgen N, Nath R, Mahajan A V and Loidl A 2007 *Phys. Rev. B* **76** 064429
[9] Kalvius G M, Noakes D R and Hartmann O 2001 *Handbook on the Physics and Chemistry of Rare Earths* ed Gschneidner Jr J A, Eyring and Lander G H **32** Ch. 206 (Amsterdam Elsevier Science) and references cited therein
[10] Sugiyama J, Månsson M, Ikedo Y, Goko T, Mukai K, Andreica D, Amato A, Ariyoshi K and Ohzuku T 2009 *Phys. Rev. B* **79** 184411
[11] Kadowaki H, Takei H and Motoya K 1955 *J. Phys.: Condens. Matter* **7** 6869
[12] Borg R J and Dienes G J 1988 *An Introduction to Solid State Diffusion* (San Diego, Academic Press)
[13] Gašparović G, Ott R A, Cho J-H, Chou F C, Chu Y, Lynn J W and Lee Y S 2006 *Phys. Rev. Lett.* **96** 046403
[14] Bobroff J, Lang G, Alloul H, Blanchard N and Collin N 2006 *Phys. Rev. Lett.* **96** 107201
[15] Soubeyroux J L, Fruchart D, Delmas C and Le Flem G 1979 *J. Magn. Magn. Mater.* **14** 159
[16] Hsieh D, Qian D, Berger R F, Cava R J, Lynn J W, Huang Q and Hasan M Z 2008 *J. Phys. Chem. Solids* **69** 3174

[17] Viciu L, Bos J W G, Zandbergen H W, Huang Q, Foo M L, Ishiwata S, Ramirez A P, Lee M, Ong N P and Cava R J 2006 *Phys. Rev. B* **73** 174104
[18] Zandbergen H W, Foo M L, Xu Q, Kumar V and Cava R J 2004 *Phys. Rev. B* **70** 024101
[19] Watanabe H, Mori Y, Yokoi M, Moyoshi T, Soda M, Yasui Y, Kobayashi Y, Sato M, Igawa N and Kakurai K 2006 *J. Phys. Soc. Jpn.* **75** 034716
[20] Huang Q, Foo M L, Lynn J W,Zandbergen H W, Lawes G, Wang Y, Toby B H, Ramirez A P, Ong N P and Cava R J 2004 *J. Phys.: Condens. Matter* **16** 5803
[21] Blangero M, Carlier D, Pollet M, Darriet J, Delmas C and Doumerc J P 2008 *Phys. Rev. B* **77** 184116

Pressure-induced new magnetic phase in $Tl(Cu_{0.985}Mg_{0.015})Cl_3$ probed by muon spin rotation

Takao Suzuki[1], Isao Watanabe[1], Fumiko Yamada[2], Motoki Yamada[2], Yasuyuki Ishii[1], Takayuki Kawamata[1], Takayuki Goto[3] and Hidekazu Tanaka[2]

[1]RIKEN Nishina Center for Accelerator-Based Science, 2-1 Hirosawa, Wako, Saitama 351-0198, Japan

[2]Department of Physics, Tokyo Institute of Technology, O-okayama, Meguro-ku, Tokyo 152-8551, Japan

[3]Faculty of Science and Technology, Sophia University, 7-1 Kioi-cho, Chiyoda-ku, Tokyo 102-8554, Japan

E-mail: suzuki_takao@riken.jp

Abstract. We carried out zero-field muon-spin-rotation (ZF-μSR) measurements in hydrostatic pressures in impurity-doped quantum spin gap system $Tl(Cu_{0.985}Mg_{0.015})Cl_3$ to investigate microscopic magnetic properties of the pressure-induced phase. The spontaneous muon spin precession, which indicates the existence of a long-range coherent order, is observed in pressures. With decreasing temperature in 3.1 kbar, the internal static magnetic field at the muon sites $H_{\rm int}$ tends to saturate to 280 Oe around 4 K, however, decreases to 240 Oe at 2.3 K. These results suggest a rearrangement of ordered spins, and we speculate that the oblique antiferromagnetic phase, which is observed in the pressure of 14 kbar on the pure $TlCuCl_3$, appears in the Mg-doped system in lower pressures.

1. INTRODUCTION

$TlCuCl_3$ is the parent material of the subject compound in this study, and is three-dimensionally coupled Cu-3d S = 1/2 spin dimer system. This system has the magnetic ground state of spin-singlet with excitation gap of $\Delta_{\rm gap}$ = 7.5 K, which originates from strong intradimer antiferromagnetic interaction [1, 2]. In the impurity-introduced $Tl(Cu_{1-x}Mg_x)Cl_3$ system, the magnetic phase transition to an ordered state is observed by magnetization and specific heat measurements, and neutron elastic scattering measurements identified that this impurity-induced ordered state is the antiferromagnetically ordered state of which the magnetic structure is the same with the case of the field-induced phase in $TlCuCl_3$ [3, 4]. The inelastic neutron scattering measurement revealed that a finite spin gap still remains below the transition temperature in $Tl(Cu_{1-x}Mg_x)Cl_3$ [5]. Imamura *et al.* reported the pressure-induced magnetically ordered phase by magnetization measurements in $Tl(Cu_{1-x}Mg_x)Cl_3$ with x = 0.012, and concluded that the change from the impurity-induced phase to the pressure-induced phase is the crossover [6]. In the case for the slightly doping of x = 0.0047, longitudinal-field muon-spin-relaxation measurements revealed that the impurity-induced magnetic moments of the Cu-3d spins slowly fluctuate at T = 20 mK, far below the magnetic phase transition temperature observed by the specific heat measurement [7]. Recently, Yamada *et al.* reported the spin reorientation phase transition at $T_R = 9.2$ K in the high pressure of 14 kbar in the pure $TlCuCl_3$, and discussed a second-order transition from the spin-flop phase to the oblique antiferromagnetic phase below T_R[8]. Magnetic orderings induced by impurity doping and by pressure in spin gap systems are still attracting much interest. And besides, it has been revealed that muons are the excellent probe to detect magnetic properties of randomness-introduced spin gap systems [9, 10, 11, 12]. The purpose of this study is to investigate the microscopic properties of the pressure-induced phase in the impurity-doped spin gap system by the zero-field muon-spin-rotation/relaxation (ZF-μSR).

2. EXPERIMENTS

Single crystals used in this study were grown from a melt by the Bridgman method. The details of crystal growth are given elsewhere [13]. The concentration of x was determined by the inductively coupled plasma atomic emission spectrometry (ICP-AES) method. Muon-spin-rotation/relaxation (μSR) measurements were performed on $Tl(Cu_{1-x}Mg_x)Cl_3$ with $x = 0.015$ at the RIKEN-RAL Muon Facility [14]. Measurements in hydrostatic pressures were carried out using a spin-polarized double-pulsed positive decay-muon beam with an incident muon momentum of 90 MeV/c. Samples are pressurized by ^{4}He gas pressure in a CuBe cell using the newly installed gas-pressurized μSR setup for the RIKEN-RAL Muon Facility [15]. Cleaved crystals were packed into the pressure cell in random directions. μSR measurements were carried out at two separate beam times. Single crystals were cleaved from the same rod for each beam time. In μSR measurements, spin-polarized muons are implanted into samples. The asymmetry was defined as follows:

$$A(t) = \frac{F(t) - \alpha B(t)}{F(t) + \alpha B(t)} \tag{1}$$

Here, $F(t)$ and $B(t)$ were total muon events counted by the forward and backward counters at time t respectively. The α is a calibration factor reflecting relative counting efficiencies between the forward and backward counters. The initial asymmetry is defined as $A(0)$. Measured time spectra were analyzed using the WiMDA computer program [16].

3. RESULTS AND DISCUSSIONS

Figure 1 shows the temperature dependence of ZF-μSR time spectra in hydrostatic pressure of 3.1 kbar. (The pressure dependence of time spectra at 2.3 K is reported in elsewhere [17].) Measurements were carried out with decreasing temperature. Below 8 K, the spontaneous muon-spin-precession is observed, which indicates the existence of a coherent long-range magnetically ordered state. The observed transition temperature in pressure is consistent with the reported temperature in the case of $x = 0.012$ deduced from the magnetization measurement[6]. In order to discuss the development of the internal static magnetic field at the muon sites which corresponds to the ordered Cu-$3d$ magnetic moment, all spectra are analyzed using the two components function as follows:

$$A(t) = A_1 e^{-\lambda_1 t} \cos(\omega t + \theta) + A_2 e^{-\lambda_2 t} G_z(\Delta, t) \tag{2}$$

The first term is the signal from the magnetically ordered region of samples, and the second term is that from the large pressure cell and the spin fluctuating region of samples, because it is quite difficult to distinguish the relaxing part $\exp(-\lambda t)$ between the pressure cell and the samples. λ_1 and λ_2 are the muon-spin-relaxation rate of each component, and ω is the muon-spin-rotation frequency. $G_z(\Delta, t)$ is the static Kubo-Toyabe function, where Δ/γ_μ is the distribution width of nuclear-dipole fields at the muon sites. γ_μ is the gyromagnetic ratio of the muon spin ($2\pi \times 13.5534$ kHz/gauss). The ratio of the amplitude A_1/A_2 is approximately

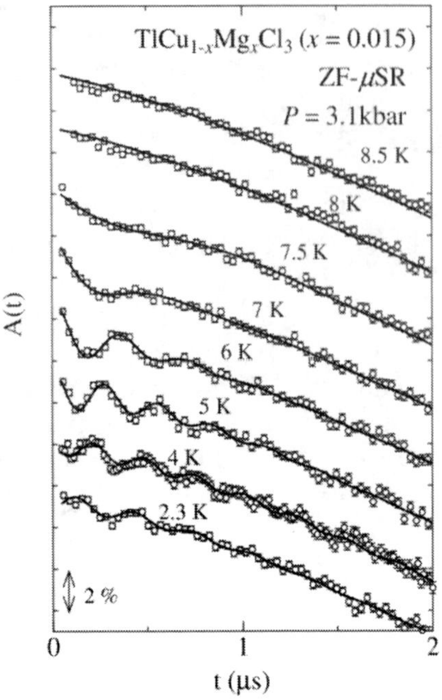

Figure 1. Time spectra of the ZF-μSR in $TlCu_{1-x}Mg_xCl_3$ with $x = 0.015$ in the hydrostatic pressure of 3.1 kbar in the temperature range from 8.5 K to 2.3 K. Solid lines are fitted results using the function (2). Each plot is shifted consecutively for clarity.

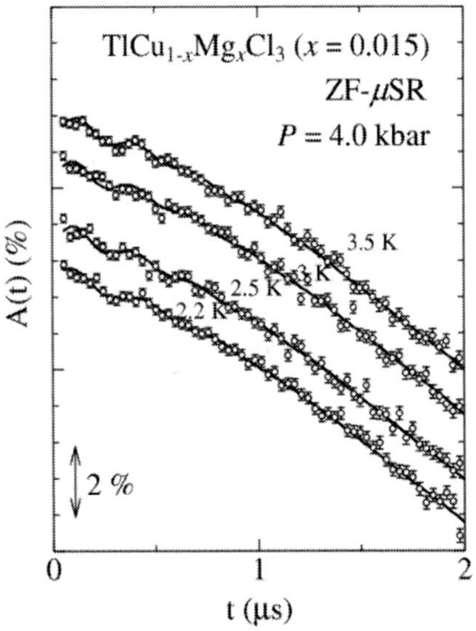

Figure 2. Time spectra of the ZF-μSR in $TlCu_{1-x}Mg_xCl_3$ with $x = 0.015$ in the hydrostatic pressure of 4.0 kbar. Solid lines are fitted results using the function (2). Each plot is shifted consecutively for clarity.

0.16 above 4 K. Fitted results are shown in Fig.1 as solid lines. From the fitted result, the internal static magnetic field $H_{\rm int}$ at the muon sites is deduced using the relation of $\omega = \gamma_\mu H_{\rm int}$ because implanted muon spins precess around the total magnetic field of the internal field and the external field at the muon sites.

Figure 2 shows the temperature dependence of ZF-μSR time spectra in hydrostatic pressure of 4.0 kbar. Spectra are analyzed in the same way with in the case for 3.1 kbar, but the ratio of the amplitude A_1/A_2 at 3.5 K is 0.04 in the case of 4.0 kbar. The small amplitude of the rotational signal is partially due to a slight discrepancy in the momentum tuning for implanting muons in samples and is partially due to the decrease of the volume fraction of the coherently ordered region at lower temperatures as discussed below. Fitted results are shown as solid lines.

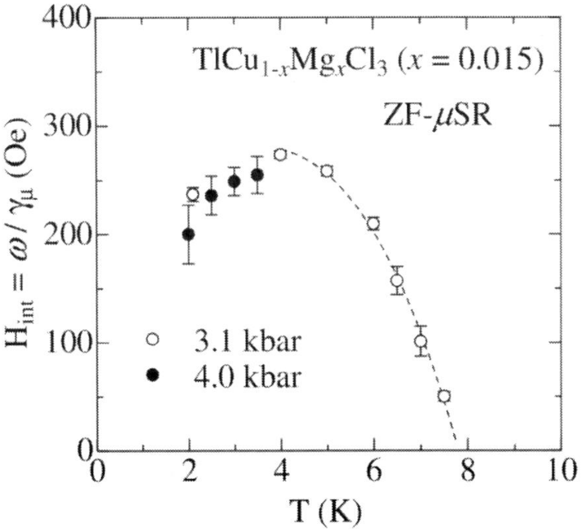

Figure 3. Temperature dependence of the internal static magnetic field $H_{\rm int}$ in 3.1 kbar (open circles) and in 4.0 kbar (closed circles). Dashed line is a guide for the eye.

Deduced internal static magnetic field $H_{\rm int} = \omega/\gamma_\mu$ at the muon sites are summarized in Fig.3. With decreasing temperature in 3.1 kbar, $H_{\rm int}$ increases monotonously, and tends to saturate to 280 Oe around 4 K. However, $H_{\rm int}$ decreases to 240 Oe at 2.3 K. Similar decrease of $H_{\rm int}$ below 4 K is observed in 4.0 kbar.

The decrease of $H_{\rm int}$ suggests an orientation of the ordered moments, because the change of the magnetic structure leads to the modification of dipole fields at the muon sites. As mentioned above, in the pure $TlCuCl_3$, the spin reorientation phase transition occurs at $T_{\rm R} = 9.2$ K in the high pressure of 14 kbar. In the pure sample, the spin gap is collapsed by pressure in $P_{\rm c} = 0.42$ kbar [18]. For the Mg-doped case, a finite spin gap still remains although the magnetic phase transition occurs [5], and it is expected that the pressure effect is more significant in the case for Mg-doped system, because the spin gap has already partially collapsed in space by the Mg-doping. In other word, the pressure-induce new phase could be appeared in lower pressures compared to the case in the pure system. Thus, we speculate that the spin reorientation phase transition occurs and that the oblique antiferromagnetic phase, which is observed in the pressure of 14 kbar on the pure $TlCuCl_3$ [8], appears in the Mg-doped system.

In the last, we discuss the decrease of the rotational amplitude in time spectra below 4 K. Generally, the amplitude of the rotational signal corresponds to the volume fraction of the coherent long-range ordered region. For that reason, the decrease of the rotational amplitude suggests that, after the spin reorientation phase transition, the spin system become less coherent due to some randomness caused by the spin reorientation. In other word, some randomness caused by the spin reorientation would disturb the coherency of the ordered state.

4. SUMMARY

Zero-field muon-spin-rotation (ZF-μSR) measurements in hydrostatic pressures were carried out in impurity-doped quantum spin gap system $Tl(Cu_{1-x}Mg_x)Cl_3$ with $x = 0.015$. The spontaneous muon spin precession, which indicates the existence of a long-range coherent order, is observed under pressures. With decreasing temperature in 3.1 kbar, the internal static magnetic field at the muon sites $H_{\rm int}$ tends to saturate to 280 Oe around 4 K, however, decreases to 240 Oe at 2.3 K. Similar decrease of $H_{\rm int}$ below 4 K is observed in 4.0 kbar. These results suggest a rearrangement of ordered spins which leads to the change of $H_{\rm int}$. We speculate that the oblique antiferromagnetic phase, which is observed in the pressure of 14 kbar on the pure

$TlCuCl_3$, appears in the Mg-doped system.

Acknowledgement

We would like to thank C. M. Goodway and all other members of high-pressure group of ISIS Facility at the Rutherford Appleton Laboratory for the technical support on μSR measurements in pressures.

References

[1] Shiramura W, Takatsu K, Tanaka H, Kamishima K, Takahashi M, Mitamura H and Goto T 1997 J. Phys. Soc. Jpn. **66** 1900
[2] Cavadini N, Henggeler Furrer W A, Güdel H -U, Krämer K, and Mutka H 1999 Eur. Phys. J. B **7** 519
[3] Oosawa A, Ono T, and Tanaka H 2002 Phys. Rev. B **66** 020405(R)
[4] Fujisawa M, Ono T, Fujiwara H, Tanaka Sikolenko H V, Meissner M, Smeibidl P, Gerischer S, and Graf H A 2006 J. Phys. Soc. Jpn. **75** 033702
[5] Oosawa A, Fujisawa M, Kakurai K, and Tanaka H 2003 Phys. Rev. B **67** 184424
[6] Imamura H, Ono T, Goto K, and Tanaka H 2006 Phys. Rev. B **74** 064423
[7] Suzuki T, Yamada F, Watanabe I, Matsuzaki T, Goto T, Oosawa A, and Tanaka H 2009 J. Phys. Soc. Jpn. **78** 074705
[8] Yamada F, Ishii Y, Suzuki T, Matsuzaki T, and Tanaka H 2008 Phys. Rev. B **78** 224405
[9] Saito T, Oosawa A, Goto T, Suzuki T, and Watanabe I 2006 Phys. Rev. B **74** 134423
[10] Goto T, Suzuki T, Kanada K, Saito T, Oosawa A, Watanabe I, and Manaka H 2008 Phys. Rev. B **78** 054422
[11] Suzuki T, Watanabe I, Oosawa A, Fujiwara T, Goto T, Yamada F, and Tanaka H 2006 J. Phys. Soc. Jpn. **75** 025001
[12] Suzuki T, Yamada F, Kawamata T, Watanabe I, Goto T, and Tanaka H 2009 Phys. Rev. B, **79** 104409
[13] Oosawa A and Tanaka H 2002 Phys. Rev. B **65** 184437
[14] Matsuzaki T, Ishida K, Nagamine K, Watanabe I, Eaton G H, and Williams W G 2001 Nuclear Instrumental Methods in Physics Research A **465** 365
[15] Watanabe I, Ishii Y, Kawamata T, Suzuki T, Pratt F L, Done R, Chowdhury M, Goodway C, Dreyer J, Smith C, and Southern M 2009 Physica B **404** 993
[16] Pratt F L 2000 Physica B **289-290** 710
[17] Suzuki T, Watanabe I, Yamada F, Ishii Y, Ohishi K, Risdiana, Goto T, and Tanaka H 2009 Phys. Rev. B **80** 064407
[18] Goto K, Fujisawa M, Ono T, Tanaka H, and Uwatoko Y 2004 J. Phys. Soc. Jpn. **73** 3254

A novel tool for detecting Li diffusion in solids containing magnetic ions; μ^+SR study on Li_xCoO_2

Jun Sugiyama[1], **Kazuhiko Mukai**[1], **Yutaka Ikedo**[2], **Hiroshi Nozaki**[1], **Martin Månsson**[3], **and Isao Watanabe**[4]

[1] Toyota Central Research & Development Laboratories Inc., Nagakute, Aichi 480-1192, Japan
[2] Muon Science Laboratory, KEK, Tsukuba, Ibaraki 305-0801, Japan
[3] LNS, ETH Zürich and Paul Scherrer Institut, CH-5232 Villigen PSI, Switzerland
[4] Muon Science Laboratory, RIKEN, 2-1 Hirosawa, Wako, Saitama 351-0198, Japan

E-mail: e0589@mosk.tytlabs.co.jp

Abstract. The diffusion coefficient of Li^+ ions (D_{Li}) in the battery material Li_xCoO_2 has been investigated by means of muon-spin relaxation (μ^+SR), because D_{Li} for positive electrode materials has not been determined correctly so far. Based on performing the experiments in zero-field and weak longitudinal-fields at temperatures up to 400 K, we determined the fluctuation rate (ν) of the fields on the muons due to their interaction with the nuclear moments. Combined with susceptibility data and electrostatic potential calculations, clear Li^+ ion diffusion was detected above $\sim$ 150 K. The D_{Li} estimated from ν was in very good agreement with predictions from first-principles calculations, and we present the μ^+SR technique as a novel and optimal probe to detect D_{Li} of unique usefulness for materials containing magnetic ions.

1. Introduction

In spite of a long research history on lithium insertion materials for Li-ion batteries [1, 2], e.g., $LiCoO_2$, $LiNiO_2$, and $LiMn_2O_4$, one of their most important intrinsic physical properties, the Li^+ ions diffusion coefficient (D_{Li}), has not yet been determined with any reliability. Although Li NMR is a powerful technique to measure D_{Li} for non-magnetic materials, it is particularly difficult to evaluate D_{Li} for materials containing magnetic ions, because the magnetic ions induce additional pathways for the spin-lattice relaxation rate ($1/T_1$), resulting in huge $1/T_1$ compared with that expected for only the diffusive motion of Li ions.

Such difficulty was clearly evident in the $1/T_1(T)$ curve for $LiCoO_2$ and $LiNiO_2$ [3, 4], and, for that reason, D_{Li} was instead estimated from the Li-NMR line width [5]. However, since the line width, i.e., the spin-spin relaxation rate ($1/T_2$) is also affected by the magnetic ions, the D_{Li} obtained by Li-NMR for $LiCoO_2$ ($= 1 \times 10^{-14}$ cm^2s^{-1} at 400 K) is approximately four orders of magnitude smaller than predicted by first-principles calculations [6]. Since lithium insertion materials always include transition metal ions, in order to maintain charge neutrality during the extraction and/or insertion of Li^+ ions, it is consequently very difficult to determine D_{Li} for these compounds unambiguously by Li-NMR.

On the other hand, the chemical diffusion coefficient (D_{Li}^{chem}), which is measured under a potential gradient, has also been determined by electrochemical measurements. Note that the relationship between D_{Li} and D_{Li}^{chem} is given by $D_{Li}^{chem} = \Theta D_{Li}$, where Θ is a thermodynamic factor. The magnitude of D_{Li}^{chem} is, however, known to be very sensitive to the measurement

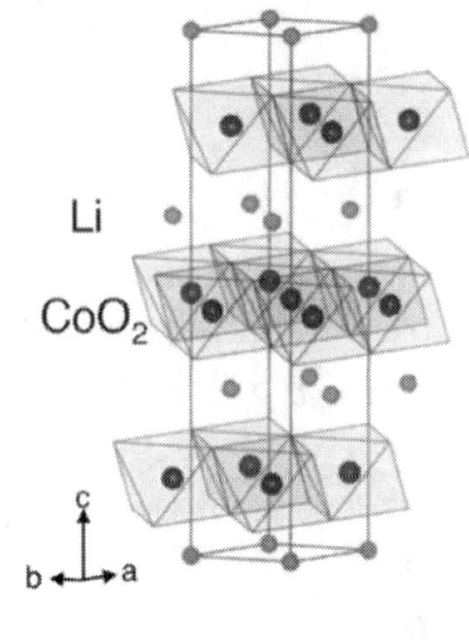

Figure 1. Crystal structure of $LiCoO_2$.

system, e.g., the electrolyte as well as the compositions of the positive and negative electrodes. As a result, the reported D_{Li}^{chem} for Li_xCoO_2 ranges from 4×10^{-8} to 10^{-10} cm^{2}s^{-1} for powder samples [7–11] and from 2.5×10^{-11} to 2×10^{-13} cm^{2}s^{-1} for thin films [12–15] at ambient T. In order to profoundly understand the physics behind the operation principle of battery materials, it is imperative to have a reliable probe to measure D_{Li} for all the components of the battery as a function of both Li content as well as T. This is at present a key issue for further development of Li-ion batteries, and in particular for future fabrication of solid-state batteries.

In contrast to NMR, the effect of localized moments in a paramagnetic (PM) state on the μ^+SR signal is very limited at high T, because the PM fluctuation is usually too fast to be *visible* by μ^+SR. Furthermore, although μ^+SR is very sensitive to the local magnetic environment, whether it is due to electronic or nuclear spins, an electronic contribution is, in principle, distinguishable from a nuclear contribution by weak longitudinal field measurements. In addition, since the μ^+SR timescale is different from the NMR's one, μ^+SR is expected to provide unique information on nuclear magnetic fields. Indeed, μ^+SR studies of Li_xCoO_2 at low T indicate that μ^+s *feel* a nuclear magnetic field caused by Li and ^{59}Co even at 1.8 K [16, 17]. This offers a possibility to determine D_{Li} of Li_xCoO_2 by μ^+SR, if μ^+s locate in the vicinity of the O^{2-} ion and make a stable μ^+-O^{2-} bond in the Li_xCoO_2 lattice. Here, we report our initial work on Li_xCoO_2 (x=0.73 and 0.53) to estimate their D_{Li} and establish μ^+SR as a novel tool to probe Li-ion diffusion.

2. Experimental

A powder sample of $LiCoO_2$ was prepared at Osaka City University by a solid-state reaction technique using reagent grade $LiOH{\cdot}H_2O$ and $CoCO_3$ powders as starting materials. A mixture of the two powders was heated at 900°C for 12 h in air. Powder X-ray diffraction (XRD) analysis showed that the $LiCoO_2$ sample was single phase with a rhombohedral system of space group $R\overline{3}m$ (a_H = 0.2814 nm and c_H = 1.4049 nm in hexagonal setting, see Fig 1). The Li-deficient samples were prepared by an electrochemical reaction using Li|$LiPF_6$-ethylene carbonate-diethyl carbonate|$LiCoO_2$ cells. The $LiCoO_2$ powder was pressed into a disc with 15 mm diameter and 0.4 mm thickness, and the disc was then used as a positive electrode. The Li_xCoO_2 disc was removed from the cell in a glove-box and packed into a sealed powder cell just before the μ^+SR measurement. In order to assess possible changes in the Li-deficient samples during the μ^+SR measurement, each sample was afterwards returned to the original cell to check their voltage

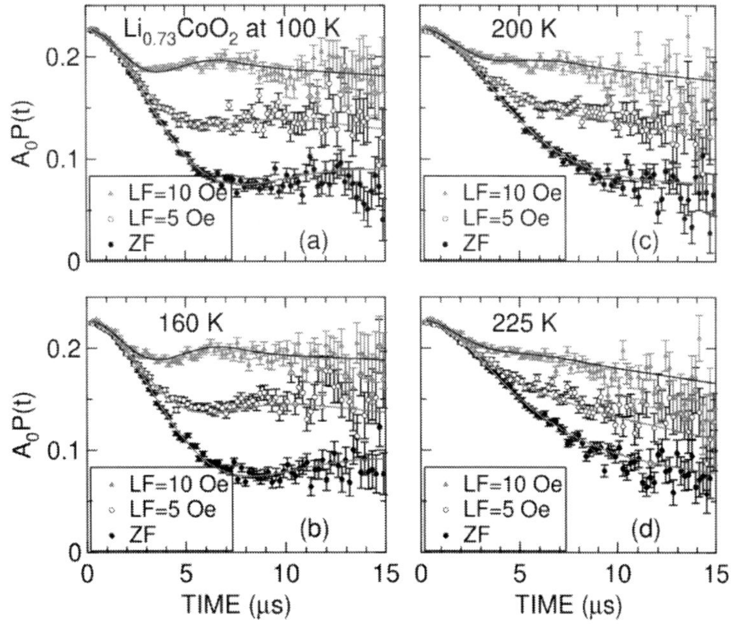

Figure 2. ZF- and two LF-μ^+SR spectra for $Li_{0.73}CoO_2$ measured at (a) 100 K, (b) 160 K, (c) 200 K, and (d) 225 K. The magnitude of LF was 5 and 10 Oe. Solid lines represent the fit result using Eq. (1).

(vs. a Li electrode). No significant change in the voltage was observed before and after the measurements. Their structures were subsequently confirmed by powder XRD, and, finally, their compositions were checked by an inductively coupled plasma atomic emission spectral analysis. The above procedure is essentially the same as that of our previous μ^+SR work on Li_xCoO_2 [17] and Li_xNiO_2 [18].

The μ^+SR spectra were measured at the surface muon beam line of the RIKEN-RAL Muon Facility at ISIS (U.K.) using a liquid-He flow-type cryostat in the T range between 10 and 400 K. The experimental techniques were described elsewhere [19]. χ was measured using a SQUID magnetometer (MPMS, Quantum Design) in the T range between 5 and 200 K under a magnetic field of H = 100 Oe.

3. Results and Discussion

Figure 2 shows the zero field (ZF-) and longitudinal field (LF-) μ^+SR spectrum for the $Li_{0.73}CoO_2$ sample obtained at 100, 160, 200, and 225 K. At 100 K, the ZF-spectrum exhibits a typical Kubo-Toyabe (KT) behavior with a minimum at $t \sim 6$ μs, meaning that the implanted muons *see* the internal magnetic field ($H_{\rm int}$) due to the nuclear magnetic moments of ^{7}Li, ^{6}Li and ^{59}Co. The applied LF clearly reduces the relaxation rate, i.e., the time slope, by *decoupling* $H_{\rm int}$. Although the ZF-spectrum still shows KT behavior at 200 and 225 K, the relaxation rate is smaller than at 100 K.

In order to estimate the KT parameters precisely, the ZF- and two LF-spectra were fitted simultaneously by a combination of a dynamic Gaussian KT function $[G^{\rm DGKT}(\Delta, \nu, t, H_{\rm LF})]$ and an offset background (BG) signal from the fraction of muons stopped mainly in the sample

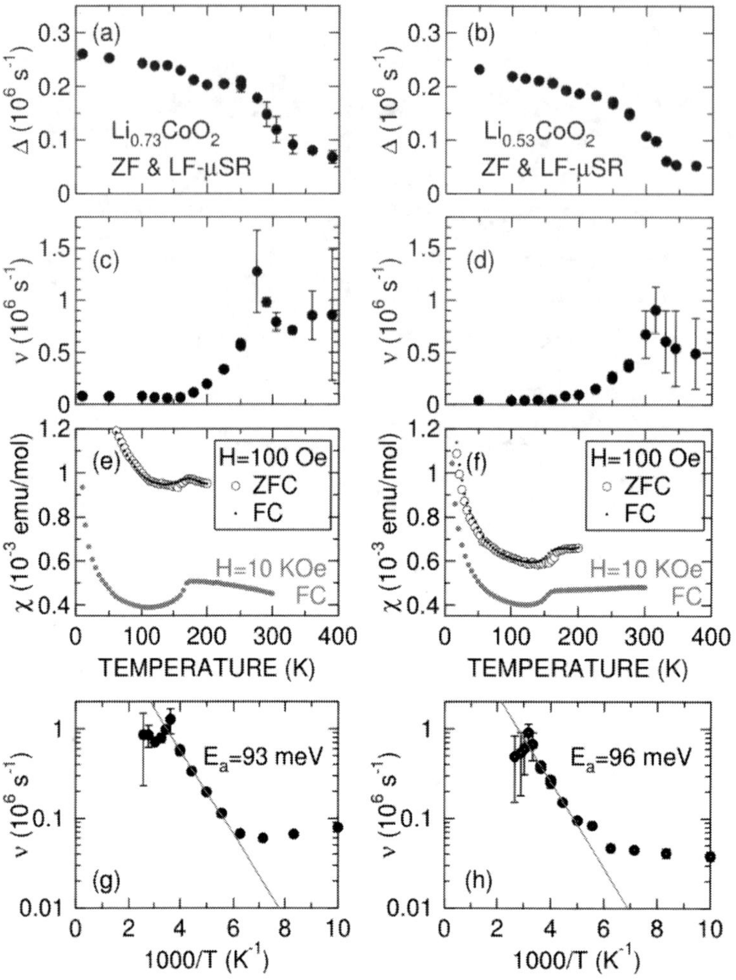

Figure 3. For $Li_{0.73}CoO_2$ and $Li_{0.53}CoO_2$ respectively, we show T dependences of (a),(b) field distribution width (Δ), (c),(d) field fluctuation rate (ν), (e),(f) susceptibility (χ), and (g),(h) the log(ν)-vs.-$1/T$ curve. Δ and ν were obtained by fitting both ZF- and LF-spectra using Eq. (1). χ was measured in both field cooling (FC) and zero field cooling (ZFC) mode with H=100 Oe. In (e),(f), the χ data [17] measured in FC mode with $H = 10$ kOe for $Li_{0.75}CoO_2$ and $Li_{0.52}CoO_2$ were also plotted for comparison. The straight lines in (g) and (h) show the activated diffusive behaviour discussed in the text.

holder, which is made of high-purity aluminum;

$$A_0 P_{\rm LF}(t) \;=\; A_{\rm KT} G^{\rm DGKT}(\Delta, \nu, t, H_{\rm LF}) + A_{\rm BG} \qquad (1)$$

where A_0 is the initial ($t = 0$) asymmetry, $A_{\rm KT}$ and $A_{\rm BG}$ are the asymmetries associated with the two signals. Δ is the static width of the local field distribution at the disordered sites, and ν is the field fluctuation rate. When $\nu = 0$ and $H_{\rm LF} = 0$, $G^{\rm DGKT}(t, \Delta, \nu, H_{\rm LF})$ is the static Gaussian KT function $G_{zz}^{\rm KT}(t, \Delta)$ in ZF. At first, we fitted all the ZF-spectra using common $A_{\rm KT}$ and $A_{\rm BG}$ in Eq. (1). The "global fit" provided that $A_{\rm KT} = 0.164056 \pm 0.000011$ (0.16889 ± 0.00018)

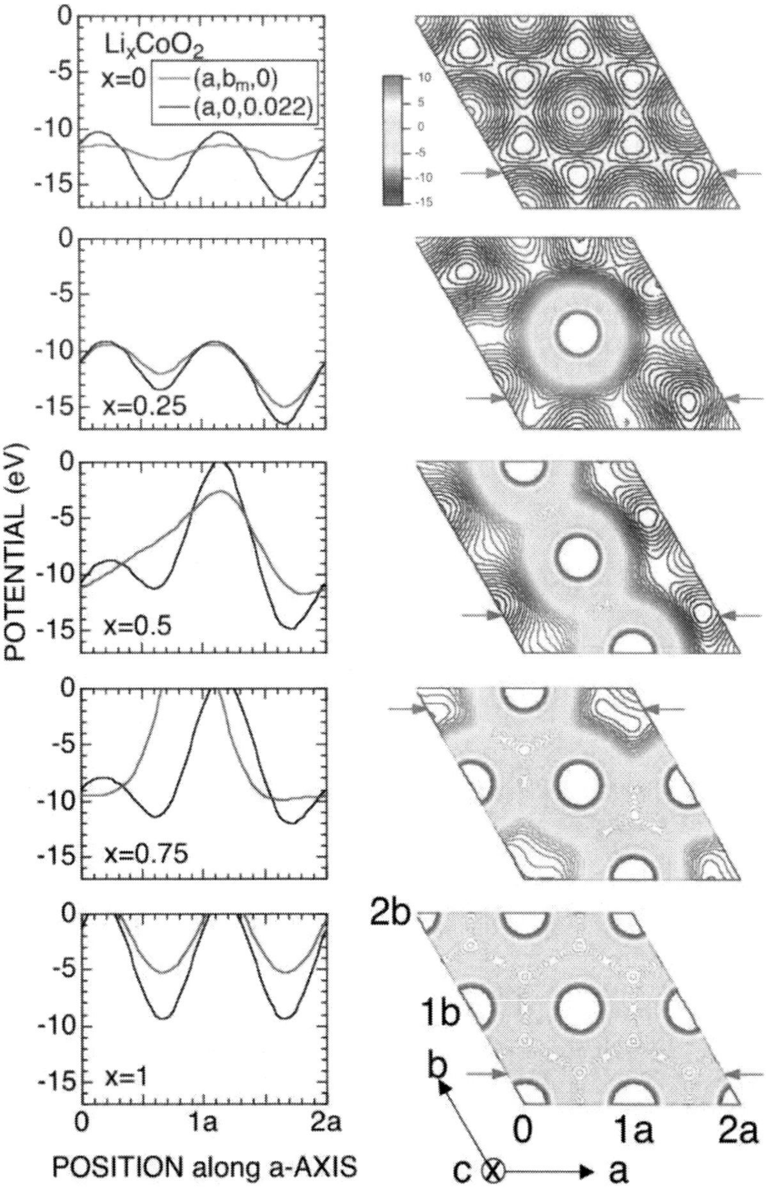

Figure 4. The variation of electrostatic potential (ϕ_E) along the a-axis at $b = b_m$ and $c = 0$ (on the Li plane) and $b = 0$ and $c = 0.022$ (1 Å away from the O^{2-} ions) in the hexagonal lattice of Li_xCoO_2 with $x = 0$, 0.25, 0.5, 0.75, and 1 (from top to bottom). Here, b_m denotes the b, at which electrostatic potential exhibits a minimum. The right panels show the corresponding distribution of ϕ_E in the Li plane. Arrows on the b-axis in the right panels represents b_m.

and $A_{BG} = 0.06350 \pm 0.00002$ (0.0692 ± 0.0002) for $Li_{0.73}CoO_2$ ($Li_{0.53}CoO_2$). Then, using the obtained A_{KT} and A_{BG}, one ZF- and two LF-spectra were global-fitted using common Δ and ν at each T.

Figure 3 shows the T dependencies of both Δ and ν for the two samples together with χ measured in a 100 Oe magnetic field. For $Li_{0.73}CoO_2$, Δ is almost constant for $5 \leq T \leq 250$ K, indicating that the μ^+ are most probably stable in the crystal lattice until $\sim$ 300 K. Δ then rapidly decreases, but levels off again for $T \geq 325$ K. The $\nu(T)$ curve is almost T-independent up to 150 K, starts to increase at $\sim$ 150 K, and exhibits a maximum at 275 K. Above 275 K, ν decreases to 0.7×10^6 s^{-1} at $\sim$ 325 K, and finally becomes almost T-independent above 350 K. The increase in ν between 150 and 275 K is well explained by a thermal activation process [Figs. 3(g) and 3(h)], which signals the onset of diffusive motion of either Li^+ or μ^+ above 150 K. The clear decrease in Δ at $\sim$ 300 K also suggests an additional diffusion of Li^+ or μ^+.

The $\chi(T)$ curve exhibits a small anomaly around 150 K with a thermal hysteresis of $\sim$ 10 K, while there is no indication of any magnetic anomalies in the T range between 200 and 300 K [Figs. 3(e) and 3(f)] [16, 17]. This suggests that the change in the μ^+SR parameters around 150 K is caused by an intrinsic change in Li_xCoO_2, but the change around 300 K is visible only by μ^+SR. The increase in ν above 150 K is, thus, most unlikely due to μ^+ diffusion but it is rather due to Li^+ diffusion, i.e., either a freezing of the Li^+ motion or an order-disorder transition of the Li^+ ions occurs below around 150 K. This is also supported by a recent ^{7}Li-NMR experiment [20], in which the NMR line width -vs.-T curve for $Li_{0.6}CoO_2$ exhibits a step-like decrease with T around 150 K by motional narrowing due to Li^+ diffusion. Since such diffusion naturally increases a local structural symmetry, it is reasonable that Δ slightly decreases with T around 150 K. On the other hand, both Li^+ and μ^+ are inferred to be diffusing above 300 K, resulting in the large decrease in Δ caused by motional narrowing. Actually, because $\Delta \leq 0.1\nu$ above 300 K, Eq. (1) is roughly equivalent to an exponential relaxation function $[\exp(-\lambda t)]$, and it is difficult to estimate Δ and ν precisely at high T.

The result for $Li_{0.53}CoO_2$ sample is very similar to that of $Li_{0.73}CoO_2$, although $\Delta_{T\to 0}(Li_{0.53}CoO_2) < \Delta_{T\to 0}(Li_{0.73}CoO_2)$ due to the decrease in the number density of Li^+ ions, as reported previously [16, 17]. Also, the magnitude of ν of $Li_{0.53}CoO_2$ is smaller than ν of $Li_{0.73}CoO_2$ in the whole T range measured, but the $\nu(T)$ curve for both samples show a clear increase with T above 150 K and a maximum around 300 K.

In order to predict the muon site(s) and to confirm the reliability of the above assumption that Li^+ ions diffuse above 150 K whereas μ^+ diffuse only above 300 K, we performed electrostatic potential (ϕ_E) calculations for the Li_xCoO_2 lattice using a point-charge model and the program DipElec [21]. As seen in Fig. 4, the site in the vicinity of the O^{2-} ions is more stable for μ^+ than the site in the Li plane for the whole x range between 1 and 0. This means that μ^+'s are bound to the O^{2-} ions so as to make a stable μ^+-O^{2-} bond in Li_xCoO_2. This is a common situation in oxides, as for example in the case for the high-T_c cuprates [22] and other cobalt oxide compounds [23, 24]. Since μ^+s are assigned as an ideal point charge, such μ^+-O^{2-} bond should be purely ionic. In fact, dipole field calculations for the site in the vicinity of the O^{2-} ions provide that $\Delta_{\rm calc} = 0.43 \times 10^6$ s^{-1} (0.35×10^6 s^{-1}) for Li_xCoO_2 with $x = 3/4$ (1/2). Furthermore, $\Delta_{\rm calc}$ is found to be comparable to Δ measured at low T in the whole x range for Li_xCoO_2 [17], if we consider the reduction of Δ by the electric field gradient effect on the nuclear moments with $I \geq 1$ [25, 26]. This suggests that the point-charge model is acceptable for determining the muon site(s) in Li_xCoO_2. As a result, it is clarified that, as T increases from 5 K, the Li^+ ions start to diffuse above 150 K ($= T_d^{\rm Li}$) and then the μ^+ diffuse above 300 K ($= T_d^{\mu}$), in spite of the mass difference between μ^+ and Li^+ ($m_{\rm Li^+}/m_{\mu^+} \sim 63$) because the muons form a hydrogen-like bond with oxygen.

Finally, we estimate $D_{\rm Li}$ using the obtained fluctuation rate ν as directly measuring the jump rate. Figure 5(a) shows the possible jump paths for the Li ions. That is, the direct jump to the nearest (vacant) Li site (path No. 1) and the jump to the interstitial site in the center of the oxygen tetrahedron (path No. 2). Assuming that ν corresponds to the jump rate of the Li ions

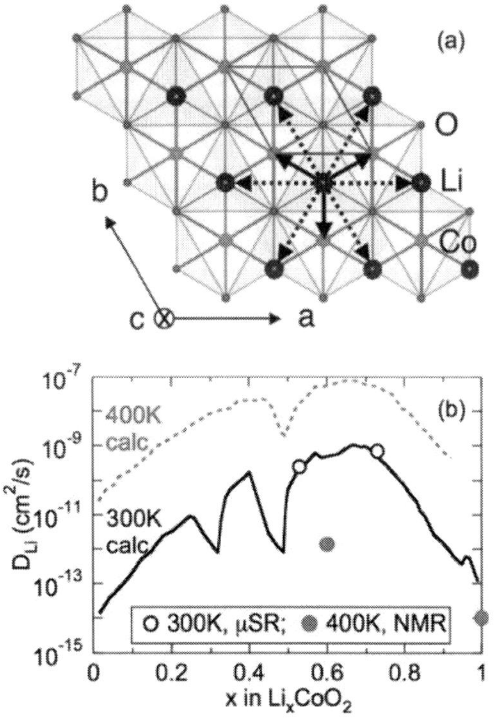

Figure 5. (a) Possible jump paths for Li ions. Broken arrows represent the direct jump to nearest (vacant) Li site (path No. 1), whereas solid arrows the jump to an interstitial site in the center of the oxygen tetrahedron (path No. 2). (b) The relationship between $D_{\rm Li}$ and x in $\rm Li_xCoO_2$ as extracted from our μ^+SR experiment (open circles). Solid and dashed lines represent the predictions by first-principles calculations [6] at $T = 300$ and 400 K, respectively, when the effective vibration frequency is a typical value (10^{13} s^{-1}). Sharp minima in the predicted curve (at $x = 1/3$ and $1/2$) are caused by Li-ordering. The ^{7}Li-NMR result [5, 20] for $T = 400$ K is also plotted (solid dot) for comparison.

between the neighboring sites, $D_{\rm Li}$ is given by [27];

$$D_{\rm Li} = \sum_{i=1}^{n} \frac{1}{N_i} Z_{v,i} s_i^2 \nu, \tag{2}$$

where N_i is the number of Li sites in the i-th path, $Z_{v,i}$ is the vacancy fraction, and s_i is the jump distance. Here, we naturally restrict the path to lie in the c-plane, i.e., along the 2D channel, because it is most unlikely that the Li ions jump across the $\rm CoO_2$ plane to an adjacent Li plane. Therefore, $N_1 = 6$, $N_2 = 3$, s_1 is equivalent to the a-axis length, $s_2 = a/\sqrt{3}$, $Z_{v,1} = 0.27$ for $\rm Li_{0.73}CoO_2$ (0.47 for $\rm Li_{0.53}CoO_2$), and $Z_{v,2} = 1$. As a result, we obtain $D_{\rm Li} = (7{\pm}2)\times10^{-10}$ cm^2/s [$(2.5 \pm 0.8) \times 10^{-10}$ cm^2/s] for $\rm Li_{0.73}CoO_2$ [$\rm Li_{0.53}CoO_2$] at 300 K. Here, ν(300 K) for $\rm Li_{0.73}CoO_2$ was estimated from the extrapolation of the linear relationship between log[ν] and T^{-1} [see Fig. 3(g)]. The estimated $D_{\rm Li}$ is found to be very consistent with the prediction by first-principle calculations [6], as seen in Fig. 5(b). Note that the jump paths used in Eq. (2) are the same to those for the first-principle calculations. This means that there is no ambiguous factor for estimating $D_{\rm Li}$ by μ^+SR. Since μ^+SR detects ν ranging from $\sim 0.01\Delta$ to $\sim 10\Delta$, it is applicable for materials with $D_{\rm Li} = 10^{-12} - 10^{-9}$ cm^2/s, when $N = 10$, $Z_v = 1$, $s = 1$ nm, and $\Delta = 0.1 \times 10^6$ s^{-1}.

In conclusion, we have been able to determine the Li diffusion coefficient, D_{Li}, of $\mathrm{Li}_x\mathrm{CoO}_2$ from the fluctuation rate of the field experienced by the muons in interaction with the nuclear moments of the diffusing ions. The value was found to be in good agreement with theoretical predictions. Consequently, we would like to suggest μ^+SR as a novel probe to investigate Li diffusion, especially for materials containing transition metal ions.

4. Acknowledgments

This work was performed at the RIKEN-RAL Muon Facility at ISIS, and we thank the staff for help with the μ^+SR experiments. We appreciate T. Ohzuku and K. Ariyoshi for sample preparation and K. Yoshimura for discussion. JS and YI are supported by the KEK-MSL Inter-University Program for Overseas Muon Facilities. This work is also supported by Grant-in-Aid for Scientific Research (B), 19340107, MEXT, Japan. The image involving crystal structure was made with VESTA.

References

[1] Mizushima K, Jones P C, Wiseman P J and Goodenough J B 1980 *Mater. Res. Bull.* **15** 783
[2] Ohzuku T and Brodd R 2007 *J. Power Sources* **174** 449
[3] Tomeno I and Oguchi M 1998 *J. Phys. Soc. Jpn* **67** 318
[4] Nakamura K, Yamamoto M, Okamura K, Michihiro Y, Nakabayashi I and Kanashiro T 1999 *Solid State Ionics* **121** 301
[5] Nakamura K, Ohno H, Okamura K, Michihiro Y, Nakabayashi I and Kanashiro T 2000 *Solid State Ionics* **135** 143
[6] der Ven A V and Ceder G 2000 *Electrochem. Solid-State Lett.* **3** 301
[7] Mizushima K, Jones P C, Wiseman P J and Goodenough J B 1981 *Solid State Ionics* **3-4** 171
[8] Yao C Y, Kao T H, Cheng C H, Chen J M and Hurng W M 1995 *J. Power Sources* **54** 491
[9] Pyun S I and Choi Y M 1997 *J. Power Sources* **68** 524
[10] Dokko K, Mohamedi M, Fujita Y, Itoh T, Nishizawa M, Umeda M and Uchida I 2001 *J. Electrochem. Soc* **148** A422
[11] Dokko K, Nakata N and Kanamura K 2009 *J. Power Source* **189** 783
[12] Striebel K A, Deng C Z, Wen S J and Cairns E J 1996 *J. Electrochem. Soc* **143** 1821
[13] Rho Y H and Kanamura K 2004 *J. Electrochem. Soc* **151** A1406
[14] Xia H, Lu L, Meng Y S and Ceder G 2007 *J. Electrochem. Soc* **154** A337
[15] Tang S, Lai M and Lu L 2008 *J. Alloys Compounds* **449** 300
[16] Sugiyama J, Nozaki H, Brewer J H, Ansaldo E J, Morris G D and Delmas C 2005 *Phys. Rev.* B **72** 144424
[17] Mukai K, Ikedo Y, Nozaki H, Sugiyama J, Nishiyama K, Andreica D, Amato A, Russo P L, Ansaldo E J, Brewer J H, Chow K H, Ariyoshi K and Ohzuku T 2007 *Phys. Rev. Lett.* **99** 087601
[18] Sugiyama J, Mukai K, Ikedo Y, Russo P L, Nozaki H, Andreica D, Amato A, Ariyoshi K and Ohzuku T 2008 *Phys. Rev.* B **78** 144412
[19] Kalvius G M, Noakes D R and Hartmann O 2001 *Handbook on the Physics and Chemistry of Rare Earths* vol 32 (Amsterdam: Elsevier Science) pp 55–451
[20] Nakamura K, Ohno H, Okamura K, Michihiro Y, Moriga T, Nakabayashi I and Kanashiro T 2006 *Solid State Ionics* **177** 821
[21] Kojima K M, Yamanobe J, Eisaki H, Uchida S, Fudamoto Y, Gat I M, Larkin M I, Savici A, Uemura Y J, Kyriakou P P, Rovers M T and Luke G M 2004 *Phys. Rev.* B **70** 094402
[22] Adams T R, Lichti R L and Gibson T L 1994 *Hyperfine Interact.* **86** 561
[23] Sugiyama J, Nozaki H, Ikedo Y, Mukai K, Andreica D, Amato A, Brewer J H, Ansaldo E J, Morris G D, Takami T and Ikuta H 2006 *Phys. Rev. Lett.* **96** 197206
[24] Sugiyama J, Nozaki H, Ikedo Y, Russo P L, Mukai K, Andreica D, Amato A, Takami T and Ikuta H 2008 *Phys. Rev.* B **77** 092409
[25] Hayano R S, Uemura Y J, Imazato J, Nishida N, Yamazaki T and Kubo R 1979 *Phys. Rev.* B **20** 850
[26] Kaiser C T, Verhoeven V W J, Gubbens P C M, Mulder F M, de Schepper I, Yaouanc A, de Réotier P D, Cottrell S P, Kelder E M and Schoonman J 2000 *Phys. Rev.* B **62** R9236
[27] Borg R J and Dienes G J 1988 *An Introduction to Solid State Diffusion* (San Diego: Academic Press) p 000

Muon spin relaxation studies on magnetic properties of organic conductors $Pd(dmit)_2$ salts and α-$(BETS)_2I_3$

S. Ohira-Kawamura[1], M. Tamura[2], R, Kato[3], K. Hiraki[4], T. Takahashi[4] and M. Iwasaki[5]

[1] J-PARC Center, Japan Atomic Energy Agency, Tokai, Ibaraki 319-1195, Japan
[2] Department of Physics, Tokyo University of Science, Noda, Chiba 278-8510, Japan
[3] Condensed Molecular Materials Laboratory, RIKEN, Wako, Saitama 351-0198, Japan
[4] Department of Physics, Gakushuin University, Toshima-ku, Tokyo 171-8588, Japan
[5] Advanced Meson Science Laboratory, RIKEN, Wako, Saitama 351-0198, Japan

E-mail: seiko.kawamura@j-parc.jp

Abstract.
Magnetic properties of organic conductors β'-$Et_2Me_2Sb[Pd(dmit)_2]_2$ (Et = C_2H_5, Me = CH_3), $P2_1/m$ crystal of $EtMe_3P[Pd(dmit)_2]_2$ and α-$(BETS)_2I_3$ were investigated by muon spin relaxation (μSR) method. It has been suggested that $Et_2Me_2Sb[Pd(dmit)_2]_2$ exhibits a charge separation phase below ~70 K and that $EtMe_3P[Pd(dmit)_2]_2$ exhibits a spin-Peierls-like state below ~25 K although it has a quasi two-dimensional structure. Our zero-field μSR results indicate that these $Pd(dmit)_2$ salts show quite similar behavior to an organic spin-Peierls system although changes in the μSR time spectra of the $EtMe_3P$ salt was ambiguous. For α-$(BETS)_2I_3$, a metal-insulator transition is expected to occur at $T_{MI} \sim 50$ K. We observed a slight enhancement of the muon spin relaxation below T_{MI} and it gradually became remarkable at lower temperatures. The present result may reveal the existence of a spin-singlet ground state showing similar temperature dependence of the muon spin relaxation rate to that in the $Pd(dmit)_2$ salts. However, the possibility that the relaxation in this system originates from magnetic fluctuations is also still conceivable.

1. Introduction

In organic materials classified in strongly correlated electron systems, the electric conductivity, including superconductivity, has been believed to be significantly related to their magnetism. In the case of some β'-$X[Pd(dmit)_2]_2$ (dmit= 1,3-dithiol-2-thione-4,5-dithiolate, X is a counter cation) which has a two-dimensional (2D) distorted triangular lattice structure of the $Pd(dmit)_2$ dimers, it has been systematically revealed that degree of the anisotropy on the triangle is the most important parameter to their magnetism and conductivity [1]. The $X[Pd(dmit)_2]_2$ systems having a relatively large anisotropy, such as X = Me_4P salt (Me = CH_3), show a long-range antiferromagnetic (AFM) order below $T_N \sim 40$ K and no metallic ground state at any hydrostatic pressure since the collinear spin structure could be stabilized by releasing its frustration [1, 2]. In contrast to them, the $EtMe_3Sb$ salt (Et = C_2H_5) having a nearly isotropic triangular lattice is expected to show no long-range magnetic order down to low temperature and metallic behavior under hydrostatic pressure[3, 4]. On the other hand, it has been recently reported that some $Pd(dmit)_2$ salts exhibit a spin-singlet ground state. Magnetic susceptibility and resistivity measurements and X-ray analysis on β'-$Et_2Me_2Sb[Pd(dmit)_2]_2$ revealed that the system exhibits a first-order transition to a charge separation phase at ~70 K [5, 6]. Furthermore, a valence bond solid (VBS) state in $P2_1/m$ crystal of $EtMe_3P[Pd(dmit)_2]_2$ was suggested from a rapid but continuous decrease in the susceptibility below 25 K toward zero [7, 8]. X-ray analysis showed modulation of intermolecular interactions along the column direction as a spin-Peierls state below the transition temperature though this system is 2D [8]. α-$(BETS)_2I_3$ (BETS = Bis(ethylenedithio)tetraselenafulvalene) is another material whose conductivity and magnetism have been of interest. This system is expected to exhibit a metal-insulator (MI) transition at T_{MI} = 50 K [9]. However, the magnetic property below T_{MI} has not been understood yet. The ^{77}Se NMR and magnetic susceptibility measurements have indicated nonmagnetic state below T_{MI} [10], while the existence of magnetic interactions has been suggested by large negative magnetoresistance at low temperatures [11]. In order to investigate detailed magnetic properties of the systems $Et_2Me_2Sb[Pd(dmit)_2]_2$, $EtMe_3P[Pd(dmit)_2]_2$ and α-$(BETS)_2I_3$, we have performed muon spin relaxation (μSR) measurements on these systems. Our zero-field (ZF-) μSR measurements indicate the spin-singlet features in the two $Pd(dmit)_2$ salts showing a gradual increase in the muon spin relaxation rate below the transition temperatures, although the change in the μSR time spectra of the $EtMe_3P$ salt was very small. Similar behavior was also observed in α-$(BETS)_2I_3$, but the MI transition at T_{MI} was not so clear. The possibility that the ground state is magnetic also still remains.

2. Experimental procedure

ZF-μSR measurements on $Et_2Me_2Sb[Pd(dmit)_2]_2$, $EtMe_3P[Pd(dmit)_2]_2$ and α-$(BETS)_2I_3$ were carried out at ARGUS spectrometer installed at the RIKEN-RAL Muon Facility in Rutherford Appleton Laboratory. 150 mg polycrystalline samples of $Et_2Me_2Sb[Pd(dmit)_2]_2$, 24 mg of $EtMe_3P[Pd(dmit)_2]_2$ and 60 mg of α-$(BETS)_2I_3$ were prepared. The samples of each system were packed into a silver foil with 25 μm in thickness to mount onto a sample holder. In the measurements on $Et_2Me_2Sb[Pd(dmit)_2]_2$ and α-$(BETS)_2I_3$, the sample was cooled down to ~1.5 K using a ^{4}He cryostat. $EtMe_3P[Pd(dmit)_2]_2$ was measured down to ~5 K using a mini-cryostat. Spin-polarized surface muons with a momentum 27 MeV/c were used. Environmental magnetic field was compensated within 30 mG during the ZF-μSR measurements.

3. Results and discussion

Figures 1(a)-(c) show ZF-μSR time spectra of $Et_2Me_2Sb[Pd(dmit)_2]_2$, $EtMe_3P[Pd(dmit)_2]_2$ and α-$(BETS)_2I_3$, respectively. In each system, muon spin depolarization due to the random and static nuclear dipole field was observed above the expected transition temperatures (open circles in the figures). The muon spin relaxation was gradually enhanced as temperature is lowered. In Figs. 1(a) and (c), the shapes of the time spectra were changed, namely they seem to include a Gaussian component above the transition temperature while they are exponential at the lowest temperature (triangles in the figures). Such a clear change was not observed in Fig. 1(b). A strong enhancement of the muon spin relaxation or a drop of the initial asymmetry, which indicates the appearance of magnetic order, was not observed in each system.

The time spectra were fitted with the following function

$$A(t) = A_0 \exp\left[-(\sigma t)^2\right] \exp(-\lambda t) + A_{bg}, \qquad (1)$$

where A_0 is the initial asymmetry for the muon spin relaxation signal from the sample, and A_{bg} is constant background. σ approximately describes the internal field distribution width. λ is the muon spin relaxation rate which reflects electronic spin fluctuations. σ should be constant because it corresponds to the field distribution width due to temperature-independent nuclear

Journal of Physics: Conference Series **225** (2010) 012041 doi:10.1088/1742-6596/225/1/012041

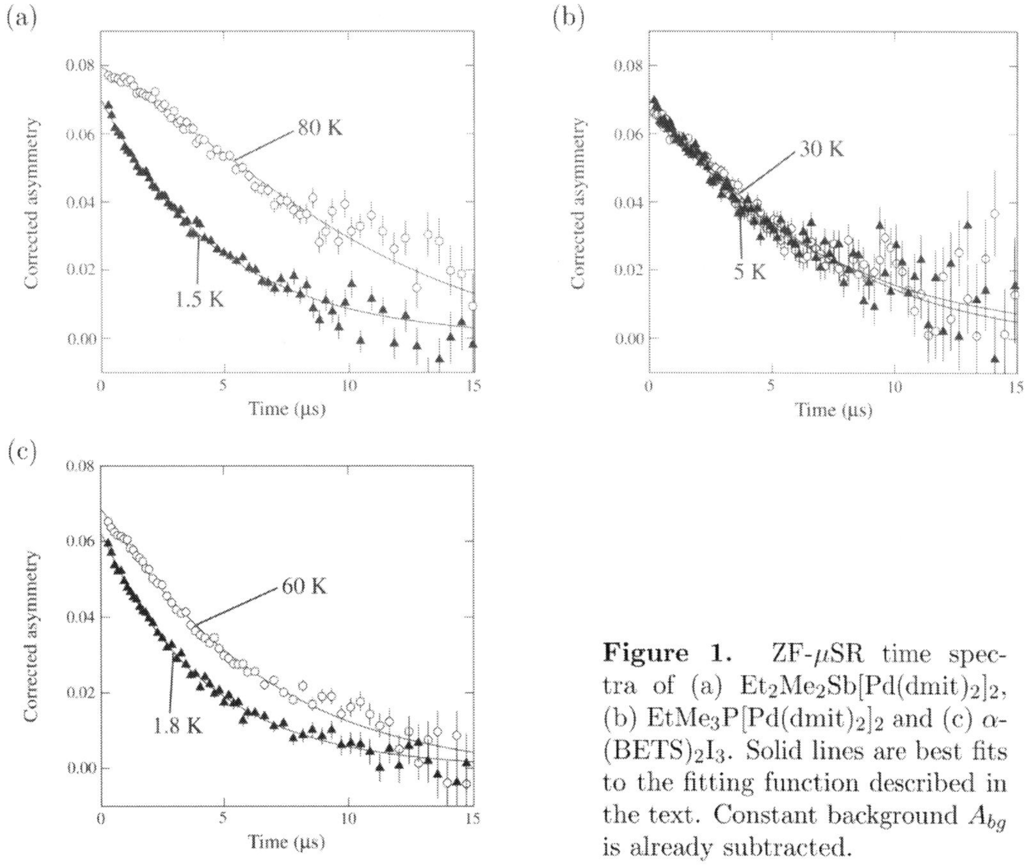

Figure 1. ZF-μSR time spectra of (a) $Et_2Me_2Sb[Pd(dmit)_2]_2$, (b) $EtMe_3P[Pd(dmit)_2]_2$ and (c) α-$(BETS)_2I_3$. Solid lines are best fits to the fitting function described in the text. Constant background A_{bg} is already subtracted.

dipoles. We however used free σ to fit the data so that the function well reproduces the result even at low temperatures. The temperature dependences of the field distribution width σ and the muon spin relaxation rate λ for each system are shown in Figs. 2(a)-(f). Here we notice some common features observed in all the systems. σ and λ are constant above the expected transition temperature. This is reasonable because the nuclear dipole field is independent of temperature and the electronic spin fluctuation is faster than the μSR time scale. λ starts to increase at around the transition temperature and possibly reaches a constant value. Then σ seems to drop at a low temperature concomitantly with the time spectra taking exponential form. Though detailed behavior is different from each other, it is confirmed that three all systems show no remarkable behavior to indicate magnetic order down to the lowest temperatures. In $Et_2Me_2Sb[Pd(dmit)_2]_2$, λ increases below $\sim$80 K, which is roughly consistent with the charge-separation transition temperature. This implies that the correlation time of the electronic spin fluctuations continues to be slow enough and becomes observable with the μSR time scale as temperature is lowered. In $EtMe_3P[Pd(dmit)_2]_2$, λ starts to increase at the reported transition temperature for the second-order phase transition to the VBS state. λ shows a small cusp at a lower temperature and then becomes constant, simultaneously with a drop of σ. The temperature dependence of λ does not show so obvious changes. α-$(BETS)_2I_3$ exhibits slight increases in both σ and λ below $T_{MI} \sim 50$ K. Because it was difficult to distinguish the exponential component from the Gaussian one for this system, we retried fitting by fixing σ in the temperature region where σ was almost constant. The results are displayed in Figs. 2(e) and (f) with open circles. Only a slight increase in λ was observed at T_{MI}. λ shows a remarkable increase at lower temperatures rather than T_{MI}.

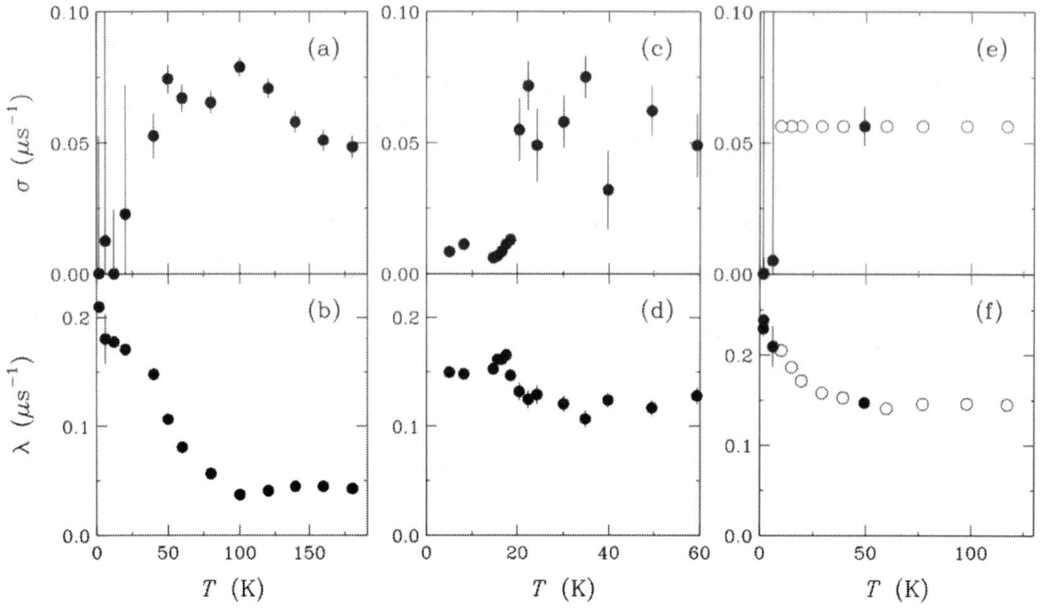

Figure 2. Temperature dependences of the field distribution width σ and the relaxation rate λ for (a)(b) $Et_2Me_2Sb[Pd(dmit)_2]_2$ [12], (c)(d) $EtMe_3P[Pd(dmit)_2]_2$ and (e)(f) α-$(BETS)_2I_3$, respectively.

When the spin-singlet state forms, muon should be affected by no electronic fluctuations. However, the muon spin relaxation is often enhanced below the transition temperature. In the spin-singlet system, the increase in λ starts at the transition temperature, in contrast to the magnetically ordered systems, where λ increases above the transition temperature and then shows a peak. We have interpreted for $Et_2Me_2Sb[Pd(dmit)_2]_2$ that the increases in λ is attributed to magnetic excitation across the gap just below the transition temperature [12]. Similar behavior of λ was also observed in $EtMe_3P[Pd(dmit)_2]_2$. The origin of the relaxation is expected to be the same as the Et_2Me_2Sb salt. However, the temperature dependence of λ is not so obvious compared to the Et_2Me_2Sb salt. As this system shows relatively large λ values above the transition temperature, spin fluctuations might exist to make the change in λ indistinct. It might be also because the transition to the spin-Peierls-like VBS state is of second-order. In this system, λ also shows temperature-independent behavior at lower temperatures. The relaxation in this temperature region should be caused by different origin. The electrons forming a spin-singlet state do not contribute to the muon spin relaxation. On the other hand, some defect creates dilute spins there, and they might be dominant to cause the exponential relaxation. Based on these consideration, it is strongly suggested that the system has a spin-singlet ground state. This may be also supported by the fact that quite similar temperature dependences of σ and λ are observed in an organic spin-Peierls system $MEM(TCNQ)_2$ [13, 14]. In α-$(BETS)_2I_3$, the observed behavior may overall resemble the two spin-singlet $Pd(dmit)_2$ systems, as far as the fitting results are compared to them. However, the MI transition at 50 K is ambiguous and the increase in λ gradually becomes noticeable at lower temperatures rather than T_{MI}. This tendency was emphasized in longitudinal field measurements under 30 G (not shown). No saturation of λ was observed down to the lowest temperature. It is implied that some temperature-dependent fluctuation becomes effective below T_{MI}, and remains down to

1.5 K. It is noted that the magnetoresistance measurement suggests the existence of magnetic interactions at low temperatures [11]. Indeed, the obtained temperature dependence of λ cannot be distinguished from a critical slowing down observed in magnetic systems. Thus, at this stage, we cannot exclude the possibility that the origin of the increase in λ is due to the magnetic interactions. More detailed μSR studies under longitudinal fields are necessary to understand the ground state.

4. Conclusion

We performed ZF-μSR measurements on organic conductors, $Et_2Me_2Sb[Pd(dmit)_2]_2$, $EtMe_3P[Pd(dmit)_2]_2$ and α-$(BETS)_2I_3$. In $Et_2Me_2Sb[Pd(dmit)_2]_2$, formation of the spin-singlet state accompanying with the charge separation was observed below ~80 K. In $EtMe_3P[Pd(dmit)_2]_2$, the field distribution width σ and the muon spin relaxation rate λ showed similar temperature dependences to those in the Et_2Me_2Sb salt, though the change in λ was not so obvious. λ further showed temperature-independent behavior in a low temperature region which might reflect spin fluctuations due to some defect. α-$(BETS)_2I_3$ also exhibited roughly similar behavior to the other spin-singlet systems. However, only a small anomaly was observed at $T_{MI} = 50$ K, and the muon spin relaxation was gradually enhanced as temperature was lowered without showing the constant-λ behavior. For this system, the possibility that the ground state is magnetic cannot be excluded at present.

References

[1] M. Tamura and R. Kato, J. Phys.: Condens. Matter **14** (2002) L729.
[2] S. Ohira, M. Tamura, R. Kato, I. Watanabe and M. Iwasaki, Phys. Rev. B **70** (2004) 220404(R).
[3] T. Itou, A. Oyamada, S. Maegawa, M. Tamura and R. Kato, J. Phys.: Condens. Matter **19** (2007) 145247.
[4] T. Itou, A. Oyamada, S. Maegawa, M. Tamura and R. Kato, Phys. Rev. B **77** (2008) 104413.
[5] M. Tamura, A. Tajima and R. Kato, Synth. Met. **152** (2005) 397.
[6] A. Nakao and R. Kato, J. Phys. Soc. Jpn. **74** (2005) 2754.
[7] R. Kato, A. Tajima, A. Nakao and M. Tamura, J. Am. Chem. Soc. **128** (2006) 10016.
[8] M. Tamura, A. Nakao and R. Kato, J. Phys. Soc. Jpn. **75** (2006) 093701.
[9] M. Inokuchi et al., Bull. Chem. Soc. Jpn. **68** (1995) 547.
[10] K. Hiraki et al., unpublished.
[11] N. Tajima et al., Synth. Met. **86** (1997) 1981.
[12] S. Ohira, M. Tamura, R. Kato and M. Iwasaki, Physica B **374-375** (2006) 122.
[13] S. J. Blundell, F. L. Pratt, P. A. Pattenden, M. Kurmoo, K. H. Chow, S. Takagi, Th. Jestädt and W. Hayes, J. Phys.: Condens. Matter **9** (1997) L119.
[14] B. W. Lovett, S. J. Blundell, F. L. Pratt, Th. Jestädt, W. Hayes, S. Takagi and M. Kurmoo, Physica B **289-290** (2000) 145.

High-resolution Auger-Electron Spectroscopy Induced by Positron Annihilation on Fe, Ni, Cu, Zn, Pd, and Au

C. Hugenschmidt, J. Mayer, and K. Schreckenbach

FRM II and E21 Physics Department, Technische Universität München, Lichtenbergstrasse 1, D-85747 Garching

E-mail: hugen@frm2.tum.de

Abstract. Positron annihilation induced Auger electron spectroscopy (PAES) enables almost background free, non-destructive surface analysis with high surface selectivity. The Auger-spectrometer at the high intense positron source NEPOMUC now allows to record positron annihilation induced Auger spectra within a short data acquisition time of 10-80 minutes. With a new hemispherical electron energy analyzer and due to the exceptional peak to noise ratio, we succeeded to measure Auger-transitions such as the $M_{2,3}VV$ double peak of nickel with high energy resolution. The relative Auger-electron intensities are obtained by the analysis of the recorded positron annihilation induced Auger spectra for the surfaces of Fe, Ni, Cu, Pd and Au. It is demonstrated, that high-resolution PAES allows to determine experimentally the relative surface core annihilation probability of various atomic levels.

1. Introduction

In positron annihilation induced Auger-electron spectroscopy (PAES) the sample is irradiated with low-energy positrons with a kinetic energy of a few 10 eV which is well below the threshold for impact ionization of core electrons. The implanted positrons thermalize very rapidly within a few picoseconds. Due to the low implantation energy the vast majority of the positrons diffuses back to the surface where they can be trapped in the surface potential before they annihilate together with an electron (see e.g. [1]). In contrast to conventional AES, where core electrons are removed by photo-ionization with X-rays (XAES) or by impact-ionization with keV-electrons (EAES), at PAES core electrons are removed by positron-electron annihilation for the initiation of the Auger-process. Since only the annihilation with core electrons leads to the emission of Auger-electrons the element dependent core annihilation probability is an important number to estimate the intensity of the relevant Auger transitions. Jensen et al. [2] calculated core annihilation probabilities for various materials which typically ranges between 1 and 10 %.

The completely different initiation of the Auger-transition leads to several advantages of PAES over conventional AES. The positron annihilation induced Auger-spectra show no secondary electron background in the range of higher Auger-transition energies due to the low positron energy. The annihilation of positrons, which are trapped in a surface state, with electrons in the topmost atomic layer leads to the extreme surface sensitivity of PAES. At PAES the emission of the Auger-electron is decoupled from the impact position of the positron, and hence the probed atom is neither perturbed by the impinging primary particle nor by the detached core electron. However, the main drawback of PAES is the very long measurement time of up to a few weeks for one spectrum, if usual lab beams based on β^+-emitters are used. For this reason, up to now only a few research groups [1, 3, 4] benefit from PAES' advantages for surface studies.

To overcome the long data acquisition times, we developed and installed a PAES-setup at the positron beam facility NEPOMUC (**NE**utron induced **PO**sitron source **MU**ni**C**h) at the research reactor Heinz-Maier-Leibniz FRM II. Both, the high intensity of the low-energy positron beam of more than 10^7 moderated positrons per second, and a new state-of-the-art hemispherical electron energy analyzer with a large acceptance angle of up to $\pm 13°$ lead not only to an enormous reduction of the measurement time but also to an improved signal-to-noise ratio. Although, the reactor-based positron source NEPOMUC delivers an intense remoderated positron beam its low-energy positron flux is in the order of 10^{-11} A. Hence, it is still six orders of magnitude lower than the electron current from conventional electron guns. However, we succeeded to demonstrate that the signal-to-noise ratio is about one order of magnitude higher at PAES than at EAES. Within this work, we present high-resolution PAES experiments on pure Fe, Ni, Cu, Zn, Pd, and Au. Besides the determination of the peak energy of the respective Auger-transitions, the relative Auger intensities are discussed in terms of core annihilation probabilities.

2. Experimental

2.1. The Auger-spectrometer at the positron beamline NEPOMUC

The positron beam facility NEPOMUC at the research reactor FRM II delivers a high-intensity 1 keV positron beam of nearly 10^9 moderated positrons per second [5]. This primary positron beam is remoderated with a tungsten single crystal in back reflection geometry for brightness enhancement [6]. The voltage of the remoderation crystal can be adjusted between 15 V and 200 V, and was set to 17 V for all presented measurements. Taking into account the positron work function of the W crystal of -3.0 eV, the kinetic energy of the remoderated beam amounts to 20 eV with respect to ground potential. The low-energy positron beam is adiabatically guided in a longitudinal magnetic field of typically 6-8 mT via a beam switch to the PAES apparatus.

The positron beam is non-adiabatically released by a magnetic field termination out of μ-metal at the entrance of μ-metal shielded Auger analysis chamber into the electrostatic guiding and focusing unit. A tungsten shielded NaI-detector is directed toward the center of the analysis chamber, and hence allows the detection of the annihilation radiation, i.e. the positron intensity, at the sample position. An electron gun enables the measurement of conventional EAES. A movable MCP-detector with phosphor screen is used for the positron beam and for electron beam adjustment. A sample lock and a sample preparation chamber equipped with an effusion cell, an electron beam evaporator, a piezo-oscillator as thickness monitor, and an argon ion sputter gun for surface cleaning are connected to the Auger-chamber [7]. The electron spectra and hence the emitted Auger-electrons are recorded via a hemispherical electron analyzer using a MCP/CCD read-out system.

2.2. Sample preparation

For the presented measurements high purity polycrystalline samples of Fe (purity >99.99 %), Ni (>99.99 %), Cu (>99.999 %), Zn (>99.99 %) Pd (>99.994 %), and Au (>99.99 %) were used. First, all samples were cleaned by acetone and rinsed by deionized water. After mounting on the sample holder, the sample surface was sputtered in the UHV-sample preparation chamber with 1 keV Ar^+-ions for typically 20-30 minutes.

2.3. Measurement

The PAES-measurements were carried out at room temperature and at a pressure of typically $< 5 \cdot 10^{-9}$ mbar inside the Auger-chamber. The kinetic energy of the incident positron beam on to the grounded sample is defined by the remoderation stage and was set to 20 eV for all measurements. Consequently, the secondary electron background ends at about 20 eV and hence well below the energies of the expected Auger transitions. The beam intensity at the sample position was about $9(3) \cdot 10^6$ moderated positrons per second.

The typical data acquisition time for the positron annihilation induced Auger-spectra presented here was 80 minutes. The entrance slit amounted to 7 mm, and the exit slit is given by the discretization of the CCD-camera read-out which amounts to 1.6 mm. The step width of the detected electrons was set to 0.25 eV and the fixed retard ratio (FRR) of the analyzer was set to 1.0 for Ni and Au resulting in an energy resolution of $\Delta E/E = 0.025$. For all other samples the step width was set to 0.5 eV, the FRR was 0.5, and the energy resolution amounted to $\Delta E/E <= 0.049$ for the PAES measurements presented here.

3. Results and discussion

For all samples Fe, Ni, Cu, Pd, Zn and Au Auger-spectra were recorded with PAES which are plotted in figure 1. All spectra show an exceptional signal-to-noise ratio which is in the case of Cu as high as S/N=11:1. Comparison with EAES showed that the signal-to-noise ratio at PAES is at least one order of magnitude higher. The secondary electron spectrum ends at about 20 eV. However, the background, which is still observable, is mainly attributed to detected (Compton- and photo) electrons which are produced by the annihilation radiation in the sample. In addition, scattered Auger-electron of Auger transitions at higher energy might also contribute to the background. The dark current of the MCP detector of the electron analyzer is negligible.

The double peak structure of certain Auger transitions, which is particularly clearly visible for Ni, correspond to the respective energy levels of core electrons with different total angular momentum quantum number, e.g. $p_{1/2}$ and $p_{3/2}$ for the $M_{2,3}$ levels. The spectrum obtained for Pd shows a significant second peak according to an additional Auger transition at about 80 eV. Moreover, at Au, three distinct Auger peaks emerge which were attributed to the O_3VV, O_2VV and $N_{6,7}VV$ transitions. Note, that with increasing electron binding energy of the annihilating electron, i.e. increasing energy of the Auger-electron, the peak intensity decreases due to the accordingly declining core annihilation probability.

For all spectra the peak energies were found to agree with the theoretical values. For example at Cu the experimentally determined separation of the $M_{2,3}VV$ transition was determined to 2.19(0.18) eV, which is in excellent agreement with the theoretical value of 2.2 eV [8]. However, mainly due to the large secondary background at EAES significant peak shifts of the Auger transitions could not be observed, which might be expected due to the completely different ionization process of the probed atom in both techniques.

The data shown in figure 1 was analyzed not only with respect to the absolute energy of the Auger-transitions but also with respect to their intensities. In order to estimate the relative core annihilation probabilities of the respective Auger transitions the Auger-intensities are plotted as a function of the binding energy of the annihilated electron for Fe, Ni, Cu, Zn, Pd, and Au (see figure 2).

With the following assumptions the variation of the observed Auger-intensities can be correlated to the relative changes of the core annihilation probability. Since the positron flux at the sample position is considered to remain constant the time normalized spectra can be compared directly. The deexcitation of the excited electron shells is governed by the Auger-process. This assumption is reliable since the effective core potential seen by the outer electrons is low and characteristic X-ray emission is not probable at the considered excitation energies. The different positron reflexion, which increases with higher nuclear charge, is not taken into account.

The calculated values of the core annihilation probabilities show a large straggling [2] but can be very roughly approximated by the formula given in [9]: $p = 600 \cdot N(E_B) \cdot E_B^{-1.6}$ with p as

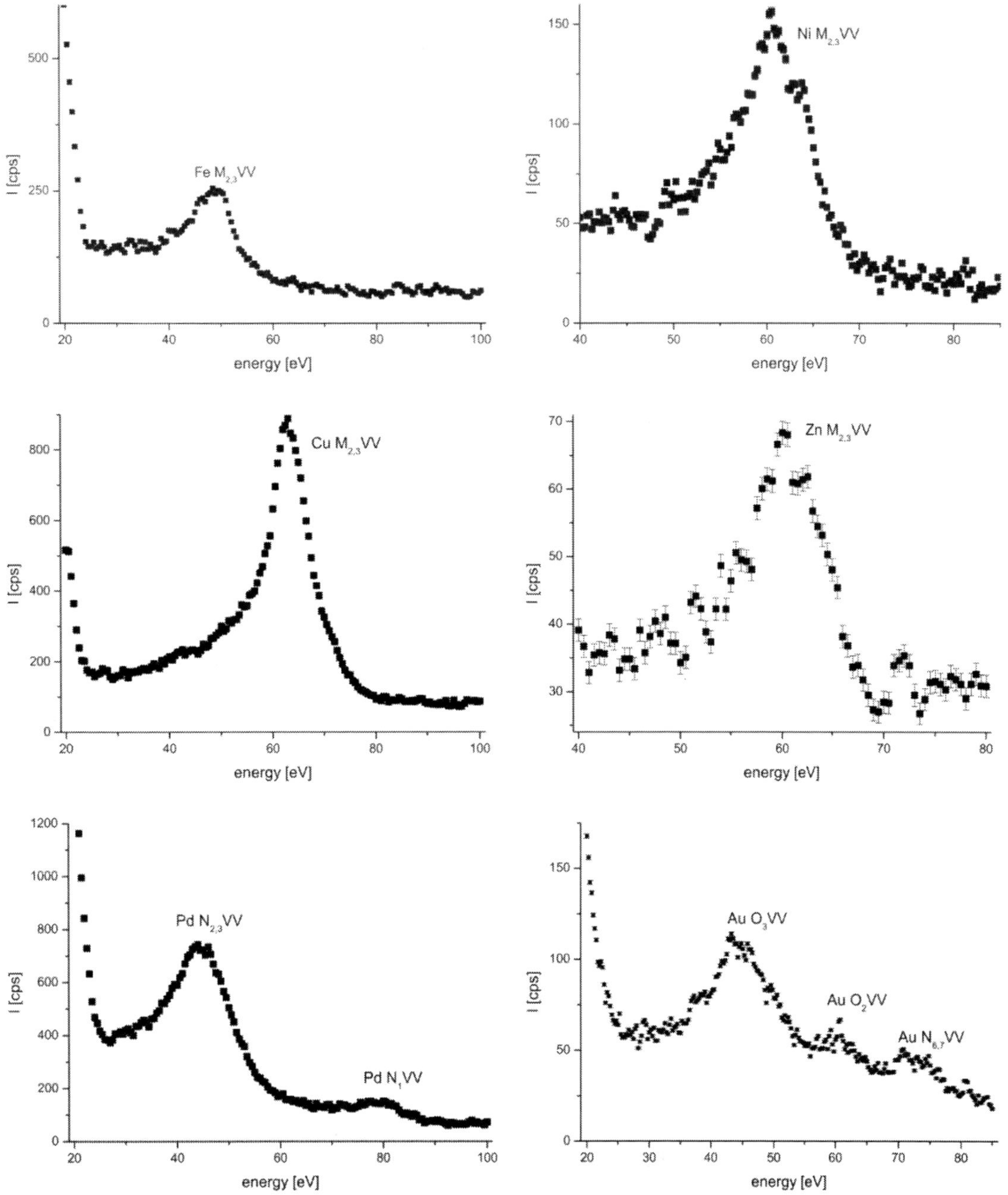

Figure 1. Positron annihilation induced Auger-electron spectra for Fe, Ni, Cu, Zn, Pd, and Au.

the core annihilation probability in %, E_B the binding energy in eV and $N(E_B)$ the number of electrons with the binding energy E_B. As shown in figure 2 the slope of the data clearly shows a trend to higher Auger intensity at lower electron binding energy which reflects qualitatively the given formula for p. This observation is attributed to the respective core annihilation probability which is lower for deeper bound electrons due to the repulsive Coulomb potential of the nucleus.

The variation of the experimentally determined relative differences of the Auger intensities is influenced by several aspects. The amount of positrons that leave the surface epithermally or moderated (provided a negative positron work function, e.g. at Cu or Ni) would vary a little at the different elements but their contribution to different surface core annihilation probability would be low. The positronium formation probability at the surface is expected to be similar at metal surfaces but might change drastically if oxygen is absorbed at the surface. This cannot be excluded although each sample was cleaned by Ar sputtering immediately before the PAES measurement. Therefore, a possible surface contamination would lead to a lower measured Auger intensity due to both, a larger positronium formation probability and to less unscattered Auger-electrons from the metal. In addition, influences due to differences in the surface topology can not be considered. In order to estimate the influence of the surface structure such as steps, adsorbates or defects on an atomic scale, a complementary tool such as scanning tunneling microscopy would be very helpful.

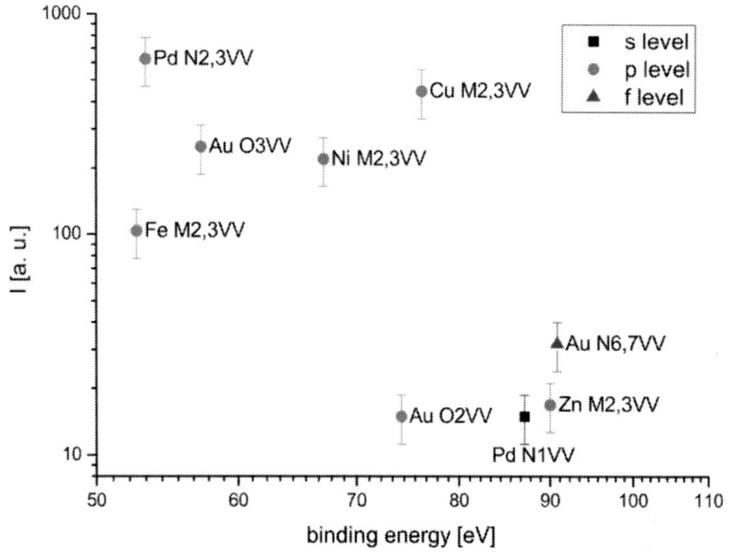

Figure 2. The relative Auger intensities obtained from the spectra shown in figure 1 as a function of the binding energy of the annihilated electron for Fe, Ni, Cu, Pd, and Au.

4. Summary and Outlook

Within the present work we demonstrated that PAES studies of surfaces can be performed within a measurement time of approximately one hour using the high-intensity low-energy positron beam at NEPOMUC. In these measurements on pure metallic surfaces the substructure of the Auger transitions, and e.g. at Au three distinct Auger peaks, could be observed due to the high resolution of the applied electron analyzer with high peak-to-noise ratio. It was shown that the Auger intensities recorded with PAES allows the estimation of the relative core annihilation probabilities experimentally in spite of influences such as positron reflexion, surface topology or possible surface contamination. In order to anneal possible surface defects introduced by Ar sputtering it is planned to implement a heatable sample holder. In addition, future studies will be focused on sub-monolayer layers on metallic substrates in order to benefit from the high surface sensitivity and the elemental selectivity of PAES.

References

[1] A. Weiss, R. Mayer, M. Jibaly, C. Lei, D. Mehl, and K. G. Lynn. *Phys. Rev. Lett.*, 61:2245, 1988.
[2] K. O. Jensen and A. Weiss. *Phys. Rev. B*, 41(7):3928–3936, 1990.
[3] R. Suzuki and T. Ohdaira and T. Mikado and H. Ohgaki and M. Chiwaki and T. Yamazaki *Applied Surface Science*, 100-101:297 – 300, 1996.
[4] C. Hugenschmidt, J. Mayer, and K. Schreckenbach. *Surf. Sci.*, 601:2459–2466, 2007.
[5] C. Hugenschmidt and B. Löwe and J. Mayer and C. Piochacz and P. Pikart and R. Repper and M. Stadlbauer and K. Schreckenbach. *Nucl. Instr. Meth. A*, 593(3):616–618, 2008.
[6] C. Piochacz, G. Kögel, W. Egger, C. Hugenschmidt, J. Mayer, K. Schreckenbach, P. Sperr, M. Stadlbauer, and G. Dollinger. *Appl. Surf. Sci.*, 255(1):98 – 100, 2008.
[7] J. Mayer, K. Schreckenbach, and C. Hugenschmidt. *Phys. Status Solidi C*, 4:3928–3931, 2007.
[8] J. Mayer, K. Schreckenbach, and C. Hugenschmidt. *Phys. Status Solidi C*, 11:2468-2470, 2009.
[9] P. Coleman. *Positron Beams and Their Applications.* World Scientific, 2000.

Journal of Physics: Conference Series **225** (2010) 012008 doi:10.1088/1742-6596/225/1/012008

Structure and phase transition of low-dimensional metals on Si(111) surfaces studied by reflection high-energy positron diffraction

Y Fukaya, M Hashimoto, A Kawasuso and A Ichimiya

Advanced Science Research Center, Japan Atomic Energy Agency, 1233 Watanuki, Takasaki, Gunma 370-1292, Japan

fukaya.yuki99@jaea.go.jp

Abstract. We investigated the surface phase transitions of Si(111)-4×1 and Si(111)-√3×√3-Ag surfaces using reflection high-energy positron diffraction (RHEPD). We measured the rocking curves of RHEPD at above and below their phase transition temperatures. By means of the intensity analyses based on the dynamical diffraction theory, we discuss the structural changes upon the phase transitions.

1. Introduction

Phase transition is one of the main topics in the condensed matter physics. The surface phase transition takes place within a few layers of the crystal surface. For instance, clean Si(001)-2×1 superstructure undergoes a c(4×2)-2×1 phase transition below 200 K [1]. This phase transition is confirmed as the order-disorder phase transition [1]. That is, asymmetric dimer atoms are thermally flipping above 200 K, leading to the 2×1 periodicity. By depositing foreign atoms on a crystal surface, it is possible to fabricate various kinds of low-dimensional structures. The low-dimensional structure shows various intriguing properties such as Peierls instability, quantum well state, and Rashba effect. To explore these properties, it is very important to study atomic coordinates of the low-dimensional structure. In particular, the structure analyses of the topmost surface are needed.

Reflection high-energy positron diffraction (RHEPD) is a surface sensitive tool to study the atomic coordinates and thermal vibrational amplitudes of the topmost surface [2,3]. In 1998, we developed RHEPD apparatus and succeeded in the observation of RHEPD patterns [4]. Since positrons have a positive charge, the total reflection takes place at grazing incidence [2]. At the total reflection condition, the incident positron is mostly reflected at the topmost surface. Figure 1 shows the penetration depth of the incident positron beam into the Si surface as a function of the glancing angle. The penetration depth was calculated by solving the simple Schrödinger equation using a step-like potential. In this case, the critical angle (θ_c) of the total reflection condition can be determined via the Snell's equation [2]:

$$\theta_c = \arcsin(V_0/E)^{1/2}, \tag{1}$$

where V_0 and E are the inner potential of the crystal and the accelerating voltage of the incident positron beam. When V_0 = 12 V for the Si crystal and E = 10 kV, the critical angle of the total reflection is estimated to be 2.0° using equation (1). In particular, in the total reflection region, the penetration depth is quite small, which is less than 2 Å. Thus, the RHEPD is very sensitive to the topmost surface structure.

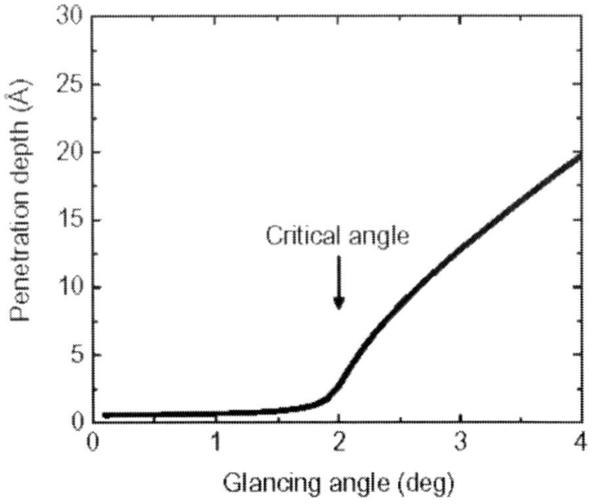

Figure 1. Penetration depth of the positron beam into the Si(111) surface as a function of the glancing angle. The critical angle of the total reflection region is estimated to be 2.0° when the accelerating voltage of the incident positron beam is 10 kV.

In this study, using the RHEPD, we investigated the surface structures at above and below the phase transition temperature. We will show two recent results concerning the phase transitions of one- and two-dimensional metals of In/Si(111) and Ag/Si(111) surfaces.

2. Experimental procedure

Samples with a dimension of 5×25×0.5 mm^3, which were cut from a mirror-polished *n*-type Si(111) wafer (resistivity: 1-10 Ωcm), were used. After transferring them into a ultra-high vacuum (UHV) chamber with a base pressure of 5×10^{-8} Pa, they were heated at 673 K for several hours and then flashed at 1473 K for a few seconds several times to obtain clean 7×7 reconstructed surfaces. To make Si(111)-4×1-In surfaces, 1 monolayer (ML: 1 ML = 7.83×10^{14} cm^{-2}) of In atoms were deposited on the Si(111)-7×7 surface at 673 K using an electron beam evaporator. To prepare Si(111)-√3×√3-Ag surfaces, 1 ML of Ag atoms were deposited on the Si(111)-7×7 surface at 773 K. The substrate temperature (T) was calibrated with an infrared radiation thermometer and a thermo couple attached near the sample.

Experiments were carried out in the UHV chamber equipped with a ^{22}Na positron source (3.7GBq) and electromagnetic lens system [5]. The positron beam was accelerated at 10 kV. The glancing angle (θ) of the incident positrons was adjusted by rotating the sample. The diffraction patterns were observed using a micro channel plate with a phosphor plane and a CCD camera. The sample was cooled down by using a cryostat.

3. Results and discussion

3.1. 8×2-4×1 phase transition of In/Si(111) surface

The 4×1 structure is composed of the zigzag chains of In atoms at room temperature [6], as represented in figure 2(a). Since angle-resolved photoemission spectroscopy (ARPES) measurements show the electronic band dispersions characteristic of one-dimensional systems, the 4×1-In structure is known as a quasi-one-dimensional atomic chain [7]. When the substrate temperature is reduced, the 4×1 structure is transformed to a 8×2 structure [8]. The phase transition temperature was determined to be 130 K using a high-resolution electron energy loss spectroscopy [9]. From the ARPES [8] and surface electrical conductivity measurements [10], it was found that the surface undergoes the metal-insulator transition at 130 K according to the 8×2-4×1 phase transition. Although the phase transition has been extensively investigated by many researchers, the atomic configuration of the 8×2 phase is still in debate. Especially, the atomic positions are not confirmed experimentally. Therefore, the phase transition mechanism is not fully clarified.

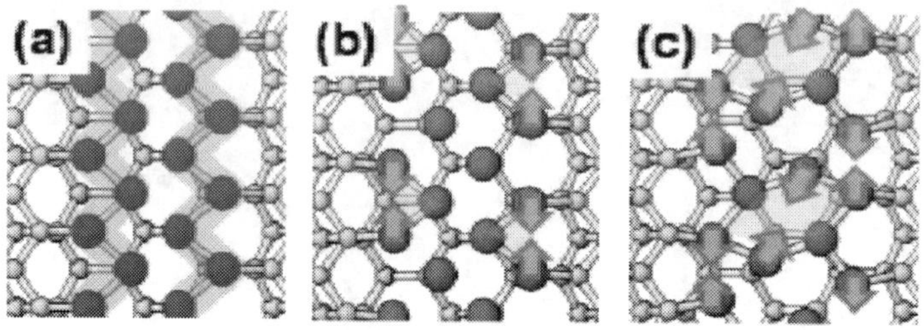

Figure 2. Structure models of (a) zigzag chain, (b) trimer, and (c) hexagon for In/Si(111) surface. Red and gray circles denote the In and Si atoms, respectively. Arrows indicate the possible displacements of In atoms from the zigzag chain to trimer or hexagon structures.

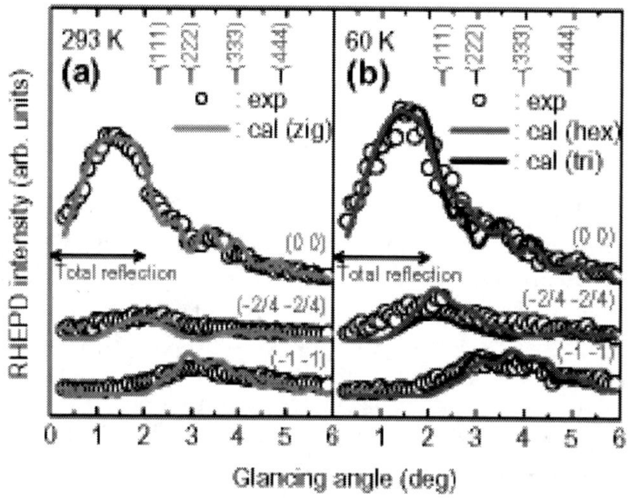

Figure 3. Rocking curves from (a) the Si(111)-4×1 surface at 293 K and (b) 8×2 surface at 60 K. The open circles and solid lines denote the measured and calculated curves, respectively. The accelerating voltage of the positron beam is set at 10 kV. The incident azimuth of the positron beam corresponds to the $[1\bar{1}0]$ direction.

Figures 3(a) and 3(b) show the RHEPD rocking curves from the Si(111)-4×1 surface at 293 K and 8×2 surface at 60 K along the $[1\bar{1}0]$ direction, respectively. As for the Si(111)-4×1 surface, small peaks are observable in the shoulder of the rocking curve for the (0 0) spot in the range of $\theta = 2.5°$-

4.5°. The intensity for the (-2/4 -2/4) spot is relatively weak. When the 4×1 structure is transformed to the 8×2 structure, the total reflection peak becomes sharp. The small peaks in the shoulder disappear. Furthermore, the intensity for the (-2/4 -2/4) spot increases. These changes indicate the atomic displacements of In atoms according to the phase transition.

We investigated the atomic positions of the 4×1 and 8×2 structures from the rocking curve analyses. We calculated the rocking curves based on the dynamical diffraction theory [11]. As for the 4×1 structure, we used the atomic positions of the zigzag chain model determined by surface x-ray diffraction [6] in the calculations. The solid lines in figure 3(a) show the calculated rocking curves using the zigzag chain structure. The calculated curves can well reproduce the small peaks in the shoulder of the rocking curve for the (0 0) spot and the relatively weak intensities for the (-2/4 -2/4) spot. The calculated curve is in good agreement with the measured curve for each spot. Therefore, we also confirmed that the 4×1 structure is composed of the zigzag chains of In atoms.

As for the 8×2 phase, two different structure models, trimer (figure 2(b)) [12,13] and hexagon (figure 2(c)) [14,15] structures have been proposed theoretically. First of all, we calculated the RHEPD intensity based on the dynamical diffraction theory using the trimer model as represented by the black lines in figure 3(b). The small peaks apparently appear in the shoulder of the rocking curve for the (0 0) spot in the range of θ = 2.5°-4.5°. The feature is not consistent with the measured curve. Then, we optimized the displacements of In atoms to minimize the difference between the measured and calculated curves based on the trimer configurations, as denoted by the arrows in figure 2(b). As a result, the measured curves for the Si(111)-8×2 surface cannot be reproduced by considering the trimer model.

Similarly, we optimized the displacements of In atoms to minimize the difference between the measured and calculated curves based on the hexagon configurations, as denoted by the arrows in figure 2(c). The blue lines in figure 3(b) show the calculated rocking curves using the optimized hexagon model. The calculated curve is in good agreement with the measured curve for each spot. In particular, the calculated curve can well reproduce the shape of the rocking curve in the θ range between 2.5° to 4.5°. We found that the optimum atomic positions are close to the values calculated by González *et al.* [14,16,17]. Therefore, we confirmed experimentally that the 8×2 structure is composed of the hexagons of In atoms. The electronic band structures calculated using the 8×2 structure of the optimized hexagons is found to be semiconducting, as calculated by González *et al.* [14,17]. The band gap of the optimized hexagon structure is estimated to be 60-202 meV. This value is in good agreement with the previous experimental results [18]. Very recently, it was found from the reflection anisotropy spectroscopy [19] and first-principles calculations [20] that the spectrum of the 8×2 phase at low temperature can be well explained by considering the hexagon model.

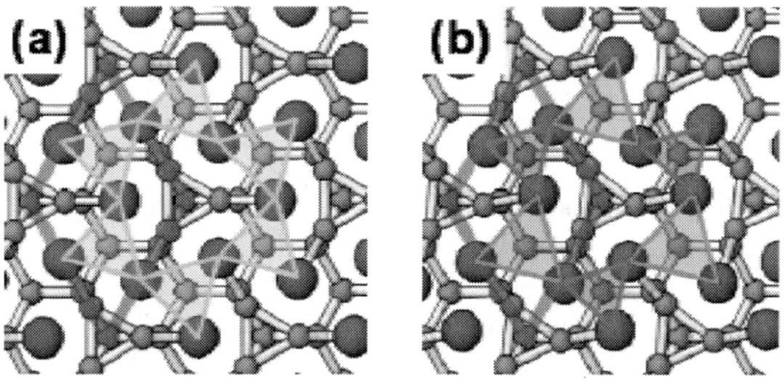

Figure 4. Schematic drawings of (a) honeycomb chained triangle (HCT) and (b) inequivalent triangle (IET) models for the Si(111)-√3×√3-Ag surface. Red and gray circles denote the Ag and Si atoms, respectively. Parallelogram indicates the √3×√3 unit cell.

3.2. √3×√3-√3×√3 phase transition of Ag/Si(111) surface

The Si(111)-√3×√3-Ag surface is one of the most extensively studied surfaces by using several surface techniques. Nowadays it becomes a typical prototype as two-dimensional metal/semiconductor systems [21]. In 1988, the framework of √3×√3-Ag superstructures has been revealed as the honeycomb chained triangle (HCT) by surface x-ray diffraction (see figure 4(a)) [22]. In 1999, this structure was slightly modified. The inequivalent triangle (IET) structure has been confirmed as the grand state structure by using the first-principles calculations and low-temperature scanning tunneling microscopy (figure 4(b)) [23]. In the IET model, the topmost Ag atoms are slightly shifted about 0.15 Å from that in the HCT model. The IET structure contains small and large Ag triangles in the unit cell of √3×√3. According to the mirror symmetry along the $[11\bar{2}]$ direction, two types of the IET structures, i.e., IET(+) and IET(-), exist. There is no difference of the surface energy between the IET(+) and IET(-) structures. When the temperature is elevated above 120 K, the STM image is changed, having the mirror symmetry along the $[11\bar{2}]$ direction [24]. To explain the change in the STM image with temperature, two different mechanisms of the phase transition have been reported, displacive phase transition from the IET to HCT structure [25,26] and order-disorder phase transition of the IET structure [27-31]. The phase transition mechanism of the Si(111)-√3×√3-Ag surface is still controversial.

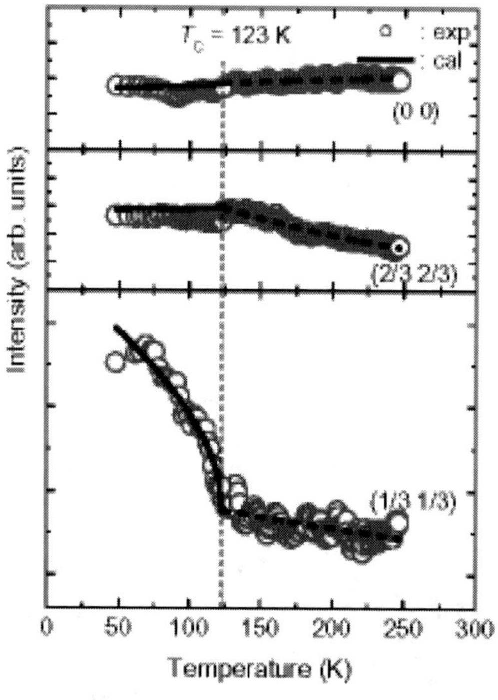

Figure 5. Temperature dependence of the RHEPD intensities from the Si(111)-√3×√3-Ag surface. The azimuthal angle of the incident positron beam corresponds to 1.5° away from the $[11\bar{2}]$ direction. The glancing angle is set at 2.0°, which satisfies the total reflection condition. The open circles denote the measured intensities. The solid lines indicate the calculated intensities using equation (3).

Figure 5 shows the temperature dependence of the RHEPD intensity from the Si(111)-√3×√3-Ag surface in the temperature range from 48 K to 247 K. The RHEPD intensities of (0 0), (1/3 1/3), and (2/3 2/3) spots were measured at 1.5° away from the [$11\bar{2}$] azimuth under the total reflection condition (θ = 2.0°). Above 123 K, the intensities for the (0 0), (1/3 1/3), and (2/3 2/3) spots slightly decrease with increasing the temperature. The intensity change with temperature follows the Debye-Waller like attenuation. However, below 123 K, the (0 0) and (2/3 2/3) spot intensities slightly decrease with decreasing the temperature. Very interestingly, the (1/3 1/3) spot intensity drastically increases with decreasing the temperature. These intensity changes cannot be explained by the displacive phase transition from the HCT to IET structure. We found that these intensity changes can be reproduced by considering the order-disorder phase transition of the IET structure, as described below.

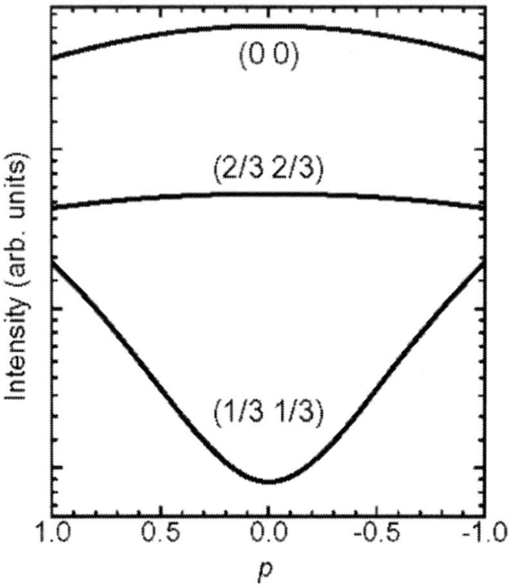

Figure 6. Calculated RHEPD intensities for (00), (1/3 1/3) and (2/3 2/3) spots as a function of the order parameter (p). p of 0.0 and 1.0 or −1.0 mean the completely disordered and completely ordered states, respectively. The calculations were done based on the dynamical diffraction theory.

Assuming n_+ and n_- to be the fraction of the IET(+) and IET(-) phases, respectively, the order parameter (p) is expressed by

$$p = \frac{n_+ - n_-}{n_+ + n_-}. \qquad (2)$$

In the completely ordered state, p = 1 or p = -1. On the other hand, in the completely disordered state, p = 0. By considering these states, we calculated the RHEPD intensity based on the dynamical diffraction theory as a function of p. The detail of the RHEPD calculation was described in the previous paper [32]. Figure 6 represents the RHEPD intensities for the (0 0), (1/3 1/3), and (2/3 2/3) spots with p. When p changes from 0 to 1 or -1, the intensities for the (0 0) and (2/3 2/3) spots decrease slightly. On the contrary, the intensity for the (1/3 1/3) spot drastically increases with p. Assuming that p depends on the substrate temperature, p is given by

$$p \propto \left|1-\frac{T}{T_c}\right|^{\beta}, \tag{3}$$

where T_c and β denote the phase transition temperature and critical exponent, respectively. From the fitting between the measured and calculated intensities, we obtained $\beta = 0.28$. The value of β is slightly larger than that (0.125) expected from the two-dimensional Ising model. The result suggests that the Si(111)-√3×√3-Ag surface system cannot be expressed by the simple two-dimensional Ising model. The calculated intensities using $\beta = 0.28$ can well reproduce the measured intensities, as shown by the solid lines in figure 5. Therefore, we found that this phase transition can be explained by the order-disorder phase transition.

References

[1] Tabata T, Aruga T, and Murata Y 1987 *Surf. Sci.* **179** L63
[2] Ichimiya A 1992 *Solid State Phenom.* **28&29** 143
[3] Fukaya Y, Kawasuso A, Hayashi K and Ichimiya A 2004 *Phys. Rev.* B **70** 245422
[4] Kawasuso A and Okada S 1998 *Phys. Rev. Lett.* **81** 2695
[5] Kawasuso A, Ishimoto T, Maekawa M, Fukaya Y, Hayashi K and Ichimiya A 2004 *Rev. Sci. Instrum.* **75** 4585
[6] Bunk O, Falkenberg G, Zeysing J H, Lottermoser L, Johnson R L, Nielsen M, Berg-Rasmussen F, Baker J and Feidenhans'l R 1999 *Phys. Rev.* B **59** 12228
[7] Abukawa T, Sasaki M, Hisamatsu F, Goto T, Kinoshita T, Kakizaki A and Kono S 1995 *Surf. Sci.* **325** 33
[8] Yeom H W, Takeda S, Rotenberg E, Matsuda I, Horikoshi K, Schaefer J, Lee C M, Kevan S D, Ohta T, Nagao T and Hasegawa S 1999 *Phys. Rev. Lett.* **82** 4898
[9] Sakamoto K, Ashima H, Yeom H W and Uchida W 2000 *Phys. Rev.* B **62** 9923
[10] Tanikawa T, Matsuda I, Kanagawa T and Hasegawa S 2004 *Phys. Rev. Lett.* **93** 016801
[11] Ichimiya A 1983 *Japan. J. Appl. Phys.* **22** 176
[12] Kumpf C, Bunk O, Zeysing J H, Su Y, Nielsen M, Johnson R L, Feidenhans'l R and Bechgaard K 2000 *Phys. Rev. Lett.* **85** 4916
[13] López-Lozano X, Krivosheeva A, Stekolnikov A A, Meza-Montes L, Noguez C, Furthmüller J and Bechstedt F 2006 *Phys. Rev.* B **73** 035430
[14] González C, Flores F and Ortega J 2006 *Phys. Rev. Lett.* **96** 136101
[15] Stekolnikov A A, Seino K, Bechstedt F, Wippermann S, Schmidt W G, Calzolari A and Buongiorno Nardelli M 2007 *Phys. Rev. Lett.* **98** 026105
[16] Hashimoto M, Fukaya Y, Kawasuso A and Ichimiya A 2009 *e-J. Surf. Sci. Nanotech.* **7** 436
[17] Fukaya Y, Hashimoto M, Kawasuso A and Ichimiya A 2008 *Surf. Sci.* **602** 2448
[18] Yeom H W, Horikoshi K, Zhang H M, Ono K and Uhrberg R I G 2002 *Phys. Rev.* B **65** 241307(R)
[19] Chandola S, Hinrichs K, Gensch M, Esser N, Wippermann S, Schmidt W G, Bechstedt F, Fleischer K and McGilp J F 2009 *Phys. Rev. Lett.* **102** 226805
[20] Wippermann S and Schmidt W G 2009 *Surf. Sci.* **603** 247
[21] Hasegawa S 2000 *J. Phys.: Condens. Matter* **12** 463
[22] Takahashi T, Nakatani S, Okamoto N, Ishikawa T and Kikuta S 1988 *Japan. J. Appl. Phys.* **27** L753
[23] Aizawa H, Tsukada M, Sato N and Hasegawa S 1999 *Surf. Sci.* **429** L509
[24] Sato N, Nagao T and Hasegawa S 1999 *Surf. Sci.* **442** 65
[25] Tajiri H, Sumitani K, Nakatani S, Nojima A, Takahashi T, Akimoto K, Sugiyama H, Zhang X and Kawata H 2003 *Phys. Rev.* B **68** 035330
[26] Nakahara H, Suzuki T and Ichimiya A 2004 *Appl. Surf. Sci.* **234** 292
[27] Sasaki N, Watanabe S, Aizawa H and Tsukada M 2001 *Surf. Sci.* **493** 188
[28] Kakitani K, Yoshimori A, Aizawa H and Tsukada M 2001 *Surf. Sci.* **493** 200

[29] Nakamura Y, Kondo Y, Nakamura J and Watanabe S 2001 *Surf. Sci.* **493** 206
[30] Matsuda I, Morikawa H, Liu C, Ohuchi S, Hasegawa S, Okuda T, Kinoshita T, Ottaviani C, Cricenti A, D'angelo M, Soukiassian P and Le Lay G 2003 *Phys. Rev.* B **68** 085407
[31] Sakamoto K, Suzuki T, Mawatari K, Kobayashi K, Okabayashi J, Ono K, Ueno N and Oshima M 2006 *Phys. Rev.* B **73** 193303
[32] Fukaya Y, Kawasuso A and Ichimiya A 2007 *Phys. Rev.* B **75** 115424

Analysis of positron annihilation spectra and reemitted positrons from clean and hydrogen adsorbed Ni(111) surfaces

S Komagata[1], H Suzuki[1], K Hirota[1], I Kanazawa[1], K Fukutani[2], K Nozawa[3] and F Komori[3]

[1]Department of Physics, Tokyo Gakugei University,
4-1-1, Koganei-shi, Tokyo 184-8501, Japan
[2]Institute of Industrial Science, University of Tokyo,
4-6-1, Meguro-ku, Tokyo 153-8505, Japan
[3]Institute of solid state Physics, University of Tokyo,
5-1-5, Kashiwa-shi, Chiba 277-8581, Japan

komagata.s@gmail.com

Abstract. The positron annihilation γ-ray spectra emitted from clean and hydrogen adsorbed Ni(111) surfaces are measured. We have discussed the disordered structures of adsorbed-hydrogens on Ni(111) surface studied by the positron annihilation spectra and reemitted slow positrons. The present experiments suggest the presence of the positronium hydride, PsH, on hydrogen-saturatedly adsorbed Ni(111) surface.

1. Introduction

The adsorption mechanism of hydrogens on transition-metal surfaces has been investigated extensively, because it is very important to application such as catalytic reaction. Especially the H/Ni(111) system is a typical example. Its phase diagram of adsorbed hydrogens for the order-disorder transition has been obtained by Christmann et al. [1] from the temperature variation of the low-energy electron diffraction (LEED) half-order spot intensity. The hydrogen atoms occupy threefold sites of a hexagonal (so-called honeycomb) structure, and the saturation coverage is around $\theta = 1$ monolayer [1]. 2×2-H phase contains 2H atoms per unit cell and this corresponds to a coverage of $\theta = 0.5$. The order-disorder transition temperature is highest for the structure with $\theta = 0.5$, and strongly decreases with disorder being introduced. This phase diagram clearly reveals an asymmetry with respect to $\theta = 0.5$. It is much wider on the low coverage side than on the high coverage one, which can be explained by the amount of disorder introduced by more vacancies and interstitial H atoms. The disordered structure on the high-coverage side is not known. On the hydrogen-chemisorbed Ni(111) surface, two temperature-programmed-desorption (TPD) peaks, β_1 and β_2, have been observed [1]. The structure, which corresponds to the β_2 state of the hydrogens on Ni(111) surface is (2×2)-2H superstructure. On the other hand, the structure of the disordered state of adsorbed hydrogens of the β_1 state is not yet known. Reemitted slow-positron spectroscopy [2] will provide important information on adsorbed light atoms on surfaces. In this study, we have measured the changes in positron annihilation spectra and reemitted slow-positrons from hydrogen-adsorbed Ni(111) surface due to different adsorbing conditions of hydrogens on Ni(111) surface. The origin of the changes in reemitted slow positrons due to different adsorbing conditions will be discussed.

2. Experiment

The positron beam intensity is ~3000 counts/s. We installed the LEED and Auger electron spectroscopy (AES) optics in the experimental chamber. The mesh of the ground level is installed in front of the sample. By changing the sample bias, information for the energy spectra of positive particles emitted from a surface of the sample can be obtained. When 35.0 V is imposed on the sample, positive particles from the surface are accelerated in the direction of the channeltron. Three meshes are installed in front of the channeltron. By changing the voltage of the center mesh and fixing the ground level on other two meshes, information for the energy distribution of positive particles can be obtained. When -20 V is imposed on the sample, a small number of positive particles (about 3 counts/600 s) are emitted from the sample. This means that the measured counts contain ~3% as reflected positrons and relatively highly energetic hydrogen ions (above 20 V). From the previous study [3], it is known that the ratio of positive ions to positrons is very low. So it is thought that most positive particles in the present study correspond to positrons. Positron annihilation spectra are measured with high-purity germanium detector. The sample is suspended by W wires and can be annealed by direct current through the wires. The highly pure hydrogen and argon gases can be introduced to the UHV chamber through a variable leak valve. The base pressure of the UHV chamber was 2.6×10^{-10} Torr. The specimen surface is cleaned by repeated argon ion bombardment. No contamination was checked by the Auger electron spectra. Then cleaned Ni(111) surface was annealed for 30 min at 1100 K to eliminate defects in 2.0×10^{-9} Torr. After the sample cools to room temperature, a sharp (1×1) LEED pattern was observed. After the sample was cooled by liquid nitrogen and then hydrogen gas at some conditions was introduced at 220 K, typical (2×2)-2H LEED pattern was observed. Reemitted slow positrons from hydrogen-adsorbed Ni(111) surface are measured in some adsorbing conditions. Furthermore, the Doppler broadening spectra of the positron annihilation from hydrogen-lightly adsorbed Ni(111) surface and hydrogen-saturatedly adsorbed one are measured.

3. Results and Discussion

On all low-index Ni surface, H adsorbs dissociatively. The Ni(111) surface remains unreconstructed under H exposure. Hydrogen seems to absorb in the hollow site [1], and the saturation coverage is $\theta\sim$ 1 monolayer. The hydrogen coverage is estimated qualitatively from hydrogen exposure (1 L = 1.0×10^{-6} Torr×1 s) and LEED patterns. At a coverage $\theta = 0.5$, the Ni(111) surface exhibits a well-defined (2×2)-2H superstructure. However, adsorbed hydrogens in the saturation coverage have the disordered structure, which is not yet known. Here, we can observe only diffraction spots of Ni(111) surface structure. We cannot observe diffraction spots of adsorbed hydrogens due to the disorder structure. The number of thermalized positrons that diffuse back to the surface is given by

$$L = \int_0^\infty P(z)N(z)dz\,, \tag{1}$$

where $P(z)$ is a positron stopping profile in the case of incident energy of 1.97 keV in Ni, and z is the distance from the surface. $N(z)$ is the total flux through the plane of the surface due to the initial δ-function profile at z, and is represented as

$$N(z) = \frac{\exp\left[-z/(D_+\tau_{\mathrm{eff}})^{1/2}\right]}{1+\beta/(D_+\tau_{\mathrm{eff}})^{1/2}}, \tag{2}$$

where D_+ and τ_{eff} are the positron diffusion constant ~0.7 cm^2/s [4] and the positron-annihilation lifetime ~100 ps [5] in Ni, respectively. β is a coefficient that takes account of internal reflection at the surface. Since the positron work function is ~-1.3 eV in Ni [6], thermalized positrons which diffuse back to the surface are emitted from the surface to vacuum as reemitted slow positrons. Oishi et al. [7] have observed that the number of reemitted slow positrons from hydrogen-saturatedly adsorbed Ni(111) surface is reduced much in comparison with that from the clean Ni(111) surface. Recently we have measured reemitted slow positron yields from hydrogen-adsorbed Ni(111) surface on different adsorbing conditions [8]. Work function of Ni(111) surface and positron-annihilation lifetime are changed by hydrogen adsorption on Ni(111) surface. Figure 1 shows normalized counts/20 min of reemitted slow positrons from hydrogen-adsorbed Ni(111) on hydrogen-introduction conditions. It is seen that the reemitted slow positron yield is maximum in the (2×2)-2H superstructure on Ni(111) surface. It is suggested strongly that increase of the reemitted slow positron yield in the (2×2)-2H superstructure is related to increase of the electron work function due to the hydrogen adsorption. An interesting experimental result is that the electron work function changes remarkably owing to chemisorption of hydrogen on the Ni(111) surface. The electron work function increases linearly by 165 meV up to $\theta \sim 0.5$ followed by a slight decrease as the β_1 chemisorption state is filled [1]. And the collision with highly dense adsorbed hydrogens induce more trapping of reemitted slow positron on hydrogen saturatedly adsorbed Ni(111) surface.

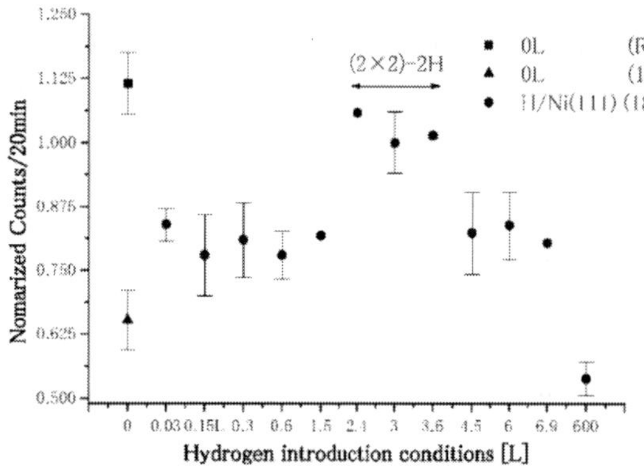

Figure 1. Normalized counts/20 min of reemitted slow positrons from hydrogen-adsorbed Ni(111) on hydrogen-introduction conditions.

Figure 2 shows the total counts of the Doppler broadening spectra of the positron annihilation γ-ray emitted from the hydrogen adsorbed Ni(111) surface in some adsorbed conditions (from clean Ni(111) surface to hydrogen-saturatedly adsorbed one). This result is consistent with reduction of reemitted slow positrons from the hydrogen-saturatedly adsorbed Ni(111) surface [7]. Since the escaping of reemitted slow positrons and ortho-positroniums from hydrogen-saturatedly adsorbed Ni(111) surface is less than that of hydrogen-lightly adsorbed one, the total count of the Doppler broadening spectrum emitted from hydrogen-saturatedly adsorbed Ni(111) surface increases in comparison with that from hydrogen-lightly adsorbed one. It is seen that the Doppler broadening spectra counts is not minimum in the (2×2)-2H superstructure on Ni(111) surface. Figure 3 shows *S*-parameter of the Doppler broadening spectra of the positron annihilation γ-ray emitted from the hydrogen adsorbed Ni(111) surface in some adsorbed conditions (from clean Ni(111) surface to hydrogen-saturatedly adsorbed one). It is of significance that the *S*-parameter of the positron annihilation spectrum of hydrogen-lightly adsorbed Ni(111) surface is higher than that of hydrogen-saturatedly adsorbed Ni(111) surface. This result suggests that relative contribution of para-positronium in the spectrum seems to increase on the hydrogen-lightly adsorbed Ni(111) surface, because escaping of the ortho-positronium from the hydrogen-lightly adsorbed Ni(111) surface increases in comparison with that from the hydrogen-saturatedly one. The total counts and *S*-parameter of Doppler broadening spectra (the positron-annihilation spectra) seem not to show the maximum or minimum at the coverage of ~0.5 ML. This might be due to that the Doppler broadening spectra contain many effects such as annihilation of para-positroniums, pick-off annihilation of ortho-positroniums, and positron annihilation in the near-surface. These experimental results show that three-annihilation γ-rays from ortho-positronium around the hydrogen-lightly adsorbed Ni(111) surface are reduced in comparison with those around the hydrogen-saturatedly adsorbed Ni(111). Now we might be able to explain the origin of the reduction of the reemitted slow-positron by adsorbed hydrogen atoms in saturated coverage as follows, that is, the interaction between highly dense adsorbed hydrogen atoms and reemitted slow-positron might disturb strongly emission of thermalized slow positrons into vacuum, and as a result positron annihilation of two γ-rays emitted from the surface increases. Furthermore the collision between highly dense adsorbed hydrogen atoms and ortho-positroniums might induce more trapping of ortho-positoniums on hydrogen-saturatedly adsorbed Ni(111) surface. Trapping of ortho-positroniums might suggest the presence of positronium hydride, PsH, on hydrogen-saturatedly adsorbed Ni(111) surface. The positronium hydride, PsH, has been observed in hydrogen-adsorbed graphite-alkali metal intercalation compounds [9].

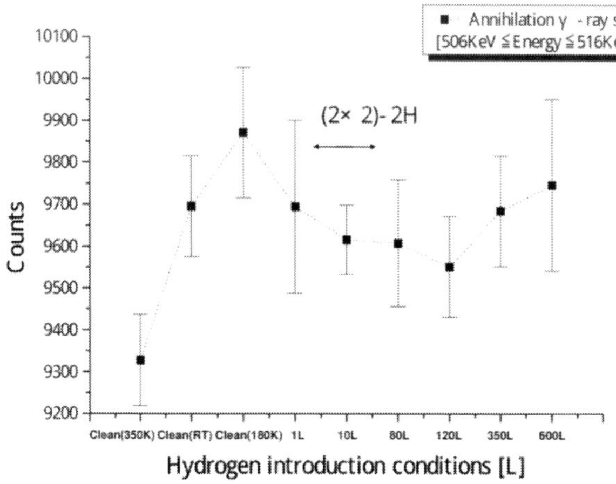

Figure 2. The total counts of the Doppler broadening spectra of the positron annihilation γ-ray emitted from the hydrogen adsorbed Ni (111) surface in some adsorbed conditions (from clean Ni (111) surface to hydrogen-saturatedly adsorbed one).

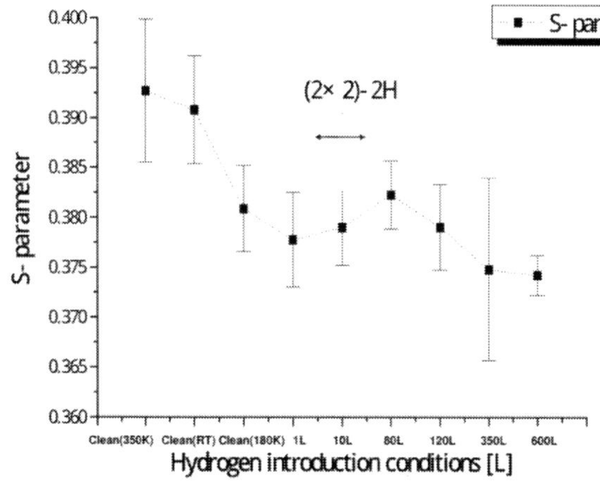

Figure 3. S-parameter (the ratio of the counts of the peak to the total counts of the spectrum) of the Doppler broadening spectra of the positron annihilation γ-ray emitted from the hydrogen adsorbed Ni (111) surface in some adsorbed conditions (from clean Ni (111) surface to hydrogen-saturatedly adsorbed one).

4. Conclusion

We have measured reemitted slow-positron yields from hydrogen-adsorbed Ni(111) surface on different adsorbing conditions. It is seen that the reemitted slow-positron yield is maximum in the (2×2)-2H superstructure on Ni(111) surface. It is suggested strongly that increase of the reemitted slow-positron yield in the (2×2)-2H superstructure is related to increase of the electron work function due to the hydrogen adsorption. And we have measured the positron annihilation γ-ray spectra from hydrogen-adsorbed Ni(111) surface on different adsorbing conditions. It is suggested that the collision with highly dense adsorbed hydrogens induces more trapping of reemitted slow positrons and ortho-positroniums on hydrogen saturatedly adsorbed Ni(111) surface. This might suggest the presence of the adhesive state of ortho-positroniums, such as the positronium hydride, PsH, in adsorbed hydrogens on hydrogen-saturatedly adsorbed Ni(111) surface.

Acknowledgements
We would like to thank Prof. Y. Murata for helpful discussions.

References

[1] Christmann K, Behm R J, Ertl G, Van Hove M A and Weinberg W H 1979 *J. Chem. Phys.* **70** 4168
[2] Kanazawa I, Wada T, Oishi Y, Terashima Y, Fukutani K, Murata Y, Ito Y, Nozawa K and Komori F 2005 *Nucl. Instrum. Methods Phys. Res.* B **232** 299
[3] Terashima Y, Arai R, Wada T, Kanazawa I, Ito Y, Fukutani K, Murata Y and Komori F 2002 *J. Phys.: Condens. Matter* **14** L349
[4] Vehanen A, Lynn K G, Schultz P J, Cartier E, Güntherodt H-J and Parkin D M 1984 *Phys. Rev.* B **29** 2371
[5] Mackenzie I K, Jackman T E and Thrane N 1975 *Phys. Rev. Lett.* **34** 512
[6] Gullikson E M, Mills A P Jr, Crane W S and Brown B L 1985 *Phys. Rev.* B **32** 5484
[7] Oishi Y, Wada T, Kanazawa I, Fukutani K, Murata Y, Ito Y, Nozawa K and Komori F 2005 *Appl. Surf. Sci.* **241** 169
[8] Komagata S, Hirota K, Suzuki H, Osawa M, Arii S, Kanazawa I, Fukutani K, Nozawa K and Komori F 2008 *Appl. Surf. Sci.* **255** 227
[9] Murakami H, Sano M, Kanazawa I, Enoki T, Kurihara T, Sakurai Y and Inokuchi H 1985 *J. Chem. Phys.* **82** 4728

Applications of Imaging Plates to the Study of Positron and Muon Research

M Doyama[1], Y Kogure[1], M Inoue[1], Y Miyake[2], K Nishiyama[2], K Shimomura[2], K Nakahara[2] , N Kawamura[2], P Strasser[2] and T Kurihara[3]
[1] Teikyo University of Science and Technology, Utsunomiya, Tochigi, 320-8551, Japan
[2] Muon Science Laboratory, High Energy Accelerator Research Organization (KEK), Tsukuba, Ibaraki, 305-0801, Japan
[3] Slow Positron Science Laboratory, High Energy Accelerator Research Organization (KEK), Tsukuba, Ibaraki, 305-0801, Japan

E-mail: doyama@ntu.ac.jp

Abstract. A muon beam intensity distribution perpendicular to the beam has been obtained using imaging plates which are commonly used in X-ray science, transmission electron microscopy, bio and medical analysis. The resolution of imaging plates depends on the resolution of an imaging plate reader. We have used BAS-TR2025 imaging plates and Bio-Imaging Analyzer BAS-2500 or BAS-3000. The positional resolution is 50 μm x 50 μm. When the beam is one kind, the coincidence measurement is not necessary so that the muon distribution perpendicular to the beam is taken in a very short time and the spatial resolution is 50 μm x 50 μm. Such a resolution cannot be obtained by usual muon measuring method. This is quite convenient to adjust muon beams. When FLA-8000 or FLA-9000 is used the resolution can be improved to 10 μm x 10 μm. Such intensity resolution of muon beams could not be imagined without using imaging plates. Transmission muon images obtained are also presented. This can be used as a non-destructive test.

1. Introduction

Imaging plates (IP) are widely used in X-ray science, medical analysis and in transmission electron microscopy. A plate is only a card which can be easily placed into a vacuum chamber without darkness. Darkness is required only during exposure and transfer imaging plate to an analyzer (reader). The images can be erased and imaging plates can be used many times. Imaging intensity is cumulative, it is not good for coincidence measurements. However, when the kind of particles arriving is known and particularly particles are one kind, not mixed, imaging plates are quite conveniently and usefully used. The sensitivity is quite high and proportional to the fluence in six digits [1]. Dead times in coincidence or anti-coincidence circuits are not problems in imaging plates. Imaging plates can be cut to the size we want. When a muon hits an electron in the valence band, the electron is excited to the conduction band of the fluorescent layer. The electron is trapped at an $F^+/F(Br)$ level or $F^+/F(F^-)$ level. The rapped electron at the $F^+/F(F^-)$ is excited to the conduction band by the laser light during reading. This emits fluorescent light which is called PSL. PSL [2] is used for the unit of irradiation of particles on imaging plates and this is not an absolute unit. In this process positrons are not created. The contamination by positrons in the original beam has been eliminated.

2. First muon beam at J-PARC [3]

At a very early stage when the muon beam was first introduced in muon research facility at J-PARC, imaging plates (BAS-TR4025) were inserted in a plastic envelope coated with aluminium spattering to avoid penetration of light. The first positive muon intensity distribution of the beam [3] taken in September, 2008 is shown in Fig. 1.

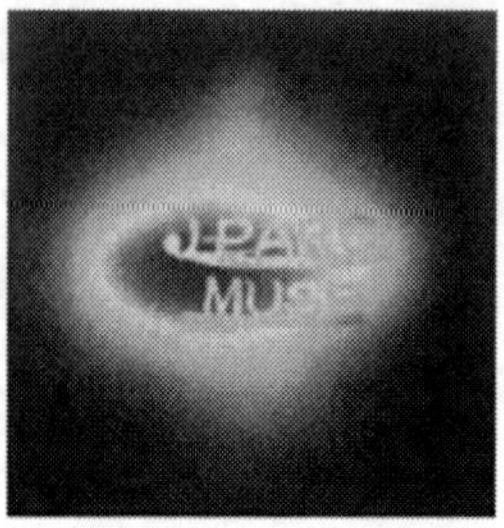

Figure 1. The first positive muon intensity distribution of the beam taken in September, 2008.

3. Quantitative muon beam at J-PARC

Figure 2 shows an example of the intensity distribution of positive muon beam on Feb. 14, 2009 at J-PARC. The effect of positrons in the muon beam has been eliminated using a DC separator. (a) is a black and white image, but (b) is originally colored but changed to black and white according to the muon intensity grade. The picture size of Fig. 2 is about 15 cm x 15 cm. The wiggling line in Fig. 2(b) is PSL along the horizontal straight line. The intensity distribution can be expressed roughly by an error function. The positive muon beam was close to an ellipse.

(a) (b)

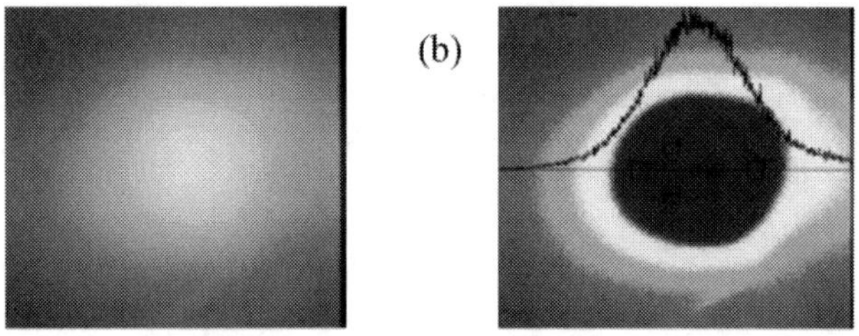

Figure 2. Positive muon image of positive muon beam intensity distribution. (a) IP image. (b) Color region depends on the value of PSL. The wiggly line shows PSL on the straight line. These are taken on Feb. 14, 2009.

Figure 3 is a blank beam on May 31, 2009 at J-PARC, eliminated positron contamination using a DC separator and defocused. Figure 3(a) is a black and white image, (b) is originally colored but changed to black and white according to the muon intensity grade. According to Figs. 3(a) and (b), the blank muon intensities in the muon beam look like uniform, but adjusting the contrast, the beam is not uniform but right upper side is the strongest as shown in Fig. 3 (c). The wiggly line shows PSL on the straight line.

To measure the absorption coefficient of metal foils, tungsten foils (40 micron-m thick), copper foils (40 micron-m thick), aluminum foils (12 micron-m thick) and nickel foils (30 micron-m thick) were taped on an imaging plate and inserted in an envelope with sputtered aluminum. The set was exposed in a positive muon beam at J-PARC on May 31, 2009. The beam was defocused. Figure 4 shows an image on an IP.

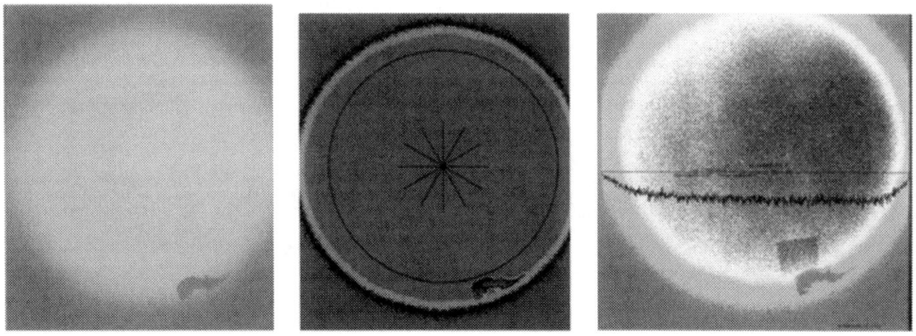

(a) (b) (c)

Figure 3. Positive muon image of positive muon beam intensity distribution on May 31, 2009, defocused. (a) IP image. (b) Color region depends on the value of PSL. (a) and (b) look like the beam is uniform. (c) The same as (a) and (b) but the color separation grade is different. The wiggly line shows PSL on the straight line. The right side upper part is the strongest in the positive muon beam.

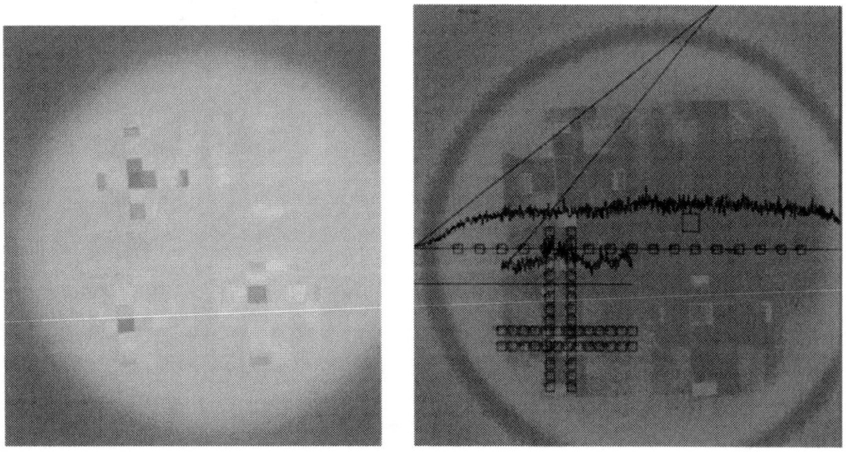

(a) (b)

Figure 4. (a) Transmission positive muon image on IP. (b) With the intensity of PSL.

The strongest part of the muon beam was near the right upper corner. Adjusting the beam intensity distribution, figure 4(b) is obtained. The wiggling curve shows the PSL along the horizontal line. The PSL values from the left except the first one are 38.5, 39.2, 37.3 and 39.2. The average PSL under two sheets is about 39.0. Similarly the average PSL under 1 sheet was 44.2.

4. Transmission muon bio-images

Transmission positive muon bio-images were taken at KEK, Tsukuba. These images are shown in this section.

Journal of Physics: Conference Series **225** (2010) 012006 doi:10.1088/1742-6596/225/1/012006

Figure 5 (a) shows a transmission muon image of a wing of a cicada. Figure 5(b) is a transmission positron image of the same wing of the cicada. The muon beam was restricted to the darker area. Figure 6 (a) shows transmission muon images of small shrimps. Figure 6 (b) shows the same specimen image by positrons taken at KEK, Tsukuba. Transmission positron images have lower contrast, but muon images have higher contrast compared with positron images, in general. Details of the imaging particles on imaging plates are not clear at present.

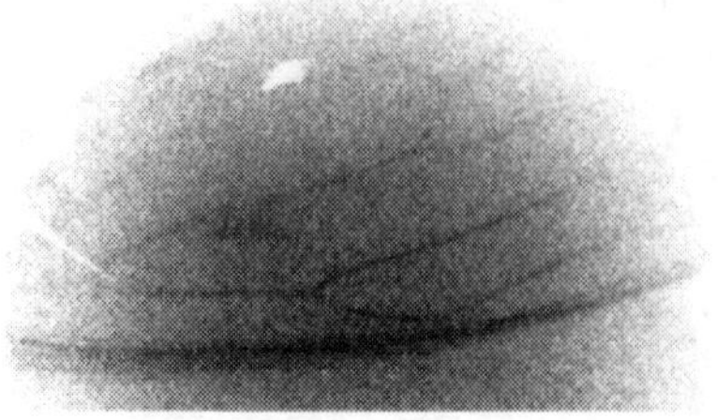

I--------------------I
5 cm
(a)

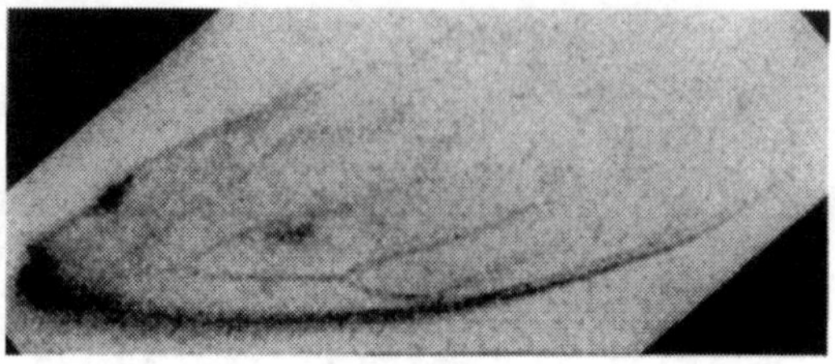

I-----------------------I
5 cm
(b)

Figure 5. (a). Transmission muon image of a wing of a cicada. 50 μm x 50 μm/pixel was used. Fig. 5 (b) transmission positron image of the same wing of a cicada.

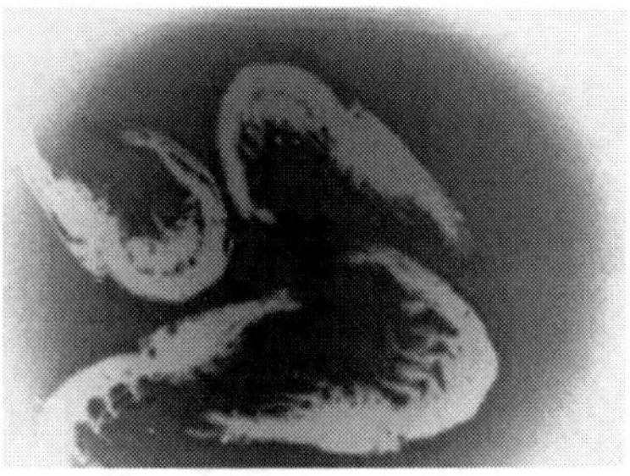

I-------------------------I
5 cm
(a)

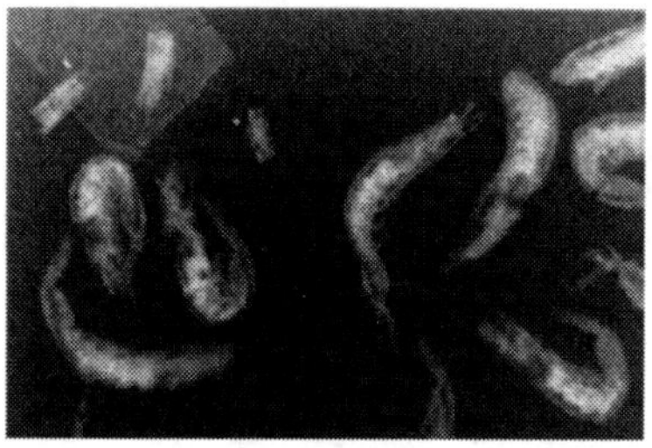

I-----------------I
5 cm
(b)

Figure 6. (a) a transmission muon image of small shrimps. The contrast of the image is high. (b) a transmission positron image. The contrast of the image is lower than that of muons.

5. Most interesting transmission muon images

Figure 7 shows a transmission muon image of copper (a) and tungsten (b) foils (40 μm thick each), taken at KEK, Tsukuba. At the cross part, the thickness is 80 μm. The light circle is the muon beam. The place where muons do not arrive, that is, outside the circle is black, no muons. The exposure time was 30 minutes. In Fig. 7(a), the place (Region 1) where muon passed a foil of 40 μm is lighter than the background of the circle. The place where muon passed two sheets (Region 2) of copper foils (80 μm) is the lightest in the circle. PSL is the highest in Region 2, nest in Region 1, and the least in Region 0, which is the beam. In Region 1, however, less muons should have arrived because muons should have been absorbed in copper foil. The Region 2 is the lightest (whitest), least muons should have arrived because muons passed two sheets. This is probably due to the fact that Bragg Peak is near the exit surface or the muon energies small.

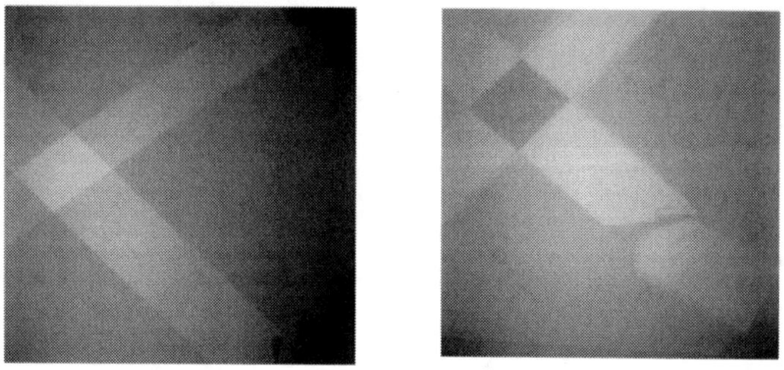

Figure 7. Transmission Muon image of copper (a) and tungsten (b) foils (each thickness is 40 μm). 50 μm x 50 μm/pixel was used. Double layer in (a) is the lightest, but the double layer of (b) is the darkest.

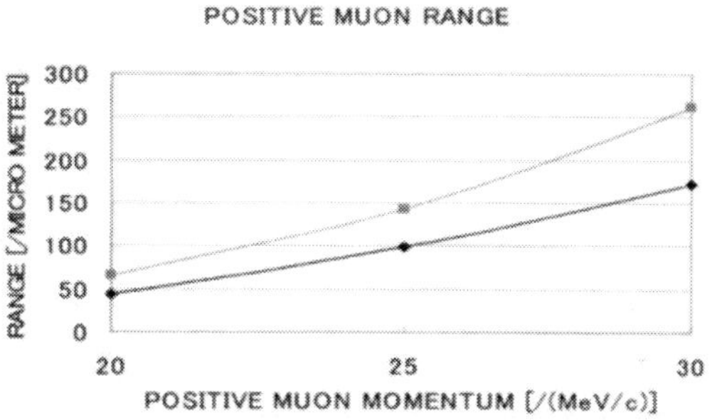

Figure 8. Relation between momenta of positive muons and ranges. Upper curve is for copper and lower curve is for tungsten.

Figure 7(b) shows a transmission muon image of tungsten foils (each foil is 40 μm thick). At the cross part, the thickness is 80 μm (Region 2). At the place where muon passed tungsten foil of 40 μm (Region 1), the image is whiter than the background of the circle. This is the same reason for copper foils. At the place where muons passed two tungsten foils (80 μm thick) (region 2) the image is the darkest within the circle. This is reasonable because the least muons arrived in Region 2.

There are three reasons to be considered. One is due to the absorption of muons, the second is the Bragg peak and the other is due to the energies of muons arriving at the particle sensitive layer in IP. Muons with high energies pass the sensitive layer but muones with lower energies are trapped in the sensitive layers of imaging plates. In the case of positrons, the relation between PSL and positron energies has been studied 2. The relation between positive muon momentum and the range is shown in Fig. 8. Muons passed through a black envelope lost some of their energies. A tentative explanation of Fig. 7 is as follows: Near the center, the doubled zone (Region 2) is the lightest in Fig. 7(a) (copper). IP sensed the most in the circle. It looks more positive muons are reached to the imaging plate, but this is probably due to the peak of the Bragg curve of muons is near the exit surface or muon energies are lowest and IP sensed most. The place where no samples exist (Region 0), muon energies are high and some muons may have passed through the particle sensitive layer of the imaging plate with less fluorescent reaction. After muons passed one sheet of copper or tungsten foil, the kinetic energies of muons are lowered and absorbed more than in Region 0. Slower muons are trapped in the particle sensitive layer of the imaging plate. All the momentums used here are the momentums calculated from the tuned voltages.

In Fig. 7(b), the double layer part of tungsten foils (region 2) is whiter than the background. If the same positive muons with about 23 MeV/c momentums are arriving at the surface of the tungsten foil, the muon range is about 65 μm which is smaller than the thickness of two tungsten sheets, i.e. 80 μm. Therefore less muons reach to the imaging plate. More quantitative results will be published near future.

Imaging plates can be used to see the details of muon beams and inside of materials to positional resolution of 50 μm. Wide applications in the future can be expected. For detail explanations, more experiments are required.

Figure 9(a) is a transmission muon image of a plug, taken at KEK, Tsukuba. Positive muons penetrated the plug. The image shows the center hole. (b) is a picture by a digital camera.

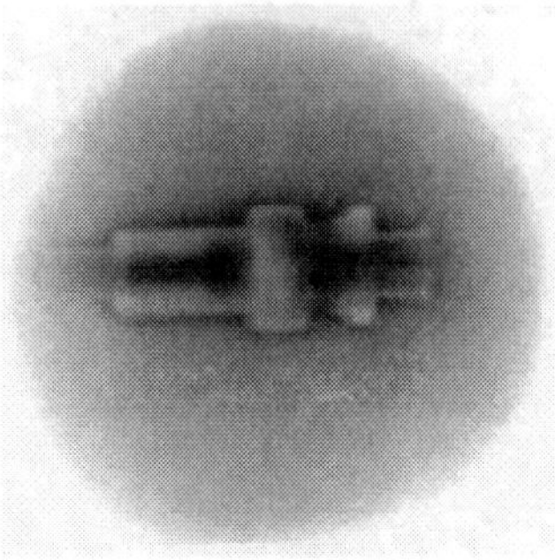

I--------------------------I
5 cm
(a) (b)

Figure 9. (a) A digital camera image of a plug. (b) Transmission muon image of the plug shown in (a). 50 μm x 50 μm/pixel was used.

6. Summary

Using imaging plates, muon intensity distribution of muon beam has been shown. The muon beam is not uniform. Transmission muon images of bio-materials are shown. Muons can be used for a non-destructive testing.

References

[1] Doyama M., Takano J., Inoue M., Yoshiie T., Hayashi Y., Kiritani M. and Oikawa T., 1997 *Nucl. Instr. And Meth. In Phys. Res.* A **394** 146.
[2] Doyama M., Kogre Y., Inoue M., Kurihara T., Yoshiie T., Oshima R. and Matsuya M., 2006 *Appl. Surface Sci.* **252** 3126
[3] Muon beam was created at 12:10, Sept. 26, 2008. News@KEK on Nov. 13, (2008).

Acknowledgments

Some of the present experiments were performed between December 2005 and March 2006 as a Collaborative Experiment Muon Science Facility, KEK in Tsukuba. Registration number was A02. We express our gratitude for giving us a great chance to perform the experiments just before closing the Muon Science Laboratory in Tsukuba. The rest of the muon experiments were performed at the earliest muon beam at the J-PARC under a Collaborative Research project at J-PARC. We express our gratitude for giving us a great chance to perform muon experiments at the earliest stage.

Surface plasmon excitation at metal surfaces studied by reflection high-energy positron diffraction

Y Fukaya, A Kawasuso and A Ichimiya

Advanced Science Research Center, Japan Atomic Energy Agency, 1233 Watanuki, Takasaki, Gunma 370-1292, Japan

fukaya.yuki99@jaea.go.jp

Abstract. We investigated the excitation process of surface plasmon at metal surfaces by fast positrons using an energy-filtered reflection high-energy positron diffraction. We measured the positron and electron energy loss spectra for the specular spots from Al(111)-1×1 and Bi(001)-1×1 surfaces at grazing angles. The mean excitation number of surface plasmon by positrons is approximately twice as large as that by electrons. Owing to the existence of the total reflection, the surface plasmons are much excited by positrons as compared with electrons. The element dependence of the surface plasmon excitation process is not significant from the comparison between the Al(111)-1×1 and Bi(001)-1×1 surfaces.

1. Introduction

When high-energy electrons are incident on crystal surfaces, various energy loss processes such as phonon, plasmon, and core-level electron excitations occur. Among the excitations, the surface plasmon excitation is known as a dominant process at grazing incidence [1]. The surface plasmon is a collective excitation of electrons at crystal surfaces. Its energy is reduced by a factor of $\sqrt{2}$ as compared with the volume plasmon [2]. Generally, the excitation process of surface plasmon has been studied by using (low-) energy electron loss spectroscopy (EELS). Recently, the energy loss process of electrons due to surface plasmon excitations has been extensively investigated through an energy-filtered reflection high-energy electron diffraction (RHEED) [3-6]. On the other hand, the energy loss process of positrons at surfaces has not been well-studied so far.

Reflection high-energy positron diffraction (RHEPD) is a surface-sensitive tool for analyzing the surface structures owing to the occurrence of the total reflection [7-10]. Since positrons have a positive charge, opposite to electrons, they are totally reflected at the first surface layer when the glancing angle is small enough. In the total reflection region, the incident positron beam does not penetrate the bulk. This is different from the case of electrons. Therefore, under the total reflection condition, the energy loss process by positrons at surfaces is expected to be different from that by electrons.

Very recently, we fabricated an energy-filtered RHEPD to measure the energy loss spectrum from a Si(111)-7×7 surface [11]. We found that under the total reflection condition the mean excitation number of surface plasmon by positrons is larger than that by electrons [11]. In this study, to investigate the excitation process of surface plasmon by positrons at metal surfaces, we measured the positron energy loss spectra from Al(111)-1×1 and Bi(001)-1×1 surfaces under the total reflection condition. To compare the excitation process by electrons, we also measured the electron energy loss spectra from the Al(111)-1×1 and Bi(001)-1×1 surfaces.

2. Experimental procedure

Substrates (15×5×0.5 mm^3) were cut from a mirror-polished *n*-type Si(111) wafer with a resistivity of 1-10 Ωcm. They were introduced into a ultra-high vacuum (UHV) chamber with a base pressure less than 6×10^{-8} Pa. They were annealed at 400 °C and flashed at 1200 °C in a few seconds several times to produce a 7×7 reconstruction. 1/3 monolayer (ML) of Al atoms were deposited on the Si(111)-7×7 surface at 670 °C using a crucible to produce a √3×√3-Al structure (1 ML corresponds to 7.83×10^{14} cm^{-2}). To prepare Al(111)-1×1 surfaces, 3 ML of Al atoms were deposited on the √3×√3-Al structure at 350 °C [12]. Since the mean inner potential (V_0) of Al crystals is 12 eV [13], the critical angle of the total reflection for the incident positron energy (E) of 10 keV is estimated to be 2.0° via Snell's equation [7]. Bi(001) surfaces were prepared by the deposition of 8 BL Bi atoms on the Si(111)-7×7 surface at room temperature (1 BL corresponds to 1.14×10^{15} cm^{-2}). Subsequently, they were annealed at about 100 °C to produce well-ordered Bi(001)-1×1 surfaces [14]. In the case of Bi, the critical angle of the total reflection for $E = 10$ keV and $V_0 = 20$ eV [15] is estimated to be 2.6°.

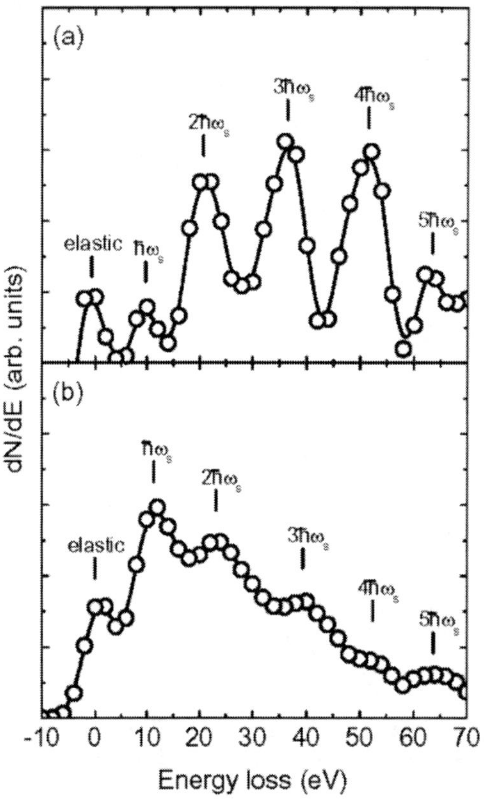

Figure 1. Energy loss spectra of the specular spots for (a) positrons and (b) electrons from the Al(111)-1×1 surfaces. The glancing angle of the positron beam is set at 1.0° (total reflection condition). The glancing angle of the electron beam is 1.5°. The incident azimuths of the positron and electron beams correspond to the [11$\bar{2}$] direction and 7.5° away from the [11$\bar{2}$] direction, respectively.

The positron beam was generated with a ^{22}Na positron source (370 MBq) and electrostatic lenses. The details of the apparatus were described elsewhere [11,16]. Electron energy loss spectra were measured using a conventional electron gun. The incident energies were 10 keV in both cases. The retarding-field-type energy analyzer was installed into the UHV chamber to measure the energy loss spectra. The energy resolution of the analyzer was estimated to be 4.6 eV. All the measurements were conducted at room temperature.

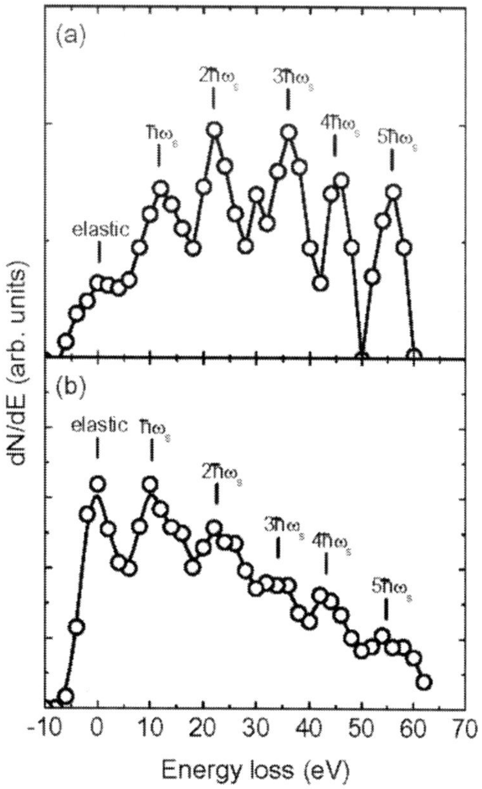

Figure 2. Energy loss spectra of the specular spots for (a) positrons and (b) electrons from the Bi(001)-1×1 surfaces. The glancing angle of the positron beam is set at 2.0° (total reflection condition). The glancing angle of the electron beam is 1.3°. The incident azimuths of the positron and electron beams correspond to 7.5° away from the $[11\bar{2}]$ direction.

3. Results and discussion

3.1. Positron energy loss spectrum

Figures 1(a) and 2(a) show the energy loss spectra (dN/dE) of the specular spots for positrons from the Al(111)-1×1 and Bi(001)-1×1 surfaces, respectively. The energy loss means the difference between the incident beam and retarding energies. The glancing angles of the incident positron beams for the Al(111)-1×1 and Bi(001)-1×1 surfaces are 1.0° and 2.0°, respectively, which satisfy the total reflection condition. The incident azimuths of the positron beams for the Al(111)-1×1 and Bi(001)-1×1 surfaces correspond to the $[11\bar{2}]$ direction and 7.5° away from the $[11\bar{2}]$ direction [17], respectively. In this energy loss region, five prominent loss peaks are observed in both spectra. For the Al(111)-1×1 surface, the intervals of the loss peak positions are not constant. According to the previous study using the electron energy loss spectroscopy, the energy loss peaks due to the surface plasmons of Al were observed from thin Al(111) films fabricated on the Si(111) surface [12]. Thus, the deviation of the intervals is not considered to be due to the effect of Al(111)/Si(111) interfaces. Although the loss peak positions are fluctuated due to the energy resolution, the averaged interval of the peak positions is estimated to be 12 eV for the Al(111)-1×1 surface and 11 eV for the Bi(001)-1×1 surface. Since the volume plasmon energy for Al crystals is 16 eV, the surface plasmon energy is estimated to be approximately 11 eV from the relation $\hbar\omega_s = \hbar\omega_v/\sqrt{2}$, where $\hbar\omega_s$ and $\hbar\omega_v$ are the surface and volume plasmon energies, respectively. In a similar way, the surface plasmon energy for Bi is estimated to be 10 eV. Thus, these peaks in figures 1(a) and 2(a) are assigned to a sequence of the surface plasmon losses. In both cases, the peak intensity of two- or three-fold surface plasmon losses is larger than the others. These results indicate that the totally reflected positron multiply excites surface plasmons.

3.2. Electron energy loss spectrum

Figures 1(b) and 2(b) show the energy loss spectra (dN/dE) of the specular spots for electrons from the Al(111)-1×1 and Bi(001)-1×1 surfaces, respectively. The glancing angles of the incident electron beams for the Al(111)-1×1 and Bi(001)-1×1 surfaces are set at 1.5° and 1.3°, respectively. The incident azimuth corresponds to 7.5° away from the $[11\bar{2}]$ direction [17]. Similar to the case of positrons, five loss peaks are observed in figures 1(b) and 2(b). These peaks in the energy loss spectra by electrons can be also assigned to a series of the loss peaks due to the surface plasmon excitations of Al and Bi. In the case of electrons, on the contrary, the peak intensity resulting from the elastic or single surface plasmon excitation ($\hbar\omega_s$) is higher than the other loss peaks. These features are in contrast to those for positrons. Therefore, multiple surface plasmon excitations by electrons are suppressed as compared with positrons.

3.3. Excitation number of surface plasmon

The intensity distributions in the energy loss spectrum due to the surface plasmon excitations can be well expressed by the Poisson distribution. From the analysis of the loss peak intensities in figures 1(a) and 2(a), we determined the mean excitation numbers as 2.8 for the Al(111)-1×1 surface and 2.4 for the Bi(001)-1×1 surface, as listed in Table 1. These values are close to that from the Si(111)-7×7 surface [11]. Therefore, it is considered that under the total reflection condition, the excitation processes of surface plasmon for positrons from the Al(111)-1×1 and Bi(001)-1×1 surfaces are nearly the same as the Si(111)-7×7 surface. Similar to it, the mean excitation numbers of surface plasmon for electrons are determined to be 1.8 for the Al(111)-1×1 surface and 1.4 for the Bi(001)-1×1 surface. The values are compatible to the previous study on the Si(111)-7×7 surface [11]. In a similar way of positrons, the excitation process of surface plasmon by electrons does not almost depend on the elements consisting of the surface structures.

Table 1. Mean excitation numbers of surface plasmon by positrons and electrons.

	Positron	Electron
Al(111)	2.8	1.8
Bi(001)	2.4	1.4
Si(111)[a]	2.6	1.4

[a] Ref. [11].

The mean excitation number (n_s) of surface plasmon is given in the form

$$n_s = t/l, \tag{1}$$

where l and t denote the inelastic mean free path of positrons or electrons due to the surface plasmon excitation and the nominal interaction length with the crystal surface, respectively. In the high-energy region, the mean free paths of positrons and electrons are nearly the same [18]. In the case of the Al(111)-1×1 surface, the mean free path is estimated to be 190 Å using the so-called TPP-2M formula [19] and $\hbar\omega_s$ = 11 eV. Thus, the nominal interaction length of positrons with the Al(111)-1×1 surface is determined to be 530 Å. Similar to it, the interaction length of electrons is determined to be 340 Å. In the case of the Bi(001)-1×1 surface, the mean free paths of positrons and electrons are evaluated to be 190 Å. Thus, the nominal interaction lengths of positrons and electrons for the Bi(001)-1×1 surface are determined as 460 Å and 270 Å, respectively. In both cases, the difference of the nominal interaction lengths between positrons and electrons is 190 Å. When positrons and electrons approach the surface from the vacuum region, both of them can excite the surface plasmon. Thus, the nominal interaction lengths of positrons and electrons are the same because the excitation process of surface plasmon is similar to each other. Positrons under the total reflection condition channel the first surface layer. On the other hand, electrons penetrate the first surface layer. Positrons channeling in the first surface layer excite the surface plasmon. However, electrons penetrating the first surface layer hardly excite the surface plasmon. Therefore, the difference of the interaction lengths between positrons and electrons can be explained by the channeling length of positrons in the first surface layer.

Acknowledgments

This work was supported by a Grant-in-Aid for Scientific Research (Grant No. 19540349) from the Ministry of Education, Culture, Sports, Science and Technology of Japan.

References

[1] Lucas A A and Šunjić M 1971 *Phys. Rev. Lett.* **26** 229
[2] Kittel C 1996 *Introduction to Solid State Physics* (New York: Wiley)
[3] Horio Y, Hashimoto Y, Shiba K and Ichimiya A 1995 *Japan. J. Appl. Phys.* **34** 5869
[4] Horio Y and Hara T 2002 *Japan. J. Appl. Phys.* **41** L736
[5] Nakahara H, Hishida T and Ichimiya A 2003 *Appl. Suf. Sci.* **212-213** 157
[6] Tanishiro Y 2003 *Hyomen Kagaku* **24** 166
[7] Ichimiya A 1992 *Solid State Phenom.* **28&29** 143
[8] Kawasuso A and Okada S 1998 *Phys. Rev. Lett.* **81** 2695
[9] Fukaya Y, Kawasuso A, Hayashi K and Ichimiya A 2004 *Phys. Rev.* B **70** 245422
[10] Fukaya Y, Kawasuso A and Ichimiya A 2007 *Phys. Rev.* B **75** 115424
[11] Fukaya Y, Kawasuso A and Ichimiya A 2009 *Phys. Rev.* B **79** 193310
[12] Li S-T, Hasegawa S, Yamashita N and Nakashima H 1989 *Appl. Surf. Sci.* **41/42** 118
[13] Radi G 1970 *Acta Crystallogr.* A **26** 41
[14] Nagao T, Sadowski J T, Saito M, Yaginuma S, Fujikawa Y, Kogure T, Ohno T, Hasegawa Y, Hasegawa S and Sakurai T 2004 *Phys. Rev. Lett.* **93** 105501
[15] Berger H, Kulyupin Yu A, Nepijko S A, Obuchov I A and Shamonya V G 1980 *Z. Physik* B **37** 23
[16] Kawasuso A, Fukaya Y, Hayashi K, Maekawa M, Okada S and Ichimiya A 2003 *Phys. Rev.* B **68** 241313(R)
[17] Ichimiya A 1987 *Surf. Sci.* **192** L893
[18] Oliva J 1980 *Phys. Rev.* B **21** 4909
[19] Tanuma S, Powell C J, Penn D R 1994 *Surf. Interface Anal.* **21** 165

International Symposium on Advanced Science Research 2009 IOP Publishing
Journal of Physics: Conference Series **225** (2010) 012033 doi:10.1088/1742-6596/225/1/012033

Evaluation of stainless steel under tensile stress using positron microbeam

M Maekawa, Y Yabuuchi and A Kawasuso
Advanced Science Research Center, Japan Atomic Energy Agency
1233 Watanuki, Takasaki, Gunma, 370-1292, Japan

E-mail: maekawa.masaki@jaea.go.jp

Abstract. Defect evolution near the crack tip in stainless steel (SUS 304) under tensile stress has been observed through *in-situ* positron microbeam measurements. It was found that the S parameter of the Doppler broadening of annihilation radiation spectrum increases near the crack tip. Upon the corrosion treatment, the crack progress occurred preferentially in the high S parameter region. The further increase of S parameter due to the complementary influences of tensile stress and corrosion environment was observed at the region away from the progressed crack. These results suggest excess that vacancy type defects play an important role in the stress corrosion cracking of stainless steel.

1. Introduction

Stainless steel is widely used as materials for nuclear power plants because of its excellent corrosion resistance. The corrosion resistance depends on toughness of chromium oxidization film formed on the surface. However, by inappropriate heat treatment, the oxidization film is degraded because of chromium precipitation. When tensile stress and corrosion environment exist under such condition, stress corrosion cracking (SCC) steel happens. Although the SCC is an important issue for the safety of nuclear power plants, its mechanism has not yet fully been clarified. To elucidate the SCC mechanism, it is important to understand structural change at crack tips in an atomic scale. Recently, fine cracks (tight cracks) of a few nano-meter are found at crack tips using an analytical transmission electron microscope (ATEM) [1]. The introduction of such tight cracks cannot be explained by the conventional theory [2, 3]. In the tight crack model, vacancy-type defects are thought to play an important role in the crack progress. Actually previous positron microbeam observation implies the introduction of vacancy-type defects around crack tips [4]. Li reports that the crack progress is affected by dislocations introduced by the plastic deformation from the *in-situ* TEM observations of SCC on the thin stainless film [5, 6].

In this work, we performed the positron microbeam observation of stainless film under the tensile stress during continuous corrosion. We will discuss the possibility of contribution of vacancy type defects to the crack progress.

2. Experimental

Figure 1 schematically shows the apparatus of our positron microbeam. Sodium-22 of 55 MBq was deposited into a 2 mm-diameter hole and sealed by a cap with a Ti window of 5 μm thick. The source is cooled down to 4 K and a solid neon film is grown on the source window as a moderator [7]. A slow positron beam with energy of 20 keV is generated by an electrostatic way [8]. The positron beam is focused onto sample surface using the commercial scanning electron microscope optics. The positron beam is laterally scanned by moving the sample. The Doppler-broadening of annihilation radiation (DBAR) is measured using a Ge detector. The obtained DBAR spectra are characterized by so-called S parameter, which is defined as the peak intensity of the spectrum.

The sample used in this study was SUS304 stainless steel of 20 μm thick. After the solution treatment at 1150 °C for 2 hours, for the sensitisation the sample was heated at 650 °C for 24 hours and finally quenched into water. The specimen was pre-cracked in the air at room temperature. The sample was fixed by a special holder made of stainless steel as shown in figure 2. Using the tension nuts, tensile stress of ~ 300 MPa was loaded to the sample. This stress is enough to introduce the SCC failure. The corrosion environment, treatments were carried out using boiled $MgCl_2$ solution along with the Japanese Industrial Standards method (JIS G0576). After the corrosion under the tensile stress, optical microscope observations and positron microbeam observations were alternatively conducted.

3. Results and discussions

First we show the results of the observations without the corrosion treatment. Figures 3(a), (b) and (c) shows the optical image around the pre-crack, the distribution of S parameter around the same crack

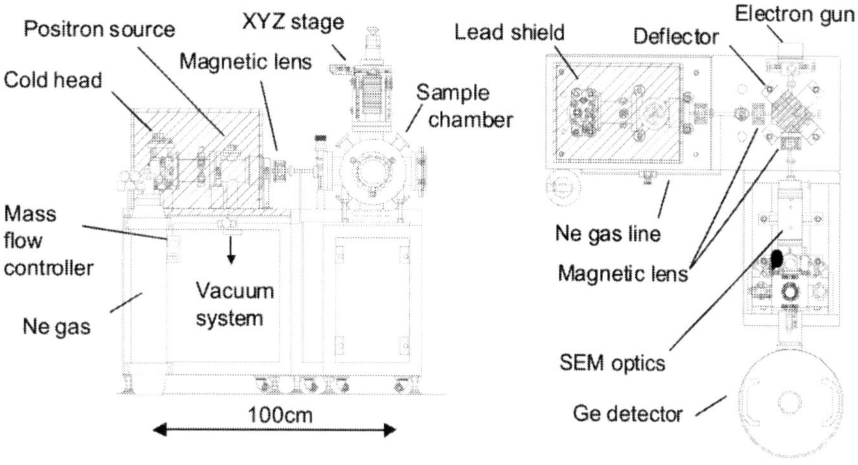

Figure 1. Schematics of the positron microbeam in JAEA

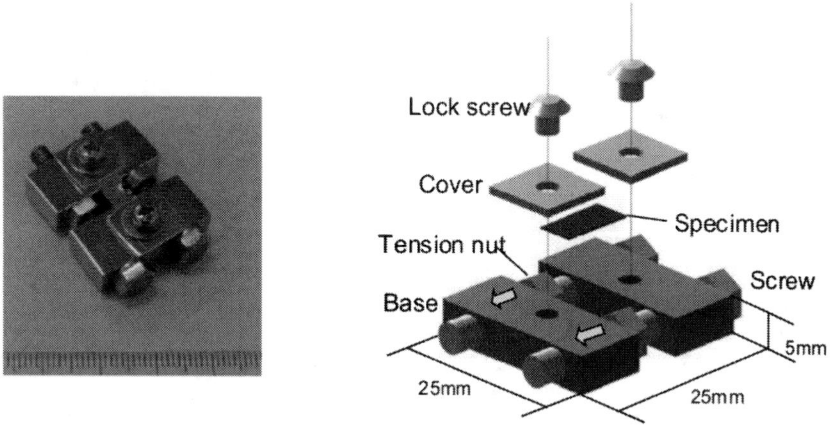

Figure 2. Schematics of the sample holder

Journal of Physics: Conference Series **225** (2010) 012033 doi:10.1088/1742-6596/225/1/012033

tip and their superimposed image, respectively. Although no morphological change is observed by the optical microscope, S parameter increases at the region of ~200 μm away from the crack tip. The relative increase of S parameter to the bulk is ~1 %. Such a small increase of S parameter may be attributed to the vacancy-type defects, dislocations and/or volume expansion.

Figures 4(a), (b) and (c) show the similar images to figures 3(a), (b) and (c) obtained after exposuring the sample to the $MgCl_2$ solution for 24 hours. From the optical image, the crack progress is observed from the pre-cracked position denoted by the dotted line on figure 4(a). The crack seems to

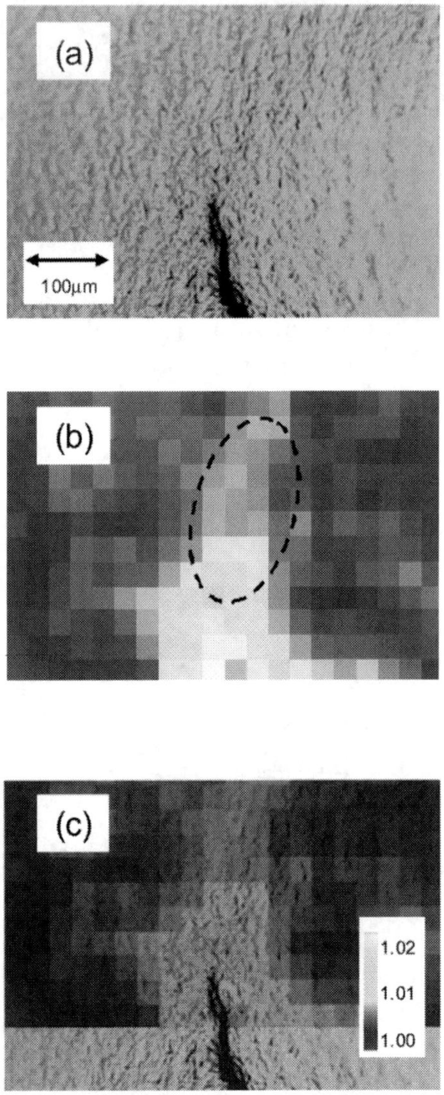

Figure 3. (a) Optical image of the pre-crack before the corrosion, (b) positron image around the crack tip and (c) superimposed image of (a) and (b).

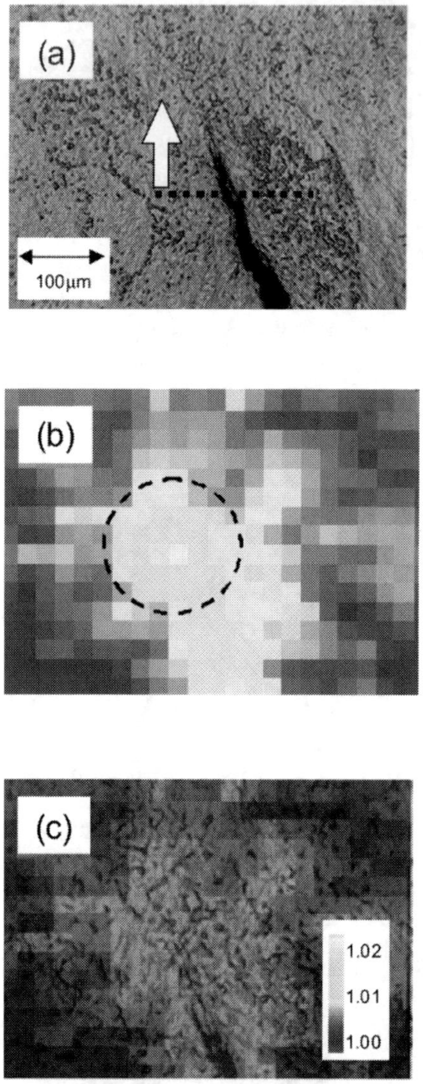

Figure 4. (a) Optical image of the SCC crack after the corrosion for 24 hours in boiled $MgCl_2$ solution, (b) distribution of S parameter around the crack tip and (c) superimposed image of (a) and (b). Dotted line on (a) shows the position of crack tip before corrosion.

progress preferentially to the direction where S parameter increased in figure 3(b). The further increase of S parameter (~2 %) is observed at the region away from the progressed crack tip. Such increase of S parameter is hardly attributed to the dislocations and/or volume expansion. This suggests that excess vacancy type defects are introduced due to the additional corrosion treatment. However, S parameter does not increase near the pre-crack region that is free from the tensile stress even after the corrosion treatment. From the above results, S parameter increases as the consequence of complementary influences due to tensile stress and corrosion environment.

In the tight crack model, some physical and chemical reasons, for example, preferential dissolutions at elementally atoms of stainless steel, dislocation movements, surface oxidation, stress at crack tip singularity and so on, are proposed as candidates for the source of vacancies. Our results seem to be consistent with the tight crack model that requires the introduction of vacancy defects.

4. Summary

We have carried out positron microbeam observations of stainless steel before and after the corrosion treatment. Applying the tensile stress, the increase of S parameter was observed near the crack tip. This suggests that vacancy-type defects are introduced by the stress concentration. The crack progress by the corrosion occurs preferentially in the region where vacancy type defects are introduced. Again, vacancy type defects are introduced near the progressed crack tip. By the further studies focusing on the vacancy type defects by the positron microbeam, the SCC mechanism will be clarified in detail.

Acknowledgements

Present study is the result of "In-situ analysis of crack tip of stainless steel under tensile stress" entrusted to "JAEA" by the Ministry of Education, Culture, Sports, Science and Technology of Japan (MEXT) and is partly supported by the MEXT Grant-in-Aid for Young Scientists (B), 2008, 20760598.

References

[1] Staehle R W 2006 Proc. Int. Conf. on water chemistry of nuclear reactor systems
[2] Swann P R 1963 Corrosion **19** 3
[3] Smith T J and Staehle RW 1967 Corrosion **23** 117
[4] Maekawa M, Kawasuso A, Hirade T and Miwa Y 2009 Mater. Sci. Forum Vol. **607** 266
[5] Li J X, Chu W Y, Wang Y B and Qiao L J 2003 Corrosion science **4** 1355
[6] Gao K W, Chu W Y, Gu B, Zhang T C, Qiao L J 2000 Corrosion **6** 515
[7] Mills Jr A P and Gullikson E M 1986 Appl. Phys. Lett. **49** 1121
[8] Maekawa M, Yu R S, Kawasuso A 2007 Phys. Stat. Sol (c) **4** 4016

A Positron Annihilation Study on the Microstructure of the Interpenetration Polymer Networks of Cyanate ester Resin/Epoxy Resin

Qi Chenze [1], Li Chunqing [1], Zeng Minfeng [1], Wang Baoyi [2], and Zhang Jian [1]

[1] Institute of Applied Chemistry, Shaoxing University, Shaoxing 312000, P. R. China;

[2] Institute of High Energy Physics, The Chinese Academy of Science, Beijing 100049, P.R. China

Email: Qichenze@usx.edu.cn

Abstract. Cyanate ester (CE) resin was blended with epoxy resin (EP) at different mass ratios (CE/EP: 100/0, 90/10, 70/30, 50/50, 30/70, 10/90, 0/100). The free volume size of CE/EP IPNs has been determined by positron annihilation lifetime spectroscopy (PALS). The size decreased as the epoxy resin content increased. The PALS results are consistent with the chemical structure changes for the copolymerizing between CE and EP. The crosslinking units of curing products (oxazoline, oxazolidinone, and polyether network) of the blends are all smaller in size than those of triazine ring structure from neat CE. Therefore, the free volume size of the blends decreases with increase of EP content. Examination of the mechanical properties, thermal stability, and morphology of the blend systems showed that addition of epoxy resin resulted in improved toughness but a little sacrifice in thermal stability when compared with pure CE. The correlations between the free volume properties and physical properties (thermal stability and mechanical properties) have been discussed.

1. Introduction

Cyanate ester resin (CE) is one of the most important kinds of thermosetting resins and has received more and more attentions for its superior mechanical properties, low water absorptivity, low outgassing in curing, high environmental resistance, and excellent dielectric properties. It has been widely used as adhesives and matrixes for composites. However, like most other thermosetting resins, it has a drawback of brittleness. The improvement of the toughness of CE while retaining its high thermal stability, modulus and dielectric properties has been an important issue in the past few decades. Current and prior attempts to improve toughness have been made through transforming the basic rigid-triazine structure into flexible structures by blending with thermoplastics and flexible rubbers, or copolymerization with other thermosetting resins like epoxy resin (EP), bismaleimides, etc [1-9]. Among the blends, CE/EP has been explored as polymer matrix in high performance composites for aerospace and electronic applications. Blending these two resins allows for modification of properties such as glass transition temperature, mechanical properties and adhesion, while reducing the cost. For a CE-

[1] To whom any correspondence should be addressed

rich system (EP as modifier), the addition of EP monomer can improve the processability and toughness of CE matrix. For an EP-rich system (CE as modifier), the addition of CE monomer can improve the thermal stability and modulus of EP matrix. Each of them can be utilized as latent catalysts to the other component.

Positron annihilation lifetime spectroscopy (PALS) is an innovative nuclear technique which has been carried out with the aim of characterizing free volume in polymeric materials during the last two decades. Free volume defined as the difference between the total volume and hypothetically occupied volume, has been considered as an intrinsic parameter that will correlate with microstructure and mechanical performance [10]. In the present study, we present the results of the free volume properties of a real CE/EP blend system with different mass ratios obtained from PALS. The differences in the the free volume properties must effectively reflect from the differences in the crosslinking network. The main purpose of this study is to use PALS as a probe to measure the free volume properties of the CE/EP blend and then examine the relationship between free volume holes and physical properties of the blend system.

2. Experimental

2.1. Materials

A phenolic-based CE resin (melting point: 76°C) was supplied by Jinan Special Structure Institute of China Aero-Industry (Jinan, China). Bisphenol A type epoxy resin (E-51) was purchased from Wuxi Huili Resin Factory (Jiangsu, China). Curing agent 4, 4-diamino diphenyl methane (DDM) was purchased from Jiangying Huifeng Chemical Company (Jiangsu, China).

2.2. Preparation of the composites

CE/EP mixtures with different mass ratios (100/0, 90/10, 70/30, 50/50, 30/70, 10/90) were thoroughly blended at 90 °C for 30 min with constant stirring. After complete mixing, each mixture was quickly poured into a preheated steel mold coated with the mold release agent. After degasification under vacuum for 30 min at 110 °C, the mold temperature was kept in the following curing procedure: 130 °C/ 1hr +150 °C/ 1hr + 170 °C/ 1hr + 190 °C/ 1hr + 210 °C/ 1hr + 230 °C/ 1hr + 250 °C/ 1hr to form the specimens. Neat EP specimens were prepared with the process: An appropriate amount of DDM (curing agent, 10 wt%) was added to EP at 90 °C, stirred for several minutes, and quickly poured into a preheated steel mold coated with the release agent. After degasification at 90 °C for about half an hour, the sample was allowed to cure at 90 °C for 2 hr and then at 150 °C for 5 hr.

2.3. Characterization of the composites

The unnotched impact strength was measured with a Charpy impact tester (XCJ-5, Chengde Testing Machine Co. Ltd., China). The tests were according to the Chinese national standard methods for impact resistance of resin casting body (GB/T 2571-1995) at 23±2 °C. The size of the samples for impact testing is about: 80 mm × 10 mm ×5 mm. The tensile properties were measured with a universal materials testing machine (SANS Testing Machine Co. Ltd., Shenzheng, China) with a crosshead speed of 2 mm/min. The tensile measurements were according to the Chinese national standard methods for tensile properties of resin casting body (GB/T 2568-1995) at 23±2 °C. The sample for tensile testing is dumbbell shaped and the scale distance is 50 mm. Five specimens of each group were prepared and tested.

Thermal gravimetric analysis (TGA) was performed using a NETZSCH STA 449C (NETZSCH Company, German) thermal analyzer. The samples were heated from 30 °C to 800 °C at a heating rate of 20 °C/ min under nitrogen atmosphere.

2.4. PALS measurement

Positron annihilation lifetime spectroscopy (PALS) was measured with an EG&G ORTEC fast-fast lifetime spectrometer (ORTEC Co., Tennessee, USA) with a FWHM=190 ps for a ^{60}Co prompt peak

Journal of Physics: Conference Series **225** (2010) 012044 doi:10.1088/1742-6596/225/1/012044

of 1.18 MeV and 1.33 MeV γ rays. A 6×10^5 Bq of positron source (^{22}Na) was deposited between two Kapton films (3 μm in thickness), which was sandwiched between two identical composite samples. All PALS measurements were performed at room temperature. Every spectrum contained about 10^6 counts. The resulting spectra were consistently modeled with a three-component fitting program PATFIT.

3. Results and Discussions

3.1. Free volume properties of the blend systems

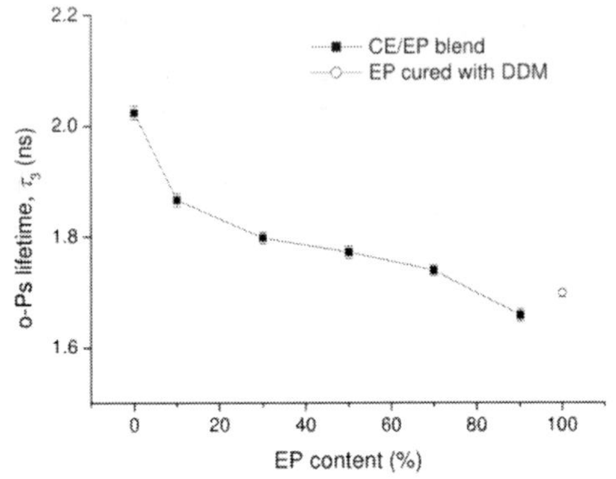

Figure 1. The *o*-Ps lifetime τ_3 of CE/EP IPNs with different EP content.

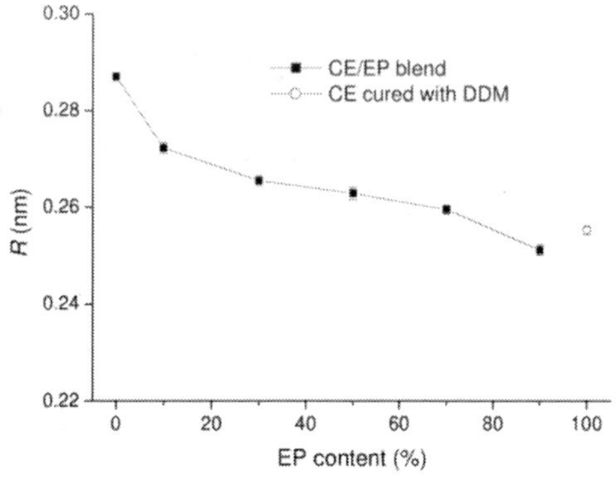

Figure 2. The calculated free volume radius *R* of CE/EP IPNs with different EP content.

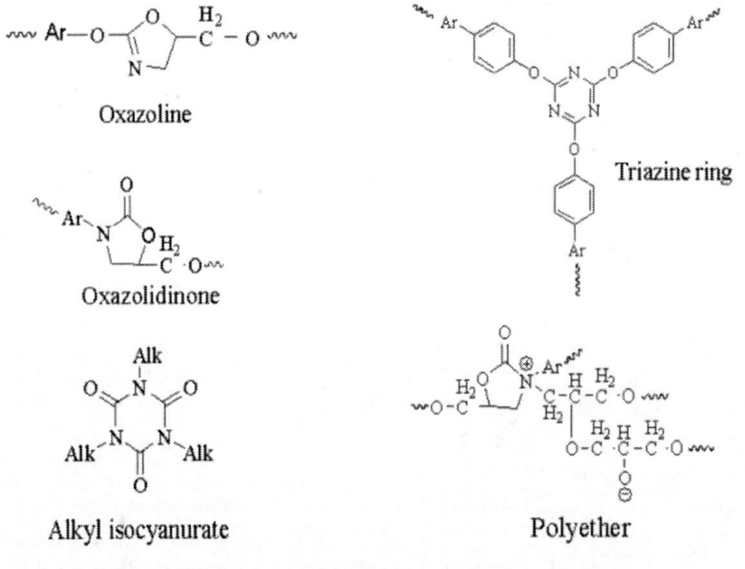

Figure 3. The curing products structure of CE/EP IPNs.

The PALS spectra for all samples were analyzed with PATFIT for three or more components of lifetime. For all samples a three-component fit was much better with variances all below a value of 1.2. Usually, the shortest lifetime τ_1 is attributed to *p*-Ps annihilation. The intermediate lifetime τ_2, is due to the free positron annihilation. The longest lifetime τ_3, is assigned to *o*-Ps annihilation. The EP content dependence of the *o*-Ps lifetime τ_3 is shown in Figure 1. The average radius *R* (as shown in Figure 2) of free volume holes has been calculated with a quantum mechanical model developed by Tao [11] as follow equation:

$$1/\tau_3 = 2\{1 - R/(R+\Delta R) + \sin[2\pi R/(R+\Delta R)]/2\pi\} \quad (1)$$

where R is the radius of the free volume hole, ΔR (=0.1656 nm) is derived from fitting the observed *o*-Ps lifetimes in molecular solids with known hole sizes [12].

Previous studies [13, 14] have showed that in the case of thermosetting-based composites, chains are restricted in the rigid network structure (in which chain units connect with chemical bonds) and their mobility depends much more on the properties of the network, such as chemical structure and crosslinking density. Our recent works [15, 16] have shown that the mass ratio of CE/EP has great influences on the reaction routines of the interpenetration polymer networks (IPNs). In a CE-rich blend system, the main curing products are triazine ring networks and oxazoline, together with a small amount of oxazolidinone. In an EP-rich blend system, the main curing products are polyether network, oxazolidinone and isocyanurate, whereas the triazine ring network is rare for the lack of CE (as see in Figure 3).

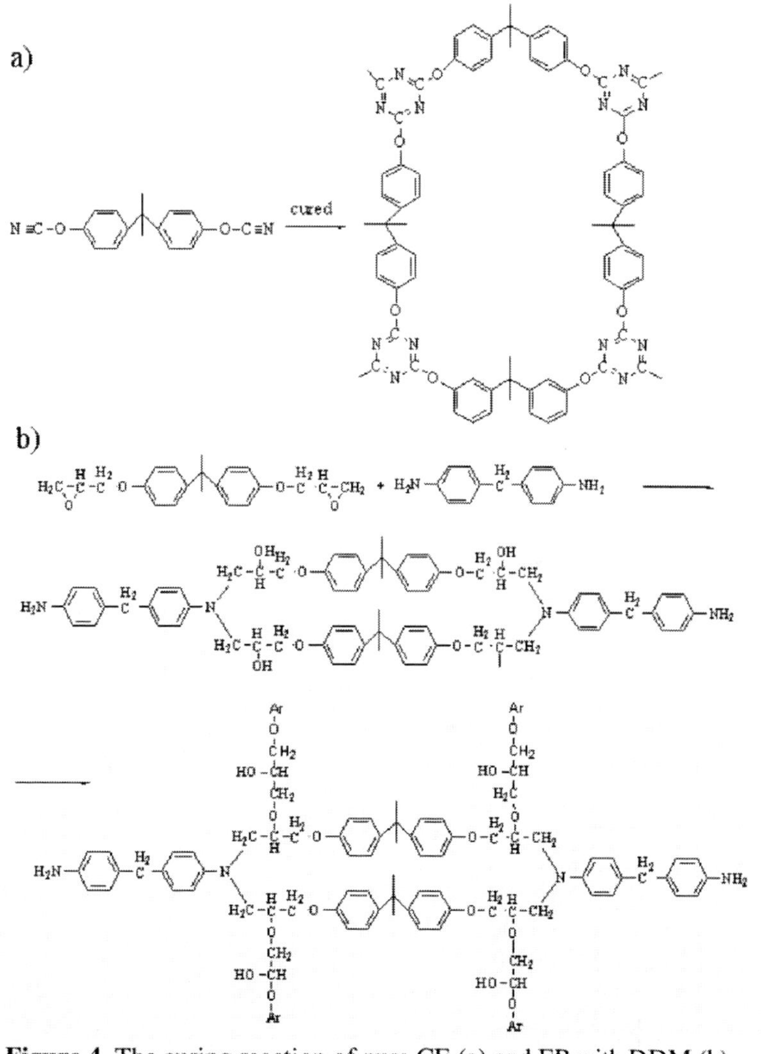

Figure 4. The curing reaction of pure CE (a) and EP with DDM (b).

The changes of R are similar with changes of *o*-Ps lifetime of the composites. It is observed that the mean free volume hole size of the blend system decreases with the increase of EP content. Before a discussion on this phenomenon, the difference of the free volume size of both crosslinked CE and EP (cured with DDM) should be mentioned. The free volume size of the matrix is determined by the crosslinking degree and the size of the crosslinking unit. As shown in Figure 4, pure CE forms a highly crosslinked network of triazine during the curing process. For comparing, the curing process of EP with DDM is also shown in Figure 4. It can be easily deduced from the structure of both curing products that the crosslinking unit of triazine is larger than that of cured EP. The smaller size of the free volume holes in EP cured with DDM should be mainly attributed to the smaller crosslinking unit.

Therefore, for CE/EP blend system, the decrease of the free volume size is due to the loss of the triazine structure in the blends with the increase of EP. As discussed above (Figure 3), cured products of oxazoline, oxazolidinone and polyethers increase with the EP content. The sizes of the crosslinking units of these curing products are all smaller than the crosslinking units of triazine ring structure. Therefore, the free volume size of the blend decreases with the EP content. In the case of CE/EP (10/90), its free volume size is even smaller than EP cured with DDM. This is caused by the different curing mechanism of EP curing process. In the case of CE/EP (10/90), the catalytic curing agent (oxazolidinone or CE) induces anionic ring-open addition polymerizations to form crosslinked polyether network (Figure 3). The catalytic curing agent does not appear in the structure of the curing products, whereas in EP cured with DDM, DDM has appeared in the structure of the curing products (as shown in Figure 4). This may be the main reason for the smaller free volume size in the case of CE/EP (10/90).

3.2. Thermal stability of the blend systems

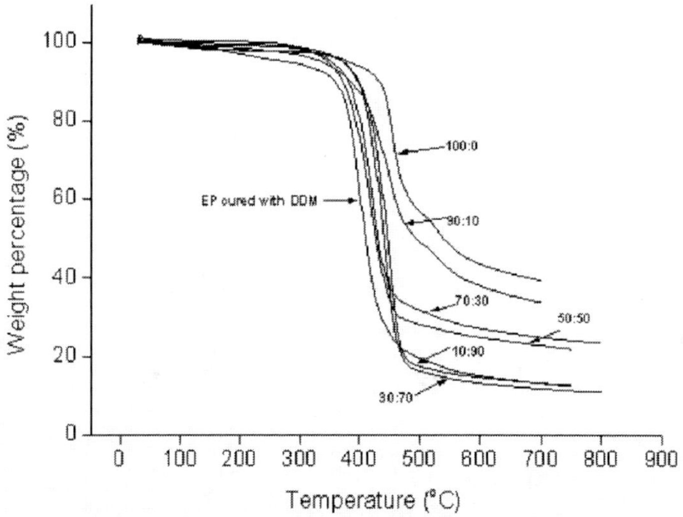

Figure 5. The thermal stability of CE/EP IPNs with different EP content.

Figure 5 shows the thermal gravimetric (TG) curves for the cured CE/EP blends with different mass ratios. Neat CE has much higher thermal stability than EP cured with DDM. The thermal stability of all the blends is better than EP cured with DDM but worse than neat CE. The blend mass ratio has great influences on the thermal stability of the blends. It is believed that CE has more thermal stability than EP because of the triazine ring structure. For the cured CE/EP blends, the number of triazine unit decreases as the EP content increases. It seems that the thermal stability of CE/EP blend system should be reduced as the EP content increases. However, in our TG study, the results show that the EP-rich systems have higher thermal stability than CE-rich systems especially in the early stage of the tests. For example, the temperature of 10% weight loss: 100/0>10/90>30/70>90/10>70/30>50/50>EP (with

DDM). Similar results have been reported by Walters NR and Lyon RE [17] who found that CE/EP (2/8) and CE/EP (4/6) had higher thermal stability than CE/EP (8/2) and CE/EP (6/4) in the early stage of the TG tests. To explain this phenomenon, we should make a careful examination of the crosslinking products of the blend system as the blend mass ratio changes. The significant differences of the blend in CE-rich systems and EP-rich systems are more oxazoline structures and triazine ring structures but fewer oxazolidinone, isocyanurate and polyether network structures formed in CE-rich systems. It is believed that the oxazoline structures degrade first and have an inferior effect on the composite [17]. Oxazolidinone, isocyanurate, triazine ring and polyether network structures have higher thermal stability than oxazoline. Therefore, an EP-rich system even has higher thermal stability in the early stage of the tests than a CE/EP (90/10) system. The higher thermal stability of the polyether network than the oxazoline structure may be attributed to its higher crosslinking density and compactness of each crosslinking unit, which is in accord with the PALS results that the free volume size of CE/EP (10/90) blend is the smallest.

3.3. Mechanical properties of the blend systems

The impact strength and tensile strength of the blend systems have been shown in Figure 6 and Figure 7. A significant increase in the impact strength of all the samples added with EP is observed when compared to that of neat CE. The tensile strength of the blend system decreases firstly and then increases as the EP content increases. Pure CE has highly symmetrical triazine ring crosslinking backbone and exhibits low toughness and high stiffness. For the blends, the highly symmetrical triazine ring crosslinking backbone has been partially destroyed and replaced by some oxazoline and oxazolidinone structure which is more thermoplastic.

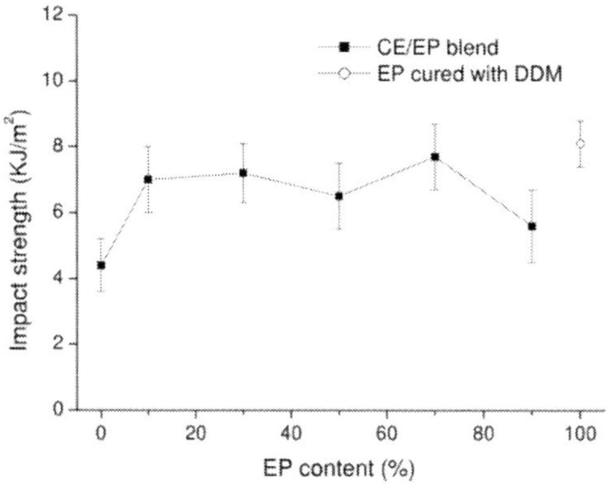

Figure 6. Impact strength of CE/EP IPNs with different EP content.

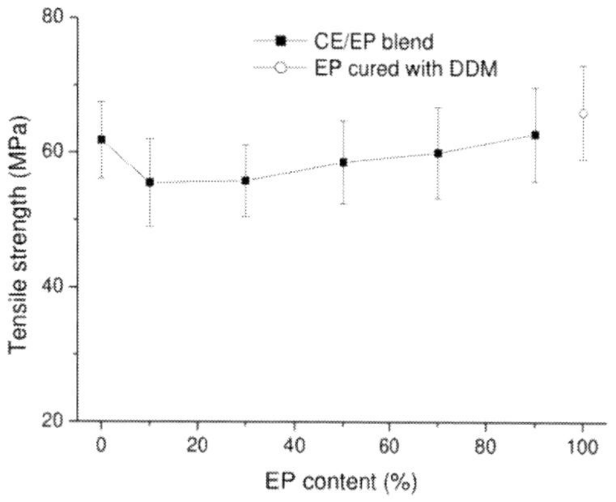

Figure 7. Tensile strength of CE/EP IPNs with different EP content.

Pure CE has highly symmetrical triazine ring crosslinking backbone and exhibits low toughness and high stiffness. For CE-rich blends, the highly symmetrical triazine ring crosslinking backbone has been partially destroyed and replaced by some oxazoline and oxazolidinone structure. The oxazoline and oxazolidinone structures exhibit more thermoplastic character and reduce the tensile strength [18]. Therefore, they have a higher toughness but lower stiffness than neat CE. For EP-rich blends, the main backbone of the crosslinking network is polyether network, together with some oxazolidinone structure and rare triazine ring structure. The increase in the tensile strength of the blends as EP content increases (EP content >30%) is due to the increase of the formation of polyether network with high crosslinking density. In the case of CE/EP (10/90), relatively lower toughness and higher stiffness is due to the highest crosslinking density of polyether network. The higher toughness of EP cured with DDM is mainly attributed to its bigger free volume size (DDM has appeared in the structure of the curing products), which can provide more room for the segmental motion.

4. Conclusions

CE resin was blended with EP with different mass ratios. Investigation of the mechanical properties, thermal stability, and morphology showed that addition of epoxy resin resulted in improved toughness but a little sacrifice in thermal stability than neat CE. The free volume properties of the blend system determined by PALS are consistent with the chemical structure changes for the co-curing reactions between CE and EP. In addition, physical properties of the blend system can be estimated from the changes in the free volume properties of the blends.

Acknowledgements: This work was supported by National Natural Science Foundation of China (grant number: 10875079), and Natural Science Foundation of Zhejiang Province, China (grant number: Y4080448).

References

[1] Ileana B R, Dagmar R, Rosa G L, Clara M G, 2005 *Euro. Polym. J.* **41** 2635
[2] Iijima T, Kaise T, Tomoi M, 2003 *J. Appl Polym. Sci.* **88** 1
[3] Mondragon I, Solar L, Nohales A, Vallo C I, Gomez C M, 2006 *Polymer* **47** 3401
[4] Hayes B S, Seferis J C, Parker G A 2000 *Polym. Eng. Sci.* **40** 1344
[5] Borrajo J, Riccardi C C, Williams R J J, Cao Z Q, Pascault J P 1995 *Polymer* **36** 3541
[6] Feng Y, Fang Z P, Gu A J 2004 *Polym. Adv. Technol.* **15** 628
[7] Feng Y, Fang Z P, Gu A J 2005 *Polym. Int.* **54** 369
[8] Dinakaran K, Kumar R S, Alagar M 2005 *Mater. Manuf. Process.* **20** 299
[9] Bartolomeo P, Chailan J F, Vernet JL 2001 *Eur. Polym. J.* **37** 659
[10] Ito K, Kobayashi Y 2003 *Appl. Phys. Lett.* **82** 654
[11] Tao S J 1972 *J. Chem. Phys.* **56** 5499
[12] Nakanishi H, Wang S J, Jean YC 1988 *In positron annihilation in fluids*. Sharma SC Ed; World Sci. Pub. Singapore.; 292-298
[13] Yu D H, Wang B, Feng Y, Fang Z P 2006 *J. Appl. Polym. Sci.* **102** 1509
[14] Wang B, Qi N, Gong W, Li X W, Zheng Y P 2007 *Radiat. Phys. & Chem.* **76** 146
[15] Zeng M F, Lu C Y, Qi C Z *J. Macromol. Sci., Part B Phys.* in press
[16] Sun X D, Zeng M F, Lu C Y, Yan F Y *Polymer-Plastics Tech. & Eng.* Submitted
[17] Walters R N, Lyon R E 2003 *Fire Mater.* **27** 183
[18] Dinakaran K, Alagar M, 2003 *Polym. Adv. Technol.* **14** 574

International Symposium on Advanced Science Research 2009 IOP Publishing
Journal of Physics: Conference Series **225** (2010) 012038 doi:10.1088/1742-6596/225/1/012038

Free volume and gas permeation in ion-exchanged forms of the Nafion® membrane

Hamdy F M Mohamed[1*], Y Kobayashi[2], C S Kuroda[1] and A Ohira[4]
[1] FC-Cubic, National Institute of Advanced Industrial Science and Technology (AIST), Tokyo 135-0064, Japan
2 NMIJ, National Institute of Advanced Industrial Science and Technology (AIST), Tsukuba, Ibaraki 305-8565, Japan

E-mail: hamdy-mohamed@aist.go.jp

Abstract. Variations of free volume and gas permeability of the Nafion® membrane upon ion-exchange of H^+ with Cs^+ or Pt^{2+} was studied as a function of temperature. Free volume was quantified using the positron annihilation lifetime technique. Our results showed that the free volume ($V_{FV,Ps}$) of the dried membrane is enlarged by thermal expansion. It was found that the ion-exchange significantly expands the free volume and at the same time decreases the permeabilities of O_2 and H_2. Good linear correlations between the logarithm of permeabilities of O_2 and H_2 at different temperatures and $1/V_{FV,Ps}$ for the ion-exchanged forms of Nafion® in the dried state suggest an important role played by the free volume in gas permeation. Considerable downward deviation of the correlations for the ion-exchanged ionomers from the H^+-form suggested the importance of polymer stiffening in gas permeation.

1. Introduction

The polymer electrolyte fuel cell (PEFC) has been considered as one of the most promising power sources for stationary and portable applications due to their high energy conversion efficiency and near-zero pollutant emission [1-4]. As a key component of PEFCs, the polymer electrolyte membrane (PEM) plays dual roles of ion conductor and gas separator [5]; the latter role is important because the crossover of H_2 and O_2 between the anode and cathode in PEFC decreases the efficiency of PEFC. DuPont's Nafion® membranes with the chemical structure shown in Fig. 1 possesses high proton conductivity and excellent chemical stability and are considered as the state of-the-art PEMs. According to the free volume model of gas permeation, free volume, present as sub-nanometer scale holes in the amorphous domain of the PEM, provides a pathway for diffusion of gases [6]. Hence the model states that the structure with less open volume is more amenable to lower permeability of small gas molecules such as O_2 and H_2.

Nanometer sized free volume holes in a polymer can be probed by the positron annihilation lifetime (PAL) spectroscopy, as long as a reasonable fraction of the positrons form positronium (Ps), the positron-electron bound state [7-10]. Ps is formed by a reaction between a positron and one of the electrons, ionized and liberated from the polymer molecules by the energetic positron itself, in the terminal positron spur [11]. The positrons that have formed Ps annihilate either from short-lived *para-*

* Visiting from Physics Department, Faculty of Science, Minia University, P.O. Box 61519 Minia, Egypt.

positronium (*p*-Ps) or long-lived *ortho*-positronium (*o*-Ps). Thus the positron annihilation lifetime spectrum of a Ps-forming substance typically contains three components, which are due to *p*-Ps with lifetime τ_1 ~ 125 ps and relative intensity I_1, free positrons with lifetime τ_2 ~ 500 ps and intensity I_2, and *o*-Ps with lifetime τ_3 ~ 1-10 ns and intensity I_3 [11, 12]. The annihilation of *o*-Ps in a polymer is governed by the "pick-off" process, in which the positron in *o*-Ps annihilates with one of the surrounding spin opposite electrons [12-14]. Approximating the free volume hole as a spherical potential well having infinitely high walls, Tao [15] and Eldrup *et al.* [16] derived the following relationship between the pick-off annihilation lifetime (τ_3) and the radius (R) of the free volume hole as;

$$\tau_3 = 0.5\left\{1-\frac{R}{R_o}+\frac{1}{2\pi}\sin\left(\frac{2\pi R}{R_o}\right)\right\}^{-1} \text{ (ns)}, \qquad (1)$$

where $R_o = R + \Delta R$ and ΔR = 0.166 nm is the thickness of the homogenous electron layer in which the positron in *o*-Ps is postulated to annihilate [17]. The free volume size $V_{FV,Ps}$ in nm^3 is given as

$$V_{FV,Ps} = 4\,\pi R^3/3. \qquad (2)$$

The above equations have been successfully used for the study of the free volume in various polymers [18-20].

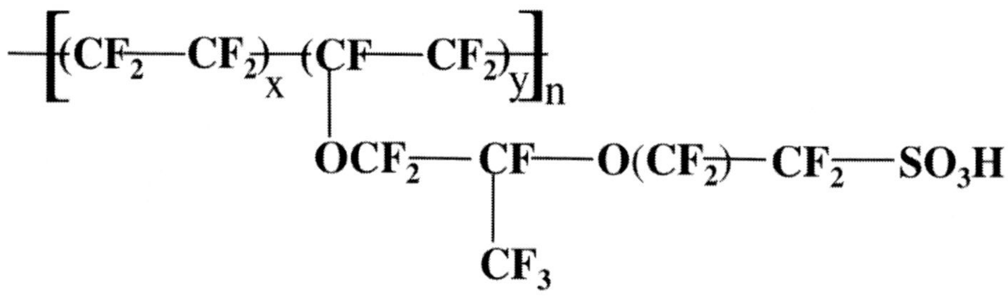

Figure 1. Chemical structure of Nafion®.

Previously, we investigated the free volume in Na^+- and K^+-forms of Nafion® [21,22]. The comparison of experimental data collected by different techniques revealed that not only free volume but also polymer stiffness plays an essential role in O_2 permeation. In this study, we extend our work on the mechanism of gas permeation to the Cs^+- and Pt^{2+}-forms of Nafion®. The study of the free volume in ion-exchanged Nafion® is of practical significance not only because an ion-exchanged form of Nafion® is sometimes used to fabricate PEFC [23,24] but also because contamination of such ions as Cs^+ and Pt^{2+} is possible during fuel cell operation from humidified feed gases or catalyst layer [25-27]. Free volume was characterized by using the positron annihilation lifetime (PAL) technique and the permeabilities of O_2 and H_2 were measured as a function of temperature. The information on the crystallinity and mechanical properties was obtained by wide angle x-ray diffraction (WAXD), dynamic mechanical analysis (DMA).

2. Experimental

The H^+-form of Nafion® was supplied from Du Pont Japan as NR212 (formally NRE212) with an ionic exchange capacity (IEC) of 0.92 meq/g and a thickness of 50 μm. The Cs^+- and Pt^{2+}-forms were prepared by exchanging H^+ with Cs^+ and Pt^{2+}, respectively following the standard procedure [28]. Briefly as-received NR212 was soaked for 1 day in aqueous 1 mol/dm^3 solutions of CsCl and $Pt(NH_3)_4(OH)_2(H_2O)_n$. The soaked films were removed from the solutions, washed several times with deionized water and dehydrated in an oven at 70 °C for 3 h. All the chemicals used were of reagent grade (Wako Pure Chemical Industries).

The PAL measurements were performed using a fast-fast coincidence system with a time resolution of 280 ps (full width at half maximum, FWHM) [29]. A positron source (^{22}Na), sealed between thin foils of Kapton, was mounted in a sample (stack of films with a total thickness of 1 mm)-source-sample sandwich. The channel width was 20.7 ps, and the data were recorded over 4096 channels. Before the measurements, the samples had been dried by evacuation for one day at room temperature (~25 °C) and then at 80 °C for 10 h under vacuum. Lifetime spectra were recorded as a function of temperature up to 100 °C. Each lifetime spectrum containing a total number of 2.5 x10^6 counts was analyzed using the PALSfit program [30] into three lifetime components to deduce lifetime τ_3 and relative intensity I_3 of the longest-lived *o*-Ps component. No source correction was applied in the data analysis. The validity of our measurement and data analysis was confirmed with certified reference materials (NMIJ-CRM 5601-a, No.07 and NMIJ-CRM 5602-a, No.01) [31,32].

Gas permeability was measured by the equal pressure method with a GTR-Tech 30XFST apparatus equipped with a Yanaco G2700T gas chromatograph by monitoring the amounts of O_2 and H_2 that permeated across the membrane from one side to the other [21,22,33]. The temperature dependence of the permeation coefficient for O_2 and H_2 (P_{O2} & P_{H2}) was measured up to 90 °C for dried samples using He and Ar as carrier gases of permeated O_2 and H_2, respectively, to the gas chromatograph. A 50 μm thick membrane was set in a cell with a gas inlet and outlet, where the temperature was controlled. Both O_2 and H_2 were supplied at a flow rate of 30 mL/min.

Dynamic mechanical analysis (DMA) was performed in the tensile mode on a Triton DMA analyzer (Tritec 2000 - Shimadzu) for the dehydrated samples. Loss tangent (tan δ) was measured at a fixed dynamic displacement of 0.5 % and a frequency of 1 Hz in air for a sample typically of 9.0 x 9.0 x 0.05 mm^3. Elongation was measured with a static load of - 0.10 N. The samples were heated with a rate of 3 °C/min from 20 °C to 150 or 330 °C.

Wide angle x-ray diffraction (WAXD) was measured in the θ-2θ mode for the dehydrated samples by Ultima IV (Rigaku Corporation) with a 285 mm goniometer. A parallel monochromatic x-ray beam obtained by reflection of Cu $K\alpha$ radiation (λ = 1.542 Å) from a fine focus x-ray tube operated at 40 kV and 40 mA by a parabolic multi-layer membrane mirror was used. Scattering profiles were recorded with a scintillation one-dimension position-sensitive detector over a range of $5° < 2\theta < 65°$ with a step of 0.025° and at a rate of 0.5°/ min.

3. Results and discussion

Figure 2A shows temperature dependence of the *o*-Ps lifetime τ_3 for dry Cs^+- and Pt^{2+}-forms of Nafion®. The right hand ordinate of the figure represents the free volume hole size quantified by Equations (1) and (2). It is seen that the free volume of all the samples is considerably enlarged by thermal expansion. Both Cs^+- and Pt^{2+}-forms of Nafion® have larger free volumes than the H^+-form over the entire temperature range studied. Temperature dependence of the free volume for the H^+-form shows an abrupt change of the thermal expansion coefficient around 20 °C due to β-transition caused by the glass transition of Nafion®. According to Osborn *et al.* [34], this glass transition of the H^+-form of Nafion® is attributed to principally main-chain (backbone) motions within the framework of a static physically cross-linked (hydrogen-bonding) network. The absence of any abrupt change in the thermal expansion coefficient of the free volume for the Cs^+- and Pt^{2+}-forms suggests that their β-transition occurs above 80 °C. Figure 3 shows the tan δ measured by dynamic mechanical analysis (DMA) as a function of temperature for dehydrated Nafion® containing H^+, Cs^+ and Pt^{2+}. A pronounced peak centered at 108 °C for the H^+-form is due to α-relaxation, which may be associated with the onset of large-scale motions of the ionomer [35]. The peak is shifted to 230 and 295 °C for the Cs^+- and Pt^{2+}-forms, respectively, so that the relaxation temperature significantly rises upon ion-exchange as reported by other authors [36,37]. The increase of the relaxation temperature by the exchange of H^+ with Cs^+ or Pt^{2+} is consistent with our result in Fig. 2A.

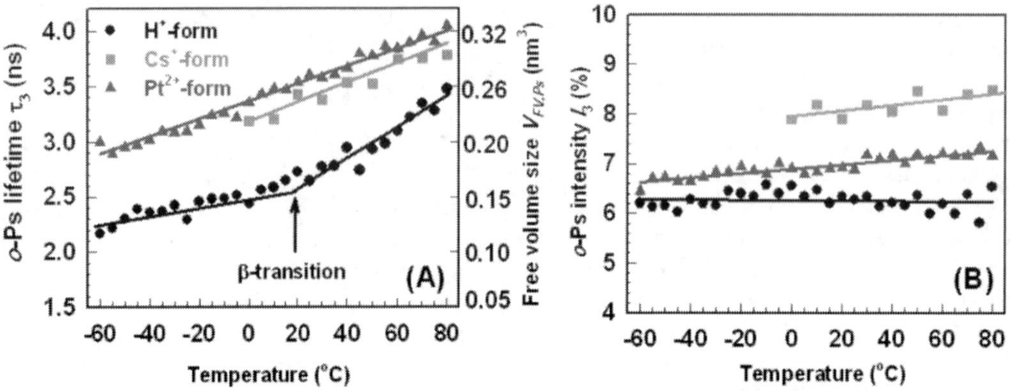

Figure 2. Lifetime τ_3 and its intensity I_3 of *o*-Ps as a function of temperature for dried Nafion® of the H^+, Cs^+- and Pt^{2+}-forms. The scale of the right ordinate in (A) is the size of free volume hole $V_{FV,Ps}$ deduced from Equations (1) and (2).

The *o*-Ps intensity I_3 for all the samples varies only slightly with temperature, indicating that Ps formation does not significantly depend on temperature as shown in Fig. 2B. Regardless of temperature, exchange of H^+ with Pt^{2+} substantially enhances I_3 and that with Cs^+ makes I_3 even higher. Since free volumes favorable for Ps formation are in the amorphous region of a polymer, Ps formation is expected to be enhanced with decreasing crystallinity [38,39]. However, all the Nafion® samples have the *o*-Ps intensity I_3 less than half the typical value of PTFE (~ 19.4 %), in spite of their consistently lower crystallinity [29,40]. Moreover, Ps formation is significantly different between the Cs^+- and Pt^{2+}-forms. Figure 4 displays WAXD patterns for dehydrated Nafion® in the H^+-, Cs^+- and Pt^{2+}-forms recorded at room temperature. A sharp peak around $2\theta = 17^\circ$ is superimposed on an amorphous halo over a range of $2\theta = 12$ - 22° in agreement with Laporta *et al.* [41]. The peak can be indexed as the (100) reflection of the hexagonal structure of polytetrafluoroethylene (PTFE)-like crystallites in Nafion® [42,43]. The crystalline reflection around 17° is weakened in the Cs^+- and Pt^{2+}-forms in comparison with the H^+-form, suggesting that in ion-exchanged Nafion® large alkali ions make the regular arrangement of the molecular chains difficult and the crystallinity is decreased. Further the crystallinity of the Cs^+- and Pt^{2+}-forms of Nafion® is similar to each other although there *o*-Ps intensities are significantly different. Therefore, the variations of Ps formation upon ion-exchange can not be fully accounted for in terms of crystallinity. Apart from crsytallinity, a chemical structure or compound that captures an electron-positron pair, which would otherwise recombine with each other to Ps, can inhibit the formation of Ps [40,44,45]. Comparison of the chemical structure between Nafion® and PTFE suggests that the reduced Ps formation in the ionomer may be due to the inhibition effect of the electron-accepting SO_3X (X = H^+, Pt^{2+}, Cs^+). If this is the case, our results in Fig. 2B suggest that SO_3H and SO_3Cs are, respectively, strong and weak inhibitors of Ps formation.

Figures 5 and 6 show O_2 and H_2 permeabilities in dried H^+-, Cs^+- and Pt^{2+}-forms of Nafion® at various temperatures, respectively. It is seen that both the permeablities increase by orders of magnitudes with increasing temperature from 20 to 90 °C. At all the temperatures studied, H_2 is much more permeable than O_2. Regardless of temperature both O_2 and H_2 are less permeable in the ion-exchanged ionomers than the H^+-form of Nafion®.

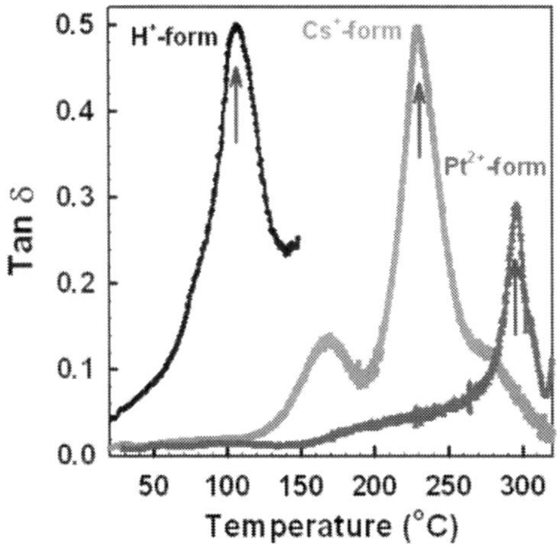

Figure 3. Loss tangent (tan δ) as a function of temperature for Nafion® containing H^+, Cs^+ and Pt^{2+}. The arrow indicates the peak due to α-relaxation.

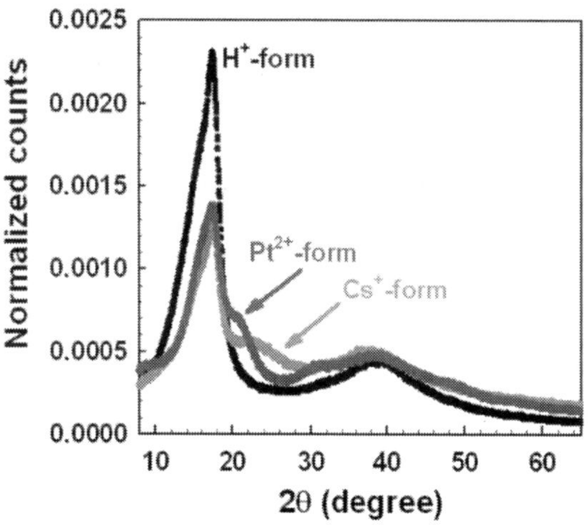

Figure 4. Wide angle x-ray diffraction patterns for Nafion® in the H^+-, Cs^+- and Pt^{2+}-forms.

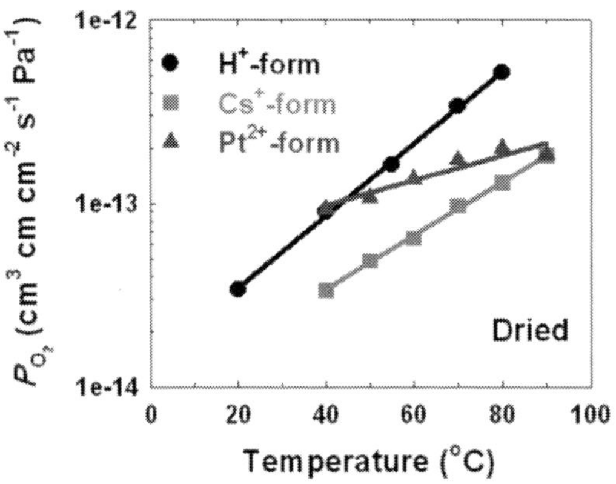

Figure 5. O_2 permeability as a function of temperature for dried Nafion® in the H^+-, Cs^+- and Pt^{2+}- forms. The straight lines are drawn just for guiding the eyes.

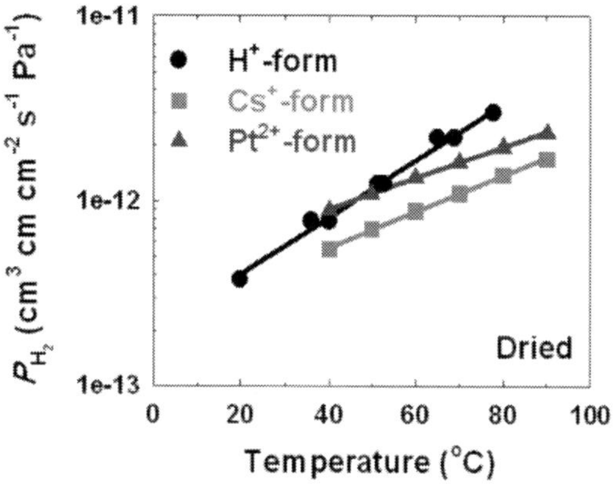

Figure 6. H_2 permeability as a function of temperature for dried Nafion® in the H^+-, Cs^+- and Pt^{2+}- forms. The straight lines are drawn just for guiding the eyes.

According to the sorption-diffusion model, gas permeation coefficient P is given as the product of permeant solubility S and diffusion coefficient D [6],

$$P = S * D. \tag{3}$$

The permeant can diffuse when it finds large enough free volume around itself to make a jump to one of the neighbouring sites. Therefore, the diffusivity can be related to the fractional free volume V_f by [46-48]

$$D = AR_gT \exp(-B / V_f), \tag{4}$$

where A and B are constants depending on the permeant, R_g is the gas constant and T is the absolute temperature. Combination of Equation (3) with Equation (4) yields the following relationship between the permeability and free volume;

$$P = SAR_gT \exp(-B / V_f). \tag{5}$$

To correlate the free volume obtained by the PAL technique to gas diffusivity and permeability, the following equations have been proposed by replacing V_f in Equations (4) and (5) with $V_{FV,Ps}$

$$D = AR_gT \exp(-b / V_{FV,PS}), \tag{6}$$

$$P = SAR_gT \exp(-b / V_{FV,Ps}) = CT \exp(-b / V_{FV,Ps}), \tag{7}$$

where b is a constant depending on the permeant and C is another constant depending on solubility and chain stiffness. Figures 7 and 8 show the plots of log P/T versus $1/V_{FV,Ps}$ for the permeation of O_2 and H_2, respectively, in dried Nafion® at different temperatures before and after ion exchange with Cs^+ and Pt^{2+}. Reasonably good correlations in line with Eq. (7) are observed, indicating the important role played by the free volume in gas permeation. However, the correlation for the Cs^+- and Pt^{2+}-forms of Nafion® is considerably shifted downward relative to the H^+-form. This shift is due to increase in C. The rise of transition temperatures upon ion-exchange in Fig. 3 indicates the increased chain stiffness, which is supposed to cause the increase of C and the shift of the correlation between the permeabilities and free volume [21].

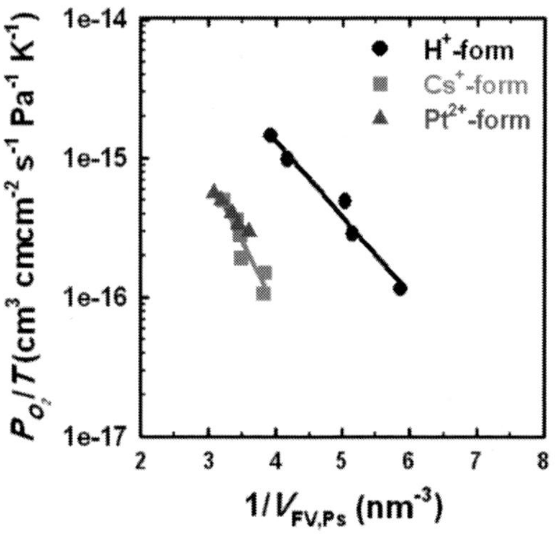

Figure 7. Correlation between the permeability of O_2 and reciprocal free volume hole size for dry ion-exchanged forms of Nafion® at different temperatures.

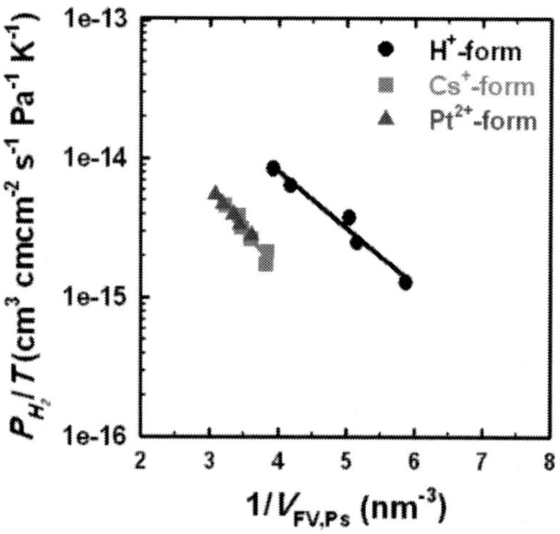

Figure 8. Correlation between the permeability of H_2 and reciprocal free volume hole size for dry ion-exchanged forms of Nafion® at different temperatures.

4. Conclusions

In the present work we studied free volume and gases permeation in Nafion® before and after ion exchange with Cs^+- and Pt^{2+}. The exchange of H^+ with Cs^+ or Pt^{2+} enlarged the free volume but significantly reduced the gas permeabilities. As evidenced by good correlations between the permeability and free volume, we confirmed the important role played by the free volume in gas permeation in dry Nafion® (H^+-, Cs^+- and Pt^{2+}-forms). The smaller gas permeabilities in the Cs^+- and Pt^{2+}-forms of Nafion® than the H^+-form may be attributed to chain stiffness enhanced by the ion exchange.

Acknowledgments
This work was financially supported by the Ministry of Economy, Trade and Industry (METI) and the New Energy and Industrial Technology Development Organization (NEDO).

References

[1] Wei Z D, Ji M B, Hong Y, Sun C X, Chan S H and Shen P K 2006 *J Power Sources* **160** 246.
[2] Vielstich W, Lamm A and Gasteiger H A 2003 *Handbook of fuel cells: Fundamentals, technology, and applications*, New York, Wiley.
[3] Ralph T R 1999 *Platinum Met. Rev.* **43** 14.
[4] Steele B C H and Heinzel A 2001 *Nature* **414** 345.
[5] Luo Z P, Li D X, Tang H L, Pan M and Yuan R Z 2006 *Int. J. Hydrogen Energy* **31** 1831.
[6] Vegt A K 2002 *From polymers to plastics* The Netherlands, Delft University Press pp. 159.
[7] Okamoto K, Tanaka K, Katsube M, Sueoka O and Ito Y 1993 *Radiat. Phys. Chem.* **41** 497-502.
[8] Kobayashi Y, Haraya K, Hattori S and Sasuga T 1994 *Polymer* **35** 925–928.
[9] Oka T, Ito K, He C, Dutriez C, Yokoyama H and Kobayashi Y 1996 *J. Phys. Chem. B* **112** 12191–12194.
[10] De Sitter K, Andersson A, D'Haen J, Leysen R, Mullens, Maurer F H J and Vankelecom I F J 2008 *J. Membr. Sci.* **321** 284-292.
[11] Mogensen E 1995 *Positron annihilation in chemistry* Berlin Heidelberg, Germany Springer-Verlag.
[12] Kobayashi Y 1991 *J. Chem. Soc. Faraday Trans.* **87** 3641-3645.
[13] Jean Y C, Mallon P E and Schrader D M 2003 *Principles and applications of positron & positronium chemistry* Hong Kong, World Scientific.
[14] Kobayashi Y, Haraya K, Kamiya K and Hattori S 1992 *Bull. Chem. Soc. Jpn.* **65** 160-163.
[15] Tao S J 1972 *J. Chem. Phys.* **56** 5499-5510.
[16] Eldrup M, Lightbody D and Sherwood JN 1981 *Chem. Phys.* **63** 51-58.
[17] Nakanishi H, Wang S J and Jean Y C 1988 *Microscopic surface tension studied by positron annihilation*, in: Sharma S C (Ed.) *Positron annihilation studies of fluids* Singapore, World Scientific pp.292-298.
[18] Mohamed H F M, El-Sayed A M A and Abdel-Hady E E 1999 *J. Physics: Condensed Matter* **11** 4461-4467.
[19] Mohamed F H M and Abd El-Aziz N S 2001 *Polymer* **42** 8013-8017.
[20] Abdel-Hady E E, Mohamed H F M and Fareed S S 2007 *Radiat. Phys. Chem.* **76** 138-141.
[21] Mohamed H F M, Kobayashi Y, Kuroda C S and Ohira A 2009 *J. Phys. Chem. B* **113** 2247-2252.
[22] Mohamed H F M, Ohira A and Kobayashi Y 2009 *Mater. Sci. Forum* **607** 58–60.
[23] Xie J, Garzon F, Zawodzinski T and Smith W 2004 *J. Electrochem. Soc. 151* A1084-A1093.
[24] Wilson M S and Gottesfeld S 1992 *J. Electrochem. Soc.* **139** L28-L30.
[25] Okada T 1999 *J. Electroanal. Chem.* **465** 1-17.
[26] Okada T 1999 *J. Electroanal. Chem.* **465** 18-29.
[27] Pozio A, Silva R F, Francesco M D and Giorgi L 2003 *Electrochim. Acta.* **48** 1543-1549.
[28] Kinumoto T, Inaba M, Nakayama Y, Ogata K, Umebayashi R, Tasaka A, Iriyama Y, Abe T and Ogumi Z 2006 *J. Power Sources* **158** 1222.
[29] Mohamed H F M, Ito K, Kobayashi Y, Takimoto N, Takeoka Y and Ohira A 2008 *Polymer* **49** 3091-3097.
[30] Olsen J V, Kirkegaard P, Pedersen NJ and Eldrup M 2007 *Phys. Stat. Sol. C* **4** 4004–4006.

[31] National Metrology Institute of Japan: Certificated Reference Materials. http://www.nmij.jp/english/service/C/crm/.
[32] Ito K, Oka T, Kobayashi Y, Shirai Y, Wada K, Matsumoto M, Fujinami M, Hirade T, Honda Y, Hosomi H, Nagai Y, Inoue K, Saito H, Sakaki K, Sato K, Shimazu A and Uedono A 2008 *J. Appl. Phys.* **104** 026102-1- 026102-3.
[33] Asano N, Aoki M, Suzuki S, Miyatake K, Uchida H and Watanabe M 2006 *J. Am. Chem. Soc.* **128** 1762-1769.
[34] Osborn S J, Hassan M K, Divoux G M, Rhoades D W, Mauritz K A and Moore R B 2007 *Macromolecules* **40** 3886-3890.
[35] Kim Yu S, Dong L, Hickner M A, Glass T E, Webb V and McGrath J E 2003 *Macromolecules* **36** 6281-6285.
[36] Miura Y and Yoshida H 1990 *Thermochimica Acta* **163** 161-168.
[37] Page K A, Cable K M and Moore R B 2005 *Macromolecules* **38** 6472-6484.
[38] Kobayashi Y, Wang C L, Hirata K, Zheng W and Zhang C 1998 *Phys. Rev. B* **58** 5384-5389.
[39] Mohamed H F M, Ito Y and Imai M 1996 *J. Chem. Phys.* **105** 4841-4845.
[40] Mohamed H F M, Kobayashi Y, Kuroda C S and Ohira A 2009 *Phys. Stat. Sol. C*, in press.
[41] Laporta M, Pegoraro M and Zanderighi L 2000 *Macromol. Mater. Eng.* **282** 22-29.
[42] Park H, Kim Y, Hong W H, Choi Y S and Lee H 2005 *Macromolecules* **38** 2289-2295.
[43] Lin J, Wycisk R, Pintauro P N and Kellner M 2007 *Electrochem. Solid-State Lett.* **10** B19-B22.
[44] Wang C L, Kobayashi Y, Zheng W, Zhang C, Nagai Y and Hasegawa M 2001 *Phys. Rev.* B **63** 064204-1-064204-9.
[45] Hirata K, Kobayashi Y and Ujihira Y 1997 *J. Chem. Soc., Faraday Trans.* **93** 139-142.
[46] Cohen M H and Turnbull D 1959 *J. Chem. Phys.* **31** 1164–1169.
[47] Fujita H 1961 *Fortschr. Hochpolym. Forsch.* **Bd.3** S.1–47.
[48] Thran A, Kroll G and Faupel F. 1999 *J. Polym. Sci. Pt. B-Polym. Phys.* **37** 3344-3358.

Free-volume structure of fluoropolymer-based radiation-grafted electrolyte membranes investigated by positron annihilation lifetime spectroscopy

S Sawada[1], A Kawasuso[2], M Maekawa[2], A. Yabuuchi[2], Y Maekawa[1]

[1]Quantum Beam Science Directorate, Japan Atomic Energy Agency
1233 Watanuki, Takasaki, Gunma 370-1292, Japan
[2] Advanced Research Institute, Japan Atomic Energy Agency
1233 Watanuki, Takasaki, Gunma 370-1292, Japan

E-mail: sawada.shinnichi@jaea.go.jp

Abstract. In the field of polymer-electrolyte-membrane (PEM) fuel cell technology, the structures of free-volume holes in the PEMs are very important because they are correlated to the supplied-gas crossover phenomenon, which sometimes deteriorates the cell performance. In this study, we investigated the size and location of free-volume holes in the crosslinked-polytetrafluoroethylene (cPTFE) based radiation-grafted PEMs by positron annihilation lifetime (PAL) spectroscopy. For comparison, the base cPTFE and polystyrene grafted films were also measured. From the analysis of PAL spectra, it was found that there were free-volume holes with different radius of 0.28-0.30 nm and 0.44-0.45 nm. The smaller holes should be located in both PTFE crystallites and poly(styrene sulfonic acid) grafts, while the larger holes are considered to exist in amorphous PTFE phases.

1. Introduction

Polymer electrolyte membrane (PEM) fuel cells are electrochemical devices to directly transform hydrogen and oxygen into electric power. In a fuel cell, protons dissociated from hydrogen at an anode are transported through a PEM to a cathode, and then react with oxygen to yield water. Recently, there has been increasing interest in PEM fuel cells due to high energy-conversion efficiency and a low environmental load [1-2]. The vital component in the fuel cell is a PEM, which acts as both an electrolyte transporting protons and a separator to prevent mixing of the supplied fuel gases. At present, the most commonly used PEM material is Nafion (DuPont Co.,), consisting of hydrophobic polytetrafluoroethylene (PTFE) backbones with pendent side chains of perfluorinated vinyl ethers terminated by sulfonic acid groups [3]. Although Nafion has excellent properties (proton conductivity, chemical stability, and mechanical strength), it is still too expensive owing to the complicated production process.

During the operation of a PEM fuel cell, supplied hydrogen and oxygen gases permeate through a PEM. This so-called gas crossover phenomenon means the loss of fuel, resulting in deterioration of cell performance [4-9]. Additionally, gas crossover is said to cause a more serious problem. When oxygen permeates through a PEM from a cathode to an anode, oxygen and proton react to form hydrogen peroxide, which would degrade the PEM [9]. In general, gas diffusion phenomena in a polymer film is believed to be related to the nanometer-sized free-volume holes [10-13]. Consequently, in order to develop the advanced PEM materials with low gas crossover, it is important to obtain a deep insight into the micro-hole structures.

The positron annihilation lifetime (PAL) spectroscopy is capable of directly probing the nanometer-sized free-volume holes. This technique was developed based on the formation of positronium (Ps), the bound state between a positron and an electron. When positrons are injected into materials, there exist three states of positrons with different lifetimes. The shortest-lifetime (~ 0.125 ns) state is *para*-positronium (*p*-Ps), in which the spins are anti-parallel. The middle-lifetime (~ 0.45 ns) state is a free positron without forming a bound state. The longest-lifetime (1-10 ns) state is *ortho*-positronium (*o*-Ps) with parallel spins. The *o*-Ps lifetime has a strong correlation with the size of the free-volume holes: that is, as the holes become smaller, the *o*-Ps lifetime decreases. Therefore, the information of free-volume structures in materials can be obtained by PAL spectra.

As a low-cost process of PEM synthesis, a radiation-induced grafting method has been widely focused [14,15]. This method consists of the following three steps: irradiation to fluoropolymer base films; grafting of styrene from generated radicals; and sulfonation of grafted films. The resultant PEM possesses the fluoropolymer back bones and poly(styrene sulfonic acid) (PSSA) graft chains. In this method, one can widely control the degree of grafting (DOG) by various grafting conditions.

In this study, we investigated the micro-hole structures in the radiation-grafted PEMs by using PAL spectroscopy measurement. The target materials are the PEMs based on crosslinked-polytetrafluoroethylene (cPTFE) films developed by our research group [16]. Up to now, there have been several reports on PAL measurement of Nafion [17,18]. On the contrary, as far as the authors know, little is known about free-volume holes in the radiation-grafted PEMs. In this study, we clarify the size and location of microvoids in the cPTFE-based grafted PEMs.

2. Experimental

2.1. Synthesis of PEMs

The PEMs were synthesized as already reported [16]. We used, as a base material, a 42-μm thick film of PTFE crosslinked with an electron beam at 100 kGy. The film was pre-irradiated with 15-kGy γ-ray in an Ar atmosphere at room temperature and then immersed in a styrene monomer at 60 °C for 2-18 h. After the grafting reaction, the film was immersed in toluene to extract any excess styrene. The grafted film was then dried under vacuum at 50 °C to a constant weight. The degree of grafting (DOG) was estimated by:

$$\mathrm{DOG} = \frac{W_G - W_O}{W_O} \times 100 \qquad (1)$$

where W_G and W_O denote the weights of the grafted and base films, respectively.

For sulfonation, the grafted film was immersed in 0.2 M chrolosulfonic acid / 1,2-dichloroethane mixture at 50 °C for 6 h. Finally, the obtained PEM was rinsed with pure water and then dried in a vacuum oven.

2.2. PAL measurement

PAL measurements were carried out using conventional "fast-fast" coincidence system with a time resolution of 225 ps. A ^{22}Na positron source was prepared by depositing and drying aqueous solution of ^{22}NaCl on a Kapton thin film. This substrate (1 cm × 1 cm) was covered with another Kapton film with the same size. The edges were glued to seal activity inside.

The PEMs were measured in the dried and fully-hydrated states at 300 K. A stack of about 10 PEM samples was set on each side of the positron-source Kapton films to ensure positron annihilation only in the samples. All the PAL measurements were conducted at the room temperature. The obtained PAL spectra with about 10^6 counts were analyzed by PATFIT program. For comparison, the virgin cPTFE and styrene-grafted films were also measured in a similar way.

3. Results and discussion

3.1. Styrene-grafted films with different DOGs

The PAL spectra for all the samples were deconvoluted into four components with lifetimes, τ_i, and intensities, I_i. In three-component analysis, the accuracy of spectra fitting was not good. The detected four components were numbered 1, 2, 3, and 4 from the shortest to the longest lifetime. Components 1 and 2 are surely *p*-Ps and free-positron, respectively. Components 3 and 4 are attributed to two types of *o*-Ps in different regions.

Figure 1 and 2 shows the lifetimes (τ_3 and τ_4) and relative intensities (I_3 and I_4) of the two types of *o*-Ps in the styrene-grafted films. For the base cPTFE film (see the plots at DOG = 0), τ_3 and τ_4 were 2.3 and 4.5 ns, respectively. In order to considerate the location of *o*-Ps, we should take into account very high crystallinity of PTFE, i.e., PTFE matrices consist of both crystalline and amorphous phases. In the former, the polymer chains are regularly and densely packed with small intermolecular free-volume voids. On the other hand, in the latter, the polymer chain arrangement is random, thereby forming larger-sized free-volume spaces between chains. Accordingly, component 3 and 4 are likely *o*-Ps in the PTFE crystalline and amorphous regions, respectively. Also, in the previous report concerning PAL measurement on PTFE films (non crosslinked) [19], the shorter- and longer-lived *o*-Ps were assumed to be in crystalline and amorphous regions, respectively.

As the DOG increased, both τ_4 and I_4 exhibited declining trends, indicating the structural change in the amorphous regions. This result can be explained by considering the radiation grafting process as follows. The grafting reaction should initiate from polymer radicals generated by pre-irradiation, located in the crystalline / amorphous interfacial layers. The polystyrene graft chains would be elongated into amorphous regions, leading to a decrease of both the size and number of free-volume holes (low τ_4 and low I_4). The similar phenomenon was observed in the PAL study for the other polystyrene-grafted film based on poly(tetrafluoro-ethylene-*co*-hexafluoropropylene) [20].

On the other hand, I_3 was raised with a DOG (see Fig. 2). From this result, one may possibly suspect that PTFE crystallites became larger in the highly-grafted film. However, this is never true. In

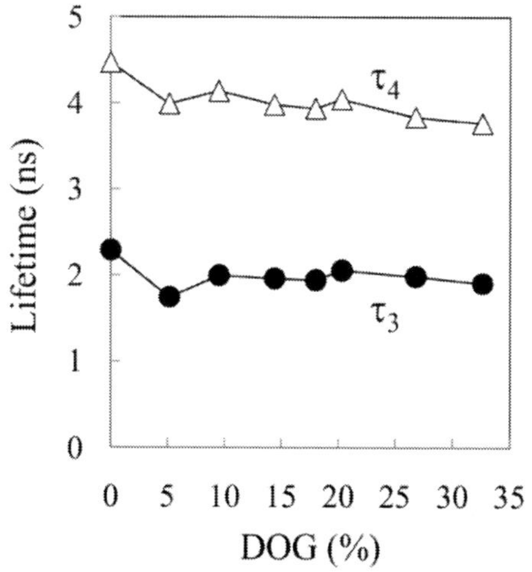

Figure 1. Lifetimes of (●) τ_3 and (△) τ_4 in the styrene-grafted cPTFE films. The plots at DOG = 0 are the results for the base cPTFE film.

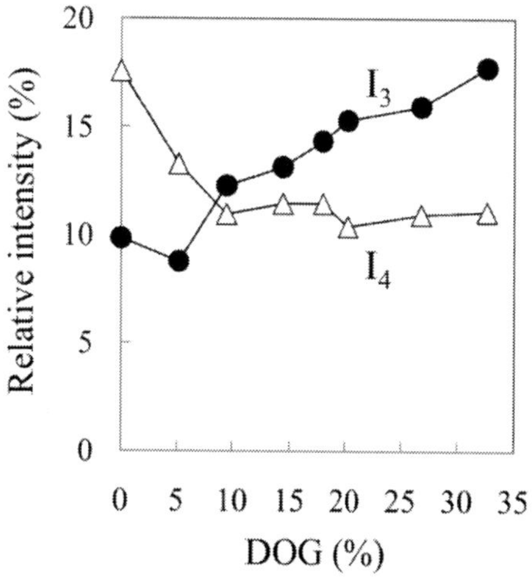

Figure 2. Relative intensity of (●) I_3 and (△) I_4 in the styrene-grafted cPTFE films. The plots at DOG = 0 are the results for the base cPTFE film.

general, the graft chain elongation leads to a disruption of polymer crystallites. Then, the relationship between I_3 and a DOG was rationalized as follows. According to the previous report [21], the *o*-Ps lifetime in polystyrene powders was about 1.9 ns, which is similar to τ_3 in the PTFE crystallites. Consequently, component 3 for styrene-grafted PTFE films is considered to be *o*-Ps not only in PTFE crystallites but also in grafted polystyrene phases. This is the reason why I_3 increased with a DOG.

3.2. Styrene-grafted films and PEMs

Figure 3 and 4 show the lifetimes and relative intensities of components 2, 3, and 4 for the grafted films and PEMs (DOG is 14% and 27%). The most significant change was observed in I_2 as shown in Fig. 3 (B) and Fig. 4 (B). Compared to grafted films, I_2 of the dried PEMs remarkably increased up to about 60%, while both I_3 and I_4 decreased. Similar observation was already reported in a previous study [22], in which the PAL measurements were performed on poly(ether ether ketone) (PEEK) films before and after sulfonation. The yield of *o*-Ps was found to be very low in sulfonated PEEK (sPEEK) compared with that in non-sulfonated one. Such a phenomenon is called inhibition of Ps formation, and is presumed to be caused by the following mechanism. Since the sulfonic acid group is an electron withdrawing component, the aromatic ring substituted by it becomes electron deficient and electron accepting. Some of the spur electrons likely tend to be captured by these electron acceptors (sulfonated aromatic rings), and hence cannot recombine with positrons to form Ps. Namely, there were not sufficient available electrons to react, which would suppress the Ps formation. In our cPTFE-based PEMs possessing the poly(styrene sulfonic acid) (PSSA) graft chains, the above-mentioned electron-scavenging process probably occurred, resulting in a decrease of both I_3 and I_4.

The τ_4 in the PEMs was higher than that in the styrene-grafted films at both DOGs. This τ_4 increase was related to the structural change in the PTFE amorphous phases, and can be accounted for as follows. In the grafted films (before sulfonation), the polystyrene graft chains should be somewhat entangled with amorphous PTFE chains. On the contrary, in the PEMs obtained by sulfonation, the repulsive interactions likely acted between the PSSA graft and PTFE chains because their chemical properties were quite different. In other words, they are completely immiscible each other (The PSSA is hydrophilic, while the PTFE is hydrophobic.). Then, the PSSA grafts may be phase-separated from the PTFE chains. Consequently, the PTFE amorphous phases, from which the PSSA chains were excluded out, should be more vacant and had more free-volume spaces (higher τ_4).

3.3. Dried and fully-hydrated PEMs

As already mentioned, the *o*-Ps lifetime has a strong correlation with the size of free-volume holes. The annihilation of *o*-Ps in the spherical holes is described by simple quantum mechanical model with infinite height potential. The semi-empirical relationship between the *o*-Ps lifetime, τ (ns), and radius of free-volume holes, R (nm), was expressed by the following Tao-Eldrup equation:

$$\tau = \frac{1}{2}\left[1 - \frac{R}{R+\Delta R} + \frac{1}{2\pi}\sin\left(\frac{2\pi R}{R+\Delta R}\right)\right]^{-1}, \qquad (2)$$

where ΔR denotes a fitted empirical electron-layer thickness (=0.166 nm). The hole volume, V (nm^3), is calculated by:

$$V = \frac{4}{3}\pi R^3. \qquad (3)$$

Table 1 and 2 show the R values calculated by equation (2), as well as the measured *o*-Ps lifetimes. The V calculated by substituting the R into equation (3) is also shown in Table 1 and 2. The V in the dried PEMs with different DOGs were almost the same each other. The V corresponding to τ_3 and τ_4 were estimated to be 0.11 nm^3 and 0.36-0.38 nm^3, respectively. The smaller holes should be in both PTFE crystallites and PSSA grafts, while the larger holes are likely in the amorphous PTFE phases. According to the previous report [23], the self-diffusion coefficient, D, of various gases including oxygen was found to be proportional to the inverse of V. From this relationship of D versus 1/V, the D of oxygen in the cPTFE PEMs was roughly evaluated to be on the order of 10^{-7} cm^2/s in the smaller holes and 10^{-4} cm^2/s in larger holes.

A final topic in this paper is the effect of hydration level on the micro-hole structures. Notice that τ_3 was a little smaller in the fully-hydrated states than in dried ones (see Fig. 3(A) and 4(A)). The calculated V values in the PEMs with DOGs of 14 and 27% were 0.10 and 0.092 nm^3, respectively (see Table 1 and 2). Although the above τ_3 decrease is within an experimental error, it implies the meaningful structural change in PEMs. When the PEMs were hydrated, PSSA graft chains should alter their conformations because absorbed water gathered around the sulfonic acid groups, which possibly made micro holes smaller. Sodaye et al. also reported the similar findings in Nafion [18]: As the water content increased, the size of free-volume holes decreased. In the near future, we are planning to

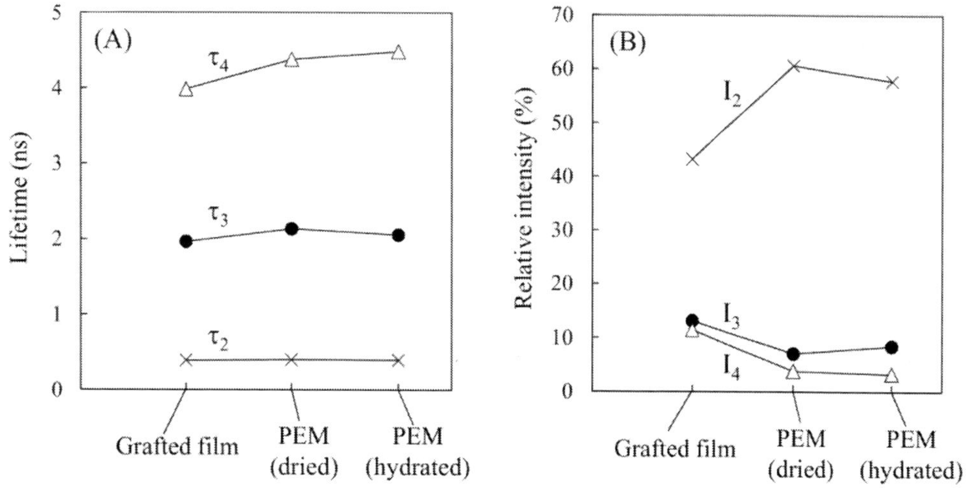

Figure 3 (A) Lifetimes and (B) relative intensities of component 1, 2, and 3 in the grafted film and PEM with a DOG of 14%.

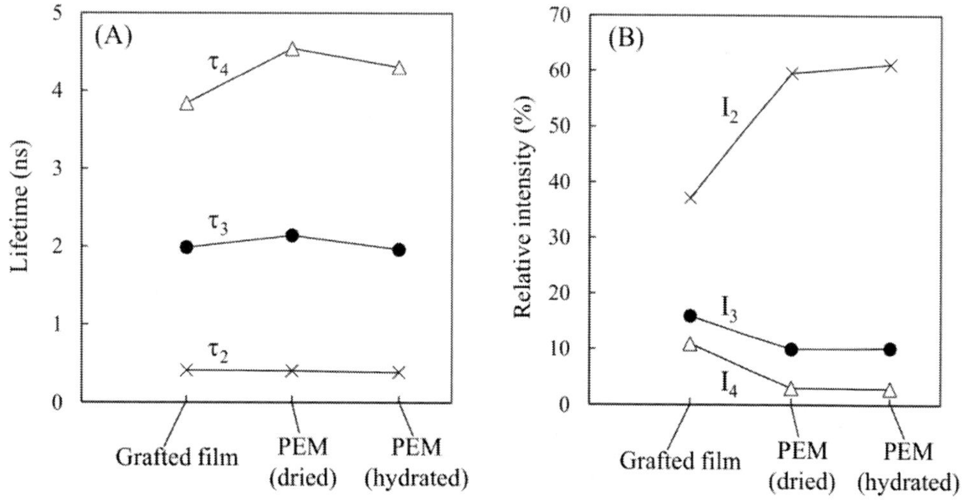

Figure 4 (A) Lifetimes and (B) relative intensities of component 1, 2, and 3 in the grafted film and PEM with a DOG of 27%.

perform the PAL measurements on PEMs under the controlled temperature and relative humidity. In addition, the permeability of hydrogen and oxygen through PEMs will be measured.

Table 1 Lifetimes of *o*-Ps, R, and V in the PEMs with a DOG of 14%

	Component 3		Component 4	
	dried	hydrated	dried	hydrated
τ_3 (ns)	2.1	2.1	4.4	4.5
R (nm)	0.30	0.29	0.44	0.45
V (nm^3)	0.11	0.10	0.36	0.38

Table 2 Lifetimes of *o*-Ps, R, and V in the PEMs with a DOG of 27%

	Component 3		Component 4	
	dried	hydrated	dried	hydrated
τ_3 (ns)	2.1	2.0	4.5	4.3
R (nm)	0.30	0.28	0.45	0.44
V (nm^3)	0.11	0.092	0.38	0.36

References

[1] Smitha B, Sridhar S and Khan A A 2005 J. Membr. Sci. **259** 10
[2] Dobrovol' skii Y A, Volkov E V, Pisareva A V, Fedotov Y A, Likhachev D Y and Rusanov A L 2007 Russ. J. Gen. Chem. **77** 766
[3] Mauritz K A and Moore R B 2004 Chem. Rev. **104** 4535
[4] Büchi F N, Gupta B, Haas O and Scherer G G 1995 J. Electrochem. Soc. **142** 3044
[5] Büchi F N, Gupta B, Haas O and Scherer G G 1995 Electrochim. Acta **40** 345
[6] Wang H and Capuano G A 1998 J. Electrochem. Soc. **145** 780
[7] Vie P, Paronen M, Strømgård M, Rauhala E and Sundholm F 2002 J. Membr. Sci. **204** 295
[8] Schmidt T J, Simbeck K and Scherer G G 2005 J. Electrochem. Soc. **152** A93
[9] Inaba M, Kinumoto T, Kiriake M, Umebayashi R, Tasaka A and Ogumi Z 2006 Electrochim. Acta **51** 5746
[10] Haraya K and Hwang S T 1992 J. Membr. Sci. **71** 13
[11] Tanaka K, Katsube M, Okamoto K, Kita H, Sueoka O and Ito Y 1992 Bull. Chem. Soc. Jpn. **65** 1891
[12] Budd P M, Mckeown N B and Fritsch D 2005 J. Mater. Chem. **15** 1977
[13] Fu Y J, Hu C C, Lee K R, Tsai H A, Ruaan R C and Lai J Y 2007 Euro. Polym. J. **43** 959
[14] Nasef M M and Hegazy E S A 2004 Prog. Polym. Sci. **29** 499
[15] Gubler L, Gürsel S A and Scherer G G 2005 Fuel Cells **5** 317
[16] Yamaki T, Kobayashi K, Asano M, Kubota H and Yoshida M 2004 Polymer **45** 6569
[17] Sodaye H S, Pujari P K, Goswami A and Manohar S B 1997 J. Polym. Sci. B: Polym. Phys. **35** 771
[18] Sodaye H S, Pujari P K, Goswami A and Manohar S B 1998 J. Polym. Sci. B: Polym. Phys. **36** 983
[19] Mohamed H F M, Abdel-Hady E E and Mohamed S S 2007 Radiat. Phys. Chem. **76** 160
[20] Sudarshan K, Rath S K, Patri M, Sachdeva A and Pujari P K 2007 Polymer **48** 6434
[21] Honda Y, Shimada T, Tashiro M, Kimura N, Yoshida Y, Isoyama G and Tagawa S 2007 Radiat. Phys. Chem. **76** 169
[22] Kobayashi Y, Mohamed H F M and Ohira A 2009 J. Phys. Chem. B **113** 5698
[23] Kobayashi Y, Haraya K, Hattori S and Sasuga T 1994 Polymer **35** 925

Diffusion of positrons in polymers studied by slow positron beam

H Suzuki[1], K Azami[1], S Komagata[1], I Kanazawa[1], R Suzuki[2], T Ohdaira[2]

[1]Department of physics, Tokyo Gakugei University, Nukuikita-machi 4-1-1, Koganei-shi, Tokyo, 183-8501, Japan

[2]National Institute of Advanced Industrial Science and Technology, 1-1-1, Umezono, Tsukuba, Ibaraki, 305-8568, Japan

E-mail: m091807x@u-gakugei.ac.jp

Abstract. Using a pulsed slow positron beam, we have measured the positron lifetimes in the systems of thin Ni films on polyethylene and polycarbonate substrates made by the molecular beam epitaxial method. We have estimated the one-dimensional diffusion lengths of positrons in polyethylene and polycarbonate.

1. Introduction

The variable-energy slow-positron beams have enabled studies of atomic vacancy-type defects at surfaces, in sub-surfaces, and in thin films [1]. The positron lifetime method with a slow-positron beam provides important information on species of vacancy defects contained in thin films [2]. The positron mobility is an important quantity for understanding the positron trapping into defects and the positron reemission from surfaces observed using slow positron beams. In the case of polymers, the positron beam allows the determination of local free-volume properties of the polymers as a function of the depth from the surface [3]. Considering the positroniums (Ps) formation, the positron mobility in polymers is especially important [4]. Since various positron spur processes relevant to the Ps formation are strongly influenced by the diffusion of positrons, its mobility plays a crucial role in determining the Ps yields in polymers.

Hirata *et al.* have reported that the diffusion lengths of Ps in polymers are very small ($D_{Ps} = 2.6 \sim 5.1 \times 10^{-6}$ cm^2/s) [5]. Diffusion mobility of free positrons in polyethylene is estimated to be $\mu = 10$cm^2/V $\cdot$ s [6], 10.3cm^2/V $\cdot$ s [7], 27.7cm^2/V $\cdot$ s [4] and 32cm^2/V $\cdot$ s [8]. The diffusion constant is given by the Einstein's relation,

$$D = \mu k_B T/e, \tag{1}$$

where k_B is the Boltzmann constant, T is temperature and e is the elemental charge. The diffusion length is then given by

$$L = \sqrt{6D\tau}. \tag{2}$$

Having the free positron lifetime of $\tau = 0.52$ nsec for polyethylene and $\mu = 32$cm^2/V $\cdot$ s, we obtained the one-dimensional diffusion length $L \equiv \sqrt{D\tau} \approx 210$ nm at room temperature.

In this study, we have determined the one dimensional diffusion lengths of positrons in polyethylene and polycarbonate based on the slow positron lifetime measurements. To avoid complicated effects (like

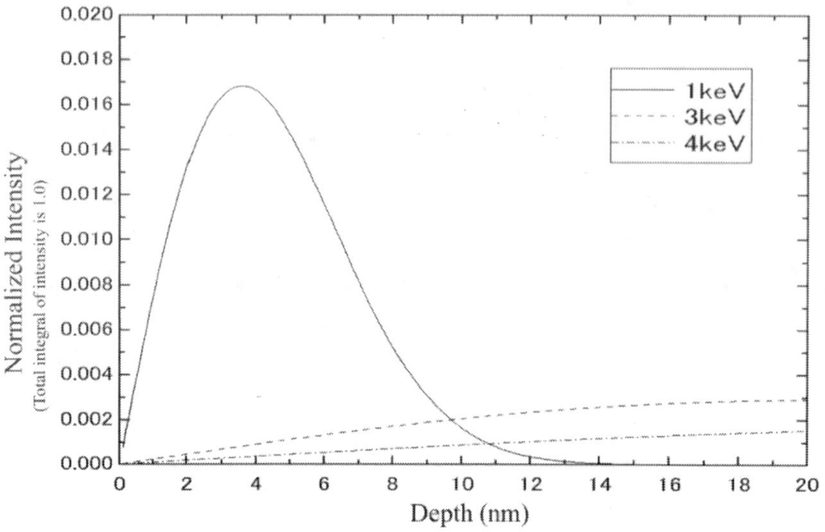

Figure 1. Positron implantation profile in Ni.

Ps states at the polymer surface), we deposited the thin Ni films on the polyethylene and polycarbonate. We compare the obtained one-dimensional diffusion lengths with those determined from the above Einstein's relation.

2. Experiment

Thin Ni films of thickness 25±5 nm and 75±5 nm were deposited on the polyethylene substrates (15mm × 15mm × 1mm, density 0.92g/cm^3 Japan Custom Co. , Ltd.) by the molecular beam epitaxial (MBE) method under a pressure of 8.5×10^{-9}Torr at room temperature. Similarly, a thin Ni film of thickness 125±5 nm was deposited on the polycarbonate substrate (15mm×15mm×1mm, density 1.20 g/cm^3, Japan custom Co. , Ltd.). The positron annihilation lifetime measurements were carried out at room temperature using an intense pulsed slow positron beam generated by the electron linac of the National Institute of Advanced Industrial Science and Technology (AIST) LINAC facility.

3. Results and Discussions

First of all, we introduce the general principle to determine the one dimensional diffusion length of positrons. The positron implantation profile is given by

$$P(z, E) = -(d/dz)\{exp[-(z/z_0)^m]\}, \tag{3}$$

where z_0 is given by

$$z_0 = AE^m/\rho/\Gamma(1 + 1/m), \tag{4}$$

where E is the incident positron energy in keV, ρ is the materials density in g/cm^3 (Ni : 8.85, polyethylene : 0.92, polycarbonate : 1.20), A, n and m are the constants: $A = 400$, $n = 1.6$ and $m = 2.0$ [10]. Figure 1 shows the positron implantation profile in Ni for E=1, 3 and 4 keV calculated by eq. (3). The fraction of positrons stopped in the i-th layer is given by

$$\eta_i(E) = \int_{a_i}^{b_i} P(z, E)dz, \tag{5}$$

Table 1. Positron lifetimes and their intensities obtained for the polyethylene substrate.

τ_1 (ns)	τ_2 (ns)	τ_3 (ns)	I_1 (%)	I_2 (%)	I_3 (%)
0.295	0.520	2.392	61.7	15.1	23.2

Table 2. Positron lifetimes and their intensities obtained for the system of the thin Ni film (25±5nm thick) on a polyethylene substrate at E=1 keV, 3 keV and 4 keV.

E (keV)	τ_1 (ns)	τ_2 (ns)	τ_3 (ns)	I_1 (%)	I_2 (%)	I_3 (%)
1	0.347	—	3.985	96.0	—	4.0
3	0.317	0.520	2.324	71.5	11.4	17.1
4	0.311	0.520	2.286	66.0	13.5	20.5

where a_i and b_i are the boundaries of the i-th layer. The forward and backward reemission yields, YF_i, YB_i, can be derived immediately from the Green's function

$$YF_i = [1/sinh(T_i/L_i)] \int_{a_i}^{b_i} sinh(z/L_i) P(z, E) dz, \tag{6}$$

$$YB_i = [1/sinh(T_i/L_i)] \int_{a_i}^{b_i} sinh(T_i - z/L_i) P(z, E) dz, \tag{7}$$

where L_i is the one-dimensional diffusion length of positrons in the i-th layer. T_i is the thickness of the i-th layer. The fraction of positrons (F_i) that annihilate in the i-th layer is given by

$$F_i = \eta_i - YF_i - YB_i + YF_{i-1} + YB_{i+1}. \tag{8}$$

Having the annihilation fraction determined from the measurements and the stopping fraction calculated by eq. (5) in the i-th layer, one can determine the one-dimensioned diffusion length, L_i, in the i-th layer using eqs.(6) through (8).

Table 1 summarizes the positron lifetimes and their intensities obtained for the polyethylene substrate [9]. The first component correspond to the mixture of annihilations of para-positronium and free positrons. The second component and third component are attributed to the annihilation of free positrons and the pick-off annihilation of ortho-positronium, respectively.

Table 2 lists the positron lifetimes and their intensities obtained for the system of the thin Ni film (25±5nm thick) on a polyethylene substrate at E=1, 3 and 4 keV. In the case of 1keV, the observed lifetime spectrum is well fitted with two lifetime components. Most positrons annihilate within the Ni-film as shown in from Fig. 1. The lifetimes τ_1 and τ_2 represent small vacancy clusters in the Ni layer and the pick-off annihilation of ortho-positronium at the Ni surface, respectively. For E=3 keV and 4keV, the observed lifetime spectra are well fitted with three lifetime components. The lifetime τ_1 corresponds to the mixture of small vacancy clusters in the Ni film and free-positron annihilation in polyethylene. The values of τ_2 and τ_3 correspond to the annihilation of free positrons in polyethylene and the pick-off annihilation of orho-positronium in polyethylene. The ratio of I_3/I_2 are 1.51 and 1.52 for E=3 keV and 4 keV, respectively. These results are consistent with those listed in table 1 [9]. Thus, from the comparision between tables 1 and 2, the annihilation fraction in the polyethylene substrate is estimated as $I_2/15.1$(=I_2 in table 1).

Table 3. Annihilation and stopping fractions in the Ni layer and the polyethylene substrate estimated for the system of the thin Ni film (25±5nm thick) on the polyethylene substrate at E=3 keV and 4 keV.

E (keV)	Ni		Polyethylene	
	F (%)	η (%)	F (%)	η (%)
3	24.7	36.5	75.3	63.5
4	10.7	16.6	89.3	83.4

Table 3 summarizes the annihilation and stopping fractions (F and η in eq. (8)) in the Ni layer and the polyethylene substrate estimated for the system of the thin Ni film (25±5nm thick) on the polyethylene substrate at E=3 keV and 4 keV. We integrated the positron stopping profile within polyethylene, taking into account of only the diffusion length Ni (near surface 6nm) [1]. Following the manner explained at the beggining of this section, we obtained the one-dimensional diffusion length as L ∼10±2 nm in polyethyene.

We have measured the positron lifetimes for the system of a Ni film (75±5nm thick) on a polyethylene substrate ate E=1, 2, 4, 6 and 8 keV. Subsequently, we obtained the one-dimensional diffusion length in the polyethylene to be L ∼10±2 nm, which is comparable to that obtained for the case of the 25±5 nm thick Ni on a polyethylene. We have further measured the positron lifetimes for the system of a thin Ni film (125±5 nm thick) on a polycarbonate substrate at E=1, 2, 6 and 8 keV. The one-dimensional diffusion length in polycarbonate was obtained to be L ∼32±2 nm.

The one-dimensional diffusion lengths obtained here (10∼30 nm) for the polyethylene and polycarbonate are much shorter as compared to the value ∼200 nm derived from the mobilities of positrons and Einstein's relation as mentioned in the introduction. The positron in polymers pushes on polymer molecules to make space for itself. This effect reduces the positron kinetic energy. Large space in polymer of sphere, however, is costly in terms of polymer free energy, so the polymer molecules push back. The balance between the positron kinetic energy and the polymer compressibility, which is relatively low in comparison with other solid materials, results in the stable structure. In the case of the positron diffusion in this state, it might not be adequate to use the Einstein's relation.

4. Conclusion

We have determined the one-dimensioned diffusion lengths in the polyethylene and polycarbonate based on the positron beam technique. The obtained diffusion lengths are one order of magnitude shorter than those reported previously using the Einstein's relation.

References

[1] Oshima N, Nakajyo T, Kanazawa I, Iwashita T, Ito Y, Soe We-Hyo and Yamamoto R 1995 *Appl. Surf. Sci.* **85** 329
[2] Takagi K, Furukawa N, Kanazawa I, Suzuki R and Ohdaira T 2002 *Surt. Sci.* **514** 298
[3] Tashiro M, Honda Y, Terashima Y, Watanabe M, Pujari P K and Tagawa S 1997 *Appl. Surf. Sci.* **116** 251
[4] Mills A P, Gullikson E M, Pfeiffer L and Rockward W S 1986 *Phys. Rev.* B**33** 7799
[5] Hirata K, Kobayashi Y and Ujihara Y 1996 *J. Chem Soc. Faraday Trans.* **92** 985
[6] Brandt W and Mourino M 1979 *Bull. Am. Phys. Soc.* **24** 72
[7] Mackenzie I K and Ghorayshi P Z 1985 *Solid State Commun.* **55** 125
[8] Zheng W, Kobayashi Y, Hirata K and Suzuki T 1998 *Radiat. Phys. Chem.* **51** 269
[9] Azami K, Habu M, Takagi K, Kanazawa I, Suzuki R and Ohdaira T 2004 *Mater. Sci. Forum* **445-446** 42
[10] Valkealahti S and Nieminen R M 1983 *Appl. Phys.* A**32** 95

International Symposium on Advanced Science Research 2009 IOP Publishing
Journal of Physics: Conference Series **225** (2010) 012059 doi:10.1088/1742-6596/225/1/012059

A study of vacancy-type defects in amorphous and crystalline FeBSi alloys after He ion irradiation

Q. Xu[1], X. Z. Cao[1], K. Sato[1], T. Iwai[2] and T. Yoshiie[1]
[1]Research Reactor Institute, Kyoto University, Osaka 590-0494, Japan

[2]Research Center for Nuclear Science and Technology,
The University of Tokyo, Ibaraki 391-1106, Japan

E-mail: xu@rri.kyoto-u.ac.jp

Abstract. In order investigate the structural dependence of defect formation during irradiation, amorphous $Fe_{79}B_{16}Si_5$ alloy and crystalline FeBSi alloy were irradiated by 300 keV helium ions at room temperature. Positron annihilation lifetimes observed for the unirradiated amorphous and the crystalline FeBSi alloys were nearly equal, i.e., 124.4± 0.3 ps and 124.6± 0.2 ps, respectively. The S parameters of the Doppler-broadening of annihilation radiation measured using a positron beam were also similar for the unirradited amorphous and crystalline FeBSi alloys. It means that the positron annihilation at the free volume in amorphous $Fe_{79}B_{16}Si_5$ alloy is not prominent. The Doppler-broadening S parameter increased after the helium irradiation. This phenomenon was more pronounced in the amorphous FeBSi alloy. It suggests that the atoms are more easily displaced in amorphous alloy than in crystalline alloy during irradiation.

1. Introduction

With advances in nuclear power technology, in particular fusion reactor technology, interest in the behavior of He in solids has increased. In a fusion reactor, helium is introduced into materials at a high rate by direct implantation, but also indirectly, such as by the nuclear transmutation reaction of (n, α). Helium, which has a strong interaction with vacancies [1, 2], forms a high density of He bubbles. In a fusion reactor, helium and He bubbles lead to surface damage such as erosion, sputtering, and blistering of the plasma-facing materials (PFMs). The impurities induced by erosion of the PFMs are a key factor in the radiative power loss of the plasma. Therefore, it is important to find materials with low helium retention and/or better resistance to He bubble formation. The configuration of atoms is one factor influencing the retention of helium in solids, and amorphous metals have a very different atomic configuration from crystalline metals. Point defects and defect clusters are produced by irradiation with high energy particles in crystalline metals. Point defects and defect clusters, especially vacancies and voids, act as helium trapping sites. Although point defects and defect clusters do not exist in amorphous metals, helium atoms are also trapped during He irradiation. It would be interesting to clarify the helium trapping sites in amorphous metals.

The positron annihilation technique is a very powerful tool in the study of fundamental microstructural features, such as small vacancy-type defects, of localized sites in condensed matter [3], because the positron annihilation rate is proportional to the electron density around the annihilation site. Of course, energy and momentum are conserved in the positron annihilation process. In the dominant decay mode of a thermal positron and electron, two gamma rays are emitted. The energies of the two photons emitted by the annihilation of a positron and an electron are different. This difference in photon energy is proportional to the longitudinal component of the electron-positron momentum in the direction of gamma emission. Measurement of the photon energies reveals information regarding the momentum distribution of valence and core electrons. In the present study, defect formation and the interaction between defects and helium under He irradiation were investigated by positron annihilation analysis.

2. Experimental procedure

The amorphous alloy $Fe_{79}B_{16}Si_5$ (Metglas 2605S-3, The Nilaco Corporation (Japan)) was obtained in 25-μm-thick sheets. The crystallization temperature (T_C) was ~788 K. Crystalline FeBSi alloy specimens were prepared in a vacuum furnace by rapidly heating amorphous ribbons to 923 K and holding for 0.5 hours. Amorphous and crystalline samples were irradiated by 300 keV He^+ ions from a Van de Graaff accelerator at the University of Tokyo to 1×10^{20} He/m^2 at room temperature. The damage induced by the ions in the matrix was not uniform, and the damage peak was about 800 nm from the irradiation surface, according to calculations using the TRIM code [4], as shown in Fig. 1. In

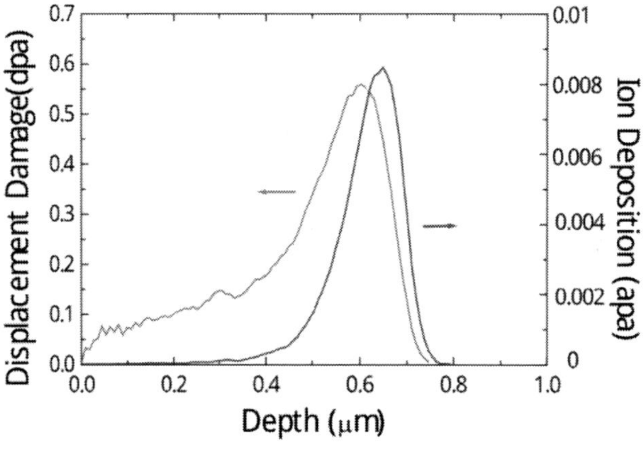

Fig. 1 Depth distribution of damage and ion deposition calculated using the TRIM code for FeBSi alloy irradiated with He ions to 1×10^{20} He/m^2.

order to investigate the irradiation depth dependence of microstructural evolution, the Doppler broadening of annihilation radiation measurements were performed using a mono-energetic positron beam apparatus combined with the Van de Graaff accelerator. Details of the apparatus used have been described previously [5]. The implantation profile of mono-energetic positrons into a material may be given by the following equation [6, 7]:

$$P(x,E) = -\frac{d}{dx}\exp(-x^m / x_0^m) \qquad (1)$$

Here, $x_0=(\alpha/\rho)E^n$, ρ is the material density in g/cm^3, E is the incident positron energy in keV, and x is the distance from the surface in cm. In the calculation of the positron implantation profile, we used the Makhovian parameters of α=4.0 μg/cm^2/keV$^{-1.6}$, m=2, and n=1.62 determined by Vehanen et al [7]. Figure 2 shows implantation profiles of positrons with typical energies of 5, 10, 15, 20, 25, and 30 keV. The implantation profile of positrons is broadened gradually with increasing beam energy. At E=20 keV, most positrons are stopped within the damaged region formed by 300 keV He^+ ions.

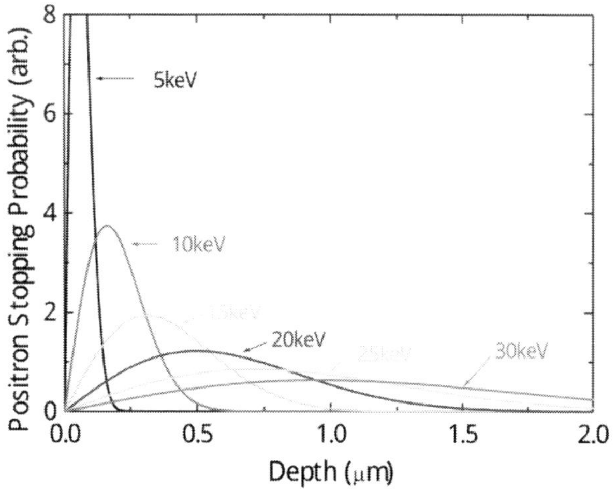

Fig. 2 Depth profile of position stopping probability for energies of 5, 10, 15, 20, 25 and 30 keV.

3. Results and discussion

The lifetime spectra were hardly decomposed into two components for the unirradiated amorphous or the crystalline FeBSi samples. Their mean lifetimes are 124.4± 0.3 and 124.6± 0.2 ps, respectively. Although free volume is introduced into amorphous alloy by rapid quench, the positron annihilation at the free volume is not prominent.

Figure 3 shows the dependence of the S parameter for the amorphous and the crystalline FeBSi samples before and after the He irradiation as a incident positron beam energy. In both the unirradiated amorphous and the crystalline FeBSi samples, the S parameter decreased with increasing positron energy up to 5 keV, and then became constant. The high S parameter at low incident positron energies comes from the positron diffusion to the surface and consequent production of ortho-positronium [8, 9]. Huomo et al. indicated that positrons with energy lower than 4 keV diffuse back to the surface in Mo [10]. The fraction of positrons diffusing back to the surface decreases at high incident positron energies. Before the He irradiation, the S parameters for the amorphous and the crystalline FeBSi samples are nearly the same when the positron energy is higher than 8 keV. This is consistent with the above-mentioned positron lifetime measurements. Thus, there are no important differences in the positron annihilation mechanisms in the amorphous and the crystalline FeBSi samples. After the He irradiation, on the other hand, the S parameters of both the amorphous and the crystalline FeBSi samples increase as the incident positron energy increases from 0 to 2 keV, and then decrease to constants with increasing the incident positron energy. The high S parameters at around E=2 keV are probably due to defects produced near the surface, which trap positrons to suppress the fraction of positrons diffusing back to the surface. The diffusion of positrons back to the surface is negligible above E=5 keV. Thus, the change of S parameter above E=5 keV reflects some structural change in the bulk. After the He irradiation, the S parameters of the crystalline FeBSi sample increases. This indicates the introduction of vacancy-type defects due to atomic displacement in the FeBSi matrix. The S parameter of the amorphous FeBSi sample further increases. Thus, vacancy-type defects that can trap positrons are also produced in the amorphous FeBSi structure [11]. Moreover, vacancy-type defects are more easily produced in amorphous FeBSi than in crystalline FeBSi by He irradiation. This may indicate that comparing the crystal phase in the amorphous phase the displacement energy is lower possibly due to the weaker binding energy between atoms.

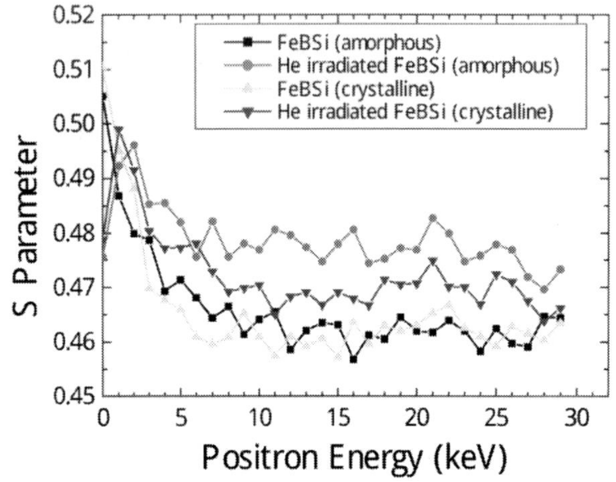

Fig. 3 Dependence of the S parameter on incident positron energy for the amorphous and crystalline FeBSi samples before and after the He irradiation.

4. Conclusion

Amorphous and crystalline FeBSi alloys irradiated with 300 keV He ion were studied by positron annihilation spectroscopy. Vacancy-type defects acting as positron trapping centers are formed in both amorphous and crystalline FeBSi alloys by He irradiation, while the production of vacancy-type defects is enhanced in the amorphous alloy.

References

[1] Klaver P., Haddeman E., Thijsse B. 1999 *Nucl. Instrum. and Meth. B* **153** 228-235.
[2] Wilson W.D., Johnson R.A. 1972 in: Gehlen P.C., Beeler J.R., Jaffee R.I. (Eds.), Interactomic Potentials and Simulation of Lattice Defects, Plenum, New York, p. 375-387.
[3] Dupasquier A., Mills A.P. 1995 Positron Spectroscopy of Solids (IOS Press, Amsterdam)
[4] Biersack J.P., Haggmark L.G. 1980 *Nucl. Instrum. and Meth.* **174** 257-269.
[5] Iwai T., Ito Y., Koshimizu M. 2004 *J. Nucl. Mater.* **329-333** 963-966.
[6] Makhov A.F. 1960 *Soviet Phys. Solid State* **2** 1942.
[7] Vehanen A., Saarinen K., Hautojarvi P., Huomo H. 1987 *Phys. Rev.* B **35** 4606-4610.
[8] Makinen J. Vehanen A., Hautojarvi P. Huomo H., Lahtinen J., Nieminen R.M. 1986 *Surf. Sci.* **175** 385-414.
[9] Brotton D.T., Rice-Evans P.C., Evans J.H. 1988 *Philos. Mag. Lett.* **57** 165-169.
[10] Huomo H., Vehanen A., Bentzon M.D., Hautojarvi P. 1987 *Phys. Rev.* B **35** 8252-8255.
[11] Xu Q., Cao X.Z., Sato K., Yoshiie T. *Philos. Mag. Lett.* **90**, DOI: 10.1080/09500830903485536.

Positron annihilation coincidence Doppler broadening measurements of irradiation-induced precipitation in Ni-Sn alloys

Koichi Sato, Qiu Xu and Toshimasa Yoshiie
Research Reactor Institute, Kyoto University, Kumatori-cho, Sennan-gun, Osaka 590-0494, Japan

E-mail: ksato@rri.kyoto-u.ac.jp

Abstract. Irradiation-induced precipitation in Ni-Sn binary alloys was examined by positron annihilation coincidence Doppler broadening (CDB) and lifetime measurements, and the correlation between precipitation and void growth was studied. *S-W* plots obtained from CDB measurements of pure Ni and Ni-Sn alloys were compared. The change in *S-W* plots in Ni-0.05at.%Sn with increasing neutron dose was nearly the same as that of pure Ni, while that of Ni-0.3at.%Sn and Ni-2at.%Sn deviated from that of pure Ni. Precipitation of Ni_3Sn was detected after annealing of Ni-2at.%Sn, and the precipitation was promoted by irradiation. In Ni-0.3at.%Sn, no precipitation was detected after an 1173 K heat treatment, but irradiation-induced precipitation did occur. The precipitation and the frequency of one-dimensional motion of interstitial clusters were closely correlated, and the clusters and vacancies annihilated each other. The precipitation suppressed void growth.

1. Introduction

Understanding the effect of solute atoms on microstructural evolution is important for the development of fusion reactor materials. Yoshiie et al. have studied damage structures in neutron-irradiated pure Ni and its binary alloys (Ni–Si, Ni–Cu, Ni–Ge, and Ni–Sn) [1–6]. The volume size factors of Si, Cu, Ge, and Sn in Ni are −5.81%, +7.18%, +14.76%, and +74.08%, respectively [7]. Well-developed dislocation networks and voids were detected in pure Ni, Ni–2at.%Cu, and Ni–2at.%Ge by transmission electron microscopy (TEM) after a dose of 0.026 dpa at 573 K , while no voids were detected and a small number of interstitial-type dislocation loops were observed in Ni–2at.%Si and Ni–2at.%Sn for the same irradiation conditions [2,4–6]. The same trend in void growth was also detected by positron annihilation lifetime (PAL) measurements [5,8].

Yoshiie et al. also reported that solute atoms of Si and Sn suppressed the one dimensional (1-D) motion of interstitial clusters, and that this was one of the main reasons for the absence of void growth [6]. They suggested that the volume size factor influenced the 1-D motion frequency. However, Sato et al. reported that the migration energy of seven interstitial clusters (I_7) required to overcome a large oversized solute atom was about 0.9 eV, and that isolated solute atoms could not suppress the 1-D motion of interstitial clusters when irradiation was performed at 573 K [9]. Segregation or precipitation is required to suppress the 1-D motion of interstitial clusters. According to the Ni-Sn phase diagram [10], 2 at.% Sn is within the solubility limit above 873 K. Below this temperature, the precipitation of Ni_3Sn should occur. In this study, irradiation-induced precipitation in Ni–Sn binary alloys was examined by positron annihilation coincidence Doppler broadening (CDB) and lifetime measurements, and the effect of precipitation on void growth is discussed.

2. Experimental procedure

The irradiated specimens were 99.99% pure Ni and Ni–Sn binary alloys with 99.99% pure Sn. The solute atom concentrations were 0.05, 0.3, and 2 at.%. In order to examine the properties of Ni_3Sn, Ni–13at.%Sn was also prepared to form a large amount of Ni_3Sn. These alloys were produced by plasma jet melting. The samples were cold rolled to a thickness of 0.1 mm, and 3-mm-diameter discs were punched out. These discs and Sn discs were annealed separately at 1173 K and 433 K, respectively, for 1 h in a vacuum of less than 10^{-4} Pa. The average grain size of the pure Ni after annealing was 30 μm. Neutron irradiation was performed at the Materials Controlled Irradiation Facility at the Kyoto University Reactor (KUR) [11] and at the Japan Materials Testing Reactor (JMTR) at the Japan Atomic Energy Agency. The neutron doses were 6×10^{-5} – 1×10^{-2} dpa at KUR and 6×10^{-3} – 0.2 dpa at JMTR. The irradiation temperatures were 573 K at KUR and 563 K at JMTR. The irradiation dose was obtained using the SPECTER code [12] with a displacement threshold energy of 24 eV. After neutron irradiation, positron annihilation CDB measurements were performed at room temperature (RT). Positron lifetime spectra were also measured using a fast digital oscilloscope and BaF_2 scintillators [13,14] at RT. The time resolution of the system was about 170 ps (full width at half maximum). Each spectrum was accumulated to a total of approximately 4×10^6 coincidence events. Lifetime spectra were analyzed using the PALSfit fitting program [15].

3. Results and discussion

Table 1 shows the positron annihilation lifetimes of unirradiated specimens used in this study. The lifetime spectra of Ni–2at.%Sn and Ni–13at.%Sn were decomposed into two components, indicating that they contained vacancy-type defects. The defects most likely arose from interface misfit between the solid solution phase and the Ni_3Sn phase, which formed during cooling after annealing (about 30 minutes from 1173 K to RT) according to the phase diagram.

Table 1. Positron annihilation lifetime of unirradiated specimens used in this study. τ_{av}, τ_1, and τ_2 denote mean lifetime, and the first and the second components in the two-component analysis, respectively. I_2 denotes the intensity of τ_2.

	τ_{av} (ps)	τ_1 (ps)	τ_2 (ps)	I_2 (%)
Pure Ni	102±1	–	–	–
Pure Sn	203±1	–	–	–
Ni–0.05at.%Sn	103±1	–	–	–
Ni–0.3at.%Sn	103±1	–	–	–
Ni–2at.%Sn	111±1	88±3	150±7	32±6
Ni–13at.%Sn	146±1	82±4	169±2	68±3

Table 2 shows the results of PAL measurements of neutron-irradiated Ni and Ni–Sn alloys. Vacancy clusters grew in pure Ni and in Ni–0.05at.%Sn. The size of the vacancy clusters in Ni–Sn alloys decreased with increasing solute atom concentration at irradiation doses of 0.1 dpa. This clearly indicated the void growth suppression effect of Sn.

Figure 1 shows typical CDB ratio curves comparing unirradiated specimens to unirradiated pure Ni. The parameters *S* and *W* were respectively defined as the ratios of the low-momentum ($|P_L| < 5\times10^{-3}$ mc) and high-momentum (20×10^{-3} mc < $|P_L|$ < 30×10^{-3} mc) areas in the Doppler-broadening spectrum to the total area, where m is the electron rest mass and c is the velocity of light. *S* represents the smaller Doppler shift resulting from the annihilation of valence electrons, and *W* represents the Doppler shift caused by the annihilation of core electrons. Interface misfit caused the increase in the *S* region of Ni–2at.%Sn and Ni–13at.%Sn.

Figure 2 shows *S*–*W* plots for (a) unirradiated specimens, (b) Ni–0.05at.%Sn, (c) Ni–0.3at.%Sn, and (d) Ni–2at.%Sn. The solid and dotted lines in Figure 3 (b)–(d) denote the variation in pure Ni and Ni–Sn alloys, respectively, with increasing neutron dose. In unirradiated specimens, the *S* parameter increased and the *W* parameter decreased with increasing Sn concentration. In irradiated pure Ni, the same trend was observed with increasing neutron dose. In Ni–0.05at.%Sn, the solid line was almost coincident with the dotted line. When the data points were compared between pure Ni and Ni–0.05at.%Sn at the same dose, those of Ni–0.05at.%Sn were closer to the data point of pure Sn than those of pure Ni. Sn atoms are 74.08% larger than Ni atoms, and strongly interact with vacancies. Therefore, complexes of the vacancy clusters and Sn atoms were formed. On the other hand, in Ni–0.3at.%Sn and Ni–2at.%Sn, the dotted line included the data point of Ni–13at.%Sn. This trend indicates the formation of Ni_3Sn. 0.3at.%Sn in Ni is below the solid solubility limit at 573 K. Irradiation-induced vacancies assisted the movement of Sn atoms, leading to the formation of Ni_3Sn. The vacancy clusters formed by irradiation combined with Sn atoms for the same reason in Ni–0.05at.%Sn. In Ni–2at.%Sn, a high concentration of irradiation-induced precipitates was formed. From PAL measurements of 1.6×10^{-4} dpa irradiated samples, the long lifetime was about 159 ps and the precipitation likely occurred at a very low irradiation dose, so that τ_2 denotes a mixed lifetime of interface misfits and vacancy clusters trapped near precipitates. Therefore, a very high intensity was obtained.

When irradiation-induced precipitation was detected, vacancy clusters did not grow. The precipitates interrupted the 1-D motion of interstitial clusters, and the clusters and vacancies annihilated each other. The interface acted as a defect sink. Precipitation is very important to prevent void growth in Ni–Sn alloys.

Table 2. Positron annihilation lifetime in neutron-irradiated Ni and Ni–Sn alloys. τ_{av}, τ_1, τ_2, and τ_3 denote mean lifetime, and the first, second, and third components, respectively. I_2 and I_3 denote the intensities of τ_2 and τ_3, respectively.

	Neutron dose (dpa)	τ_{av} (ps)	τ_1 (ps)	τ_2 (ps)	I_2 (%)	τ_3 (ps)	I_3 (%)
Pure Ni	2.7×10^{-4}	136±1	88±1	222±3	29±1	–	–
	7.6×10^{-2}	247±1	52±4	154±4	53±1	488±4	25±1
Ni–0.05at.%Sn	1.6×10^{-4}	139±1	91±1	289±4	20±1	–	–
	7.6×10^{-2}	190±1	80±14	176±6	78±4	488±20	8±1
Ni–0.3at.%Sn	1.6×10^{-4}	144±1	71±2	179±2	58±1	–	–
	7.6×10^{-2}	166±1	128±9	190±7	58±12	–	–
Ni–2at.%Sn	1.6×10^{-4}	155±1	42±9	159±1	93±1	–	–
	7.6×10^{-2}	162±1	30±16	163±1	96±1	–	–

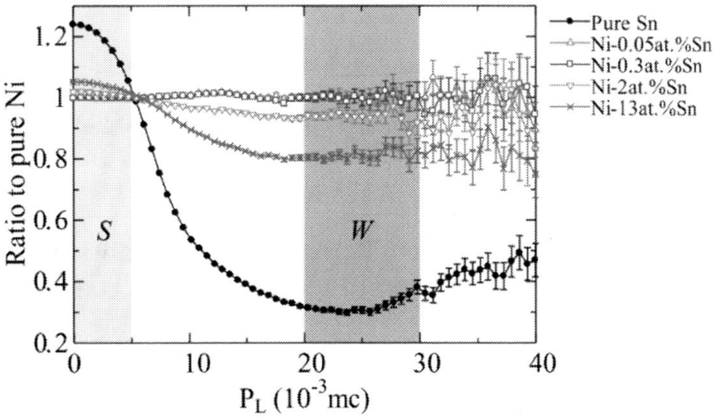

Figure 1. Typical CDB ratio curves of unirradiated specimens to unirradiated pure Ni.

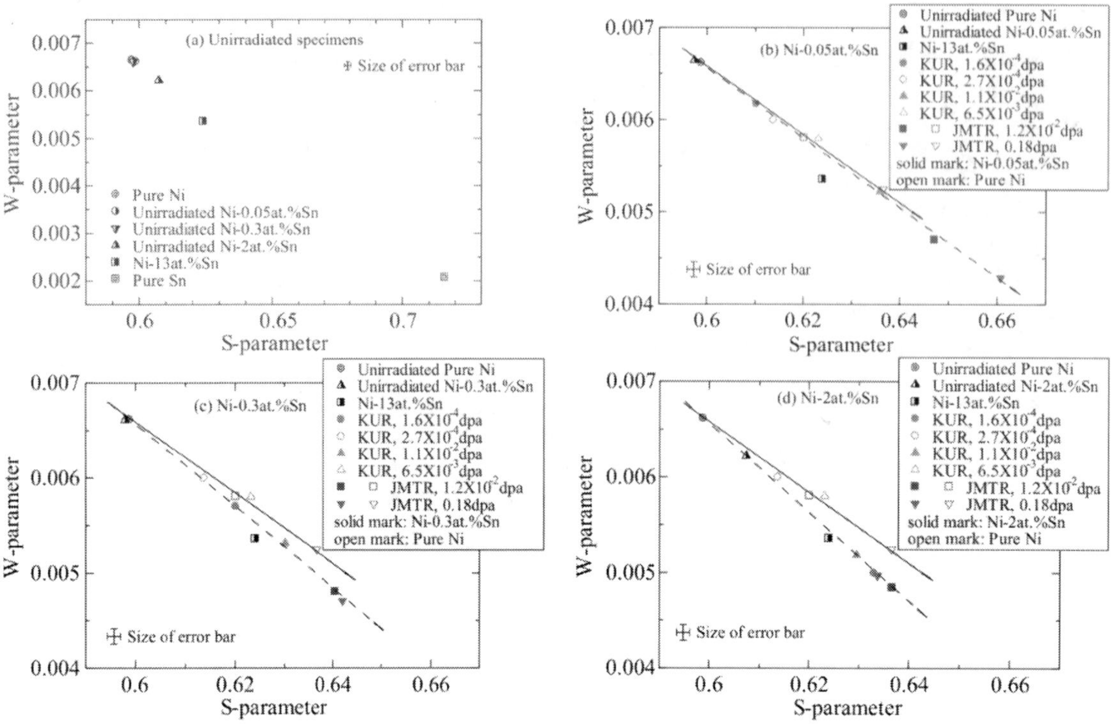

Figure 2. *S–W* plots for (a) unirradiated specimens, (b) Ni−0.05at.%Sn, (c) Ni−0.3at.%Sn and (d) Ni−2at.%Sn.

Acknowledgements
A portion of this study was performed under the sponsorship of JNES open application research project for enhancing the basis of nuclear safety. (JNES : Japan Nuclear Energy Safety Organization)

References

[1] Kojima S, Yoshiie T and Kiritani M 1988 *J. Nucl. Mater.* **115−157** 1249−1253
[2] Kiritani M, Yoshiie T, Kojima S, Satoh Y and Hamada K 1990 *J. Nucl. Mater.* **174** 327−351
[3] Yoshiie T, Kojima S, Satoh Y, Hamada K and Kiritani M 1992 *J. Nucl. Mater.* **191−194** 1160−1165
[4] Hamada K, Kojima S, Ogasawara Y, Yoshiie T and Kiritani M 1994 *J. Nucl. Mater.* **212−215** 270−274
[5] Yoshiie T, Xu Q, Satoh Y, Ohkubo H and Kiritani M 2000 *J. Nucl. Mater.* **283−287** 229−233
[6] Yoshiie T, Ishizaki T, Xu Q, Satoh Y and Kiritani M 2002 *J. Nucl. Mater.* **307−311** 924−929
[7] King H W 1966 *J. Mater. Sci.* **1** 79−90
[8] Sato K, Itoh D, Yoshiie T, Xu Q, Taniguchi A and Toyama T, to be published
[9] Sato K, Yoshiie T and Xu Q 2007 *J. Nucl. Mater.* **367−370** 382−385
[10] Massalski T B and Okamoto H *Binary Alloy Phase Diagrams* (ASM International)
[11] Yoshiie T, Hayashi Y, Yanagita S, Xu Q, Satoh Y, Tsujimoto H, Kozuka T, Kamae K, Mishima K, Shiroya S, Kobayashi K, Utsuro M and Fujita Y 2003 *Nucl. Instr. Meth. Phys. Res.* A **498** 522−531
[12] Greenwood L R 1985 *ANL/FPP/TM* **197** (Argonne, Argonne National Laboratory)
[13] Saito H, Nagashima Y, Kurihara T and Hyodo T 2002 *Nucl. Instr. Meth. Phys. Res.* A **487** 612−617
[14] Toyama T, Nagai Y, Tang Z, Hasegawa M, Almazouzi A, van Walle E and Gerard R 2007 *Acta Mater.* **55** 6852−6860
[15] Olsen J V, Kirkegaard P, Pedersen N J and Eldrup J 2007 *Phys. Stat. Sol.* C **4** 4004−4006

Behavior of free volume in ZrCuAl bulk metallic glass after irradiation

Yuka Fukumoto[1], Akito Ishii[1], Akihiro Iwase[1], Yoshihiko Yokoyama[2], Fuminobu Hori[1]

[1] Department of Materials Science, Osaka Prefecture University, 1-1, Gakuen-cho, Nakaku, Sakai,Osaka 599-8531, Japan,

[2] Institute for Materials Research, Tohoku University, 2-1-1, Katahira, Aoba-ku, Sendai 980-8577, Japan

E-mail: horif@mtr.osakafu-u.ac.jp

Abstract. Free volumes in $Zr_{50}Cu_{40}Al_{10}$ bulk metallic glasses irradiated by 100 MeV and 200 MeV Xe ions and 2 MeV electrons at room temperature have been investigated by using positron annihilation techniques, the micro Vickers hardness and X-ray diffraction measurements. No crystallization took place due to both ion and electron irradiations. The coincidence Doppler broadening spectra of annihilation gamma rays also remained unchanged by ion and electron irradiation. Meanwhile, an increase and a decrease of positron lifetimes after electron and ion irradiation, respectively, were observed. This suggests the structural modification of free volumes. The response of positron lifetime to ion and electron irradiations may be connected with the tendency of hardness.

1. Introduction

The bulk metallic glasses (BMGs) are expected to be useful for various applications due to their superior mechanical properties, hardness, strength, corrosion resistance and micro-formability [1-3]. We suggest that the properties of BMGs are further improved by irradiation. So far, thin film of metallic glass with electron and ion irradiation has been extensively studied [4-8]. Thin film metallic glasses are crystallized due to both electron [4] and ion irradiations [7-8]. However, we suspect that different irradiation indoses may appear for BMGs because of indose of thickness. In this study, we investigated the effects of ion and electron irradiations on the free volumes and mechanical properties of ZrCuAl BMGs.

2. Experimental procedure

The samples (ϕ 8 mm x 0.5 mm discs) were cut from a $Zr_{50}Cu_{40}Al_{10}$ BMGs (ϕ 8 mm x 60 mm) rod fabricated by the tilt casting method in an arc furnace [9] and polished to mirror surfaces. These were irradiated by 100 MeV or 200 MeV Xe ions with a maximum dose of about 1.0×10^{14} ions/cm^2 or 2 MeV electrons with a maximum dose of about 1.4×10^{18} e$^-$/cm^2 at room temperature. Using a ^{22}NaCl source (172 kBq) deposited on thin Kapton foils, positron annihilation lifetime (PAL) and coincidence Doppler broadening (CDB) measurements were carried out at room temperature. All the PAL spectra were analyzed by RESOLUTION program [10]. To confirm crystallinities of the samples due to irradiation, X-ray diffraction (XRD) measurements were performed using a Rigaku Ultima IV. Furthermore, micro-Vickers hardness test was performed before and after irradiation by using a Shimadzu HMV-2 with applied load of 200 and 1000 g imposed for 10 s.

3. Results and discussion

3.1. Free volume

Figure 1 shows the XRD patterns obtained for the as-prepared sample, the samples irradiated with 2 MeV electrons, 100 MeV and 200 MeV Xe ions and the sample annealed at 773 K without irradiation. The broad peak observed for the as-prepared sample indicates its amorphous state. After the heat treatment at 773 K, the peak becomes sharp suggesting the occurrence of crystallization. Similar results have been reported by Matsubara *et al.* [11]. The peak widths observed after electron and ion irradiations are nearly the same as that of the as-prepared sample. This indicates that the crystallization does not take place due to the above irradiations.

Figure 2 shows the positron lifetimes obtained after 2 MeV electron irradiation, 100 MeV and 200 MeV Xe ion irradiations as a function of dose. All the positron lifetime spectra were analyzed considering one lifetime component. The positron lifetime increases after electron irradiation. This indicates that the size of free volumes becomes large by electron irradiation. On the other hand, the positron lifetime decreases after Xe ion irradiation. The decrease of positron lifetime is more significant for 200 MeV Xe ion irradiation at the dose of 10^{14} ions/cm^2. Hori *et al.* have reported that the decrease of positron lifetime observed upon annealing is caused by the reduction of free volume size associated with the structural relaxation and crystallization [12-13]. The decrease of positron lifetime after Xe ion irradiation may be explained in the similar way to it. Also, higher energy irradiation causes more shrinking of free volume.

Figure 3 shows the CDB spectra (ratio to that of pure Al) obtained for the as-prepared sample, the samples irradiated by 2 MeV electrons and 200 MeV Xe ions. For comparison, the CDB spectra for pure Zr and pure Cu are also plotted. The CDB spectrum of the as-prepared sample does not match the spectra of Zr, Cu and Al. This means that free volumes, where positrons are localized, are not surrounded by one specific element. Ishii *et al.* found the segregation of Zr atoms around free volumes in the same BMG alloy sample [14]. That is, the fraction of Zr atoms around free volumes is greater than that of chemical composition for this alloy system. There are no significant changes in the CDB spectra at any irradiation conditions. Namely, the elemental ratio around free volumes, where positrons are localized, does not change by irradiation.

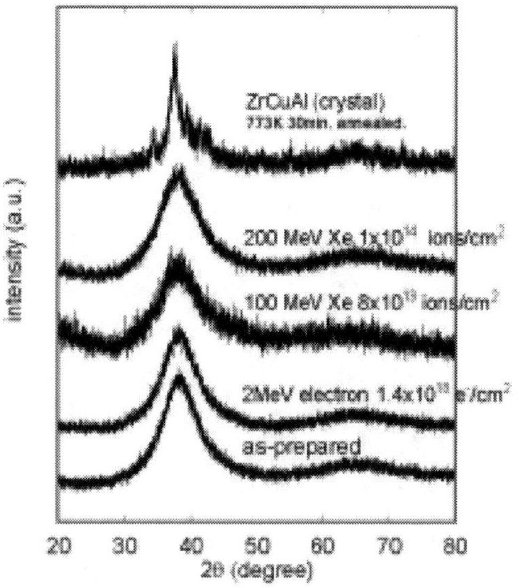

Figure 1. XRD patterns of the $Zr_{50}Cu_{40}Al_{10}$ BMGs in the as-prepared state, after 2MeV electron, 100 MeV and 200MeV Xe ion irradiations and after the heat treatment at 773 K.

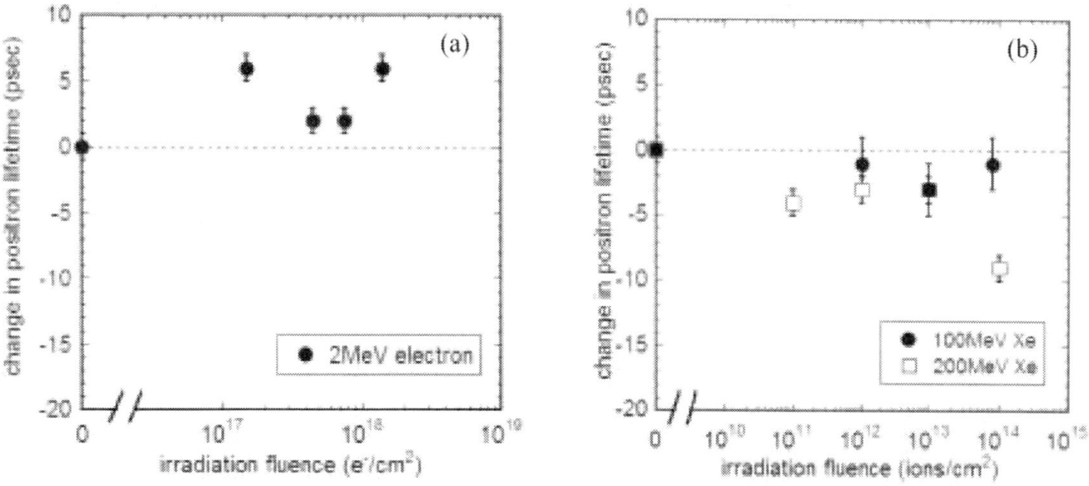

Figure 2. Positron lifetime obtained for the $Zr_{50}Cu_{40}Al_{10}$ BMGs after (a) 2MeV electron irradiation and (b) 100 and 200 MeV Xe ion irradiation as a function of irradiation dose.

Taking into account of the above results, it is concluded that electron irradiation enhance the free volume size without atomic reconstruction in $Zr_{50}Cu_{40}Al_{10}$ BMGs. Similar behavior due to structural relaxation by heat treatment was observed below the glass transition temperature [12]. While, ion irradiation reduces the free volume size without atomic reconstruction. The reduction of free volumes with atomic ordering due to crystallization were found above the glass transition temperature [13]. Therefore, it is considered that crystallization does not occur in $Zr_{50}Cu_{40}Al_{10}$ BMGs by electron and Xe ion irradiation.

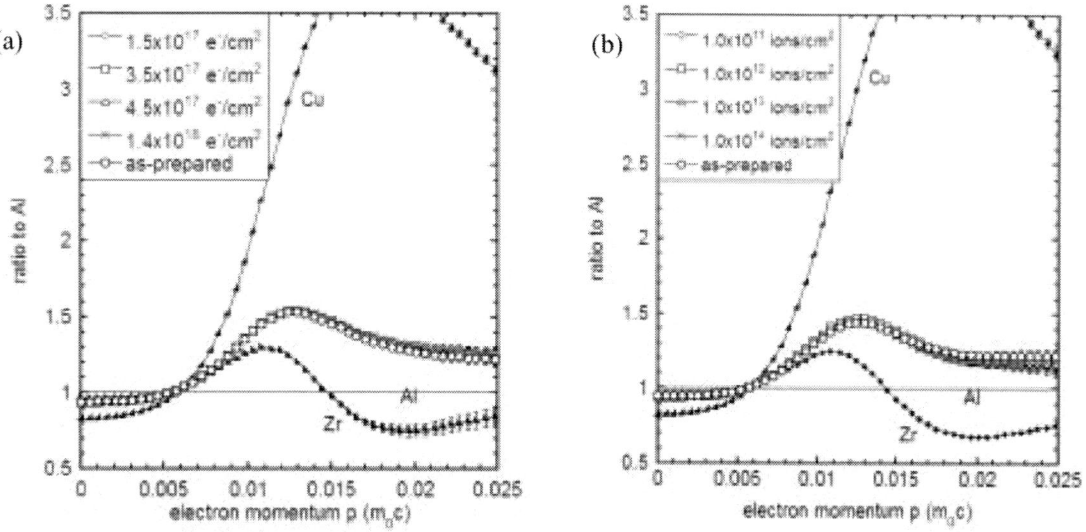

Figure 3. The CDB (ratio to that of pure Al spectrum) spectra of the samples irradiated with (a) 2MeV electrons and (b) 200 MeV Xe ions. In both figures, the CDB spectra of the as-prepared sample, pure Cu and Al metals are also plotted.

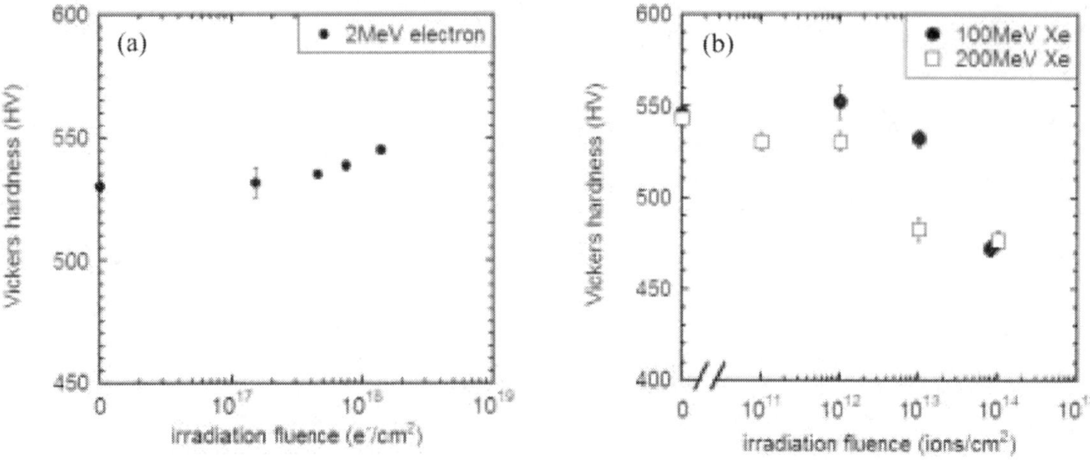

Figure 4. The Vickers micro-hardness obtained after (a) 2 MeV electron irradiation and (b)100 MeV and 200 MeV Xe irradiations as a function of irradiation dose.

3.2. Hardness

Figure 4 shows the Vickers micro-hardness (HV) measured after electron and Xe ion irradiation. The hardness of $Zr_{50}Cu_{40}Al_{10}$ BMG is changed by electron and ion irradiation. The micro-hardness increases by electron irradiation, and it decreases by Xe ion irradiation. These features seem to be qualitatively in good agreement with the change of free volume size shown in Fig. 2. Iqbal *et al.* reported that the increase of hardness of $Zr_{55}Cu_{30}Al_{10}Ni_5$ BMG by heat treatment may be related to the formation of crystalline phase [16]. Yokoyama *et al.* reported that the hardness of this alloy does not change by annealing below the glass transition temperature [15]. Consequently, it can be supposed that the change of hardness by ion irradiation is not caused by structural relaxation. However, the hardened mechanism by the electron irradiation is difficult to explain. In order to clarify the mechanism of hardness change by irradiation, the further experiments and analyses are still needed.

4. Conclusions

In this study, the structural changes of $Zr_{50}Cu_{40}Al_{10}$ BMG after electron and swift heavy ion irradiation were studied by employing the positron annihilation spectroscopy, the XRD and the hardness measurements. No crystallization after electron and Xe ion irradiations was observed, while the different structural changes were induced by electron and Xe ion irradiation. That is, the free volume size in BMG increases after electron irradiation and decreases after Xe ion irradiation without compositional changes. These features are compatible to the changes of hardness upon irradiation. Our results suggest that swift heavy ion irradiation results in more important structural modifications to the "bulk" metallic glass as compared to the electron irradiation.

The absence of crystallization upon irradiation is contradictory to the previous reports concerning the crystallization of thin film metallic glasses by irradiation [4, 7-8]. We suppose that the radiation effect for metallic glass strongly depends on the thickness and also incident ion energy. For instance, crystallization of $Zr_{55}Cu_{30}Al_{10}Ni_5$ BMG by 10 keV Ar ion irradiation with total dose of $2.7x10^{17}$ ions has been reported [16]. The further studies are still needed to elucidate the effects of sample thickness and irradiation condition on the modification of metallic glasses.

Acknowledgement

Authors would like to express cordial thanks to Dr. Takeshi Oshima and Dr. Norito Ishikawa for electron and ion irradiation at Japan Atomic Energy Agency. This work was also performed under the inter-university cooperative research program of the Advanced Research Center of Metallic Glasses, Institute for Materials Research, Tohoku University.

Journal of Physics: Conference Series **225** (2010) 012010 doi:10.1088/1742-6596/225/1/012010

References

[1] Inoue A 2000 *Acta. Mater.* **48** 279.

[2] Löffler J F 2003 *Intermetallics* **11** 529.

[3] Johnson W L 2002 *JOM Journal of the Minerals, Metals and Materials Society* **54** 40

[4] Nagase T, Hosokawa T, Takizawa K, Umakoshi Y 2009 *Intermetallics* **17** 657

[5] Petrusenko Y, Bakai A, Borysenko V, Astakhov A, Baarankov D 2009 *Intermetallics* **17** 246

[6] Sugita K, Matsumoto M, Mizuno M, Araki H, Shirai Y 2008 *J. Phys. Conf. Ser.* **106** 1

[7] Dunlop A, Jaskierowicz G, Rizza G 2003 *Phys. Rev. Rett.* **90** 1

[8] Carter J, Fu E G, Bassiri G, Dvorak B M, Theodore N D, Xie G, Lucca D A, Martin M, Hollander M, Zhang X, Shao L 2009 *Nucl. Instrum. Methods in Phys. Res.* B **267** 1518

[9] Yokoyama Y, Fukaura K, Inoue A 2002 Intermetallics **10** 1113

[10] Kirkegaard P, Eldrup M, Mogensen O E, Pedersen N J 1981 *Comp. Phys. Commun.* **23** 307

[11] Matsubara E, Ichitsubo T, Itoh K, Fukunaga T, Saida J, Nishiyama N, Kato H, Inoue A 2009 *J. Alloys. Comp.* **483** 8

[12] Ishii A, Hori F, Iwase A, Fukumoto Y, Yokoyama Y, Konno T J 2008 *Mater. Trans.* **49** 1975

[13] Hori F, Yano T, Yokoyama Y, Akeno Y, Konno T J 2007 *J. Alloys. Comp.* **434-435** 207

[14] Ishii A, Iwase A, Fukumoto Y, Yokoyama Y, Konno T J, Hori F *J. Alloys. Comp.* submitted

[15] Yokoyama Y, Akeno Y, Yamasaki T, Liaw P K, Buchanan R A, Inoue A 2005 *Mater. Trans.* **46** 2755

[16] Iqbal M, Akhter J I, Hu A Q, Zhang H F, Qayyum A, Sun W S 2007 *J. Non-cryst. Solids* **353** 2452

International Symposium on Advanced Science Research 2009 IOP Publishing
Journal of Physics: Conference Series **225** (2010) 012014 doi:10.1088/1742-6596/225/1/012014

Positron annihilation study of the hardening behavior in Al-Cu based alloy by electron and heavy ion irradiation

Fuminobu Hori[a], Ippei Kobayashi[a], Yuichi Saito[b], Norito Ishikawa[b], Takeshi Oshima[b], Akihiro Iwase[a]

[a]Department of Materials Science, Osaka Prefecture University, 1-1 Gakuen-cho, Sakai, Osaka 599-8531, Japan
[b]JAEA Tokai, Tokai-mura, Naka-gun, Ibaraki 319-1195, Japan

E-mail: horif@mtr.osakafu-u.ac.jp

Abstract. Al-Cu based alloy, which is generally called duralumin (JIS2017), was irradiated with 10 MeV Iodine ions, 200 MeV Xenon ions and 3 MeV electrons at room temperature respectively. The micro Vicker's hardness and positron annihilation coincidence Doppler broadening (CDB) measurements have been performed before and after irradiation. Only in the case of ion irradiation, the Vicker's hardness increases with increasing ion dose. Nevertheless, there was no difference in the profile CDB spectrum for before and after irradiation. On the other hand, we found that the micro hardness of this alloy, which was Xe ion irradiated and subsequently annealed at 423 K, is greater than that of age hardened alloy without irradiation. CDB ratio curve of the age hardened Duralumin is clearly different in the electron momentum range around 0.015-0.025 mc from that of the ion irradiated alloy. The results of three-dimensional atom probe (3DAP) also show that a lot of small clusters were found after ion irradiation but large precipitations have found in annealed Duralumin. These results reveal that a number of small clusters formed in this alloy after ion irradiation, and they should strongly affects the micro hardness.

1. Introduction

Radiation enhanced segregation is one of the important phenomena. For instance, it is well known that the precipitation of Cu impurity atoms in reactor pressurized vessel (RPV) steels causes their low temperature embrittlement because of the formation of fine Cu precipitates by which vacancy mechanism act as obstacle against the dislocation motions.[1-6] In other words, radiation induced precipitation formation by using supersaturated impurity atoms changes its mechanical property.[7] In the case of RPV, this is a negative effect but we start applying this phenomenon a tool for the hardness control of any alloys.

In the present study, we have tried to irradiate of high energetic particles to Al-Cu based alloy system, which is called Duralmin, in order to control and improve the mechanical properties. Moreover, we have studied the process of some clusters formation in this alloy under electron and swift heavy ion irradiation by using positron annihilation and other techniques. In this paper, therefore, we report the results of the micro Vicker's hardness, positron annihilation Doppler broadening and three-dimensional atom probe measurements for Al-Cu based alloys aged or irradiated with electron and heavy ion, such as Xenon and Iodine.

2. Experimental procedure

Al-Cu based alloy (Duralumin-JIS2017: The chemical compositions of the two alloy samples in the present study are listed in Table 1) was prepared by arc melt method. 10 mm x 10 mm x 1mm size samples were cut for micro Vickers and positron annihilation measurements. They were annealed at 793 K for 1 hour in air and quenched into ice water. I-ion and Xe-ion irradiations with an energy of 10 MeV and 200 MeV into Al-Cu based alloys were performed at room temperature by using a tandem type accelerator at the Japanese Atomic Energy Research Agency (JAEA) Takasaki, JAPAN. 3 MeV electrons irradiation have been performed for Al-Cu based alloy at room temperature by using tandem accelerator at JAEA Tokai. The maximum damage depth for 10 MeV Iodine and 200 MeV Xenon into Al-Cu based alloy was calculated to be approximately 4 and 20 μm, respectively. After irradiation isothermal annealing at 423 and 453 K and isochronal annealing were performed. Coincidence Doppler broadening (CDB) measurement for each annealed and irradiated sample was performed. CDB spectra were measured using two HP-Ge detectors, which were located at an angle of 180 degree relative to each other, and total counts of CDB spectra consist of about more than 10^8. As a positron source, we used ^{22}NaCl with an activity of 170 kBq, which was sandwiched by thin Kapton foils. Micro Vickers hardness was also measured by conventional equipment. The measuring temperature was room temperature and the applied load were 25 and 100 gf with holding time of 10 s. Applied load of these value is optimized for estimated damage peak for each ion irradiation. Atom probe analyses were performed for as prepared, aged at 423 K and I-ion irradiated samples using a three-dimensional atom probe (3DAP) equipment [8]. Needle-like atom probe specimens were prepared with a diameter of 100 nm by the micro-polishing technique.

Table 1. Chemical composition of the Al-Cu based alloy.

Cu	Mn	Mg	Si	Fe	Zn	Cr	Al
3.5-4.5	0.4-1.0	0.4-0.8	0.2-0.8	0.7	0.25	0.1	balance

(at.%)

3. Results and Discussions

Micro Vickers hardness of Al-Cu based alloy at room temperature is about 140 Hv. Age hardening of Al-Cu alloy at room temperature is well known generally, but this phenomenon was not observed for this Al-Cu based Duralmin alloy. Figure 1 shows micro Vickers hardness for Al-Cu based alloy before and after 3 MeV electron irradiation up to dose of 1.5 x 10^{18} /cm^2. This figure clearly shows that hardness does not change at all by any electron irradiation dose. Figure 2 shows

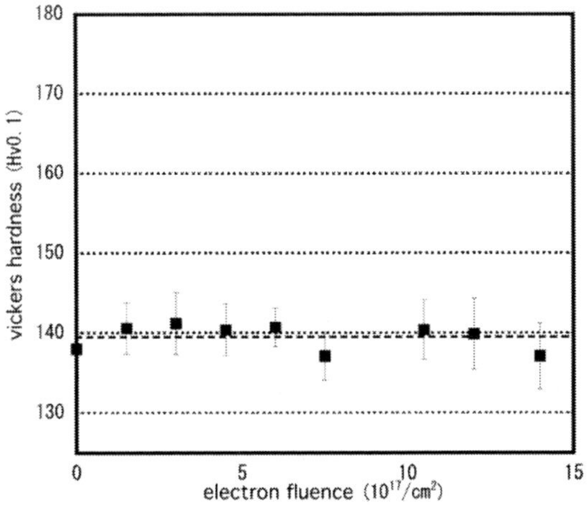

Figure 1. Change in micro Vickers hardness for Al-Cu alloy as a function of 3 MeV electron irradiation dose. (applied load is 100 gf)

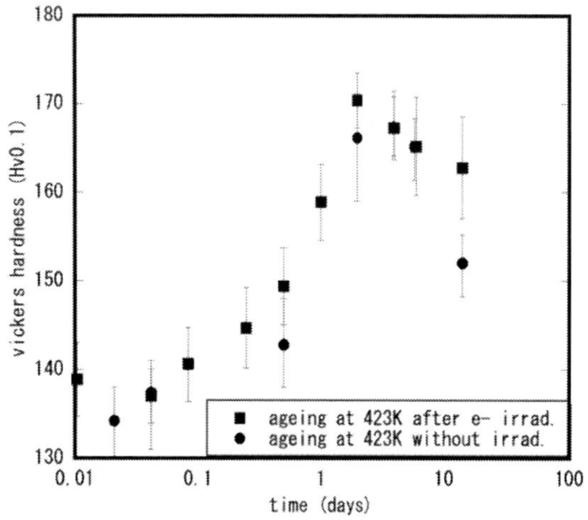

Figure 2. Change in micro Vickers hardness for Al-Cu alloy as a function of dose of 10 MeV I-ion irradiation.

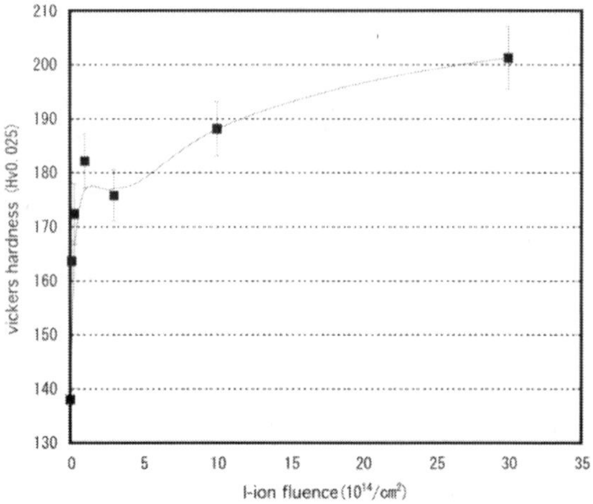

Figure 3. Change in micro Vickers hardness for Al-Cu based alloy as a function of 10 MeV I-ion irradiation dose. (applied load is 25 gf)

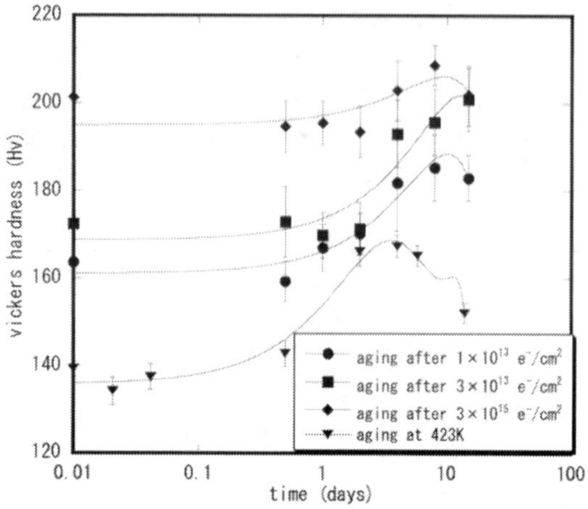

Figure 4. Change in micro Vickers hardness as a function of annealing time at 423 K for I-ion irradiated and unirradiated Al-Cu alloys.

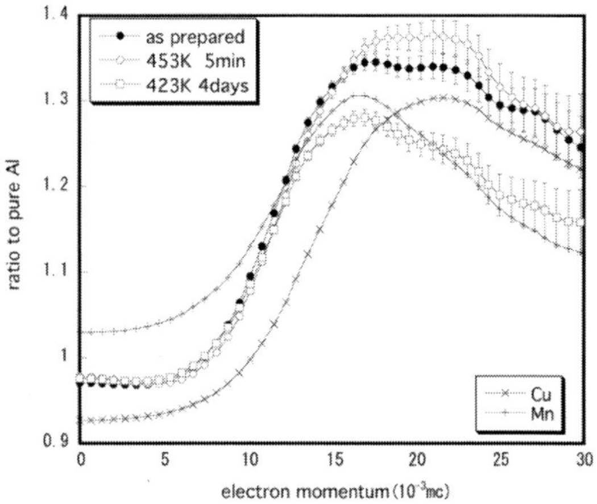

Figure 5. CDB ratio spectra for as prepared and annealed Al-Cu alloys without irradiation in the form of the ratio to pure Al as a function of electron momentum. Also shown is CDB ratio spectrum of pure Cu and Mn normalized to Al.

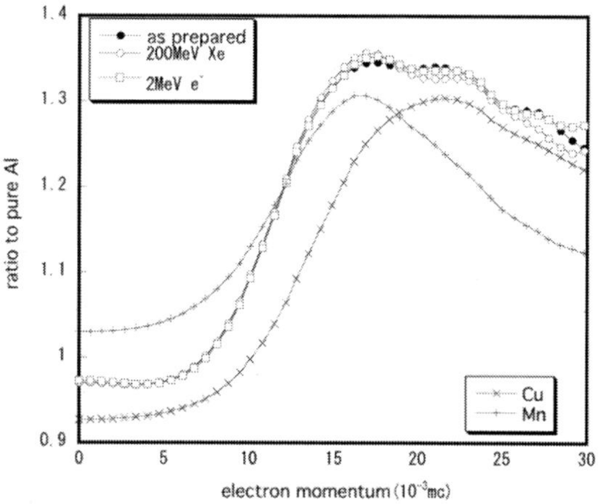

Figure 6. CDB ratio spectra for as prepared, 2 MeV electron and 200 MeV Xe-ion irradiated Al-Cu alloys as a function of electron momentum.

change in hardness with annealing before and after 10^{18} e$^-$/cm^2 electron irradiation. After electron irradiation, hardness gradually increases with increasing annealing time as same as the aging behavior without irradiation, as shown in this figure. As the result, electron irradiation does not affect hardness.

Figure 3 shows Vickers hardness for annealing after I-ion irradiation. In this figure, Vickers hardness of this alloy increases by I-ion irradiation depending on irradiation dose. The hardness of 200 MeV Xe-ion irradiated sample was also increased. The trend of change in hardness with annealing after ion irradiation is similar to that for age hardening as shown in figure 4. However, it clearly shows that the peak value of hardness after about 10 days annealing for each irradiated sample is greater than that for age hardened sample without irradiation. The indent-depth of the Vickers' load of 25 gf is about few μm and the effect of indentation extends to the depth several times larger than the indent-depth, and the depth of damage peak introduced by this irradiation is about 4μm. Therefore, measured hardness of ion irradiated Al-Cu based alloy is mean value from the surface to around damage peak. In other wards, the local hardness around damage peak might be higher than that of measured Vickers hardness. It is considered that the difference of hardness change between electron and heavy ion irradiation is mainly attribute to total amounts of dpa (displacements par atom). In fact that total dpa of electron irradiation is few orders (at least five orders at damage peak in this case) smaller than that of ion irradiation in this case. However, it is difficult to discuss the dpa dependence of hardness, because the ion irradiation induced dpa for the depth is not unique.

Figure 5 shows the CDB ratio curves for as prepared and age annealed Al-Cu based alloys, which is expressed in the form of the ratio of CDB intensity to that of pure Al. The fraction of Cu and Mn atoms for positron trapping clearly changes by the each annealing. This is supposed that some kind of new clusters are formed and/or grows larger. The results of three-dimensional atom probe measurement shows that some shape of large size precipitates (over 20 nm each), such as Mg and Si rich cluster and Al and Cu rich clusters, are observed only in case of aging without irradiation. Positron trapping at Ω precipitates including Ag and Mg atoms associated with Cu atoms in similar alloy systems is reported [9]. Therefore, certain kinds of these clusters are detected by positron in this case. The CDB ratio profiles for electron, Xe-ion irradiated and as prepared Al-Cu based alloys without irradiation are shown in figure 6. In this figure, no marked change among these samples was observed. In case of electron irradiation, this result has a good correlation with the result of unchanging trend of micro hardness after electron irradiation as mentioned above. Taking into account for these results, it might form small clusters, which are not able to detect by positron and have no effect for hardness, or no cluster formation occur in this alloy matrix by the electron irradiation. On the other hand, ion irradiation makes hardened this alloy even though no remarkable change in CDB ratio curve is observed. The estimated maximum depth of damaged area by this irradiation is about 20 μm and the positron penetration depth as a probe of defect detection is about 100 μm. Therefore, about 10 ~ 20% of positrons annihilate at the damaged area so that formed clusters by the ion irradiation may be extremely small. 3DAP result shows that large size precipitates, which are seen in age hardening heat treated alloy without irradiation, and well known plane type precipitates like a GP-zone in this alloy system are not found in case ion irradiated sample. Instead, large amounts of small size of the order of 1 nm segregations are observed in the ion irradiated sample. Hence, this fact of these small clusters formation after heavy ion irradiation is consistent with the result of CDB measurement. Moreover, 3DAP result also shows that the number density of the cluster in ions irradiated sample is greater than that in age hardening treated one without irradiation. As a consequence, large amounts of small clusters are formed in Al-Cu based alloy by swift heavy ion irradiation, and they should strongly affects the micro hardness.

4. Summary

Al-Cu based alloys of Duralmin were irradiated with electron, Xe and I ions at ambient temperature. We have found that the electron irradiation does not affect the hardness but the change in hardness due to Xe and I-ion irradiation was measured. The change in hardness with annealing at 423 K observed in Al-Cu based alloy irradiated with I ions is remarkably larger than that for age hardened Al-Cu based alloy without irradiation. From the results of positron annihilation CDB and 3DAP measurements, it was found that low density of large size precipitates are formed in aged sample without irradiation, and large amounts of small size clusters are formed in ion irradiated sample, respectively. These facts suggest that these clusters affect the mechanical properties especially the hardness and their effects strongly depend on the size and density of clusters. The present experimental results show that swift heavy ion irradiation can be used for the modification of hardness and its hardness can be improved near surface than in age hardenable Al-Cu based alloy.

Acknowledgement

This work was partly supported by a grant from Osaka Nuclear Science Association (ONSA). Authors would like to express their cordial thanks to Dr. Kaneno and Dr. Senboshi at Osaka Prefecture University for sample preparation and useful discussions.

References

[1] Iwase A, Hasegawa T, Chimi Y, Tobita T, Ishikawa N, Suzuki M, Kambara T, Ishino S 2002 Nucl. Instr. and Meth. in Phys. Res. B, **195** Issue: 3-4 309-314

[2] Ishino S, Chimi Y, Bagiyono, Tobita T, Ishikawa N, Suzuki M, Iwase A 2003 J. Nucl. Mater. **323** Issue: 2-3 354-359

[3] Hori F, Morita A and Oshima R 1999 J. Electron Microscopy 48(5) 585-589

[4] Odette G R and Lucas G E 1998 Radiation Effects of Defects in Solid **144**, 189

[5] Hori F, Tanaka S, Kuramoto E, Iwase A 2006 Nucl. Instr. and Meth. in Phys. Res. B **245** 180-183

[6] Phythian W J, Foreman A J E, English C A, Buswell J T, Hetherington M, Roberts K and Pizzini S 1992 *Effects of Radiation on Materials:15th Int. Symp., ASTM STP* **1125** 131-150

[7] Nakagawa S, Hori F, Chimi Y, Ishikawa N, Kitagawa M, Oshima R, Tobita T, Taniguchi R, Suzuki M and Iwase A 2007 Nucl. Inst. Meth. in Phys. Res. B **257** 397-401

[8] Dacosta G, Vurpilot F, Bostel A, Bouet M and Deconihout B 2005 *Rev. Sci. Instrum.* **76** 013304

[9] Honma T, Yanagita S, Hono K, Nagai Y and Hasegawa M 2004 Acta Mater. **52** 1997-2003

Free volume in Zr-based bulk glassy alloys studied by positron annihilation techniques

A Ishii[1], A Iwase[1], Y Yokoyama[2], T J Konno[2], Y Kawasuso[3], A Yabu-uchi [3], M Maekawa[3], and F Hori[1]

[1]Department of Materials Science, Osaka Prefecture University, 1-1 Gakuen-cho, Sakai, Osaka 599-8531, Japan

[2]Institute for Materials Research, Tohoku University, Katahira 2-1-1 Aobaku, Sendai 980-8577, Japan

[3]Takasaki Advanced Radiation Research Institute, Japan Atomic Energy Agency, 1233 Watanuki-machi, Takasaki, Gunma 370-1292, Japan

E-mail: horif@mtr.osakafu-u.ac.jp

Abstract. The eutectic $Zr_{50}Cu_{40}Al_{10}$ bulk glassy (BG) alloys, which have a good glass forming ability, have been investigated for various properties, such as toughness, electronic resistivity and so on. Recently, the hypoeutectic $Zr_{60}Cu_{30}Al_{10}$ BG alloys have been paid attention, because they exhibit no degradation of ductility and toughness after the isothermal annealing below the glass transition temperature (T_g). Although the toughness correlates with the free volume to be considered, the free volume of the hypoeutectic BG alloy has not been assessed in detail so far. In order to study the free volume in each BG alloy, therefore, positron annihilation lifetime and coincidence Doppler broadening (CDB) measurements have been performed for $Zr_{60}Cu_{30}Al_{10}$ (hypoeutectic) and $Zr_{55}Cu_{35}Al_{10}$ (middle-composition) and $Zr_{50}Cu_{40}Al_{10}$ (eutectic) BG alloys. The value of the positron lifetime for hypoeutectic BG alloy is virtually the same. Additionally, the CDB ratio curve for hypoeutectic alloy dose not match that for eutectic alloy in an electron momentum region around 0.015 m_0c, indicating that the fraction of Zr atoms around the free volume in hypoeutectic BG alloy is greater than that in eutectic BG alloy. Moreover, the CDB results show that the local structure around free volume for hypoeutectic BG alloy is different from that for eutectic BG alloy.

1. Introduction

Zr-based bulk glassy alloys show high tensile strength [1] and a high Chrapy impact value [2]. These alloys are recognized by good balance between high strength and high toughness. However, Zr-based BG alloys often show the degradation of mechanical properties owing to structural relaxation. Recently, our co-workers have reported that the fatigue behaviour of hypoeutectic $Zr_{60}Cu_{30}Al_{10}$ BG alloys remains unchanged after complete structural relaxation [3]. The free volume is defined as frozen excess open volume in the glassy alloys and perceived as an important factor to describe some properties for glassy alloys. Hence, the fatigue behaviour of BG alloys is also associated with free volume. However, the evaluation based on microscopic measurement of free volume for hypoeutectic BG alloys has not been carried out, so that the relation between free volume and mechanical properties has not been clarified yet. Positron annihilation lifetime can detect and estimate the size of free volume, and positron annihilation coincidence Doppler broadening (CDB) measurement provides the information on the detailed electron distribution around the free volume. Therefore, the local atomic element analysis around free volume has become possible by this technique. So far, we have reported that there is a lot of Zr fraction around free volume for eutectic BG alloy [4]. In this study, we evaluate and discuss the size of free volume and its structure for ternary eutectic, hypoeutectic and other composition of Zr-Cu-Al BG alloys by using positron annihilation techniques. We also estimate the relationship between macroscopic change in density and microscopic change in free volume size during structural relaxation for eutectic BG alloys.

2. Experimental

As a starting material for bulk glassy alloys, ternary eutectic $Zr_{50}Cu_{40}Al_{10}$ (E-BG), hypoeutectic $Zr_{60}Cu_{30}Al_{10}$ (H-BG) and another composition (middle-content of zirconium) $Zr_{55}Cu_{35}Al_{10}$ (M-BG) alloy ingots were prepared by arc melting the mixtures of pure Zr, Cu and Al metals in an argon gas atmosphere. To avoid oxygen contamination, a special crystalline Zr rod (<100 mass ppm oxygen) was used. The master alloy was completely re-melted and cast into a rod shaped cast vessel (ϕ8 mm x 60 mm), which is called the tilt casting method. E-BG alloy specimens were annealed isothermally for 5 hours at 473, 573 and 673 K below glass transition temperature (T_g). The phase characterization of the as-prepared and annealed samples was performed by X-ray diffraction (XRD) measurements. The density of the cast Zr-Cu-Al BG alloys was measured by Archimedes method [5] at room temperature. Positron annihilation lifetime spectra were measured by using a conventional fast-fast circuit with a time resolution of about 200 psec (FWHM) and slow positron beam at Japan Atomic Energy Agency (JAEA) at room temperature. As a positron source for conventional method, we used ^{22}NaCl with an activity of 286 kBq, which was sandwiched by thin Kapton foils. The energies of the positron beams were set to 15 keV. The positron annihilation lifetime spectra consist of 1.0×10^6 counts. All the positron annihilation lifetime spectra were analyzed by the POSITRONFIT program [6]. Positron annihilation coincidence Doppler broadening (CDB) measurement was carried out at room temperature. Each CDB spectrum consists of more than 1.0×10^8 counts. The CDB ratio spectrum was obtained by normalizing the momentum distribution of each spectrum to that of Al metal. It is well established that the change in low electron momentum corresponds mainly to the free volume change, and the detailed profile at higher electron momentum region reflects the local structure and their atomic elements [7].

3. Results and discussions

Figure 1 shows the results of positron annihilation lifetime and density for E-BG, H-BG and M-BG alloy. The values of positron lifetime for these alloys are nearly constant. We tried all positron lifetime spectra to decompose into two or more components, but only one component was analyzed in all samples. Generally, positron lifetime value provides the information about not only the open volume size of vacancy type defect but also its concentration. Therefore, this result shows that the same size of free volume exists in these alloys. Moreover, the result of density measurement does not correspond to that of positron lifetime. Atomic density for each alloy is not constant but depends on Zr atom content. This difference between density and positron lifetime results may be caused by the variation of free volume concentration for each alloy. This is considered that the melting temperature of Zr based BG alloy varies with Zr content, so that the amount of quenched in free volume of these alloys becomes different. Then, we estimate the local structure especially focus on Zr atom fraction around the free volume for each BG alloy from the results of CDB ratio analysis.

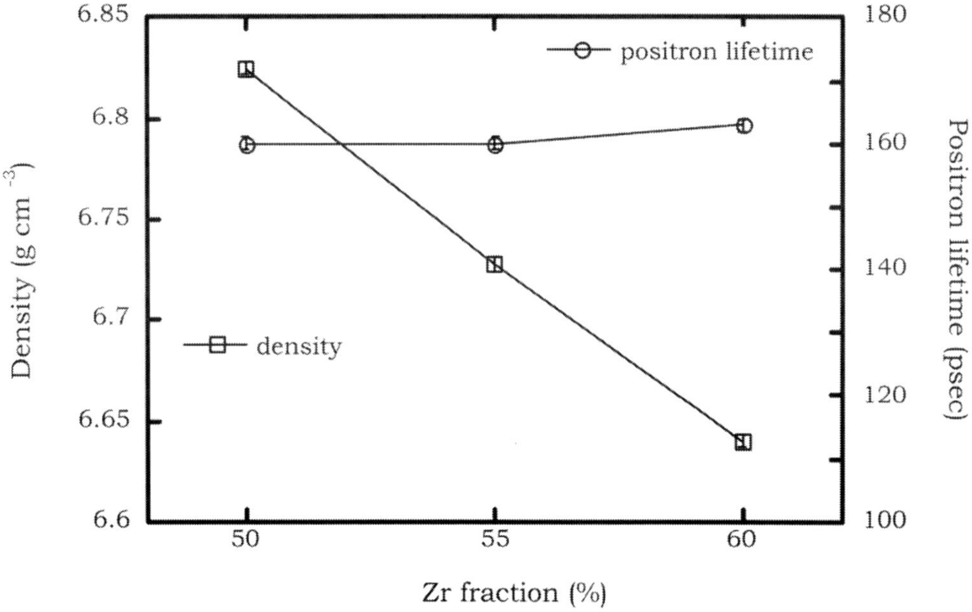

Figure 1. Positron annihilation lifetime and bulk density for ternary E-BG, H-BG and M-BG alloys.

Figure 2 shows CDB ratio curve of each BG alloy with those of Cu and Zr metals, expressed as a ratio of CDB intensity of Al. The positron affinities of A_+ for Zr, Cu and Al are -3.98, -4.81 and -4.41 eV, respectively [8]. It seems that all elements might be attracted to positron because the differences of the positron affinity ΔA_+ between A_+^{Al}, A_+^{Cu} and A_+^{Zr} are not so big. Even if Zr precipitates exist in this alloy matrix, positron may not detect them because positron affinity for Zr is lower than that for other elemental atom (Al and Cu). This result indicates that positron is not necessarily selectively annihilated with Zr atom. Nevertheless, CDB ratio curve shows a characteristic feature of Zr atom. This fact suggests that Zr atoms tend to aggregate around free volume in this system of glassy alloy. Moreover, it can be recognized from this figure that the electron momentum distribution profiles for these BG alloys depend on chemical composition. The CDB spectra indicate a marked difference between E-BG and H-BG alloy in the region of the peak around 0.015 m_0c. In other words, the spectrum of the H-BG alloy is more similar to that of Zr metal than that of the E-BG alloy. This result exhibits that the fraction of Zr atoms around free volume in H-BG and M-BG alloy is greater than that in E-BG alloy. This fact reveals that the local structure around free volume for Zr-Cu-Al ternary BG alloys strongly depends on their chemical composition. This difference of local structure may be related to the property change of brittleness for Zr-Cu-Al BG alloys [3].

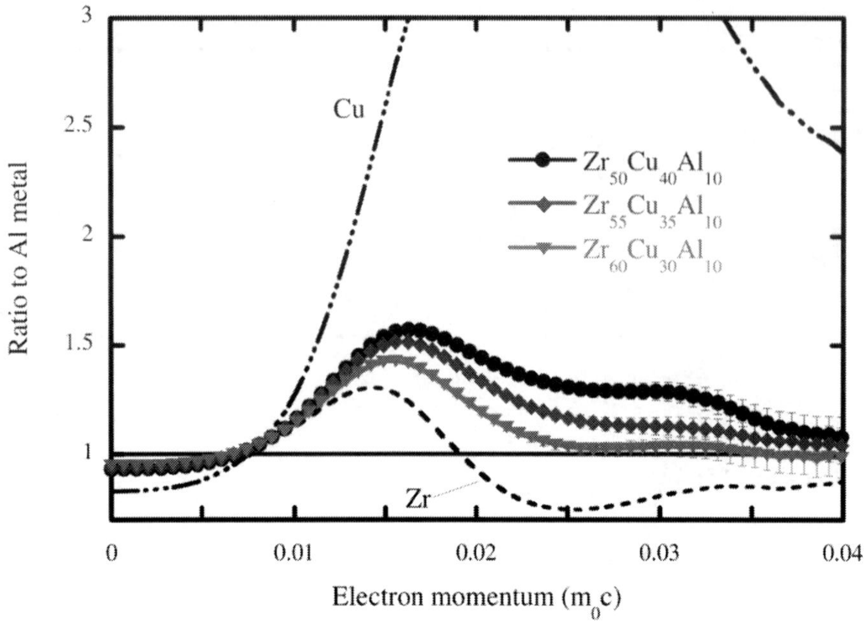

Figure 2. CDB ratio spectra of E-BG, M-BG, and H-BG alloy with those of Cu and Zr metals, expressed as a ratio of CDB intensity of Al metal.

Figure 3 shows each changing value of bulk density and positron lifetime measured by using positron from radio active source 22-Na and slow positron beam with energy of 15 keV for E-BG alloy after each isothermal annealing at 473, 573 and 673 K for 5 hours. As shown in this figure, the difference of positron lifetime comparing with as-prepared sample increases with increasing annealing temperature. This result agrees with our previous results [4,9,10]. As can be seen from this figure, density increases, and absolute value of positron lifetime decreases with increasing annealing temperature. This result means that density increases caused by the shrinkage of free volume, depending on annealing temperature.

As mentioned above, the size of free volume for each as quenched BG alloy is virtually the same, although their density depends on the Zr fraction of alloy composition. Further, Zr atom fraction around free volume, namely the local structure near by the free volume, for these alloys is essentially different from each other. Hence, it can be supposed that the relaxation process of local structure and macroscopic change of density for each alloy during annealing are not always analogous. In this point of view, other relaxation measurements for a series of chemical composition and annealing temperature are needed.

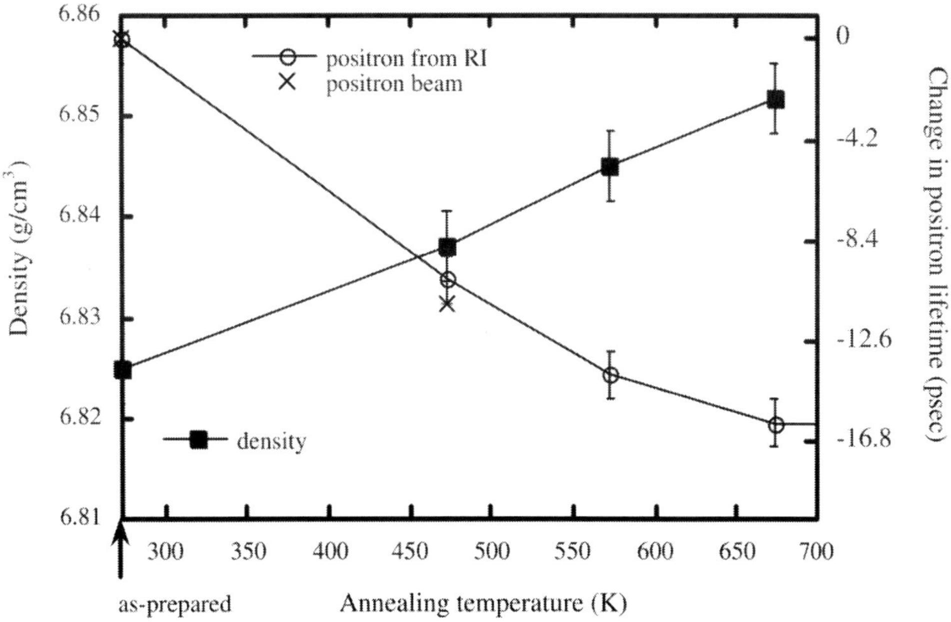

Figure 3. Change in positron lifetime measured by using positron from RI and slow positron beam and bulk density of E-BG after isothermal annealing at 473, 573 and 673 K for 5 hours.

4. Summary

The nature of free volume for various compositional Zr-Cu-Al ternary bulk glassy alloys has been studied by positron annihilation lifetime and coincidence Doppler broadening measurements, and we obtained following conclusions.

The size of free volume for Zr-Cu-Al ternaly bulk glassy alloys is essentially the same without depending on their chemical composition. The number density of free volume, which might be corresponding to bulk density, depends on alloy composition. It found that the local structure around free volume for hypoeutectic bulk glassy alloys is different from that for other composition of bulk glassy alloys. Moreover, free volume change especially around free volume after relaxation corresponds to change in density for eutectic alloy.

Acknowledgement
Authors would like to express cordial thanks to Prof. Haruyama to useful discussions for free volume theory and heat treatment. This work was financially supported by Sasakawa Scientific Research Grant from The Japan Science Society. This work was also performed under the inter-university cooperative research program of the Advanced Research Center of Metallic Glasses, Institute for Materials Research, Tohoku University.

References

[1] Zhang T and Inoue A 1998 *Mater. Trans.* **39** 1230
[2] Inoue A and Zhang T 1996 *Mater. Trans.* **37** 1726
[3] Yokoyama Y, Yamasaki T, Liaw P K and Inoue A 2008 *Acta Mater.* **56** 6097

[4] Ishii A, Hori F, Iwase A, Fukumoto Y, Yokoyama Y and Konno T J 2008 *Mater. Trans.* **49** 1975
[5] Haruyama O and Inoue A 2006 *Appl. Phys. Lett.* **88** 1319060-1
[6] Kirkegaard P, Eldrup M, Mogensen O E and Pedersen N J, 1981 *Comp. Phys. Commun.* **23** 307
[7] Hautojarvi P, Positrons in Solids, Springer-Verlag, Berlin 1979
[8] Puska M J, Lanki P and Nieminen R M 1989 *J. Phys.: Condens. Matter* 1 6081
[9] Hori F, Yano T, Yokoyama Y, Akeno Y, Konno T J 2007 *J. Alloys Compd.* **434** 207
[10] Yano T, Yorikado Y, Akeno Y, Hori F, Yokoyama Y, Iwase A, Inoue A and Konno T J 2005 *Mater. Trans.* **46** 2886

A positron beam study on vacancy formation in iron by ion beam irradiation at low temperature

T Iwai[1], K Murakami[2], Y Katano[1], T Iwata[1], T Onitsuka[3] and H Abe[4]

[1] Nuclear Professional School, School of Engineering, The University of Tokyo, 2-22 Shirakata-Shirane, Tokai-mura, Ibaraki 319-1188, Japan
[2] Department of Nuclear Engineering and Management, School of Engineering, The University of Tokyo, 7-3-1 Hongo, Bunkyo-ku, Tokyo 113-8654, Japan
[3] Nuclear Safety Research Center, Japan Atomic Energy Agency, 2-4 Shirakata, Tokai-mura, Ibaraki 319-1195, Japan
[4] Institute for Materials Research, Tohoku University, 2-1-1 Katahira, Aoba-ku, Sendai 980-8577, Japan

iwai@nuclear.jp

Abstract. This study intends to investigate cascade damage structure produced by energetic ion irradiation. Cascade damage structure is preserved at low temperature below stage I where interstitial atoms begin to migrate. Then positron beam is implanted to the irradiated surface as a vacancy probe to evaluate vacancy concentration remained in the irradiated specimens. By this method, defect production efficiency was evaluated for iron irradiated with proton and carbon ions. The defect production efficiency values indicate enhanced recombination for carbon irradiation due to primary knock-on atoms (PKA) with higher energies.

1. Introduction

Point defect production by energetic particle irradiation is a common process in materials used for nuclear systems. This phenomenon leads to many kinds of degradation in the materials exposed to radiations, especially fast neutrons. Initial defect production processes (cascade damage) are important because they control the subsequent microstructural evolution and the macroscopic property change of the materials. Many efforts, mainly computational works, have been put into this fundamental process [1,2] to obtain damage parameters such as defect production efficiency, freely migrating defect production, and in-cascade clustering. But direct experimental validation is not sufficient because of the difficulty that comes from the time scale and the size (<nm) of the processes. There is a great need for experimental validation to raise the reliability and accountability of the computational works.

In this study, aiming at providing some experimental outcome of defect production processes, positron annihilation technique is applied to iron irradiated at low temperature below stage I where point defect configuration is preserved. Pure iron is chosen not only because steels are main structural materials in nuclear systems but also because many simulation results of iron exist for further comparison. Positron annihilation spectroscopy is a well established technique to probe vacancies and their clusters in iron, and slow positron beam is suitable for the detection of vacancies in ion-irradiated iron [3]. In this paper, an experimental evaluation of cascade damage parameters was challenged by combination of ion irradiation and positron beam Doppler broadening spectroscopy at low temperature below Stage I.

2. Experimental

Samples are 99.998% pure iron sheets annealed at 1023 K for an hour in vacuum. Dimensions of each sample are 15mm x 15mm x 0.1mm. Ion irradiations were carried out with a 3.75MV single ended Van de Graaff accelerator at High Fluence Irradiation Facility, The University of Tokyo (HIT). Ion beam conditions are 1 MeV H^+ and 2.8 MeV C^+ with an angle of incidence of 30°. All irradiations were carried out at 12 K with a cryostat system (IWATANI Cryo Mini 4K Cryocooler HE05).

The variable energy positron beam has been equipped at HIT and is connected to the irradiation chamber of the Van de Graaff accelerator for the in-situ or in-chamber positron beam Doppler broadening experiment [3,4]. This uses a sealed ^{22}Na isotope of 740 MBq as a positron source and a tungsten foil as a positron moderator. Generated positrons from β^+ decay are thermalized and emitted from the moderator, and then extracted by electric field of 50 V. They are magnetically transported along the S-figure beamline with solenoid to sample chamber, where a high purity germanium detector (ORTEC GEM 20180-P) is equipped for Doppler broadening measurement. Doppler broadening spectra can be measured at controlled positron energies from 50 eV to 30 keV. In this study positron beam Doppler broadening spectra were measured by fixed energy positrons: 20 keV for 1MeV H^+ and 15 keV for 2.8MeV C^+. Depth profiles of defect production calculated by SRIM-2008 [5] and positron stopping of Makhov profile [6] are shown in Figure 1. Irradiation and Doppler measurement are sequentially repeated at 12 K to obtain detailed dose dependence of S parameter defined as the ratio of counts in the central region (511 ± 0.8 keV) to counts in the whole photo peak.

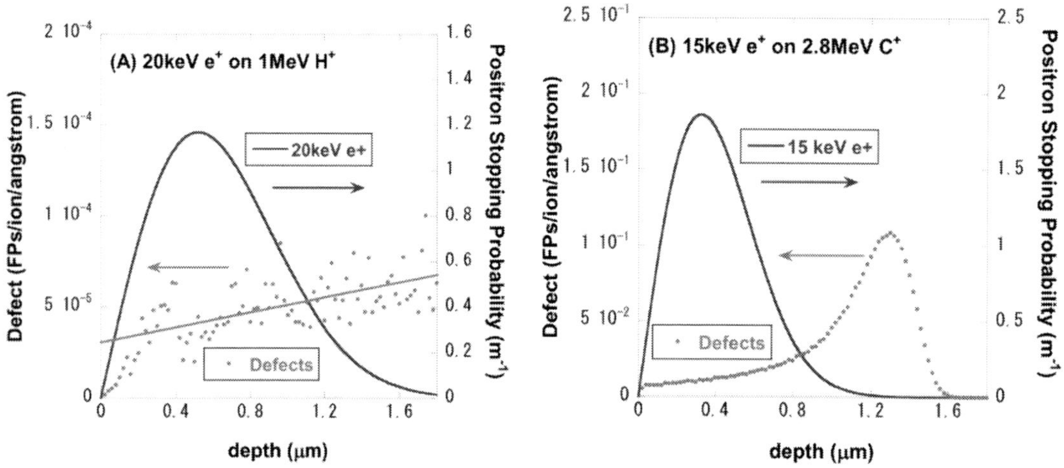

Figure 1 Depth profile of defect production and positron stopping (A) 20keV e^+ on 1MeV H^+ (B) 15keV e^+ on 2.8MeV C^+.

3. Result and discussion

Figure 2 shows a dose dependence of S parameter in iron at 12 K for H^+ and C^+ irradiation. Displacement per atom calculated from Norgett-Robinson-Torrens model (NRT model) [7] by SRIM-2008 is employed as the dose unit. This value is averaged between the surface and 1μm in depth. S parameter for both irradiations was increased with dose, then tended to saturate. When defect concentration is sufficiently high, most of positrons should be trapped at defects and measured S parameter can be treated as defect-specific S parameter S_d. In this experiment, this S parameter saturation should occur due to high concentration of defects preserved without thermal process hereafter. Then this saturated value in figure 1 can be treated as S_d. Increase in S parameter of H^+ irradiation is faster than that of C^+ irradiation, and the saturated S value is similar for each ions. This S value is much lower than S_d for larger vacancy clusters in the reference [3] of Fe^{2+}-irradiated iron at room temperature.

In simple case, measured S parameter can be generally expressed as a linear combination of specific S parameters and each positron fraction:

$$S_{measured}(E) = S_b f_b(E) + S_d f_d(E) + S_s f_s(E) \tag{1}$$

where subscript *b*, *d* and *s* denote bulk, defect and surface, respectively [8]. At 15 and 20 keV e^+ in irradiated specimens, contribution of back diffusion to surface should be minimal. Along with the two state trapping model, defect concentration (C_d) can be expressed as:

$$C_d = \frac{(S_{measured} - S_b)\lambda_b}{(S_d - S_{measured})\mu} \tag{2}$$

where μ and λ_b denote specific trapping rate for defects and bulk annihilation rate ($=9.1\times10^9\ s^{-1}$), respectively.

From computational researches, majority of vacancies exist as monovacancies, even in a cascade (only ~3% for vacancies in clusters if nearest neighbor criterion is applied [9]). Assuming that the almost all the trapping sites in this experiment are monovacancies is reasonable and consistent with these terms. Then μ value for monovacancies estimated by Vehanen et al. ($(1.1\pm0.2)\text{x}10^{15}\ s^{-1}$) [10] can be applied. By equation (2), C_v (vacancy concentration) can be estimated.

In this temperature range below stage I assigned by Takaki et al. [11], vacancies and interstitials are immobile to suppress mutual annihilation and clustering after cascade processes. Under this condition where cascade structure is preserved, the ratio of vacancy concentration to NRT dpa can be treated as defect production efficiency η. Figure 3 shows the dose dependence of the vacancy concentration along with a line which indicates that η=1. Linear increase of C_v with NRT dpa was observed up to 0.3 mdpa as expected, and the slopes are 'apparent' defect production efficiencies. They are estimated to be 0.3±0.1 for H^+ and 0.2±0.1 for C^+.

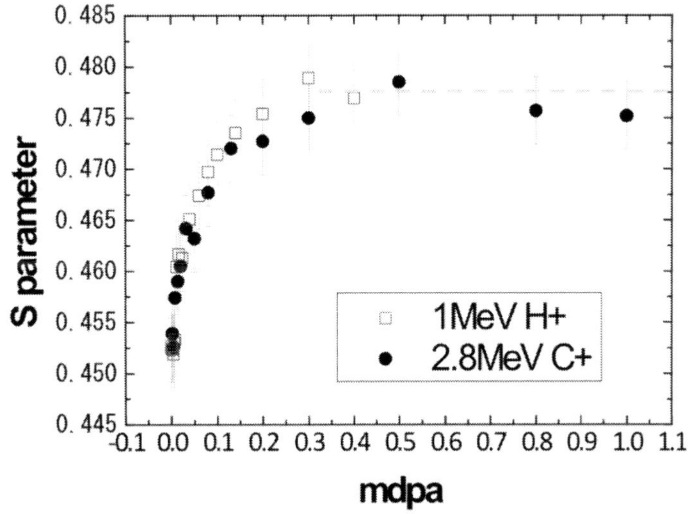

Figure 2 Dose dependence of S parameter at 12K in iron.

For comparison, defect production efficiency was roughly estimated by PKA energy spectrum from SRIM-2008 calculation and published computational efficiency values compiled by Malerba [2]. The estimated values were 0.52 for 1MeV H^+ and 0.42 for 2.8MeV C^+. Although our experimental values qualitatively represent the difference of ion species well, our values are approximately half of the computational values. These values are not necessary to coincide, but we cannot still exclude possibilities of several phenomena to reduce our experimental values. The one possibility is the positron trapping at vacancies locally concentrated near PKAs (cascade). The μ value for monovacancies estimated by Vehanen et al. was estimated by combined experiment of electric resistivity and positron annihilation spectroscopy for electron-irradiated iron. In that case, Frenkel pairs are uniformly distributed to trap positrons effectively. But locally concentrated vacancies by cascade damage might reduce positron trapping probability compared to homogeneous defect distribution due to larger defect free volume. This effect may reduce apparent vacancy concentration. Another possibility is modest migration of interstitials to occur recombination of closed Frenkel pairs, possibly caused by electronic excitation accompanied with ion irradiation.

In-cascade clustering is another important subject of cascade damage structure. However, our experimental method was not suitable for detecting clustering in cascade because considerable concentration of monovacancies around clusters effectively trap positrons to hide clusters from positrons.

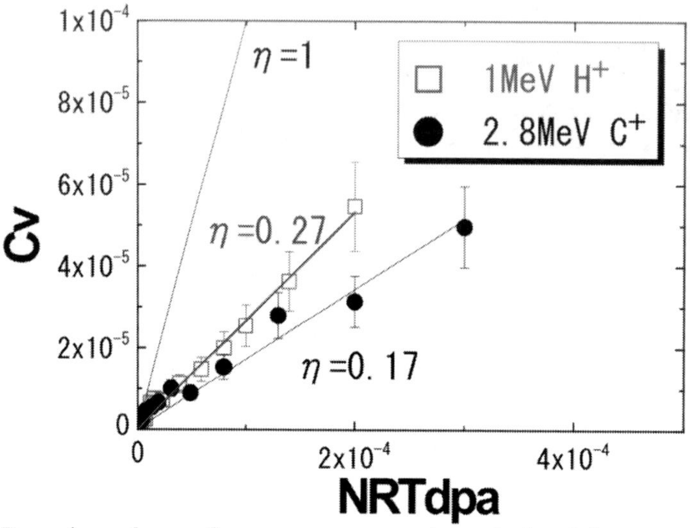

Figure 3 Dose dependence of vacancy concentration calculated from measured S parameter.

4. Conclusions

Vacancy accumulation by ion irradiations at low temperature below stage I was investigated with positron beam Doppler broadening. Formed vacancy-type defects were likely to monovacancies, independent of PKA energy spectrum. 'Apparent' defect production efficiency values qualitatively represent the difference in PKA energy spectrum of H^+ and C^+. The values were lower than simulation-based ones, possibly due to inhomogeneous distribution of vacancies caused by cascades and enhanced mutual annihilation of Frenkel pairs. Positrons are too sensitive to monovacancies nearby to probe in-cascade vacancy clustering.

Acknowledgments
This work was supported by JSPS KAKENHI 17560742 and 20560771.

References

[1] Calder A F and Bacon D J 1993 *J. Nucl. Mater.* **207** 25.
[2] Malerba L 2006 *J. Nucl. Mater.* **351** 28
[3] Iwai T, Ito Y and Koshimizu M 2004 *J. Nucl. Mater.* **329-333** 963
[4] Iwai T and Ito Y 2004 *Mater. Sci. Forum* **445-446** 120
[5] Ziegler J F, Biersack J P and Ziegler M D 2008 *SRIM The Stopping and Range of Ions in Matter* (Morrisville, NC: Lulu Press)
[6] Vehanen A, Saarinen K, Hautojärvi P and Huomo H 1987 *Phys. Rev. B* **35** 4606
[7] Norgett M J, Robinson M T and Torrens I M 1975 *Nucl. Eng. Des.* **33** 50
[8] Schultz P J and Lynn K G 1988 *Rev. Mod. Phys.* **60** 701
[9] Stoller R E 2000 *J. Nucl. Mater.* **276** 22
[10] Vehanen A, Hautojärvi P, Johansson J, Yli-Kaupilla J and Moser P 1982 *Phys. Rev. B* **25** 762
[11] Takaki S, Fuss J, Kuglers H, Dedek U and Schultz H 1983 *Rad. Eff.* **79** 87

Study of Gadolinium-doped Cerium Oxide by XRD, TG-DTA, impedance analysis, and positron lifetime spectroscopy

S. Ohta[1], T. Kosaka[1] and K. Sato[2]
[1] Department of Chemistry, Tokyo Gakugei University, Nukuikita 4-1-1, Koganei, Tokyo 184-8501, Japan
[2] Department of Environmental Sciences, Tokyo Gakugei University, Nukuikita 4-1-1, Koganei, Tokyo 184-8501, Japan

E-mail: tkosaka@u-gakugei.ac.jp

Abstract. X-ray diffraction, thermogravimetry differential thermal analysis, impedance analysis, and positron lifetime spectroscopy were conducted for Gadolinium-doped cerium oxides (GDC) prepared using oxalate coprecipitation method. XRD revealed the fluorite structure indicating that gadolinium is successfully doped into cerium oxide. Prior to sintering, the vacancy-sized free volume and nanovoid were observed at grain boundaries. The vacancy-sized free volumes shrank with increasing sintering temperatures and finally got dominant.

1. Introduction

Solid oxide fuel cells have been intensively studied due to their high power conversion efficiency. There has been a long-standing problem of extremely high operating temperature for solid oxide fuel cells. Gadolinium-doped cerium oxides are one of the most promising candidates as the electrolytes operated at intermediate temperature [1-3]. The total ionic conductivity of GDC is governed by the local atomic structure of transgranular [4], which is expected to be correlated with sintering. Song et al. [5] reported that the points defect is an important factor which influences the evolution of structure and chemical changes in hydrogen irradiated CeO_2. Zhang et al. [6] investigated the effect of transition metal oxide loading on sintering behaviour and ion conductivity of GDC. It is thus of significance to investigate the local structural change induced by sintering. In this paper, we investigated GDC prepared using oxalate coprecipitation method by XRD, TG-DTA, impedance analysis, and positron lifetime spectroscopy.

2. Experimental

Gadolinium-doped cerium oxides (GDC) were prepared using oxalate coprecipitation method [7]. The cerium and gadolinium nitrate mixed solution at a molar ratio of $Ce^{3+}/ Gd^{3+}= 4/1$ was dropped into a stirred oxalate acid solution to produce the oxalate precipitate. The precipitates were calcined at 873K for 1 h in the air to form oxides. They were compacted into a pellet by uniaxial pressing (20MPa) and sintered in the wide temperature range (873 - 1473K) for 6 h in air. The lattice constant of GDC was determined by XRD using NaCl as internal standard. The relative density was measured by Archimedes method. The Pt electrodes were formed on the both side of GDC pellet by heating at 1473 K using Pt paste (Tanaka Kikinzoku TR-7070). The ionic conductivity of the pellet was measured from 673K to 973K conducted by AC impedance analyzer (Princeton applied research corp. VersaSTAT3) in the frequency range from 100 μHz to 1MHz. Positron lifetime spectroscopy was conducted using ^{22}Na positron source at room temperature. The positron lifetime spectra were numerically analyzed using the POSITRONFIT code [8].

3. Results and discussion

Figure 1 shows the XRD patterns of GDC and CeO_2 sintered at 1073K for 6h in air. The XRD pattern indicates that GDC has a fluorite type structure without any other phases. All peaks of GDC are shifted to low angle region against to the corresponded ones of CeO_2 indicating the expansion of interplanar spacing for GDC by Gd doping. The lattice constants calculated for GDC and CeO_2 are 0.5429 nm and 0.5415 nm, respectively. This implies that it is partially substituted Ce^{4+} with Gd^{3+} and formed of an oxygen vacancy successfully.

The results of TG-DTA analysis for Ce and coprecipitated oxalate are shown in Figure 2. They lost chemically adsorbed water at 473K and began to decompose to oxides with the producing CO, CO_2 and O_2 gases at 573K. The exothermic peak in DTA curves was detected at a higher temperature for coprecipitated than for Ce oxalate.

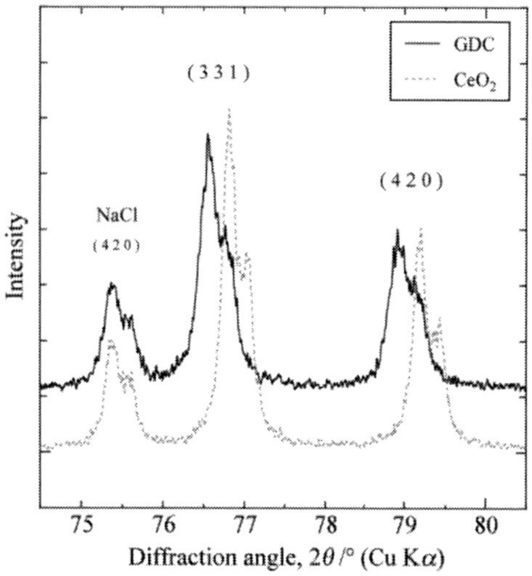

Figure 1. XRD patterns of GDC and CeO_2.

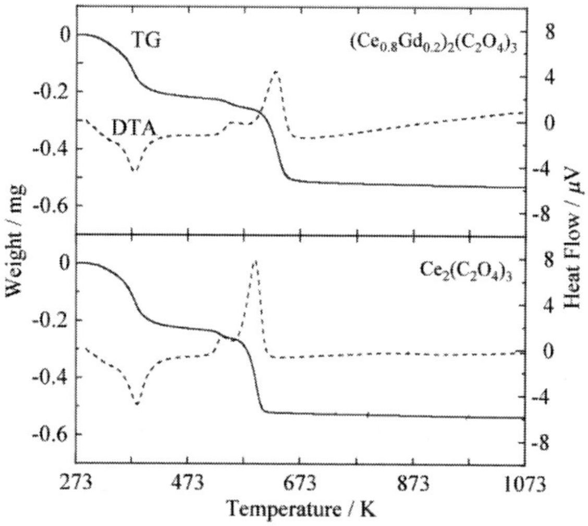

Figure 2. TG-DTA curves of Ce and coprecipitated oxalate.

Figure 3 shows the temperature dependence of the total ionic conductivity for GDC. The data was fitted with least-square method. Sameshima et al. reported that CeO_2 sintered at 1873K for 4 h in dried air shows the total ionic conductivity of 4.9×10^{-5} S / cm at 773 K lower by 20 times than the value for GDC sintered at 1473 K in the present study. The activation energy of GDC (81kJ / mol) is higher than that of CeO_2 (61.7kJ / mol) measured by Sameshima et al [9]. This indicates that the substitution of Ce^{4+} with Gd^{3+} with larger ionic radius than that of Ce^{4+} is caused to form the high potential energy barrier. The total ionic conductivity is explained by oxygen ion diffusion and is influenced by the number of oxygen vacancies existed in the samples.

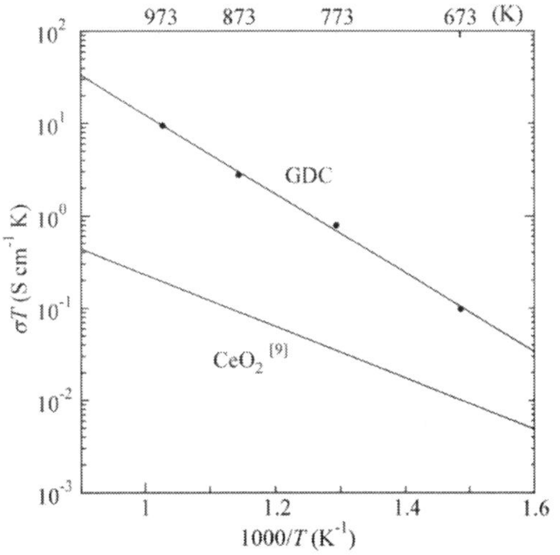

Figure 3. Arrhenius plots of total conductivity of GDC and CeO_2. Solid lines are results of the least-square fit.

Figure 4 shows the results of positron lifetime spectroscopy for GDC. Prior to sintering, two components τ_1 (~ 260 ps) and τ_2 (~ 500 ps) corresponding to a vacancy-sized free volume and a nanovoid were obtained with their relative intensities I_1 (~ 80 %) and I_2 (~ 20 %). The average size of GDC crystallite evaluated from 311 peak broadening using Scherrer's equation is 12.8 nm before sintering, which is by far smaller than the typical positron diffusion length in solids of ~ 300 nm [10]. Positrons implanted in GDC crystallite can thus efficiently diffuse out and annihilate at grain boundaries. We therefore conclude that the vacancy-sized free volumes and nanovoids detected by the present positron lifetime spectroscopy are present at grain boundaries. Similar observation was reported for yttria-stabilized zirconia by Cizek et al. [11].

Upon increasing temperatures the positron lifetime τ_1 decreases and its relative intensity I_1 increased, signifying shrinkage of vacancy-sized free volumes together with sintering. The lifetime τ_2 and its relative intensity I_2 exhibit the opposite tendency to those of vacancy-sized free volume (see right hand in Figure 4). The vacancy-sized free volumes got dominant at the sintering temperature of 1473 K, while the nanovoids almost disappear. The results suggest that a substantial increase of local electron density at grain boundaries occurs as a result of sintering and the process of sintering follows the kinetics of vacancy-sized free volumes and nanovoids at grain boundaries.

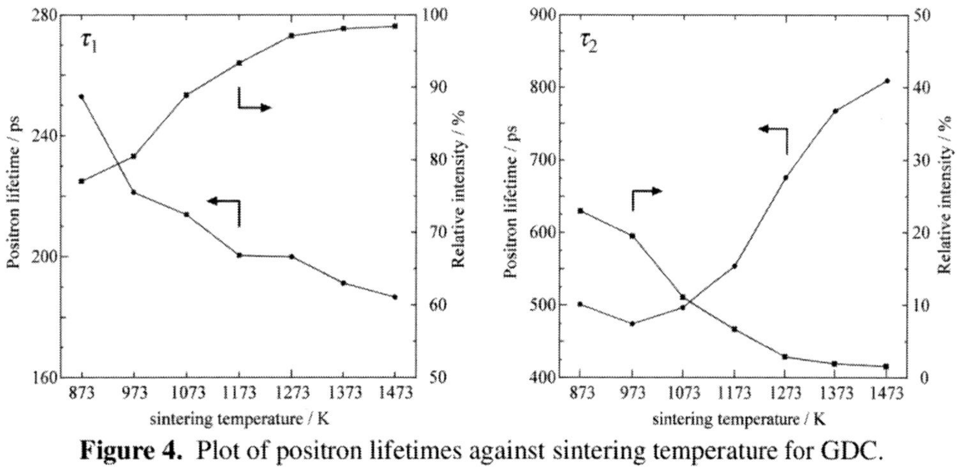

Figure 4. Plot of positron lifetimes against sintering temperature for GDC.

4. Conclusions

We prepared gadolinium-doped cerium oxide using oxalate coprecipitation method. XRD revealed the fluorite structure indicating that gadolinium is successfully doped into cerium oxide. The total ionic conductivity of GDC is 9.8×10^{-4} S / cm at 773K, which is 20 times higher than that of CeO_2 sintered at 1873K in air. This result indicated that the oxygen vacancies have an influence on the total ionic conductivity effectively. Positron lifetime spectroscopy revealed the presence of vacancy-sized free volumes and nanovoids, of which the kinetics is associated with sintering.

Acknowledgments

This work was partially supported by Grants-in-Aid for Scientific Research from the Ministry of Education, Science, Sports and Culture of Japan (Grant Nos. 20740173 and 2154317). The authors would like to thank to Dr. K. Shinozaki (Tokyo Institute of Technology) for experimental assistance in the AC impedance measurement.

References

[1] Reddy K. R. et al. 2005 *J. Electrochem.* **15** 45
[2] Kim H. N. et al. 2006 *J. Electroceram.* **17** 793
[3] Jadhav L. D. et al. 2009 *J. Alloys and Compounds* **470** 383
[4] Hara A. et al. 2008 *J. Ceram. Soc. Japan* **116** 291
[5] Song M. et al. 2002 *Nucl. Instr. and Meth. in Phys. Res. B* **191** 586
[6] Zhang T. S. et al. 2004 *Solid State Ionics* **168** 187
[7] Higashi K. et al. 1999 *Key Eng. Mater.* **159-160** 25
[8] Kirkegaard P. et al. 1974 *Comput. Phys. Commun.* **7** 401
[9] Sameshima S. et al. 2000 *J. Ceram. Soc. Japan* **108** 1060
[10] Staab T. E. M. et al. 1999 *J. Mater. Sci.* **34** 3833
[11] Cizek J. et al. 2007 *phys. stat. sol. (c)* **4** 3847

Investigation of hydrogen interaction with defects in zirconia

O Melikhova[1], J Kuriplach[1], J Čížek[1], I Procházka[1], G Brauer[2] and W Anwand[2]

[1] Charles University in Prague, Faculty of Mathematics and Physics, V Holešovičkách 2, CZ-18000 Prague 8, Czech Republic
[2] Institut für Ionenstrahlphysik und Materialforschung, Forschungszentrum Dresden-Rossendorf, Postfach 510119, D-01314, Dresden, Germany

E-mail: oksivmel@yahoo.com

Abstract. Defect studies of a ZrO_2 + 9 mol. % Y_2O_3 single crystal were performed in this work using a high resolution positron lifetime spectroscopy combined with slow positron implantation spectroscopy. In order to elucidate the nature of positron trapping sites observed experimentally, the structural relaxations of several types of vacancy-like defects in zirconia were performed and positron characteristics for them were calculated. Relaxed atomic configurations of studied defects were obtained by means of *ab initio* pseudopotential method within the supercell approach. Theoretical calculations indicated that neither oxygen vacancies nor their neutral complexes with substitute yttrium atoms are capable of positron trapping. On the other hand, zirconium vacancies are deep positron traps and are most probably responsible for the saturated positron trapping observed in yttria stabilized zirconia single crystals. However, the calculated positron lifetime for zirconium vacancy is apparently longer than the experimental value corresponding to a single-component spectrum measured for the cubic ZrO_2 + 9 mol. % Y_2O_3 single crystal. It was demonstrated that this effect can be explained by hydrogen trapped in zirconium vacancies. On the basis of structure relaxations, we found that zirconium vacancy – hydrogen complexes represent deep positron traps with the calculated lifetime close to the experimental one. In zirconium vacancy – hydrogen complexes the hydrogen atom forms an O-H bond with one of the nearest neighbour oxygen atoms. The calculated bond length is close to 1 Å.

1. Introduction

Zirconia (ZrO_2) exhibits a high thermal stability and a low thermal conductivity together with good mechanical properties (high strength and enhanced fracture toughness). Zirconia-based materials are promising for the use in a large area of industrial applications including refractory materials, ceramic glazes, oxygen sensors, solid oxide fuel cells etc [1]. Three zirconia polymorphs are known to exist at ambient pressure [2]. The high temperature cubic phase is stable at temperatures above 1380 °C. At lower temperatures zirconia transforms into a tetragonal and subsequently into a monoclinic phase.

The high temperature cubic phase can be stabilized down to room temperature by addition more than 8 mol % of trivalent yttria (Y_2O_3) [2]. Such system is then called yttria stabilized zirconia (YSZ). An embedment of Y_2O_3 into the ZrO_2 host lattice leads to a deviation from stoichiometry resulting in the creation of a high amount of native oxygen vacancies. Hence, vacancies play certainly a very important role in physical properties of YSZ.

High quality YSZ single crystals are nowadays commercially available. As vacancies strongly influence properties of YSZ, it is very important to characterize in detail vacancies in YSZ crystals. It

Journal of Physics: Conference Series **225** (2010) 012035 doi:10.1088/1742-6596/225/1/012035

this work we investigated vacancies and vacancy-like defects using positron annihilation spectroscopy (PAS), which is a well established experimental technique with a high sensitivity to open volume point defects like vacancies. In order to facilitate interpretation of experimental PAS data, we performed also theoretical modeling of defects in YSZ by state-of-art *ab initio* electronic structure calculations including ionic relaxations.

2. Experimetal

A commercially available ZrO_2 + 9 mol.% Y_2O_3 single crystal with dimensions $10 \times 10 \times 0.5$ mm^3 supplied by Crystec GmbH, Berlin was investigated in the present work. Surfaces of YSZ crystals were optically polished.

A digital positron lifetime (LT) spectrometer [3] with an excellent time resolution of 145 ps (FWHM ^{22}Na) was employed for LT studies. The decomposition of experimental LT spectra to individual components was performed by a maximum likelihood-based procedure [4].

Slow positron implantation spectroscopy (SPIS) was performed on a magnetically guided variable energy slow positron beam SPONSOR [5]. The energy of incident positrons can be varied in the range from 0.03 to 36 keV. Doppler broadening (DB) of annihilation profile was measured by high purity Ge detector with an energy resolution of (1.06 ± 0.01) keV at 511 keV. The shape of broadened annihilation profile was analyzed using the S parameter and the dependence of the S parameter on the positron energy was fitted by the VEPFIT code [6].

3. Theoretical calculations

As lattice relaxations around defects are known to be rather high in semiconductors including oxides [7], it is important to calculate positron characteristics for relaxed defect geometries. *Ab initio* calculations of atomic relaxations around defects and determination of the relaxed defect geometry having the lowest energy was performed using the Vienna *ab initio* simulation package (VASP) [8]. The calculations were performed using the generalized gradient approximation (GGA) projector augmented-wave (PAW) pseudopotential. *4s*, *4p*, *5s* and *4d* Zr electrons and *2s* and *2p* O electrons were included in the *ab initio* VASP pseudopotential calculations of electronic structure and ion relaxations. For calculations of vacancy-hydrogen complexes *1s* H electron was included in the VASP calculations as well. Defects were modeled using 96 atom-based supercells employing 2×2×2 *k*-mesh sampling centered at the Γ point. Cubic ZrO_2 (fluorite) lattice with lattice parameter a = 5.08 Å [9] was assumed in calculations.

Positron lifetimes and positron binding energies to defects were calculated in the limit of vanishing positron density within the so called 'conventional scheme' described in ref. [10], which becomes nowadays a standard for calculations of positron parameters. Note that positron binding energy to a defect is defined as

$$E_B = E_f - E_d, \quad (1)$$

where E_f is the ground state energy of a free, delocalized positron in the perfect ZrO_2 lattice and E_d is the ground state energy of a positron trapped at a particular defect.

The electron density needed for the construction of the effective positron potential was calculated using two approaches:

(i) The atomic superposition (ATSUP) method [11] which uses superimposed atomic electron densities calculated by a self-consistent relativistic atomic code [12]. The ATSUP approach *neglects* charge transfer in crystals, but it is a very fast method which can be used even for very large supercells.

(ii) The self-consistent valence electron density calculated by VASP was used in positron calculations. In this approach frozen-core electron orbitals, which are unaffected by crystal bonding, were added to the self-consistent VASP valence electron density. Hence in this approach charge transfer is *included* in calculations of positron parameters.

Two approaches were used also for the correlation part of the positron potential:

(i) The local-density approximation (LDA) using results of Arponen and Pajanne [13] parametrized by Boroński and Nieminen [14] taking into account incomplete positron screening [15] considering a high frequency dielectric constant $\varepsilon_\infty = 4.62$, and

(ii) A GGA scheme for electron-positron correlations as formulated by Barbiellini et al. [16].

Combination of the approaches described above results in four various schemes how to calculate positron parameters denoted here: ATSUP-LDA, ATSUP-GGA, VASP-LDA, and VASP-GGA.

4. Results and discussion

4.1. Experimental results

LT measurements revealed that the ZrO_2 + 9 mol.% Y_2O_3 crystal exhibits a single component spectrum (apart from the source component) with a lifetime of (175 ± 1) ps. This lifetime is remarkably longer than the calculated lifetime of for free positrons in zirconia (see Table 1). Hence, this result suggests that virtually all positrons in the studied YSZ crystal are trapped at certain defects (saturated positron trapping).

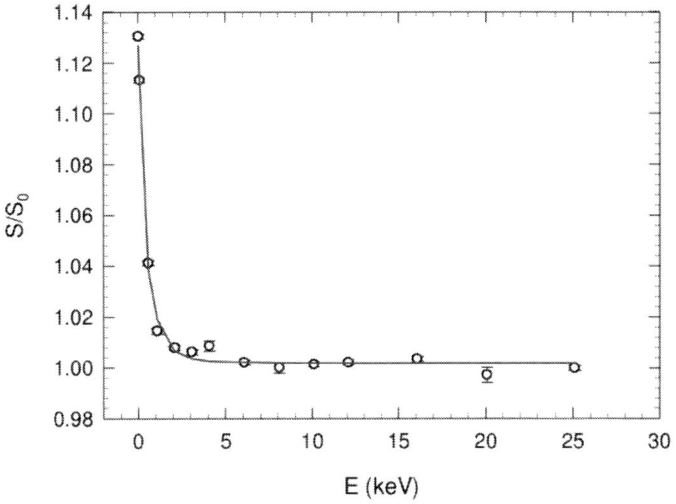

Figure 1. Dependence of the *S* parameter (open circles) on the positron energy the *E* for the ZrO_2 + 9 mol.%Y_2O_3 single crystal. The *S* parameter was normalized to the bulk value S_0 measured at *E* = 25 keV. The solid line shows a model curve fitted by the VEPFIT.

Figure 1 shows dependence of the *S* parameter on the positron energy *E* measured on the ZrO_2 + 9 mol.% Y_2O_3 crystal by SPIS. The solid line plotted in the figure is a model function calculated by the VEPFIT software assuming a single layer model. A very short positron diffusion length $L_+ = (8.6 \pm 0.4)$ nm was obtained from fitting. This gives a strong evidence that positrons are trapped at certain lattice defects. Note that similar result was obtained by Gryzspan et al. [17] on another cubic ZrO_2 crystal.

The concentration of positron traps can be estimated from SPIS results using the relation

$$c = \frac{1}{\nu \tau_B}\left(\frac{L_{B,+}^2}{L_+^2} - 1\right), \tag{2}$$

where τ_B and $L_{B,+}$ denote the free-positron lifetime and diffusion length in a perfect ZrO_2 lattice, while L_+ is the mean positron diffusion length measured in the sample. Here we used $\tau_B \approx 150$ ps calculated using the VASP-GGA approach (see Table 1) and $L_{B,+} \approx 150$ nm, which is a typical mean positron diffusion length in similar materials [18]. The symbol ν denotes the specific positron trapping rate to defects of given type. Using $\nu \approx 10^{14}$ at. s^{-1}, which is a typical value for monovacancies in most solids [10], we estimated the defect concentration in the studied crystal to be $c \approx 2$ at.%. Despite of uncertainties of input quantities, Eq. (2) implies a very high defect concentration comparable in order of magnitude with the concentration of substitutional yttrium atoms in the crystal (5.5 at.%). This testifies that the defects responsible for positron trapping in the ZrO_2 + 9 mol.% Y_2O_3 single crystal must be of a structural origin.

4.2. Theoretical modeling of vacancies

Calculated relaxed geometries of the oxygen and zirconium vacancies are shown in figures 2A and 2B, respectively. Calculated positron parameters for the defect-free crystal and various defects are listed in Table 1. One can see in figure 2A that in the case of the oxygen vacancy the nearest neighbor Zr ions and also the second nearest neighbor O ions relax towards the vacancy, i.e. the free volume of the oxygen vacancy is *reduced.* It was shown [19] that even in a rigid lattice the oxygen vacancy represents a too shallow potential well that is unable to confine a positron. The same conclusion holds for the relaxed oxygen vacancy, which due to the inward relaxation of neighbors becomes even a shallower well. Inability of the oxygen vacancy to trap positrons is demonstrated by the positron lifetime which approaches the bulk lifetime and a very low or even a negative positron binding energy (see Table 1).

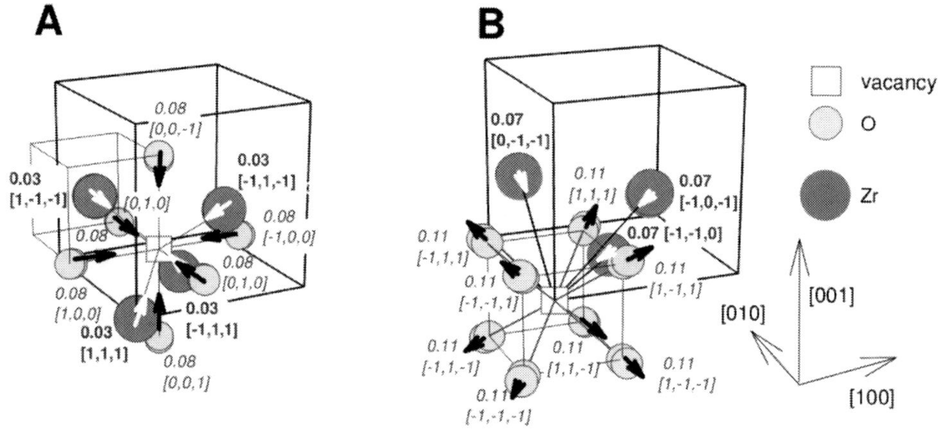

Figure 2. Relaxed defect geometries for (A) oxygen vacancy, (B) zirconium vacancy. The values given are atomic displacements in Å and displacement directions.

On the other hand, the zirconium vacancy is a deep positron trap even in the rigid ZrO_2 lattice. Figure 2B shows that the nearest neighbor oxygen anions relax outward, i.e. the open volume of the zirconium vacancy is *increased* due to relaxation. Hence the relaxed zirconium vacancy is even a deeper positron trap than the zirconium vacancy in a rigid lattice. This is reflected by a significantly increased lifetime of trapped positrons and a high positron binding energy.

One can see in Table 1 that results for the same defect may considerably differ when calculated by various approaches. It should be mentioned that the precision of the approaches used here is not the same. It is well known [16] that LDA approximation of electron-positron correlation overestimates annihilations with core electrons. This drawback is to some extend compensated if electron density is constructed by superposition of atomic densities and charge transfer is neglected [20]. Hence lifetimes calculated using VASP-LDA approach are usually shorter than experimental values and better agreement with experiment is obtained employing a simpler ATSUP-LDA approach. The GGA approximation of electron-positron correlations is more sensitive to details of electron density [16] and atomic superposition is not precise enough for the GGA scheme. As the consequence, the ATSUP-GGA approach over estimates positron localization and lifetimes calculated within this scheme are often found to differ from experimental values. On the other hand, positron lifetimes calculated within the GGA scheme with self-consistent electron density (i.e. including the charge transfer) are in very good agreement with experiment for most chemical elements [16,21]. In any case, the suitability of both the LDA and GGA approaches for ZrO_2 and other oxides needs to further investigated and we refer readers also to ref. [22] for further discussion.

Table 1. Positron lifetimes τ and positron binding energies E_B for various positron states calculated within the four various approaches described in section 3. Two different configurations of zirconium vacancy with hydrogen are described in the text.

	charge transfer neglected ATSUP-LDA		*charge transfer included* VASP-LDA		*charge transfer neglected* ATSUP-GGA		*charge transfer included* VASP-GGA	
	τ(ps)	E_B (eV)	τ(ps)	E_B (eV)	τ(ps)	E_B (eV)	τ(ps)	E_B (eV)
Bulk	138	-	133	-	147	-	151	-
O-vacancy	139	0.01	132	−0.12	147	0.02	153	0.05
Zr-vacancy	216	2.75	220	2.80	238	2.60	287	2.33
Zr-vacancy with H (i) symmetrical position	164	1.87	167	0.89	178	1.75	195	1.08
Zr-vacancy with H (ii) O-H pair	197	2.37	171	1.44	211	2.28	175	1.83

Moreover, due to the ionic nature of bonding, the effect of charge transfer in ZrO_2 is relatively important. This is demonstrated in figure 3A which compares the electron and positron density in a perfect ZrO_2 crystal calculated within the ATSUP scheme, i.e. neglecting the charger transfer, and using the VASP-GGA approach when the charge transfer was included. The electron density in the interstitial region among oxygen anions was reduced due to the charge transfer from Zr atoms to O anions. As the positron density has a maximum in these interstitial regions, the positron annihilation rate (and thereby the lifetime) is very sensitive to the electron density there and its decrease caused by charge transfer leads to a longer bulk positron lifetime calculated within the VASP-GGA approach.

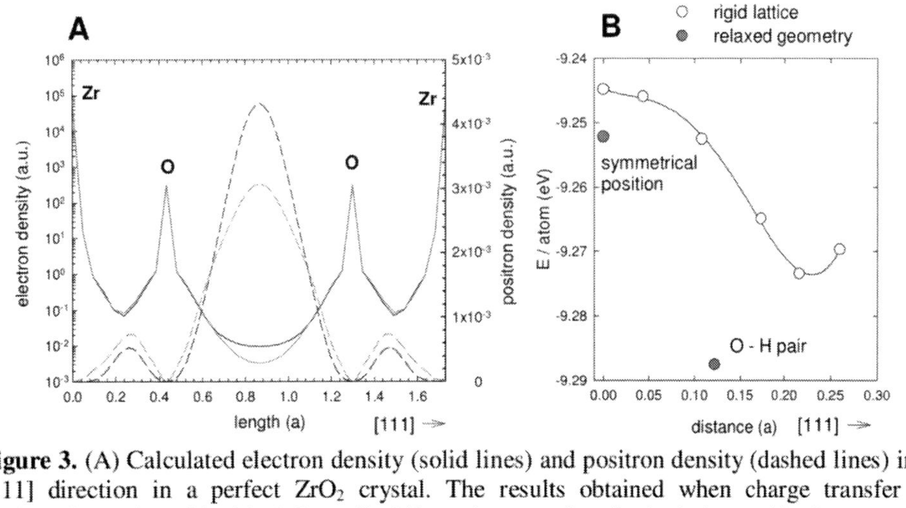

Figure 3. (A) Calculated electron density (solid lines) and positron density (dashed lines) in the [111] direction in a perfect ZrO_2 crystal. The results obtained when charge transfer was neglected are plotted by black lines. Red lines show results of calculations with charge transfer included. (B) The supercell energy per atom as a function of the hydrogen distance from the center of the vacancy. Open circles show results for hydrogen moved in the [111] direction from the vacancy center towards the nearest neighbour oxygen atom in the rigid lattice. Full red circles are energies of relaxed configurations (i) and (ii) described in the text.

Our results showed that oxygen vacancies even if present in a high concentration cannot explain the component with the lifetime 175 ps determined by LT spectroscopy. Note that complexes of oxygen vacancies with yttrium atoms were proposed by some authors as possible positron traps in YSZ, see e.g. [23,24]. A variety of these complexes, which can be expected to be formed in YSZ, were investigated in ref. [25]. However, none of them was found to be really capable of positron trapping. Thus, we can conclude that positron traps found in ZrO_2 + 9 mol.% Y_2O_3 crystal should be related to the zirconium vacancy. However, the zirconium vacancy is too deep positron trap characterized by a lifetime, which is remarkably longer than the 175 ps experimental value. Shortened lifetime of trapped positrons could be explained by some impurity associated with zirconium vacancies and reducing their free volume. Such an impurity could also reduce the formation energy of the zirconium vacancy, which is known to be rather high in pure ZrO_2 [26]. When considering possible impurities, hydrogen can be considered as the most promising candidate. First, the presence of hydrogen can hardly be avoided in any experimental technique of YSZ crystal growth. Second, hydrogen is highly mobile and can be relatively easily incorporated into ZrO_2 structure due its tendency to bind with oxygen. Indeed, it was found on the basis of our theoretical calculations that hydrogen in a perfect ZrO_2 crystal becomes bound to one of the nearest neighbor oxygen anions and forms an O-H pair with a bonding length of ≈ 1 Å.

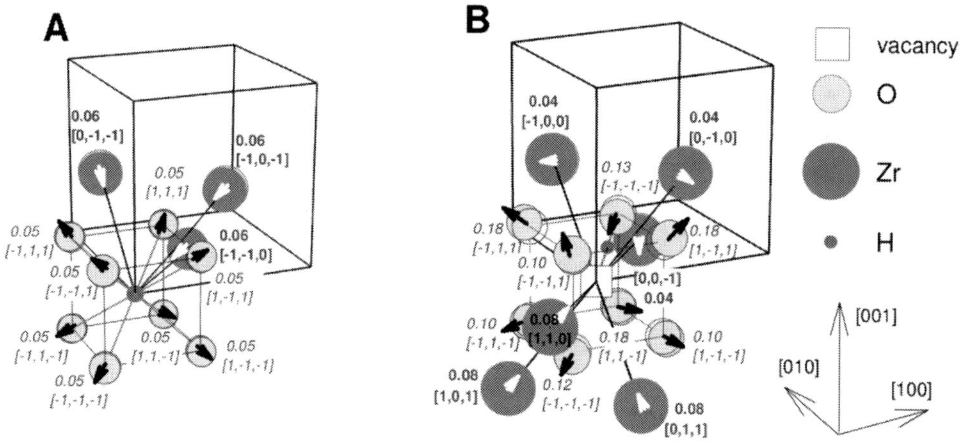

Figure 4. Relaxed defect geometries for two configurations of the zirconium vacancy containing a hydrogen atom: (A) "symmetrical case" with retained cubic symmetry and hydrogen in the center of vacancy, (B) the lowest energy geometry where hydrogen is displaced 1.13 Å from the centre of vacancy towards a nearest neighbour O atom and forms O-H pair with bonding length of 0.96 Å. The values given in the figure are atomic displacements in Å and displacement directions.

Our calculations showed that hydrogen is trapped in the zirconium vacancy. Figure 4 shows the relaxed geometry of the zirconium vacancy containing hydrogen. Two cases were considered:

(i) "symmetrical position" of hydrogen in the zirconium vacancy keeping the cubic symmetry of the ZrO_2 lattice (see figure 4A) and

(ii) the lowest energy geometry where the requirement of cubic symmetry was released. This configuration is shown in figure 4B.

In the case (i) hydrogen is located in the center of the zirconium vacancy and the nearest neighbor oxygen anions relax outward from the center of vacancy, i.e. in the similar manner as in the case of a plain zirconium vacancy only with lowered magnitude of their displacements. When the symmetry restriction is released, i.e. the case (ii), hydrogen moves towards a nearest neighbor oxygen anion and forms an O-H pair with bond length of ≈ 1 Å. The relaxation of other nearest neighbor oxygen anions is unequal and depends on the distance of oxygen anion from the O-H pair. Nevertheless in general, the nearest neighbor oxygen anions (except the one which formed the O-H pair) relax outward the center of the vacancy. To obtain a rough picture about hydrogen energies in the ZrO_2 lattice with the zirconium vacancy we calculated energy of a rigid supercell with hydrogen atom moving from the center of the vacancy towards one of the nearest neighbor oxygen anions. The calculated energy per atom is plotted in figure 3B as a function of the hydrogen distance from the center of the vacancy. Clearly the minimum in energy corresponds to hydrogen displaced from the center of the vacancy towards oxygen and the formation of the O-H pair. The red points in Figure 3B show the energy of relaxed geometry in the case (i) and (ii). Obviously the case (ii), i.e. formation of O-H pair, represents the lowest energy configuration of the zirconium vacancy containing hydrogen.

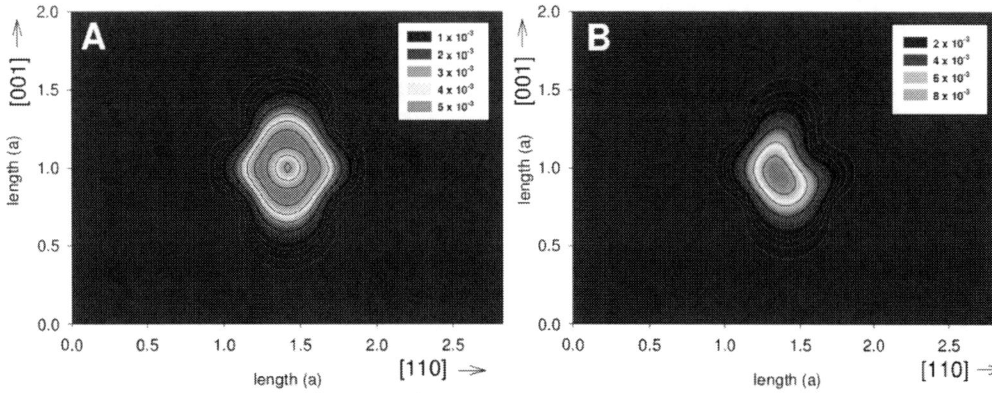

Figure 5. Positron density in (110) plane calculated using VASP-GGA approach for relaxed zirconium vacancy with hydrogen: (A) "symmetrical case" with retained cubic symmetry and hydrogen in the center of vacancy, (B) the lowest energy geometry where hydrogen forms O-H pair with a nearest neighbor oxygen anion

Figures 5 A and B show the calculated positron density for the configurations (i) and (ii), respectively, while calculated positron characteristics are listed in Table 1. One can clearly see that the zirconium vacancy containing a hydrogen atom is capable of positron trapping. Moreover, the presence of hydrogen in the zirconium vacancy leads in both cases considered here to a shortening of lifetime of trapped positrons (see table 1). Interestingly, positron lifetime calculations within the VASP-GGA scheme for the lowest energy configuration of the zirconium vacancy containing hydrogen, i.e. the case (ii), results in a positron lifetime of 175 ps which agrees very well with the lifetime measured on the YSZ single crystal. Hence, we suggest that experimental data measured on the ZrO_2 + 9 mol.% Y_2O_3 single crystal can be explained by positron trapping at zirconium vacancies containing hydrogen. We note that similar defects were recently identified in hydrothermally grown ZnO crystals [22]. This result emphasizes necessity to determine experimentally the hydrogen content in YSZ single crystals and compare it with the estimated concentration of zirconium vacancies.

5. Conclusions

Investigations of defects in ZrO_2 + 9 mol.% Y_2O_3 single crystal were performed by positron lifetime measurements combined with slow positron implantation spectroscopy. The interpretation of experimental data was made with the aid of *ab inito* modeling of vacancies and vacancy-like defects in ZrO_2. It was found that oxygen vacancies are too shallow and are unable to confine positrons. On the other hand, zirconium vacancies are deep positron traps. Experimental data testified that the ZrO_2 + 9 mol.% Y_2O_3 single crystal contains a high density of defects because virtually all positrons annihilate from the trapped state. However, the experimental lifetime 175 ps is shorter than the calculated lifetime of positrons trapped at the zirconium vacancy. Extended *ab initio* modeling of zirconium vacancy-hydrogen complexes revealed that hydrogen forms an O-H pair with one of the nearest neighbor oxygen anions. It was demonstrated that presence of hydrogen in the zirconium vacancy leads to a shortening of the positron lifetime and, in principle, can explain experimental data measured on the ZrO_2 + 9 mol.% Y_2O_3 single crystal.

Acknowledgments
We are grateful to M.J. Puska for his ATSUP code that served as a basis for further developments. This work was supported by the Academy of Science of the Czech Republic (project KJB101120906) and the Ministry of Schools, Youths and Sports of the Czech Republic through the research plan № MSM 0021620834 and the project COST OC 165.

References
[1] Ryshkewitz E and Richardson D W 1985 *Oxide Ceramics: Physical Chemistry and Technology* (Hasskell, General Ceramics)
[2] Goff J P, Hayes W, Hull S, Hutchings M T and Clausen K N 1999 *Phys. Rev. B* **59** 14202
[3] Becvar F, Cizek J, Prochazka I and Janotova J 2005 *Nucl. Instr. Meth. A* **539** 372
[4] Prochazka I, Novotny I and Becvar F 1997 *Mater. Sci. Forum* **255-257** 772
[5] Anwand W, Kissener H-R and Brauer G 1995 *Acta Phys. Polon. A* **88** 7
[6] van Veen A, Schut H, Clement M, de Nijs J, Kruseman A and Ijpma M 1995 *Appl. Surf. Sci.* **85** 216
[7] Stapper G, Bernasconi M, Nicoloso N and Parrinello M 1999 *Phys. Rev. B* **59** 797
[8] Kresse G and Hafner J 1994 *Phys. Rev. B* **49** 14251
[9] Howard C J, Hill R J and Reichert B E 1988 *Acta Crystallogr. Sect. B: Struct. Sci.* **44** 116
[10] Puska M J and Nieminen R M 1994 *Rev. Mod. Phys.* **66** 841
[11] Puska M J and Nieminen RM 1983 *J. Phys. F: Met. Phys.* **13** 333
[12] Desclaux J P 1975 *Comput. Phys. Commun.* **9** 31
[13] Arponen J and Pajanne E 1979 *Ann. Phys. (N.Y.)* **121** 343
[14] Boroński E and Nieminen R M 1986 *Phys. Rev. B* **34** 3820
[15] Puska M J, Mäkinen S, Manninen M and Nieminen R M 1989 *Phys. Rev. B* **39** 7666
[16] Barbiellini B, Puska M J, Korhonen T, Harju A, Torsti T and Nieminen R M 1996 *Phys. Rev. B* **53** 16201
[17] Grynszpan R I, Saude S, Mazerolles L, Brauer G and Anwand W 2007 *Rad. Phys. Chem.* **76** 333
[18] Staab TEM, Krause-Rehberg R and Kieback B 1999 *J. Mater. Sci.* **34** 3833
[19] Melikhova O, Kuriplach J, Cizek J, Procházka I, Anwand W, Brauer G, Konstantinova T E and Danilenko I A 2007 *phys. stat. sol. (c)* **4** 3831
[20] Puska M J, Seitsonen A P and Nieminen R M 1995 *Phys. Rev. B* **52** 10947
[21] Robles J C, Ogando E, and Plazaola F 2007 *J. Phys.: Condens. Matter* **19** 176222
[22] Brauer G, Anwand W, Grambole D, Grenzer J, Skorupa W, Cizek J, Kuriplach J, Procházka I, Ling CC and So CK 2009 *Phys. Rev. B* **79** 115212
[23] Wang Z, Chen Z Q, Zhu J, Wang S J and Guo X 2000 *Rad. Phys. Chem.* **58** 697
[24] Wang Z, Chen Z Q, Wang S J and Guo X 2000 *J. Mat. Sci. Lett.* **19** 1275
[25] Cizek J, Melikhova O, Prochazka I, Kuriplach J, Kuzel R, Brauer G, Anwand W, Konstantinova T E and Danilenko I A 2009 *Phys. Rev. B*, submitted.
[26] Foster A S, Sulimov V B, LopezGejo F, Shluger A L and Nieminen R M 2001 *Phys. Rev. B* **64** 224108

Muon spin spectroscopy of the rod-like liquid crystal 4-*n*-octyl-4′-cyanobiphenyl (8CB)

I McKenzie[1], **H Dilger**[2], **R Scheuermann**[3], **A Stoykov**[3]
[1] ISIS Facility, STFC Rutherford Appleton Laboratory, Chilton, Didcot, Oxon, OX11 0QX, UK
[2] Institut für Physikalische Chemie, Universität Stuttgart, Pfaffenwaldring 55, D-70569 Stuttgart, Germany
[3] Laboratory for Muon Spin Spectroscopy, Paul Scherrer Institute, CH-5232 Villigen, Switzerland

E-mail: iain.mckenzie@stfc.ac.uk

Abstract. Avoided level crossing muon spin resonance (ALC-μSR) spectroscopy has been used to study the four cyclohexadienyl-type radicals produced by the addition of muonium (Mu) to the rod-like liquid crystal 4-*n*-octyl-4′-cyanobiphenyl (8CB). ALC-μSR spectra have been obtained over a wide temperature range in the isotropic (I), nematic (N), smectic-*A* (Sm-*A*) and crystalline (Cr) phases. The ALC-μSR spectra of the Mu adducts of 8CB are very similar to those of the Mu adducts of 4-*n*-pentyl-4′-cyanobiphenyl (5CB). Four Δ_0 resonances were observed in the ALC-μSR spectra in the I, N and super-cooled Sm-*A* phases, from which the methylene proton hyperfine coupling constants (hfcs) of the Mu adducts of 8CB were determined as a function of temperature. The methylene proton hfcs of two of the radicals have unusual temperature dependence in the N and Sm-*A* phases and have smaller values than would be predicted from extrapolating the data in the isotropic phase. The shift of the hfcs is due to the dipolar hyperfine coupling, the ordering of the molecules along the applied magnetic field and the fluctuations of the molecules about the director. Δ_1 resonances appear just below the Cr–Sm-*A* phase transition, which indicates that this transition is associated with cessation of rotation around the long molecular axes.

1. Introduction

A liquid crystal (LC) is a material that exhibits phases in a temperature region between the solid or crystalline (Cr) phase and the liquid or isotropic (I) phase that possess some properties of both liquids and solids [1-2]. The mesophases can flow like a liquid but have anisotropic optical and electro-magnetic characteristics. LCs are characterized by the shape of the molecules with calamitic LC made up of long, rod-like molecules. Calamitic LC can form a large number of mesophases with the most common being the nematic (N) phase. In the N phase the material gains an amount of orientational order but no positional order. This reordering is thought to be due to the packing constraints of the rod-like molecules. The molecules in the N phase order along a preferred direction known as the director (*n*). Below the N phase the LC can form smectic phases where the molecules, although still forming a fluid, prefer to lie on average in layers, with each layer being essentially a 2-D nematic LC. In the smectic-*A* (Sm-*A*) phase the director of each layer is parallel to the layer normal.

The physical and chemical properties of the rod-like cyanobiphenyl (nCB) LC have been extensively studied, partly because several members of this family can exhibit mesophases near room temperature and have potential commercial applications. 4-*n*-octyl-4′-cyanobiphenyl (8CB) is the first member of this series to exhibit both N and Sm-*A* phases. The phase transition temperatures depend on the heating or cooling rate (T_{NI} = 312.2 K, $T_{\mathrm{Sm}\text{-}A\mathrm{N}}$ = 305.3 K and $T_{\mathrm{CrSm}\text{-}A}$ = 267.0 K with a cooling rate of 10 K min^{-1} [3]).

It is a longstanding goal to better understand the behaviour of the molecules within the LC. Many of the common techniques for characterizing LCs, such as polarizing optical microscopy, only provide information about the bulk properties. Magnetic resonance techniques such as nuclear magnetic resonance (NMR) and electron paramagnetic resonance (EPR) has been used to provide information about the dynamics and ordering of the 8CB molecules at the molecular level. Both NMR and EPR have studied small probe molecules, such as substituted benzenes, alkanes or stable nitroxide radicals like tempo, that become orientationally ordered due to the anisotropic environment in 8CB [4-14]. NMR has also been used to study molecular dynamics of pure 8CB [15,16].

Spin labelling involves attaching a paramagnetic group to a molecule in order to probe the molecular dynamics and local environment. Spin labels can be introduced into LC with unsaturated bonds by the addition of muonium (Mu), a light isotope of hydrogen [17]. The result is a radical, known as a muoniated radical, in which a proton has been replaced with a positive muon. Muons are sensitive probes because they are produced with spin polarisation close to 100%, the parity-violating decay of the muon provides a convenient way to monitor the evolution of their spin, and its lifetime (2.2 μs) is comparable to the lifetimes of many organic radicals. There are a series of experimental techniques collectively known as muon spin rotation, resonance and relaxation or μSR that are powerful tools for studying the structure and dynamics of muoniated radicals in soft matter, solids, liquids, and gases [18]. Avoided level crossing muon spin resonance (ALC-μSR) has been used extensively to study soft matter systems, particularly the partitioning of cosurfactants in lamellar phase dispersions [19-21].

The ALC-μSR technique involves measuring the asymmetry of the muon decay, which is proportional to the muon polarization, as a function of a magnetic field applied parallel to the initial direction of the muon spin. In high magnetic fields, the eigenstates of the radical can be approximated by pure Zeeman states, so there is no evolution of the muon's spin with time, and the asymmetry is independent of the magnetic field. At specific values of the applied magnetic field, nearly degenerate pairs of spin states can be mixed through the isotropic and anisotropic components of the hyperfine interaction. The muon polarization oscillates between the two mixing states, and this leads to a loss of time-integrated asymmetry. There are three types of ALC resonances, which are referred to as Δ_0, Δ_1 and Δ_2. The subscript refers to the difference in the total spin quantum number between the two mixing states. The Δ_2 resonance is extremely weak and narrow and rarely observed.

The Δ_1 resonance field is given by

$$B_{res}^{\Delta_1} = \frac{A_\mu}{2\gamma_\mu} - \frac{A_\mu}{2\gamma_e} \tag{1}$$

where A_μ is the muon hfc, γ_μ is the muon gyromagnetic ratio, and γ_e is the electron ratio. The Δ_1 resonance arises from mixing between spin states with the same electron and proton spins but different muon spin directions. These spin states are mixed only in the presence of anisotropy, so the Δ_1 resonance can be considered to be diagnostic of a frozen state or anisotropic motion.

The Δ_0 resonance is due to mixing between spin states that have the same electron spin but opposite muon and proton spins and is observed for muoniated radicals in the solid, liquid, or gas phases. The Δ_0 resonance field depends on both the muon hfc and the proton hfc, A_p, and is given by

$$B_{res}^{\Delta_0} = \frac{1}{2}\left[\frac{A_\mu - A_p}{\gamma_\mu - \gamma_p} - \frac{A_\mu + A_p}{\gamma_e}\right] \tag{2}$$

where γ_p is the proton gyromagnetic ratio.

The muon and methylene proton hfcs of muoniated cyclohexadienyl-type are approximately proportional to each other.

$$A_\mu = K \cdot A_p \tag{3}$$

The main contribution to K is the γ_μ/γ_p ratio, which is 3.183, and the rest results from the mass dependence of the molecule's vibrational modes. K was found to be 4.123 for the cyclohexadienyl radical in benzene at 298 K [22]. This approximation allows to determine A_p from the Δ_0 resonance field by combining equations (2) and (3) to give

$$B_{res}^{\Delta_0} = \Theta \cdot A_p \tag{4}$$

where $\Theta = 1.67 \times 10^{-2}$ T MHz^{-1} around room temperature.

The only LC to have been studied by μSR is 4-*n*-pentyl-4′-cyanobiphenyl (5CB) [23-25]. Four types of substituted cyclohexadienyl radical were formed by Mu addition to the biphenyl moiety of 5CB, and their proton hfcs were measured using ALC-μSR. The proton hfcs of two of the muoniated isomers shift from their isotropic value in the N phase and this was demonstrated to result from the ordering of the molecules, alignment of the molecules with the magnetic field and fluctuations of the molecules about the director. 8CB is the first LC to be studied by μSR that exhibits a Sm-*A* phase.

2. Experimental

8CB was purchased from Sigma-Aldrich and was used without further purification. Oxygen was removed by melting the sample in a nitrogen atmosphere and bubbling with nitrogen gas for approximately two hours. The degassed samples were sealed in brass cells with an internal volume of 17 ml and 50 μm brass foil windows.

The ALC-μSR experiments were performed using the ALC spectrometer at the πE3 beamline of the Paul Scherrer Institute in Villigen, Switzerland. The sample was initially melted, and the ALC-μSR spectra were obtained at several temperatures as the sample was slowly cooled (~0.2 K min^{-1}). The magnetic field was always greater than 1.4 T while the sample was cooling in order to produce an oriented monodomain with the director aligned along the magnetic field [1,16]. The samples were left for 15-30 min to equilibrate at a given temperature before measurements were made, and several scans were performed at each temperature (each taking at least 45 min) to obtain a sufficient signal-to- noise ratio. No changes were observed in the ALC-μSR spectra as a function of time at a given temperature, so we conclude that the sample was at thermal equilibrium at all times when the measurements were made. The ALC-μSR spectra of 8CB in the I, N and Sm-*A* phases were obtained between 1.65 and 2.15 T with a step size of 1 mT, while the ALC-μSR spectra in the Cr phase were obtained between 1.2 and 2.2 T with a step size of 2 mT.

ALC-μSR spectra have a large field-dependent background that is very sensitive to the stopping position of the muons. Each resonance was fit with a single Lorentzian function, and a fifth-order polynomial was used to model the background. The fitting was performed with the MINUIT function minimization library in the ROOT package from CERN [26]. All of the ALC-μSR spectra in the I, N and Sm-*A* phases were fit between 1.7 and 2.1 T. The corrected asymmetry is the asymmetry after the subtraction of the background.

3. Results and Discussion

3.1. ALC-μSR spectra of the isotropic, nematic and smectic A phases

Representative ALC-μSR spectra of the Mu adducts of 8CB in the Sm-*A*, N and I phases are shown in figure 1. Four resonances were observed between 1.8 and 2.0 T, which are the methylene proton Δ_0 resonances of the radicals shown in figure 2. They are known to be Δ_0 resonances as they are present in the I phase and assignment is based on a comparison of the experimental hfcs with the DFT calculated hfcs reported for the Mu adducts of 4-*n*-methyl-4′-cyanobiphenyl (1CB) [25]. Such a comparison is valid because the hfcs should be independent of the length of the alkyl chain.

No Δ_1 resonances were observed at any temperature in the Sm-*A*, N and I phases, which indicates that all of the radicals have very small dipolar hfcs. The small dipolar hfcs arise from the rapid rotation of the molecules about the long axis. The rapid rotation results in the hyperfine tensor having axial symmetry with one component parallel to the rotation axis ($D^{\parallel}$) and two components perpendicular to the rotation axis ($D^{\perp}$), with $D^{\parallel} = -2D^{\perp}$. The angle between the principal components of the hyperfine tensor and the axis of rotation results in a very small $D^{\parallel}$.

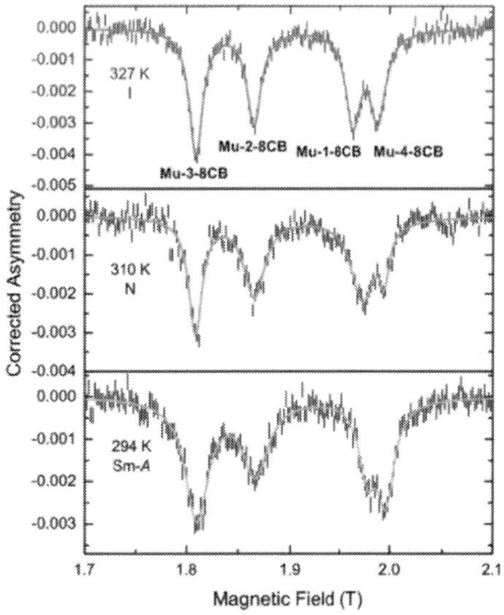

Figure 1. ALC-μSR spectra of the Mu adducts of 8CB in the isotropic (327 K), nematic phase (310 K) and smectic-*A* phases (294 K). The solid lines represent the best fits to the data. The resonances are assigned to the structures shown in Figure 2.

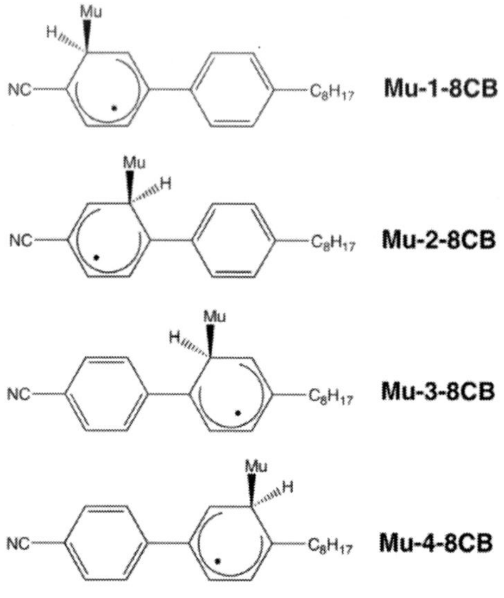

Figure 2. Structures of the four substituted muoniated cyclohexadienyl radicals formed by the addition of muonium to 8CB.

3.2. Temperature dependence of the proton hyperfine coupling constants in the isotropic, nematic and smectic-A phases

The four Δ_0 resonances broadened and shifted to higher fields as the temperature was lowered. The methylene proton hfcs of the four Mu adducts of 8CB are shown as a function of temperature in figure 3, and these increased linearly with decreasing temperature in the I phase, although dA_p/dT in Mu-2-8CB ($-(7.8\pm0.9) \times 10^{-3}$ MHz K^{-1}) and Mu-3-8CB ($-(3.9\pm0.7) \times 10^{-3}$ MHz K^{-1}) is less than that of Mu-1-8CB ($-(2.06\pm0.14)\times10^{-2}$ MHz K^{-1}) and Mu- 4-8CB ($-(1.40\pm0.09)\times10^{-2}$ MHz K^{-1}). The dA_p/dT values for the Mu adducts of 8CB are similar to the values of the corresponding values of the Mu adducts of 5CB [25]. The muon and methylene proton hfcs of cyclohexadienyl-type radicals are known to increase linearly with decreasing temperature due to the vibrational motion of the CHMu group [22]. The methylene proton hfcs of Mu-1-8CB and Mu-4-8CB increased approximately linearly with decreasing temperature in the N and Sm-*A* phases, while the methylene proton hfcs of Mu-2-8CB and Mu-3-8CB deviate from linearity below T_{NI} and are lower in the N and Sm-*A* phases than one would predict based on an extrapolation of the values measured in the I phase. The shift of the hfc,

ΔA_p, is defined to be the difference between the measured hfc and the extrapolated isotropic hfc. The temperature dependence of ΔA_p is shown in figure 4. The shift of A_p is given empirically by

$$\Delta A_p = C(T_{NI} - T)^{\beta} \tag{5}$$

The critical exponent, β, has values of 0.22±0.07 for Mu-2-8CB and 0.22±0.03 for Mu-3-8CB while the N–I phase transition temperature was determined to be 313.5±1.5 K from Mu-2-8CB and 310.2±0.9 K from Mu-3-8CB. No shift of the proton hfc was observed in Mu-1-8CB or Mu-4-8CB because the dipolar hfcs in these molecules are further averaged by torsional motion of the biphenyl moieties. The torsional barriers of Mu-1-1CB and Mu-4-1CB were shown to be much lower than those of Mu-2-1CB and Mu-3-1CB [25].

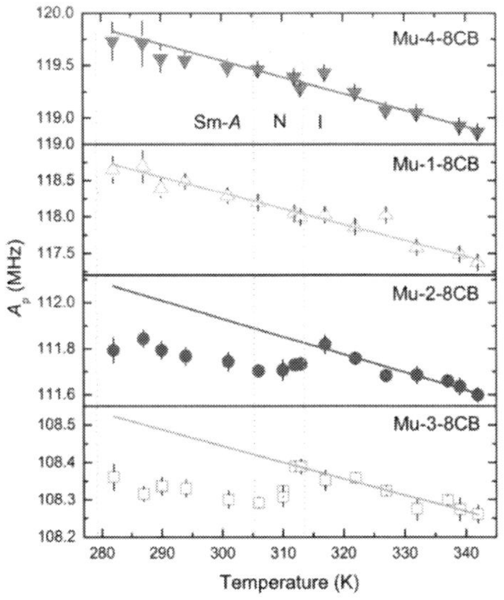

Figure 3. Temperature dependence of the methylene proton hyperfine coupling constants of the Mu adducts of 8CB: Mu-1-8CB (△), Mu- 2-8CB (●), Mu-3-8CB (□), and Mu-4-8CB (▼). The dotted vertical lines show the I—N, N—Sm-*A* and Sm-*A*—Cr phase transitions. The solid lines are the fits in the isotropic phase, extrapolated to the nematic and smectic-*A* phases.

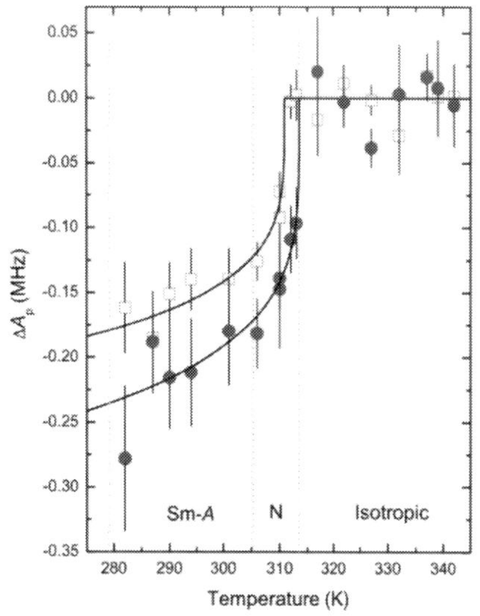

Figure 4. Temperature dependence of the hyperfine coupling constant shift, ΔA_p, for two of the Mu adducts of 8CB: Mu-2-8CB (●) and Mu-3-8CB (□). The dotted vertical lines show the I—N, N—Sm-*A* and Sm-*A*—Cr phase transitions. The solid lines are the best fits to the data using equation (5).

The hfcs of Mu-2-8CB and Mu-3-8CB shift from their isotropic values because of the ordering of the molecules, the alignment of the molecules with the magnetic field and the dipolar hyperfine interaction. We have assumed that the angular distribution of 8CB molecules in the N and Sm-*A* phases can be obtained using the Maier-Saupe (MS) theory. The normalized distribution of molecules as a function of angle, $f(\theta)$, according to the MS theory is given by [1]

$$f(\theta) = \frac{\exp\left(m \cdot \cos^2\theta\right)}{4\pi Z} \tag{6}$$

where the normalization constant, Z, is given by

$$Z = \int_0^1 \exp(mx^2)dx \tag{7}$$

The angular distribution depends on the order parameter (S), the interaction energy between molecules (U), the Boltzmann constant (k_B) and the temperature (T) since [1]

$$m = \frac{3U \cdot S}{2k_B T} \tag{8}$$

It has been previously demonstrated that for radicals with a small axially symmetric dipolar coupling aligned along the rotation axis and an angular distribution given by the MS theory the shift of the proton hfc is given by [25]

$$\Delta A_p = D_p^{\parallel}\left[1 - \frac{k_B T}{(2U \cdot S)}\right] \tag{9}$$

The proton hfcs of Mu-2-8CB and Mu-3-8CB in the N and Sm-*A* phases are lower than expected based on the isotropic values because $k_B T/(2U{\cdot}S) > 1$ in this temperature region. This can be demonstrated just below T_{NI}, where according to the MS model the order parameter is 0.44 and $k_B T_{NI}/U(T_{NI}) = 4.55$ [1]. The shift approaches $D_p^{\parallel}$ as the temperature goes to zero, which is the expected value for static molecules that are aligned along the magnetic field. The temperature dependence of ΔA_p indicates that $D_p^{\parallel}$ is also temperature dependent and goes to zero near the phase transition. $D_p^{\parallel}$ can be calculated from ΔA_p assuming that the order parameter is known. The temperature dependence of $S(T)$ for cyanobiphenyl LC can be approximated to within one percent with the following expression.

$$S(T) \approx 0.1 + 0.9\left[1 - 0.99(T/T_{NI})\right]^{0.25} \tag{10}$$

The decrease of $D_p^{\parallel}$ as the temperature is lowered is due to the fluctuations of the radicals about the local director and the fluctuations of the director decreasing in amplitude. This is responsible for the broadening of the Δ_0 resonances with decreasing temperature. The rate of change of $D_p^{\parallel}$ appears to be greater in the N phase than in the Sm-*A* phase. The dipolar coupling constant of Mu-2-8CB is consistently larger than that of Mu-3-8CB and this is in agreement with the calculated $D_p^{\parallel}$ values for the molecules only undergoing fast uniaxial rotation [25].

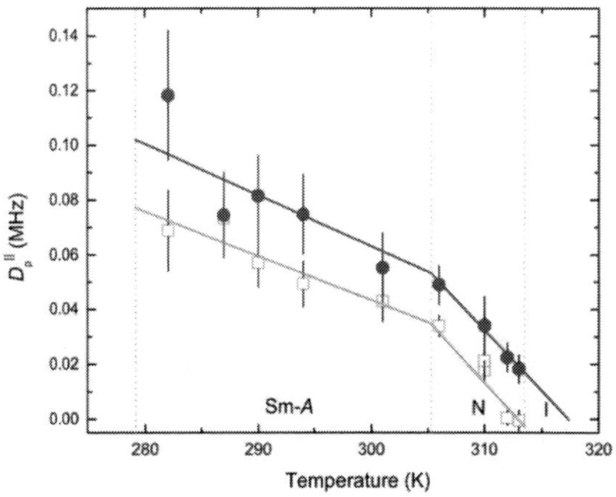

Figure 5. Temperature dependence of the dipolar hyperfine coupling constants ($D_p^{\parallel}$) of two of the Mu adducts of 8CB in the nematic phase: Mu-2-8CB (●) and Mu-3-8CB (□). The solid lines are a guide for the eyes.

3.3. ALC-μSR spectra of the crystalline phase
There was a large change in the ALC-μSR spectra when the sample was cooled below 279.2 K. This was determined to be the phase transition temperature based on the observation of a temperature spike due to the release of latent heat. This was not observed in 5CB because the crystallization enthalpy of 5CB (-3.77 kJ mol^{-1}) is much smaller than that of 8CB (-24.33 kJ mol^{-1}) [27]. The ALC-μSR spectra of the Cr phase of 8CB are very different from that of 5CB, although both contain large and broad overlapping resonances. The assignment of the resonances is based on the theoretical calculations on 1CB [25]. The resonance at 1.67 T is due to overlapping Δ_1 resonances of Mu-2-8CB and Mu-3-8CB. The resonance at 1.82 T is due to overlapping Δ_0 resonances of Mu-2-8CB and Mu-3-8CB and Δ_1 resonances of Mu-1-8CB and Mu-4-8CB. The resonance at 2.01 T is due to overlapping Δ_0 resonances of Mu-1-8CB and Mu-4-8CB. The appearance of Δ_1 resonances indicates that the magnitude of the dipolar hfc has increased considerably, which suggests that the molecules are no longer rotating rapidly about their long axis. The large amplitude of the resonances in the Cr phase compared with those in the higher temperature phases is consistent with the larger dipolar hfcs in the Cr phase.

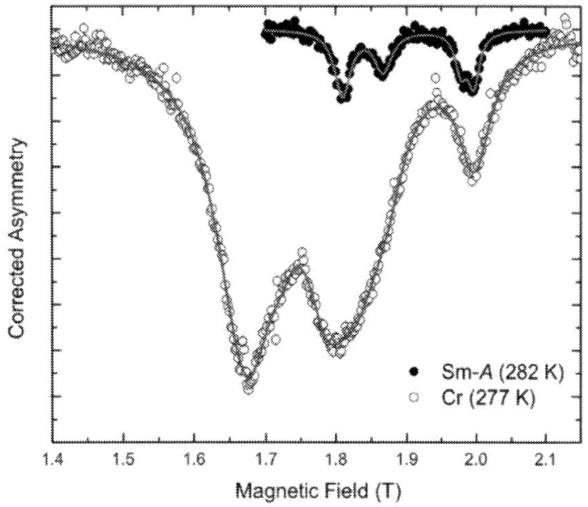

Figure 6. ALC-μSR spectrum of the Mu adducts of 8CB in the smectic-*A* phase (● – 282 K) and the crystalline phase (○ – 277 K). The solid lines represent the best fit to the data.

4. Conclusions
Four types of substituted muoniated cyclohexadienyl radical were formed by the addition of muonium to the calamitic LC 8CB. The methylene proton hfcs were measured by ALC-μSR over a wide temperature range that encompassed the I, N, Sm-*A* and Cr phases. The proton hfcs of two muoniated radical isomers shift from the isotropic values in the N and Sm-*A* phases due to the alignment of the molecules along the magnetic field, the ordering of the molecules and fluctuations of the molecules about the director. The dipolar hyperfine coupling constant is approximately zero at the N–I phase transition due to large amplitude fluctuations of the radicals about the director. The dipolar coupling increase as the temperature is lowered due to this vibrational motion decreasing in amplitude. The rate of decrease in the amplitude of the fluctuations about the director is larger in the N phase than in the Sm-*A* phase. The appearance of Δ_1 resonance below the Cr–Sm-*A* phase transition indicates that the molecules no longer rotate rapidly around the long axis in the Cr phase.

5. Acknowledgments

Support from the staff at the Laboratory for Muon Spin Spectroscopy at the Paul Scherrer Institute is gratefully acknowledged. This research project has been supported by the European Commission under the sixth Framework Programme through the Key Action: Strengthening the European Research Area, Research Infrastructures. Contract nr: RII3-CT- 2003-505925.

6. References

[1] de Gennes P G and Prost J 1993 *The Physics of Liquid Crystals* (Oxford: Oxford University Press)
[2] Collings P J and Hird M 1997 *Introduction to Liquid Crystals: Chemistry and Physics* (London: Taylor and Francis Ltd.)
[3] Sharma D, MacDonald J C and Iannachione G S 2006 *J. Phys. Chem.* B **110** 16679
[4] Yethiraj A, Sun Z, Dong R Y and Burnell E E 2004 *Chem. Phys. Lett.* **398** 517
[5] Yethiraj A, Weber A C J, Dong R Y and Burnell E E 2007 *J. Phys. Chem.* B **111** 1632
[6] Polnaszek C F and Freed J H 1975 *J. Phys. Chem.* **79** 2283
[7] Brooks S A, Luckhurst G R, Pedulli G F and Roberts J 1976 *J. Chem. Soc., Faraday Trans. 2* **72** 651
[8] Rao K V S, Polnaszek C F and Freed J H 1977 *J. Phys. Chem.* **81** 449
[9] Lin W J and Freed J H 1979 *J. Phys. Chem.* **83** 379
[10] Meirovitch E, Igner D, Igner E, Moro G and Freed J H 1982 *J. Chem. Phys.* **77** 3915
[11] Nayeem A and Freed J H 1989 *J. Phys. Chem.* **93** 6539
[12] Morsy M A, Hwang J S and Oweimreen G A 1997 *J. Phys. Chem.* B **101** 2120
[13] Morsy M A, Oweimreen G A and Al-Tawfiq A M 1998 *J. Phys. Chem.* B **102** 3684
[14] Oweimreen G A and Hwang J S 2001 *Chem. Phys. Lett.* **334** 83
[15] Magnuson M L and Fung B M 1994 *J. Chem. Phys.* **100** 1470
[16] Emsley J W, Long J E, Luckhurst G R, Pedrielli P 1999 *Phys. Rev.* E **60** 1831
[17] Roduner, E. 2005 Muonium - an ultra-light isotope of hydrogen. *Isotope Effects in Chemistry and Biology* ed A Kohen and H H Limbach (New York: CRC Press) pp 433-450
[18] McKenzie I and Roduner E 2009 *Naturwissenschaften* **96** 873
[19] Scheuermann R, Tucker I M, Creeth A M, Dilger H, Beck B and Roduner E 2002 *Phys. Chem. Chem. Phys.* **4** 1510
[20] Scheuermann R, Tucker I M, Dilger H, Staples E J, Ford G, Fraser S B, Beck B and Roduner E 2004 *Langmuir* **20** 2652
[21] Martyniak A, Dilger H, Scheuerman R, Tucker I M, McKenzie I, Vujosevic D and Roduner E 2006 *Phys. Chem. Chem. Phys.* **8** 4723
[22] Yu D, Percival P W, Brodovitch J C, Leung S K, Kiefl R F, Venkateswaran K, Cox S F J 1990 *Chem. Phys.* **142** 229
[23] Lovett B W, Stießberger J S, Blundell S J, Jestädt Th, Ardavan A, Marshall I M, Pratt F L and Reid I D 2000 *Physica* B **289- 290** 612
[24] Lovett B W, Blundell S J, Stießberger J S, Pratt F L, Jestädt Th, Hayes W, Cottrell, S. P and Reid I D 2001 *Phys. Rev.* B **63** 054204
[25] McKenzie I, Dilger H, Stoykov A and Scheuermann R 2009 *J. Phys. Chem.* B **113** 10135
[26] http://root.cern.ch (accessed October 22, 2009)
[27] Oweimreen G A and Morsy M A 2000 *Thermochimica Acta* **346** 37

Field and Temperature Dependent Charge Transport Characteristics in Regio-regular Poly(3-octylthiophene-2,5-diyl) Studied by Muon Spin Relaxation

Risdiana[1,2], Fitrilawati[2], R. E. Siregar[2], R. Hidayat[3], A. A. Nugroho[3], M. O. Tjia[3], Y. Ishii[1], I. Watanabe[1]

[1] Advanced Meson Science Laboratory, Nishina Center, RIKEN, 2-1, Hirosawa, Wako, Saitama 351-0198, Japan

[2] Department of Physics, Padjadjaran University, Jl. Raya Bandung-Sumedang km.21 Jatinangor Sumedang, Indonesia

[3] Physics of Magnetism and Photonics Research Division, Faculty of Mathematics and Natural Science, Bandung Institute of Technology, Ganesha 10 Bandung, Indonesia

E-mail: risdiana@riken.jp

Abstract. Spin dynamic of regio-regular Poly(3-octylthiophene-2,5-diyl) has been investigated with longitudinal field (LF) muon-spin-relaxation (μSR) techniques. The LF dependent muon-spin depolarization rate indicates the occurrence of dimensional crossover from one-dimensional intra-chain spin diffusion to three-dimensional inter-chain spin diffusion at 50 K.

1. Introduction

Polythiophene (PT) and its derivative are among the conducting polymers which have been of great research interest due to their chemical and thermal stability [1] as well as their broad applications and possibilities of new applications [2]. They are also easily grafted with side changes for property modifications [3]. In particular, the PT derivative of poly(3-alkylthiophene) (P3AT) has been attracted intensive study because of the effective modification of its properties by variations of the alkyl side-chain length [4]. For instance, conductivity was reported to decrease with increasing alkyl side chain length while stronger luminescence was observed in the case longer chain length [5]. The dependence of hole mobility on the alkyl chain length has also been reported, with the poly(3-hexylthiophene) showing the highest hole mobility among the series of P3AT [6].

We have studied the microscopic and intrinsic charge transport processes in both regio-regular and regio-random poly(3-hexylthiophene-2,5-diyl) (P3HT) by measuring their of longitudinal field (LF) muon-spin-relaxation (μSR) behaviors [7, 8]. The muonium, being made up of a positive muon and an electron, is readily formed as slows down to a near stop in the sample and picks up an electron from a carbon double bond in the polymer, thereby attaching itself to the carbon atom by sharing the electron. This leaves an unpaired electron at the neighbor carbon atom. Following a rapid electronic and structural relaxation of the surrounding polymer, a negative polaron, a quasiparticle composed of an unpaired electron plus its accompanying polarization field, is formed. The mobile polaron will move away from its initial site and diffuse up and down the polymer chain, giving rise to intra-chain charge transport, which may or may not be accompanied by inter-chain hopping in the conducting polymer [9, 10, 11]. The experimental results for poly(3-hexylthiophene-2,5-diyl) show that the temperature-dependent charge carrier mobility depends on their regio-regularity and exhibits abrupt change associated with the "transition" from intra-chain diffusion to inter-chain diffusion at 25 K and 50 K for the regio-regular and regio-random samples, respectively [7, 8].

Here, we report the study of intrinsic charge transport processes in the regio-regular poly(3-octylthiophene-2,5-diyl) (P3OT) along and perpendicular to the chain by LF-μSR method. The analysis of the field and temperature dependent effects on the intra-chain hopping mechanism and inter-chain coupling effect in the polymers is performed on the basis of commonly accepted theoretical model. In this connection, the possible chain length effect on the diffusive spin motion in the regio-regular P3AT is also discussed.

2. Experimental

The regio-regular Poly(3-octylthiophene-2,5-diyl)(Sigma-Aldrich) samples were pressed and wrapped in a 25 μm silver foil and mounted on a silver plate in the cryostat for muon measurements. The μSR measurements were performed at temperature varied from 10 K to 300 K in longitudinal magnetic field ranging from 0 to 350 mT using a pulsed positive surface muon beam [12, 13] at the RIKEN-RAL Muon Facility at the Rutherford-Appleton Laboratory in the UK.

The μSR function known as asymmetry parameter $A(t)$ at a time t is defined as $A(t) = [F(t) - \alpha B(t)]/[F(t) + \alpha B(t)]$, where $F(t)$ and $B(t)$ are total muon events counted by the forward and backward counters, respectively, and α is the calibration factor reflecting the relative counting efficiencies between the forward and backward counters.

3. Results

Asymmetry parameters at various longitudinal magnetic fields are shown in figure 1 for two temperatures of 10 K (a) and 300 K (b). The initial asymmetries in both cases increase monotonously with increasing field as a results of repolarization of the muonium states [14]. In each case, the dynamical variation of the asymmetry parameters display visible field dependencies, while comparison between Fig 1(a) and (b) also shows perceptible temperature effect. In order to obtain the detailed information of polaron diffusion on the polymer chains, all the time spectra were analyzed using the following two-component function:

$$A(t) = A_1 exp(-\lambda_1 t) + A_2 exp(-\lambda_2 t) \tag{1}$$

where A_1 and A_2 are the initial asymmetries and λ_1 and λ_2 are the corresponding depolarization rates associated with the fast and slow components, respectively. The solid line in Figure 1 is the best fit obtained using the diffusion model mentioned above. According to Butler *et al.* [15], the LF dependence of the dynamical depolarization rate reflects the dimensionality of the spin diffusion. That is, the depolarization rate is proportional to $a\mathrm{H}^{-0.5}$ for the one-dimensional intra-chain diffusion and while it is characterized by a $c - b\mathrm{H}^{0.5}$ curve for the three-dimensional inter-chain diffusion where H is the applied longitudinal field and a, b, c are constant.

Figure 2 shows the longitudinal-field (LF) dependence of the muon-spin depolarization rate (λ_1) in regio-regular P3OT at temperatures of 300 K, 75 K, 50 K and 10 K. At the low temperatures below 50 K, the relaxation rate displays $a\mathrm{H}^{-0.5}$ field-dependence characteristic of one-dimensional intra-chain diffusion, implying that below 50 K, the charge transport is dominated by mobility along (intra) the polymer chain. Upon increasing the temperature,

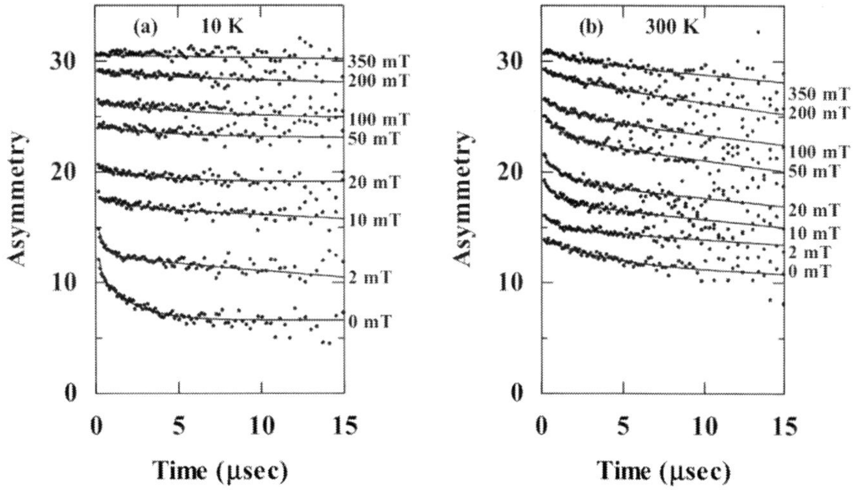

Figure 1. The Asymmetry data of regio-regular poly(3-octylthiophene-2,5-diyl) at (a) 10 K and (b) 300 K for various longitudinal magnetic field values. The solid lines are best fits to equation (1).

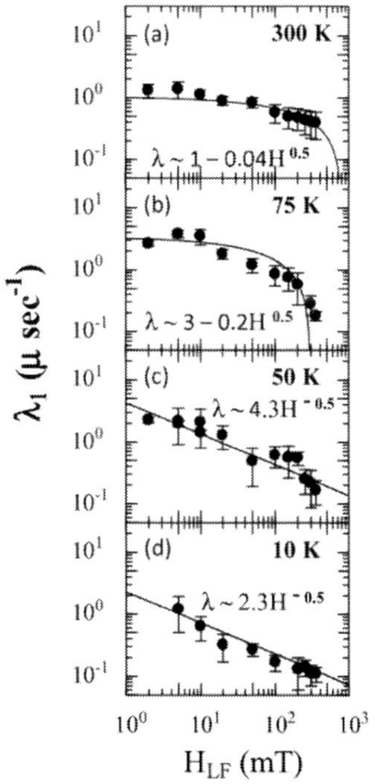

Figure 2. The longitudinal-field dependence of relaxation rate λ_1 of regio-regular poly(3-octylthiophene-2,5-diyl) at 300 K, 75 K, 50 K and 10 K. The solid line are the best fit using three-dimensional model (a, b) and one-dimensional model (c, d), respectively.

the charge carrier mobility deviates drastically from one-dimensional intra-chain diffusion and crossover completely to three-dimensional inter-chain diffusion as demonstrated by the well fitted $c - b\mathrm{H}^{0.5}$ field-dependent variation of the relaxation rate. A similar tendency of LF dependent variation of λ_2 was also observed although the values of λ_2 are two orders of magnitude smaller than that of λ_1. The present results have thus revealed the dimensional crossover of charge transport in regio-regular P3OT at 50 K.

Comparing with μSR data of regio-regular P3HT that shows dimensional crossover at 25 K [7], it is apparent that inter-chain polaron diffusion process in the present case (P3OT), requires the assistance of higher temperature. While an accurate and more thorough explanation of this difference is not available from this experiment, it may nonetheless be related to the difference side chain length in the alkyl. Clearly, longer side chain leads to larger distance between the chains which implies in turn the need of large thermal energy to support the inter-chain hopping process.

4. Conclusion

We have studied the spin dynamics in regio-regular poly(3-octylthiophene-2,5-diyl) by means of LF-μSR techniques. Our experimental results have revealed a characteristic spin diffusion in one-dimension intra-chain transport at low temperatures below 50 K and in three-dimension inter-chain diffusion at temperatures above 50 K. The present result has thus demonstrated the crossover from one-dimensional intra-chain diffusion to three-dimensional diffusion featuring strong temperature-induced inter-chain hopping effect. In addition comparison of our present result with a previous result also indicated the dependence of side chain length on the dimensional crossover of the charge transport.

Acknowledgments

We thank Dr. Ohishi and Dr. Suzuki for experimental assistance. Dr. F. L. Pratt for some discussions. Risdiana is also grateful to RIKEN for financial support of RIKEN Foreign Postdoctoral Researcher (FPR).

References

[1] Petil A. O., Heeger A. J., Wudl F., 1988 *Chem. Rev.* **88** 183
[2] Greenham N. C., Moratt S. C., Bradley D. D. C, Friend R. H., Holmes A. B., 1993 *Nature* **365** 628
[3] Hidayat R., Fujii A., Ozaki M., Yoshino K. 2001 *Jpn. J. Appl. Phys.* **40** 7103
[4] Kaneto K., Lim W. Y., Takashima W., Endo T., Rikukawa M., 2000 *Jpn.J. Appl. Phys.* **39** L872
[5] Yoshino K., Morita S., Uchida M., Muro K., Kawai T., Ohmori Y., 1993 *Synth. Met.* **55-57** 28
[6] Babel A., Jenekhewa S. A., 2005 *Synth. Met.* **148** 169
[7] Risdiana, Fitrilawati, R. Hidayat, R. E. Siregar, M. O. Tjia, I. Watanabe, to be publish in Journal of Physics: Conference Series.
[8] Risdiana, Fitrilawati, Hidayat R., Nugroho A. A., Siregar R. E., Tjia M. O., Watanabe I., to be publish in Physica B doi:10.1016/j.physb.2010.01.080
[9] Nagamine K., Ishida K., Matsuzaki T., Nishiyama K., Kuno Y., Yamazaki T., Shirakawa H., 1984 *Phys. Rev. Lett.* **53** 1763
[10] Pratt F. L., Blundell S. J., Pattenden P. A., Hayes W., Chow K. H., Monkman A. P., Ishiguro T., Ishida K., Nagamine K., 1997 *Hyperfine interactions* **106** 33
[11] Pratt F. L., Blundell S. J., Jestadt Th., Lovett B. W., Husmann A., Marshall I. M., Hayes W., Monkman A. P., Watanabe I., Nagamine K., Martin R. E., Holmes A. B., 2000 *Phisica B* **289-290** 625
[12] T. Matsuzaki, K. Nagamine, M. Tanase, M. Kato, K. Kurosawa, H. Sugai, K. Ishida, S. N. Nakamura, I. Watanabe and G. H. Eaton, Nucl. Instr and Meth. A **480** (2002) 814.
[13] K. Nagamine, T. Matsuzaki, K. Ishida, I. Watanabe, R. Kadono, G. H. Eaton, H. J. Jones, G. Thomas and W. G. Williams, Hyperfine Interact. **87** (1994) 1091.
[14] Pratt F. L., 1997 *Phil. Mag. Lett.* **75** 371
[15] Butler M. A., Walker L. R., Soos Z. G., 1976 *J. Chem. Phys.* **64** 3592

Journal of Physics: Conference Series **225** (2010) 012004 doi:10.1088/1742-6596/225/1/012004

Pilot experiment for muonium photo ionization in GaAs

K. Shimomura[1], P. Bakule[2], F. L. Pratt[2], K. Ohishi[3], K. Ishida[3], I. Watanabe[3], Y. Matsuda[4], K. Nishiyama[1], E.Torikai[5], K. Nagamine[1,3,6]
[1] Muon Science Laboratory, IMSS, KEK, Tsukuba, Ibaraki, 305-0801, Japan
[2] ISIS, Rutherford Appleton Laboratory, Chilton, Oxon OX11 0QX, U.K
[3] Advanced Meson Science Laboratory, RIKEN, Wako, Saitama 351-0191, Japan
[4] Graduate School of Arts and Science, University of Tokyo, 3-8-1 Komaba, Tokyo 153-8902, Japan
[5] Interdisciplinary Graduate School of Medicine and Engineering, University of Yamanashi, Kofu, Yamanashi, 400-8511, Japan
[6] Department of Physics and Astronomy, University of California, Riverside, California , 92521,U.S.A

E-mail: koichiro.shimomura@kek.jp

Abstract. Direct observation of muonium photo ionization in GaAs was tried for the first time, with wide range wave length from 1325nm to 800nm lasers in n-type GaAs at 15 K. Recently, Lichti *et al.* determined the energy levels in the band gap of T center muonium (as an acceptor) and BC muonium (as a donor) by reanalysis of the existing data obtained by various μSR techniques for several semiconductors like Si, Ge, GaAs, GaP etc. In these semiconductors, GaAs is the best sample to apply the muonium photo ionization method for the first time, because the energy level of T center muonium is above 0.54 eV from the valence band, therefore the ionization energy for $Mu_T^- \rightarrow Mu_T^0 + e^-$ is 0.98eV (corresponding laser wave length is 1260nm), which is within the region of present OPO laser system produced, which was installed RIKEN-RAL

1. Introduction

Hydrogen is a ubiquitous impurity in most semiconductors, including elemental (e.g., Si), compound (e.g., GaAs) one. Theoretical modeling of hydrogen impurities in semiconductors carried out over a period of many years has produced the accepted picture of H behavior in these materials [1], albeit with rather meager direct experimental evidence for the isolated atomic defect except in Si. The indirect verification of these models is mostly from results on the complexes formed by passivation reactions of mobile hydrogen with other impurities in which the addition of H satisfies various bonding mismatches and removes defect levels from the gap, thus eliminating the defect-related electrical activity. Isolated hydrogen is typically a negative-U center, where $U=E_A(-/0)-E_D(0/+)$ is the separation between donor and acceptor thermodynamic levels. Consequently, in most semiconductors, hydrogen shows compensating properties, existing as H^+ in *p*-type and H^- in *n*-type materials under equilibrium conditions.

Isolated hydrogen can exist in three charge states, H^+, H^0, and H^-, with the ionic centers residing at different interstitial sites. For predominantly covalent cubic materials, H^+ is stable at or near the center of a stretched bond [bondcentered(BC) site] and H^- is stable in the large tetrahedral void (T site) of the diamond or zinc-blende structure. Both of these sites can support a neutral center, but the electronic wave functions and the hyperfine interactions are vastly different at the two locations. When H^0 resides in the void, it is simply an interstitial impurity atom with a large isotropic hyperfine constant, somewhat reduced from the free-atom value by relatively small overlaps with near neighbors. However, in forming H_{BC}^0, hydrogen has reacted with the host to form a radical in which the unpaired electron resides in antibonding orbitals on its two nearest neighbors, producing a small anisotropic hyperfine interaction. Most of the direct experimental evidence supporting the above picture comes from studies of muonium (Mu), a light isotope of hydrogen in which the proton has been replaced by a positive muon.

Recently, Lichti *et al.* determined the energy levels in the band gap of T center muonium (as an acceptor) and BC muonium (as a donor) by reanalysis of the existing data obtained by various μSR techniques for several semiconductors like Si, Ge, GaAs, GaP etc [2]. To check these values by independently and directly, the laser photo ionization of muonium in these semiconductors is one of the best methods. And if this method works well, it is also applicable to the muonic oxygen atom, which is nitrogen atom analog in oxide semiconductors like ZnO or TiO_2, where nitrogen impurities are recently considered to play key roles in the developments.

Among several semiconductors, GaAs is the best sample to apply the muonium photo ionization method for the first time, because the energy level of T center muonium is above 0.54 eV from the valence band, therefore the ionization energy for $Mu_T^- \rightarrow Mu_T^0 + e^-$ is 0.98eV (corresponding laser wave length is 1260nm), which is within the region of existing OPO laser system produced.

2. Experiments

The experiment was conducted at Port 2 of the RIKEN-RAL Muon facility at ISIS of Rutherford Appleton Laboratory (UK) by using pulsed (a double pulse with 75 ns width and 350 ns separation with 50 Hz repetition) 4 MeV positive muon beam. The sample used was 360 μm of thickness 50.8 mm diameter single crystalline GaAs with 3×10^{16} cm^{-3} Si doping. The [001] axis was placed along muon/laser direction.

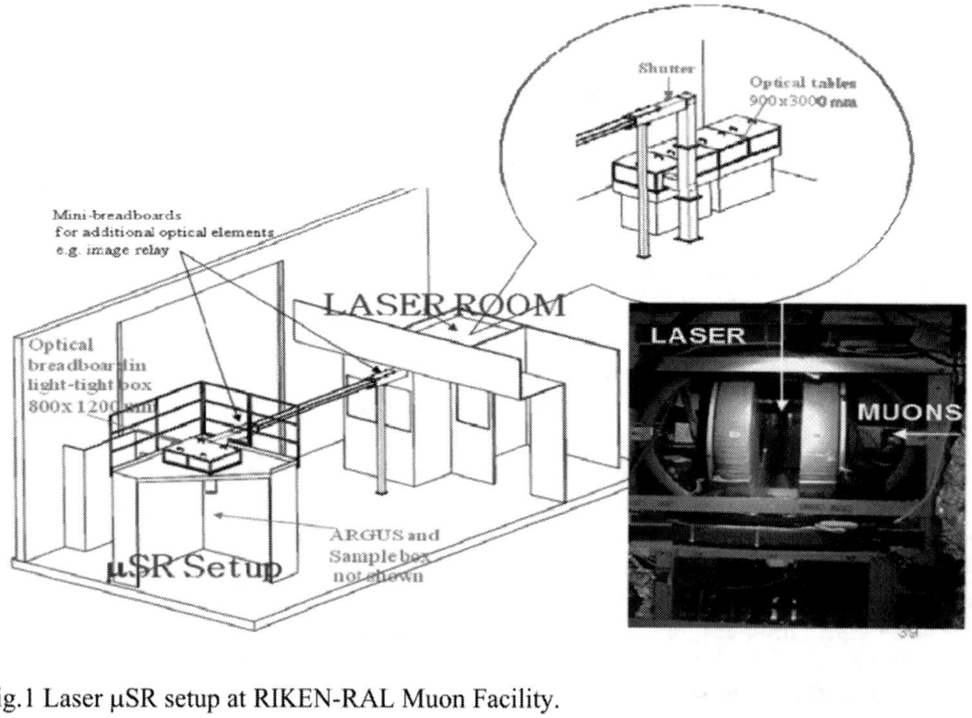

Fig.1 Laser μSR setup at RIKEN-RAL Muon Facility.

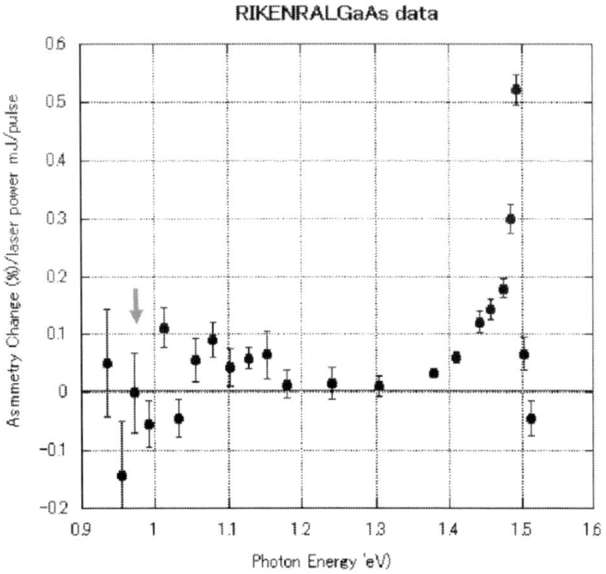

Fig.2 The photon energy dependence of the asymmetry change of Mu_T^- in GaAs under laser irradiation. Asymmetry change is normalized by laser power. The arrow shows the ionization energy of Mu_T^-, determined by Lichti *et. al.*

The tunable laser light with repetition frequency of 25 Hz was generated using a widely tunable OPO system (Continuum Panther EX OPO) pumped by 355 nm beam from Continuum Powerlite 9025 Nd:YAG laser. The linearly polarized laser output was transported through a light-tight enclosure to optical breadboard located in the muon beam port above the muon spectrometer at a distance of ~8 m using 9 silver mirrors. The laser pulse energy over the sample could be varied using a computer controlled attenuator up to 3 mJ/pulse.

At the ending part of the laser transport, a sample box with He-flow cryostat was installed. The second laser steering prism was placed right at the downstream of the muon beam so that the laser light is injected to the rear side of the sample through fused-quartz window. Sample was placed in a strain-free manner in a specially designed sample holder containing He gas. Care was taken for the muon beam not to be stopped at neither mirror nor cold plate. All measurements are performed under zero field condition, where only diamagnetic component (here Mu_T^-) is observed. At zero field, Mu_T^0 is expected to lose their polarization immediately, due to large nuclear hyperfine interaction with surrounding Ga and As.

Preliminary data is shown in Fig.2. Around 1.4 to 1.5 eV, .a large asymmetry drop were observed, these are the effect of the interactions between Mu_T^- and photo excited electron, which is already established in the previous study [3.4]. Around 1.0 to 1.2 eV, there is some bump, which cannot be explained by the above picture. The most reasonable explanation is the effect of direct photo ionization of Mu_T^- and Mu_T^0 lose their polarization. Unfortunately, at this time, laser power and quality is not enough around 0.9 to 1.1 eV, therefore we cannot say strongly there is a difference between above and below Mu_T^- ionization energy. To improve this point, we should optimize laser power at this region and prepare laser beam size adjuster.

References

[1] For reviews, Patterson B. D. 1998 *Rev. Mod. Phys.* **60**, 69; Kiefl R. F. and Estle T. L.1991 in *Hydrogen in Semiconductors* (Academic, San Diego)
[2] Lichti R.L.*et al.* 2007 *Phys. Rev. B***76**, 045221
[3] Shimomura K. *et al.* 2003 *Physica B* **326**, 151
[4] Yokoyama K. *et al.* 2009 *Physica B* **404**, 856

Defect studies with positrons: what could we learn on III-nitride heterostructures?

Filip Tuomisto[1], Jussi-Matti Mäki[1], Olli Svensk[2], Pekka T. Törmä[2], Muhammad Ali[2], Sami Suihkonen[2], and Markku Sopanen[2]

[1]Department of Applied Physics, Aalto University, P.O. Box 11100, FI-00076 Aalto Espoo, Finland

[2]Department of Micro and Nanosciences, Aalto University, P.O. Box 13500, FI-00076 Aalto Espoo, Finland

filip.tuomisto@tkk.fi

Abstract. We have applied positron annihilation spectroscopy to study 400 – 500 nm InGaN-based LED structures, as well as InGaN and AlGaN materials with varying In and Al contents. We find that the effect of adding In to GaN on the annihilation parameters obeys the Vegard's law, while in the case of AlGaN the possible effect of Al is completely screened by efficient formation of cation vacancies. The results obtained in the InGaN LED structures are indistinguishable from defect-free GaN, suggesting that the positrons annihilate preferentially in the barriers of the MQW system.

1. Introduction

In device structures fabricated out of III-nitrides (alloys of AlN, GaN and InN) the structural quality of both the semiconductor materials themselves and the heterointerfaces between them often limit significantly the output due to the unfortunate aspect of having lattice constants that differ considerably one from another. In addition to extended defects such as stacking faults or dislocations, point defects have a significant impact on the electrical and optical properties of the nitride semiconductor materials. Positron annihilation spectroscopy is a method that is particularly well suited for studying vacancy defects in semiconductors. During the past decade it has been applied in numerous studies where the identities, concentrations and characteristics of both in-grown and processing-induced vacancy defects in nitride semiconductors have been determined, see e.g. [1–8] and the references therein.

In this work we apply positron annihilation spectroscopy to study InGaN light emitting device (LED) structures (wavelengths 400 – 500 nm), as well as InGaN and AlGaN materials with different In and Al contents, respectively. We interpret the data based on earlier results [6–11] obtained in GaN, InN, and AlN and the cation (group III) vacancies in these materials.

2. Experimental details

The positron annihilation characteristics were measured in thin epitaxial films and device structures by using a mono-energetic continuous positron beam. As lifetime experiments are not possible with a continuous beam, the Doppler broadening of the annihilation radiation was monitored as a function of the beam energy, providing a depth profile of the signal. The motion of the annihilating electron-positron pair causes a Doppler shift in the annihilation radiation $\Delta E = cp_L / 2$, where p_L is the longitudinal momentum component of the pair in the direction of the annihilation photon emission. This causes the broadening of the 511 keV annihilation line. The shape of the 511 keV peak gives thus the one-dimensional momentum distribution $\rho(p_L)$ of the annihilating electron-positron pairs. A Doppler shift of 1 keV corresponds to a momentum value of $p_L = 0.54$ a.u ($\approx 3.91 \times 10^{-3}\ m_0c$).

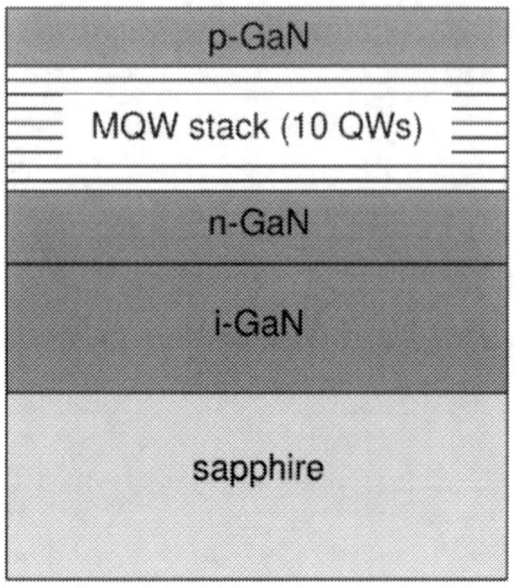

Figure 1. Schematic representation of the relevant parts of the LED structures studied in this work. The initial layer of nominally undoped GaN (i-GaN) is 2 μm thick, the n-type doped GaN layer is 1 μm thick, and the p-GaN electron blocking layer is 200 nm thick. The active MQW area consists of 10 InGaN well – GaN barrier pairs, where the well thickness is 4 nm and barrier thickness 25 nm, hence the total thickness of the MQW stack is 290 nm.

The Doppler broadening can be experimentally measured using a Ge gamma ray detector with a good energy resolution. The typical resolution of a detector is around 1–1.5 keV at 500 keV. This is considerable compared to the total width of 2–3 keV of the annihilation peak meaning that the experimental lineshape is strongly influenced by the detector resolution. Therefore, various shape parameters are used to characterize the 511 keV line. The low electron-momentum parameter S is defined as the ratio of the counts in the central region (typically $p_L < 0.4$ a.u.) of the annihilation line to the total number of the counts in the line. In the same way, the high electron-momentum parameter W is the fraction of the counts in the wing regions of the line (typically $p_L > 1.5$ a.u.). Due to their low momenta, mainly valence electrons contribute to the region of the S parameter. On the other hand, only core electrons have momentum values high enough to contribute to the W parameter. Therefore, S and W are sometimes called the valence and core annihilation parameters, respectively.

The high-momentum part of the Doppler broadening spectrum arises from annihilations with core electrons that contain information on the chemical identity of the atoms. Thus the detailed investigation of core electron annihilation can reveal the nature of the atoms in the regions where positrons annihilate. In order to study the high-momentum part in detail, the experimental background needs to be reduced. A second gamma ray detector can be placed opposite to the Ge detector and the only events that are accepted are those for which both 511 keV photons are detected [12, 13]. Depending on the type of the second detector, electron momenta even up to $p \approx 8$ a.u. ($\approx 60 \times 10^{-3}$ m_0c) can be measured with the coincidence detection of the Doppler broadening.

The nitride thin films and device structures studied in this work were grown by metal-organic vapor phase epitaxy (MOVPE) directly on sapphire substrates. For further details about the growth procedures, see Refs. [14–16]. We studied $In_xGa_{1-x}N$ and $Al_xGa_{1-x}N$ alloys with In and Al contents ranging from up to 22 % and 25 %, respectively. The device structures were realistic LED structures corresponding to wavelengths in the 400-500 nm, with In content varying from 5 % to 15 % in the quantum wells (QWs) in the active area consisting of a multi-quantum-well system of 10 well-barrier pairs. The schematic of the LED structures is shown in Fig. 1.

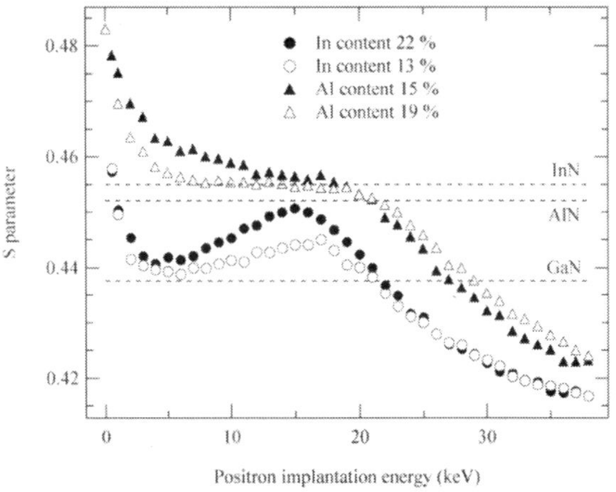

Figure 2. The S parameter measured as a function of positron implantation energy in selected $In_xGa_{1-x}N$ and $Al_xGa_{1-x}N$ layers grown by MOVPE on sapphire. The dashed lines show the characteristic S parameters of GaN, InN and AlN.

3. InGaN and AlGaN layers

Figure 2 shows the S parameter measured at room temperature as a function of positron implantation energy in selected $In_xGa_{1-x}N$ and $Al_xGa_{1-x}N$ samples. When positrons are implanted close to the sample surface with $E = 0 - 1$ keV, a relatively high S parameter of $S = 0.46$ ($S = 0.48$) is recorded in the $In_xGa_{1-x}N$ ($Al_xGa_{1-x}N$) samples. These values characterize the defects and chemical nature of the near-surface region of the samples at the depth 0 – 5 nm. The region of constant S gives the characteristic value of the $Al_xGa_{1-x}N$ layer and extends up to 20 keV in the thickest sample. In the $In_xGa_{1-x}N$ layers, the S parameter increases strongly toward the layer/substrate interface, indicating that the material is different close to the interface. The decrease of the S parameter at higher implantation energies is due to the increasing fraction of positrons annihilating in the sapphire substrate, for which the characteristic S parameter is roughly $S = 0.41$. It is clearly seen that the S parameter of the layer increases with increasing content of either In or Al in the nitride lattice.

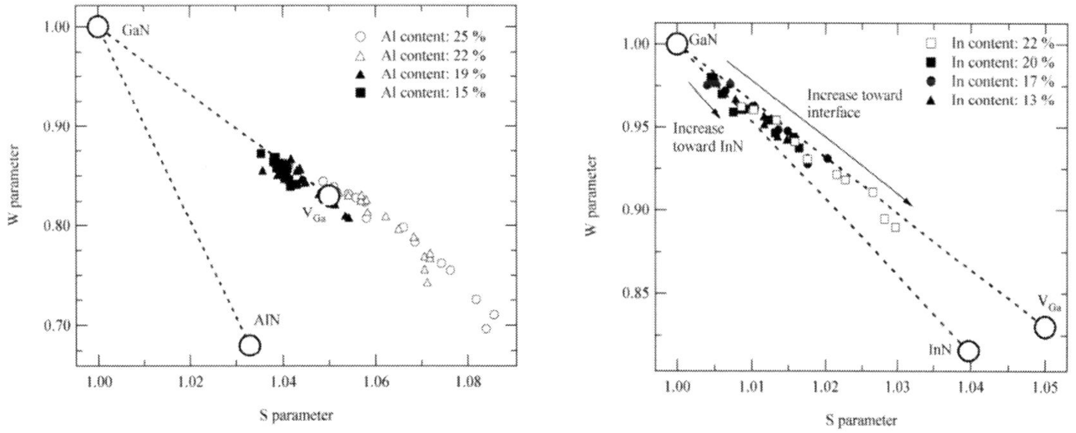

Figure 3. The relative S and W parameters (divided by the values characteristic to GaN) measured in the $Al_xGa_{1-x}N$ (a) and $In_xGa_{1-x}N$ (b) layers. Data points corresponding to the GaN, InN and AlN lattice and the Ga vacancy in GaN are also shown.

The characteristic S parameters of InN and AlN, determined from well-defined reference samples [6, 11] are significantly higher than that of GaN. However, as can be seen in figure 3a, the reason for the increase in $Al_xGa_{1-x}N$ layers is enhanced annihilation in group III vacancies. On the other hand, the increase in the S parameter far away from the interface in the $In_xGa_{1-x}N$ layers is quite close to what one would expect if the S parameter is assumed to be given by a linear combination of the S parameters of InN and GaN with the fractions given directly by the In content. The strong increase in the S parameter toward the layer/sapphire interface is due to enhanced trapping at group III vacancies abundant near the interface, as can be seen in figure 3b, and not due to an inhomogeneous In distribution. A similar effect can be observed in epitaxial vapor-phase GaN grown directly on sapphire [17,18]. Hence it would seem that there is no preference for positrons to annihilate in either locally In-rich or In-poor regions in the lattice. Interestingly, in recent state-of-the-art first principles calculations [19], it is shown that in polar nitride superlattices positrons always prefer to annihilate in pure GaN instead of areas with In or Al, while the situation is less clear in non-polar structures, even if the positron affinity is the lowest for GaN (compared to InN and AlN).

4. LED structures

In addition to the bandgap modulation that is essential for the electronic device, the use of different nitrides creates electric fields in the structure induced by spontaneous polarization due the non-inversionsymmetric wurtzite structure of the materials [20–22]. These electric fields are bound to affect the positron localization in a similar manner as the electron and hole localization. Indeed, recent theoretical calculations show that in a polar structure the positron localize at the well-barrier interface in a nitride heterostructure [19]. Figure 4 shows a schematic representation of the situation. In more detail, the positron doesn't actually localize exactly at the interface, but on the GaN side in such a way that more than 90 % of the positron density is in the GaN layer and less than ten 10 % in the In- or Al-containing layer. Further, the positron affinity is favorable in GaN (compared to InN and AlN) and hence even in non-polar structures, where the polarization-induced electric field is missing, positrons localize in the GaN layer (but the distribution is wider and centered at the center of the layer, not near an interface).

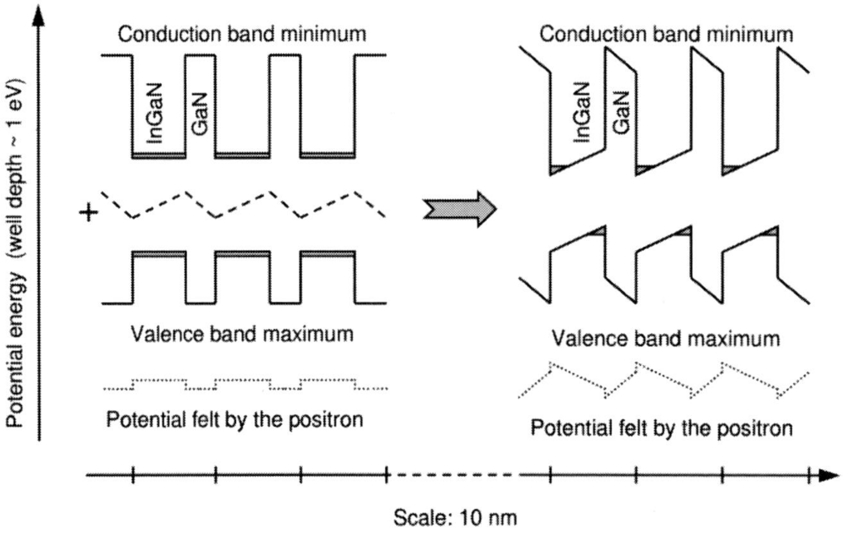

Figure 4. Schematic representation of the polarization-induced electric field on positron localization.

Figure 5 shows the S parameter measured as a function of energy at room temperature in the InGaN LED structures. The data are very clear: no traces of In are observed in the data from the MQW depths (shaded area in the figure), indicating that positrons do not annihilate in In-rich surroundings to a measurable extent. The data coincide with defect-free (i.e., no defects detected with positrons) GaN, and in fact the vacancy concentrations in the barrier GaN seem to be even lower than that in the n-

GaN on which the LED structure is grown, as the S parameter increases at higher positron implantations energies.

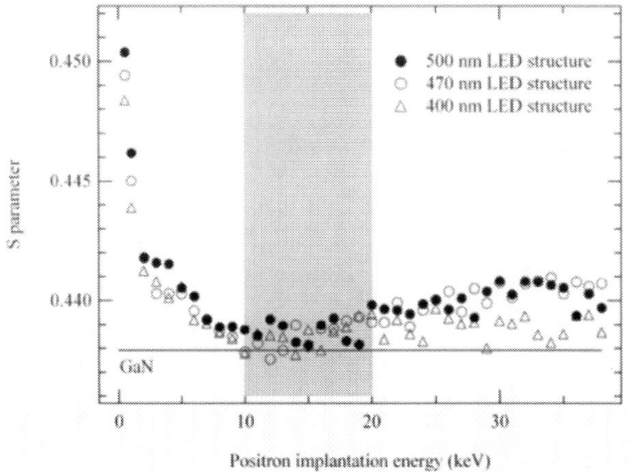

Figure 5. The S parameter measured as a function of positron implantation energy at room temperature in the LED structures. The grey area shows the MQW range of depths.

Finally, it is important to note that the above interpretations of the experimental data concerning the preference of positrons to annihilate in pure GaN or In- and Al-rich regions hold only when the crystal lattice is otherwise intact. When group III-vacancy defects are present, the sensitivity to the immediate surrounding of the positron annihilation site is lost, as the positron "feels" predominantly the nitrogen atoms immediately neighbouring the vacancy. In fact, the above-mentioned theoretical calculations also show that the preference of GaN over AlN and InN is lost when group III vacancies are introduced in the lattice, irrespective of the location of the vacancy (in any of the nitrides or at the interface), and the positrons localize at the vacancy.

5. Summary

In summary., we have applied positron annihilation spectroscopy to study 400 – 500 nm InGaN-based LED structures, as well as InGaN and AlGaN materials with varying In and Al contents. We find that the effect of adding In to GaN on the annihilation parameters is close to that what one could expect from Vegard's law, while in the case of AlGaN the effect of Al is completely screened by efficient formation of cation vacancies. The results obtained in the InGaN LED structures are indistinguishable from defect-free GaN, suggesting that the positrons annihilate preferentially in the barriers of the MQW system, in good agreement with recent theoretical predictions.

Acknowledgments

This work has been partially funded by the Academy of Finland. We acknowledge the support of the Multidisciplinary Institute of Digitalization and Energy (MIDE) of the Helsinki University of Technology.

References

[1] Saarinen K, Laine T, Kuisma S, Nissilä J, Hautojärvi P, Dobrzynski L, Baranowski J M, Pakula K, Stepniewski R, Wojdak M, Wysmolek A, Suski T, Leszczynski M, Grzegory I and Porowski S 1997 *Phys. Rev. Lett.* **79** 3030

[2] Tuomisto F, Suski T, Teisseyre H, Krysko M, Leszczynski M, Lucznik B, Grzegory I, Porowski S, Wasik D, Witowski A, Gebicki W, Hageman P and Saarinen K 2003, *phys. status solidi (b)* **240** 289

[3] Tuomisto F, Saarinen K, Lucznik B, Grzegory I, Teisseyre H, Suski T, Porowski S, Hageman P R and Likonen J 2005 *Appl. Phys. Lett.* **86** 031915

[4] Hautakangas S, Ranki V, Makkonen I, Puska M J, Saarinen K, Xu X and Look D C 2006 *Phys. Rev. B* **73** 193301

[5] Paskova T, Hommel D, Paskov PP, Darakchieva V, Monemar B, Bockowski M, Suski T, Grzegory I, Tuomisto F, Saarinen K, Ashkenov N, Schubert M 2006 *Appl. Phys. Lett.* **88** 141909
[6] Tuomisto F, Pelli A, Yu K M, Walukiewicz W and Schaff W 2007 *Phys. Rev. B* **75** 193201
[7] Tuomisto F, Paskova T, Kröger R, Figge S, Hommel D, Monemar B and Kersting R 2007 *Appl. Phys. Lett.* **90** 121915
[8] Tuomisto F, Ranki V, Look D C and Farlow G C 2007 *Phys. Rev. B* **72** 165207
[9] Oila J, Kemppinen A, Laakso A, Saarinen K, Egger W, Liszkay L, Sperr P, Lu H and Schaff W J 2004 *Appl. Phys. Lett.* **84** 1486
[10] Pelli A, Saarinen K, Tuomisto F, Ruffenach S and Briot O 2006 *Appl. Phys. Lett.* **89** 011911
[11] Tuomisto F, Mäki J-M, Chemekova T Yu, Makarov Yu N, Avdeev O V, Mokhov E N, Segal A S, Ramm M G, Davis S, Huminic G, Helava H, Bickermann M and Epelbaum B M 2008 *J. Crystal Growth* **310** 3998
[12] Alatalo M, Kauppinen H, Saarinen K, Puska M J, Mäkinen J, Hautojärvi P and Nieminen R M: 1995 *Phys. Rev. B* **51** 4176
[13] Asoka-Kumar P, Alatalo M, Ghosh V J, Kruseman A C, Nielsen B and Lynn K G 1996 *Phys. Rev. Lett.* **77** 2097
[14] Lang T, Odnoblyudov M A, Bougrov V E, Suihkonen S, Svensk O, Törmä P T, Sopanen M, Lipsanen H 2007 *J. Crystal Growth* **298** 276
[15] Suihkonen S, Svensk O, Törmä P T, Ali M, Sopanen M, Lipsanen H, Odnoblyudov M A, Bougrov V E 2008 *J. Crystal Growth* **310** 1777
[16] Svensk O, Törmä P T, Suihkonen S, Ali M, Lipsanen H, Sopanen M, Odnoblyudov M A, Bougrov V E 2008 *J. Crystal Growth* **310** 5154
[17] Oila J, Kivioja J, Ranki V, Saarinen K, Look D C, Molnar R J, Park S S, Lee S K, Han J Y 2003 *Appl. Phys. Lett.* **82** 3422
[18] Tuomisto F, Saarinen K, Bockowski M, Suski T, Paskova T and Monemar B 2006 *J. Appl. Phys.* **99** 066105
[19] Makkonen I, Snicker A, Mäki J-M, Tuomisto F and Puska M J, submitted.
[20] Bernardini F, Fiorentini V and Vanderbilt D 1997 *Phys. Rev. B* **56** R10024
[21] Lefebvre P, Morel A, Gallart M, Taliercio T, Allegre J, Gil B, Mathieu H, Damilano B, Grandjean N and Massies J 2001 *Appl. Phys. Lett.* **78** 1252
[22] Bernardini F and Fiorentini V 1998 *Phys. Rev.* B **57** R9427

Study of Cs-loaded SiO_2 glass by thermal analysis and positron annihilation spectroscopy

K. Sato[1], H. Murakami[1], K. Ito[2], K. Hirata[2] and Y. Kobayashi[2]

[1] Department of Environmental Sciences, Tokyo Gakugei University, 4-1-1 Koganei, Tokyo 184-8501, Japan
[2] National Institute of Advanced Industrial Science and Technology (AIST), Tsukuba, Ibaraki 305-8565, Japan

E-mail: sato-k@u-gakugei.ac.jp

Abstract. SiO_2 glass with Cs clusters was investigated by differential scanning calorimetry (DSC), thermogravimetric differential analysis (TG-DTA), and positron age-momentum-correlation (AMOC) spectroscopy. No significant peaks arising from Cs clusters were observed in TG-DTA and DSC curves, suggesting Cs clusters in smaller amount than the detection limit of thermal analysis. In contrast to that, AMOC spectroscopy showed *ortho*-positronium (*o*-Ps) pick-off annihilation influenced by the presence of Cs clusters. The results demonstrate the feasibility of probing an electronic state of small metal clusters loaded into a disordered system by *o*-Ps.

1. Introduction

Small metal clusters with the size of a few nano-meters have attracted attention in the last decades, since they show physical and chemical properties different from the bulk [1]. Due to simple electronic structures of alkali metals, theoretical calculations have been extensively conducted for small clusters of Na [2], K [3], and Li [4], which predict the presence of a specific magic number. Apart from randomly isolated cluster systems, alkali-metal clusters loaded into periodically-ordered cages of zeolite have provided an opportunity to design new functional materials with respect to thermochromism [5] and magnetism [6]. For example, K clusters loaded into K-type zeolite A are known to show ferromagnetic property [7]. Na-loaded Na-ion type zeolite exhibits various colors due to optical absorptions by Na clusters in the visible light region [8].

Previously, we reported that the momentum distribution of annihilating γ rays originating from *o*-Ps pick-off annihilation is sensitively influenced by the presence of small Cs clusters loaded into free volumes of SiO_2 glass [9]. This implies the feasibility of probing a local electronic state of loaded metal clusters by *o*-Ps. In this work, we further investigate Cs-loaded SiO_2 glass by AMOC spectroscopy in combination with thermal analyses as, e.g., DSC and TG-DTA.

2. Experiments

Unloaded and Cs-loaded SiO_2 glasses were studied in the present work. Conventional x-ray diffraction (XRD) measurements with Cu K-α radiation were performed at room temperature. DSC and TG-DTA measurements were performed for 10 mg samples in the temperature range of 293-773 K at a heating rate of 10 K/min.

AMOC measurements were performed by acquiring the Doppler broadening and positron lifetime data coincidently. The positron birth γ ray from ^{22}Na and one of the annihilation γ rays were detected by BaF_2 scintillators with 2" diameter × 1" thickness coupled with photomultiplier tubes (DET1 and DET2). DET1 and DET2 signaling 1.27 MeV and 511 keV γ rays were squarely aligned to each other. For the simultaneous measurement of the annihilation γ rays, a high purity Ge detector was placed collinearly to DET2. The total counts of a typical AMOC spectrum were ~ 30 million. A time resolution of 230 ps FWHM was obtained. In the present experiments, positron-age dependent Doppler broadening spectra were analyzed by the S parameter ($S(t)$), which was determined as the ratio of the central area over (-2.5 to +2.5) $\times 10^{-3}$ m_0c to the total area of the Doppler broadening peak at positron age t.

3. Results and Discussion

Figure 1 shows XRD patterns observed for unloaded and Cs-loaded SiO_2 glasses. The diffraction pattern of Cs-loaded SiO_2 glass is of typical amorphous phase as unloaded SiO_2 glass indicating that no crystallization is induced upon Cs loading.

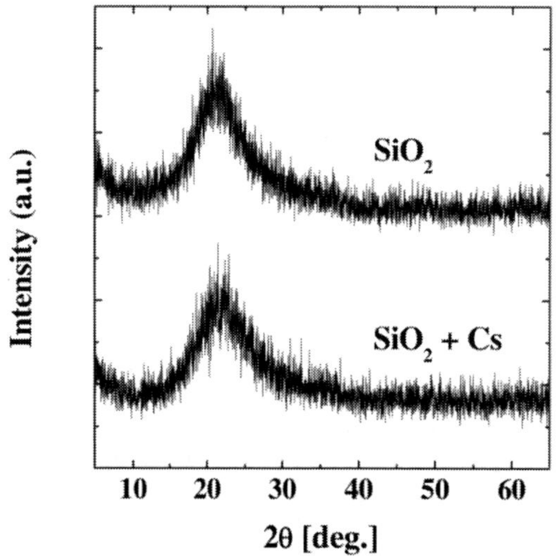

Figure 1. XRD patterns for unloaded and Cs-loaded SiO_2 glasses. The vertical axis is in arbitrary unit (a. u.). The data are adequately shifted for clarification.

Figure 2 shows DSC data for unloaded and Cs-loaded SiO_2 glasses. DSC curves are essentially identical to each other. No significant peaks arising from melting of Cs clusters as well as their structural modification were observed for Cs-loaded SiO_2 glass. Similarly to DSC data, neither unloaded nor Cs loaded SiO_2 glass exhibits observable peaks in their TG (a) or DTA (b) curves (see Figure 3 (a) and (b)). Based on the results of TG-DTA in addition to more sensitive DSC measurements, we conclude that Cs clusters less than the detection limit of thermal analysis is loaded into free volumes in SiO_2 glasses.

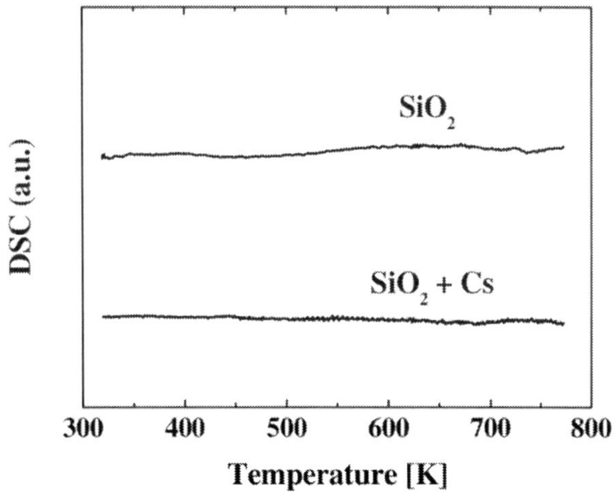

Figure 2. DSC curves for unloaded and Cs-loaded SiO_2 glasses. The vertical axis is in arbitrary unit (a. u.). The data are adequately shifted for clarification.

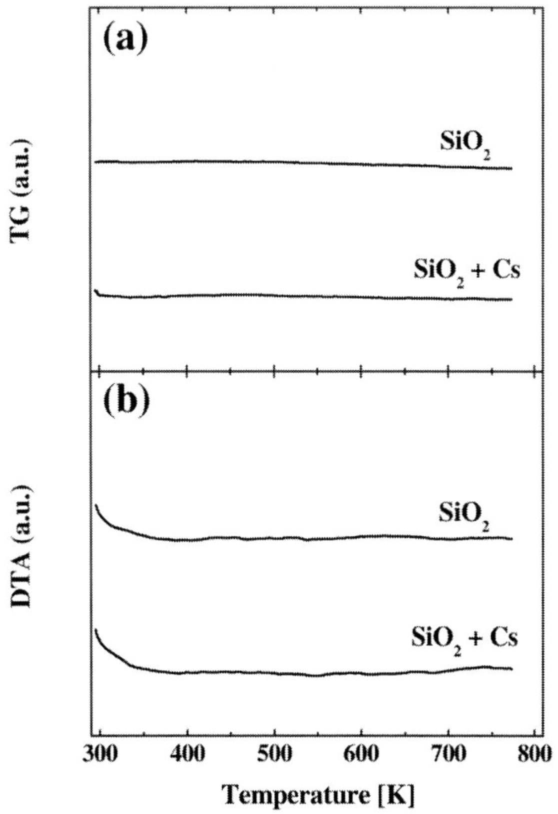

Figure 3. TG-DTA curves for unloaded and Cs-loaded SiO_2 glasses. The vertical axis is in arbitrary unit (a. u.). The data are adequately shifted for clarification.

Figure 4 shows the S parameters obtained by AMOC spectroscopy as a function of positron age ($S(t)$) for unloaded and Cs-loaded SiO_2 glasses [9]. The S parameters indicate high values around $t = 0$. They rapidly decrease with increasing positron age up to 0.5 ns and then level off above 2.0 ns. At younger positron age, *p*-Ps, *o*-Ps, and positrons all contribute to the momentum distributions of positron-electron annihilation. Around $t = 0$ ns, high values of S parameters are observed due to the narrow momentum distribution of *p*-Ps self annihilation. At the positron age above 0.4 ns, more than 90 % of the shortest-lived *p*-Ps has already disappeared. Above 1.2 ns, the positrons have mostly disappeared as well and, as a consequence, *o*-Ps dominantly contributes to the momentum distribution. The saturated values of $S(t)$ at older positron age thus reflect the momentum distributions of *o*-Ps pick-off annihilation.

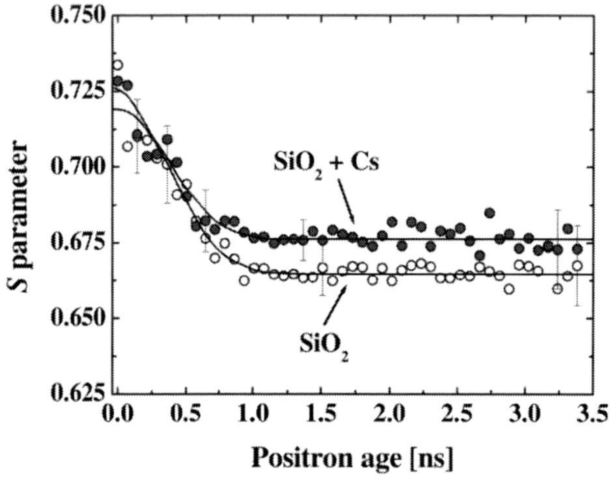

Figure 4. $S(t)$ parameters obtained by AMOC spectroscopy for unloaded and Cs-loaded SiO_2 glasses. Solid lines are drawn for guiding the eye.

The saturated values of $S(t)$ are significantly higher for Cs-loaded SiO_2 than for unloaded one. This demonstrates narrowing of the momentum distribution of *o*-Ps pick-off annihilation in the former. In contrast to DSC and TG-DTA, AMOC spectroscopy successfully detects the local electronic change induced by loading Cs clusters through the momentum distribution of *o*-Ps pick-off annihilation [9]. Upon Cs loading, free volumes in SiO_2 glass mostly surrounded by O atoms [10] get partially occupied by Cs atoms, resulting in chemical environment different from unloaded SiO_2 glass. The change of chemical environment may be caused by the electrons of Cs atoms around free volumes. Alternatively, the effect of loading Cs clusters may responsible for the change of chemical bonding of O atoms. As mentioned above, O atoms are principally located around free volumes in SiO_2 glass [10]. If loading of Cs clusters causes chemical shift in binding of electrons of O atoms, *o*-Ps pick-off annihilation could be affected by that. More comprehensive studies employing x-ray photoelectron spectroscopy (XPS) and ultraviolet photoelectron spectroscopy (UPS) in combination with AMOC spectroscopy is under way.

4. Conclusion

Cs-loaded SiO_2 glass was investigated by DSC, TG-DTA, and positron annihilation AMOC spectroscopy. No significant peaks arising from Cs clusters were observed in TG-DTA and DSC curves. AMOC spectroscopy succeeded in detecting the local electronic change induced by loading Cs clusters through the momentum distribution of *o*-Ps pick-off annihilation. This implies the feasibility of probing a local electronic state of loaded metal clusters by *o*-Ps. We discussed the reason why *o*-Ps state is sensitively influenced by a small amount of Cs clusters loaded into fee volumes in SiO_2 glass.

Acknowledgments

This work was partially supported by Grants-in-Aid for Scientific Research from the Ministry of Education, Science, Sports and Culture of Japan (Grant Nos. 20740173 and 2154317).

References

[1] for example, Yamamoto Y, Miura T, Suzuki M, Kawamura N, Miyagawa H, Nakamura T, Kobayashi K, Teranishi T and Hori H 2004 *Phys. Rev. Lett.* **93** 1168011
[2] Ishii Soh, Ohno Kaoru, Kawazoe Yoshiyuki and Louie Steven G 2001 *Phys. Rev. B* **63** 1551041
[3] Saito S, Zhang S B, Louie S G and Cohen M L 1990 *J. Phys.: Condens. Matter* **2** 9041
[4] Broyer M, Chevaleyre J, Dugourd Ph, Wolf J P and Wöste L 1990 *Phys. Rev. A* **42** 6954
[5] Harrison M R, Edwards P P, Klinowski J and Thomas J M 1984 *J. Solid State Chem.* **54** 330
[6] Duan Truong Cong, Nakano Takehito and Nozue Yasuo 2007 *J. Magn. Magn. Mater.* **310** 1031
[7] Kodaira T, Nozue Y, Ohwashi S, Goto T and Terasaki O 1993 *Phys. Rev. B* **48** 12245
[8] Maniwa Yutaka, Masubuchi Shin-ichi, Minamisawa Toshifumi, Kira Hiroshi, Kodaira Tetsuya and Kazama Shigeo 2006 *Chem. Phys. Lett.* **424** 97
[9] Sato K, Murakami H, Ito K, Hirata K and Kobayashi Y 2009 *Macromolecules* **42** 4853
[10] Zachariasen W H 1932 *J. Am. Chem. Soc.* **54** 3841

Characterization of helium bubbles in Si by slow positron beam

M Maekawa and A Kawasuso
Advanced Science Research Center, Japan Atomic Energy Agency
1233 Watanuki, Takasaki, Gunma, 370-1292, Japan

E-mail: maekawa.masaki@jaea.go.jp

Abstract. High-dose He-ion-implanted Si has been studied by positron annihilation method. As well known, due to post-implantation annealing at 1173 K, the positron lifetime and the Doppler broadening *S* parameter dramatically increased due to the formation of microvoids. Meanwhile, the post-implantation annealing at 573K resulted in a significant decrease in the positron lifetime and an apparent modification of the Doppler broadening spectrum near the momentum region of p=0.01 m_0c. From the first principles calculation, we confirmed the positron trapping at He bubbles formed in large vacancy clusters.

1. Introduction

It is known that well-ordered microvoids (~10nm) are formed in Si by the high-dose He implantation and subsequent heat treatment [1-4]. Such microvoids have been investigated concerning the gettering issues of Si. As the precursor states of microvoids, helium bubbles with comparable sizes to those of microvoids may be formed. However, the details of helium bubble in Si have not yet been clarified.

Positron annihilation spectroscopy (PAS) is a powerful tool to study defects in materials. Using a slow positron beam, a depth profiling of defects with typical depth of several microns is possible. PAS has been applied to low-dose He ion irradiated Si [5, 6]. For the high-dose He implanted Si, the microvoid formation is mainly reported [7]. However, the positron trapping at helium bubbles in Si has not yet been reported. In this study, we attempted to investigate the fundamental aspects of interaction between positrons and irradiation defects and/or the He bubbles in Si induced by the high-dose implantation of He ions.

2. Experimental

Samples used in this study were cut from a phosphorus-doped *n*-type Si wafer with the resistivity of 48~67 Ω cm. The samples were subjected to the He ion implantation with the energy of 50~200 keV to a dose of 1×10^{17} cm^{-2} at room temperature using a 400 keV ion implanter. The implanted region is estimated to be at 0.5~1.5 μm from the surface by the 'TRansport of Ion in Matter' (TRIM) simulation [8]. After the He implantation, the samples were annealed in argon ambient from 373 K to 1173K.

The Doppler broadening of annihilation radiation (DBAR) measurements were carried out as a function of incident positron energy (*E*). The obtained DBAR spectra were characterized by *S* and *W* parameters, which are defined as the peak and tail intensities, respectively. *S* (*W*) parameter tends to increase (decrease) when positrons are trapped at vacancy defects because of the narrowing of DBAR

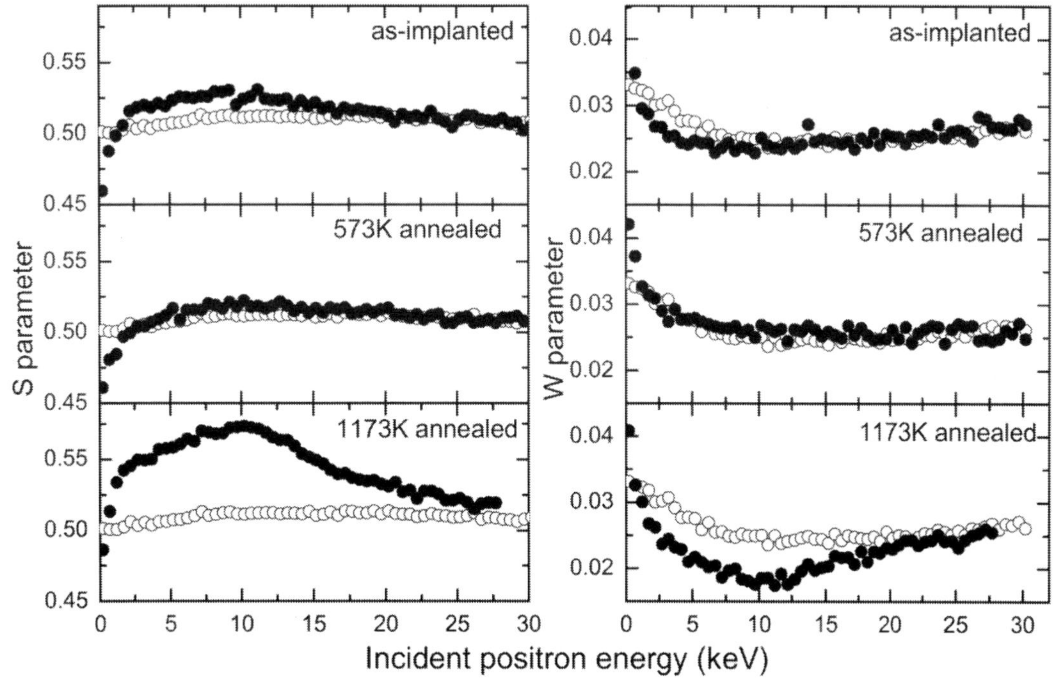

Figure 1. S and W parameters as a function of incident positron energy obtained after annealing at various temperatures. Open circles denote *S* and *W* parameters obtained for the unimplanted state.

spectrum due to the reduction of core electron annihilation rate. All the *S* and *W* parameters are normalized to those for the unimplanted state. To investigate detailed changes in the DBAR spectrum, the coincidence Doppler broadening (CDB) measurement was performed. Positron annihilation lifetime measurements were also carried out using the pulsed positron beam. After subtraction of the background components, the lifetime spectra were analyzed using the PATFIT program [9].

3. Theoretical calculation

To interpret the experimental results, the positron annihilation characteristics were calculated for various model structures based on the first principles manner [10, 11]. The valence electron momentum distribution was calculated using the conventional scheme:

$$\rho_V(\mathbf{p}) = \pi r_e^2 c^2 \sum_n \left| \int e^{-i\mathbf{pr}} \Psi_+(\mathbf{r}) \Psi_n(\mathbf{r}) \sqrt{\gamma(\mathbf{r})} d\mathbf{r} \right|^2 , \quad (1)$$

where r_e is the classical electron radius, c is the speed of light, $\Psi_+(\mathbf{r})$ is the positron wave function, $\Psi_n(\mathbf{r})$ is the valence electron wave function of the n-th band, and $\gamma(\mathbf{r})$ is the enhancement factor. The summation was performed over all the occupied states. The core electron momentum distribution was calculated following the Alatalo's method:

$$\rho_C(\mathbf{p}) = \pi r_e^2 c^2 \sum_{i,nlm} \left| \int e^{-i\mathbf{pr}} \Psi_+(\mathbf{r}) \Psi_{i,nlm}(\mathbf{r}-\mathbf{R}_i) \sqrt{\gamma(\mathbf{r})} d\mathbf{r} \right|^2 , \quad (2)$$

where $\Psi_{i,\,nlm}(\mathbf{r}-\mathbf{R}_i)$ represents the core electron wave function specified by the principal, azimuthal, and magnetic quantum numbers (nlm) for the i-th atom and $\mathbf{R}_i$ denotes the position vector. The total momentum distribution is given by $\rho(\mathbf{p})=\rho_V(\mathbf{p})+\rho_C(\mathbf{p})$. The DBAR spectrum was obtained after integrating $\rho(\mathbf{p})$ in two directions and subsequent convolution with the Gaussian resolution function having a FWHM of 3.92×10^{-3} m_0c. The positron lifetime was obtained by integrating the momentum density in all momentum space.

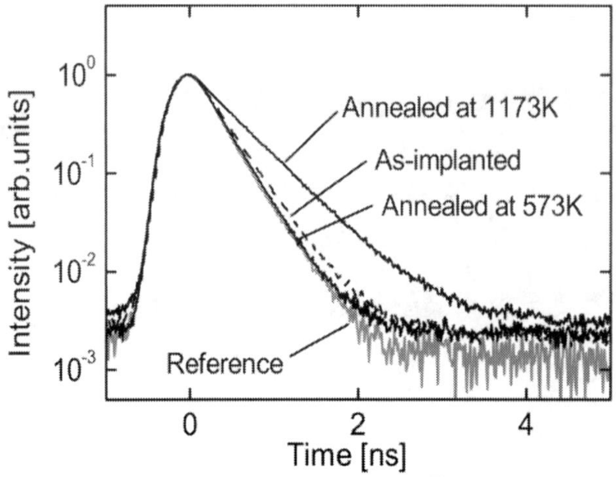

Figure 2. Positron lifetime spectrum for reference, as-implanted state, annealed at 573K and at 1173K.

A supercell with 3× 3× 3 conventional unit cell containing 216 Si atoms was constructed as a perfect Si crystal. For calculating vacancy states, relevant atoms were removed from this supercell. Also, for He bubbles or He vacancy complexes, the optimum positions of He atoms were determined through the molecular dynamics simulation. The valence electron wavefunctions were calculated based on the norm-conserving pseudo-potential method using the ABINIT4.1.4 code [12]. For the perfect Si crystal, the *k*-point mesh was 2× 2× 2. The cut-off energy of the plane wave basis set was 24 Ryd. The core electron wavefunctions were represented by the Slater function parameterized by Clementi and Roetti [13]. A self-consistent positron wavefunction was calculated based on the two-component density functional theory in order to minimize the energy functional [14]. The Boróński–Nieminen enhancement factor was adopted.

4. Results and Discussion

Figure 1 shows the *S* and *W* parameters as a function of incident positron energy (*S*-*E* and *W*-*E* curves) obtained after annealing at various temperatures. After the implantation, *S* (*W*) parameters at E=5~12 keV corresponding to the He-implanted region increase (decrease) suggesting the introduction of vacancy defects. After annealing at 573 K, *S* (*W*) parameters decrease (increase) to the values for the unimplanted state and subsequently increase (decrease) dramatically with increasing annealing temperature. Figure 2 shows the positron lifetime spectra measured before and after the He implantation and subsequent annealing at 573 K and 1173 K at E=8 keV. The prolonged positron lifetime due to implantation is shortened after annealing at 573 K. After annealing at 1173 K, the positron lifetime is sufficiently prolonged.

Detailed annealing behaviour of *S* (*W*) parameters and positron lifetime are plotted in Fig. 3. After annealing at 573 K, *S* parameter decreases to 1.023, while *W* parameter exceeds the level of unimplanted state. Upon the further annealing up to 1173 K, *S* (*W*) parameters reach 1.13 (0.70). The positron lifetime shows similar annealing behaviour to that of *S* parameter. The positron lifetime for the as-implanted state is 270 ps and decreased to 230 ps that is only 10 ps longer than that for the unimplanted state (220 ps). After annealing at 1173 K, the average positron lifetime reaches 500 ps.

It is known that well-ordered microvoids are formed after He implantation and subsequent annealing at 1000-1200 K. The dramatic increases in the *S* parameter and positron lifetime at 1173 K shown in Figs. 1 and 2 are explained as the positron trapping at microvoids as reported in the previous studies [15-17].

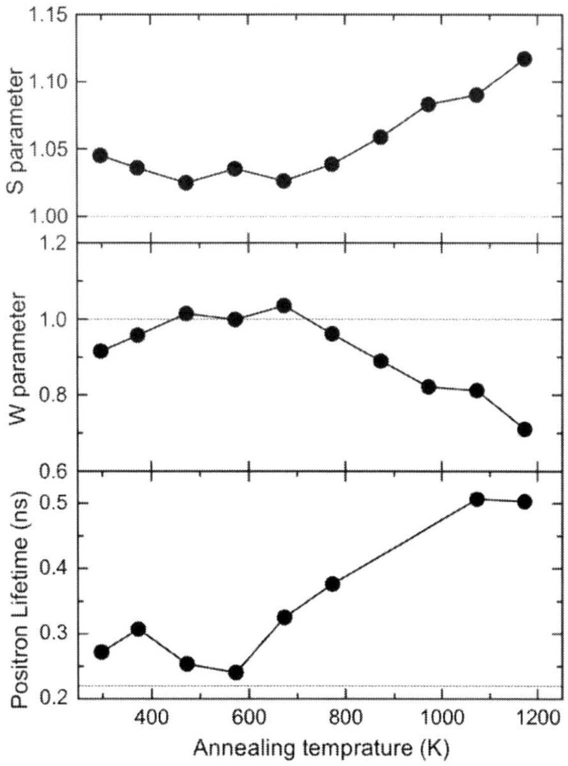

Figure 3. Annealing behaviour of S and W parameters and positron lifetime.

Here, we focus on the decreases in *S* parameter and positron lifetime and the increase in *W* parameter at 573 K. Figure 4 shows the DBAR spectrum measured after annealing at 573 K in the coincidence mode. (To enhance the change of spectrum, the spectrum is differentiated by that of the unimplanted state.) The suppressed intensity in the low momentum region ($0 \sim 3 \times 10^{-3}$ m_0c) corresponds to the decrease in *S* parameter shown in Figs. 1 and 3. A characteristic bump is seen at around 10×10^{-3} m_0c. This corresponds to the increase in *W* parameter in Figs. 1 and 3. We examine the DBAR spectrum at 573 K and positron lifetime from the aid of the first principles calculation as follows.

Major vacancy defects after the He implantation may be divacancies since monovacancies are mobile at room temperature in Si. It is inferred that due to annealing at 573 K divacancies are occupied by mobile He atoms. Alternatively, He bubbles may be formed as a consequence of He agglomeration. In this case, relatively large open volumes are filled with He atoms. As shown in Fig. 4, by introducing He atoms into divacancies the calculated spectrum becomes to be analogous with the experiment, but the overall spectrum shapes are not compatible. Meanwhile, when a large vacancy cluster (V_{18}) with an approximate diameter of 1 nm is filled with He atoms the experimental spectrum is better explained.

Figure 5 shows the calculated positron lifetimes for divacancy with He atoms and V_{18} cluster with He atoms. The positron lifetime decreases with increasing the number of He atoms. The positron lifetimes for divacancy plus two He atoms and divacancy plus four He atoms are 235 ps and 224 ps, respectively. That for V_{18} plus 36 He atoms is 233 ps. These calculated lifetimes are comparable to the

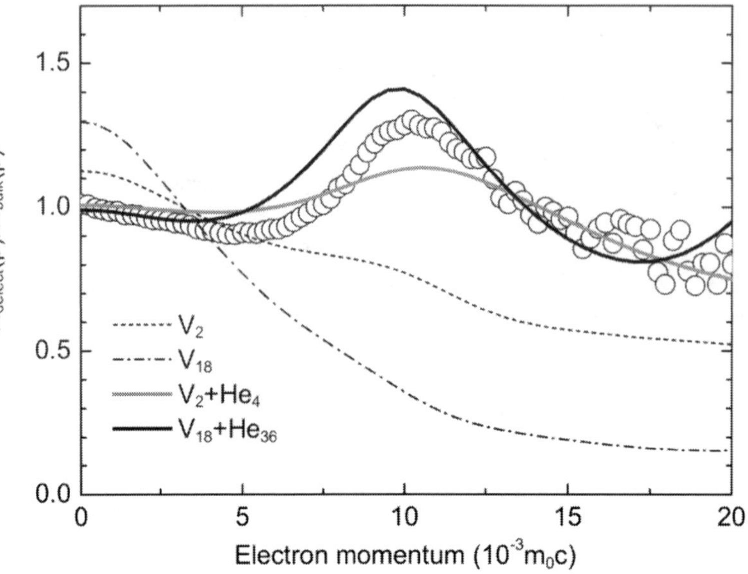

Figure 4. DBAR spectrum obtained after the He-implantation and subsequent annealing at 573 K (open circles) in the coincidence mode. The spectrum is differentiated by that for the unimplanted state to enhance the structure. Calculated DBAR spectra for a divacancy (broken line), a V_{18} cluster (chained line), a divacancy-He_4 complex (gray line) and a V_{18}-He_{36} complex (black solid line) are also shown.

experimental value obtained after the He implantation and subsequent annealing at 573 K as shown in Fig. 3.

The possibility of positron trapping at He bubbles with a few nm radius is also anticipated through a more conventional way as follows: The annihilation rate of positrons (the inverse of lifetime) may be given by

$$\lambda = \pi r_0^2 c Z_{eff} n_{He} + \lambda_{void} , \tag{1}$$

where r_0 is the classical electron radius, n_{He} is the He density in bubbles, λ_{void} is the positron lifetime in empty voids and Z_{eff} is given as 3.94 for He. Having the annihilation rates at 573 K (4.35 ns^{-1}) and 1173 K (2 ns^{-1}) where He bubbles and microvoids are formed, we obtain n_{He}=8× 10^{22} cm^{-3}. This is comparable to that assumed in the above model calculation for V_{18} plus He_{36} (n_{He}=7× 10^{22} cm^{-3}). Combining the gas equation $P=n_{He}kT$, where P is pressure, k is the Boltzmann constant and T is temperature, and the surface tension equation $P=2\gamma/r$ [18], where γ is the surface energy and r is the bubble radius, we obtain

$$r = \frac{2\gamma}{n_{He}kT} . \tag{2}$$

Assuming γ=6.24× 10^{18} eV/m^2 for Si [19], the bubble radius is estimated to be approximately 3 nm. Thus, it is concluded that He bubbles with a few nm radius are formed in the low temperature annealing process and positrons are trapped there [6, 20].

Figure 5 shows the cross-sectional transmission electron microscope (TEM) images obtained after the He implantation and subsequent annealing at 573 K and 1173 K. Many small black spots with a few nm radius are observed after annealing at 573 K. After annealing at 1173 K, these black spots vanish and microvoids with 1~40 nm radius appeared. The black spots observed after annealing at 573

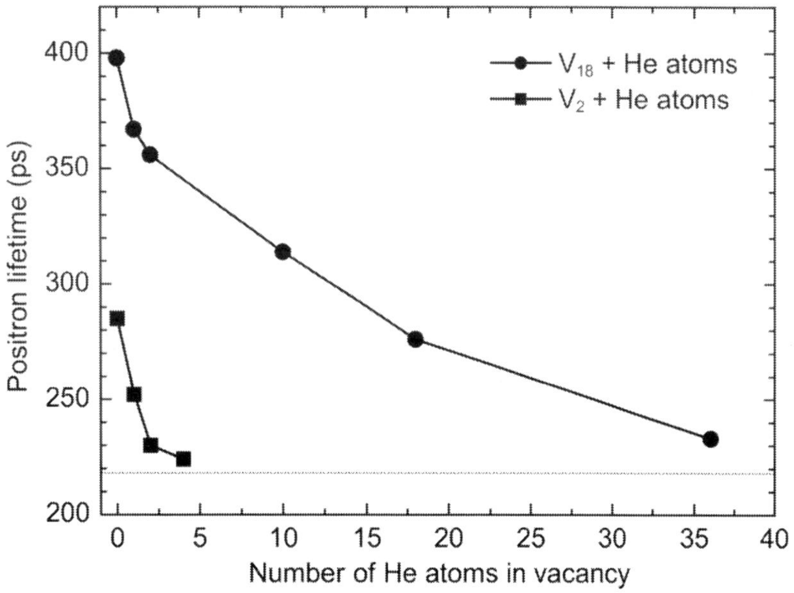

Figure 5. Calculated positron lifetimes for divacancy and V_{18} cluster filled with He atoms as a function of number of He atoms.

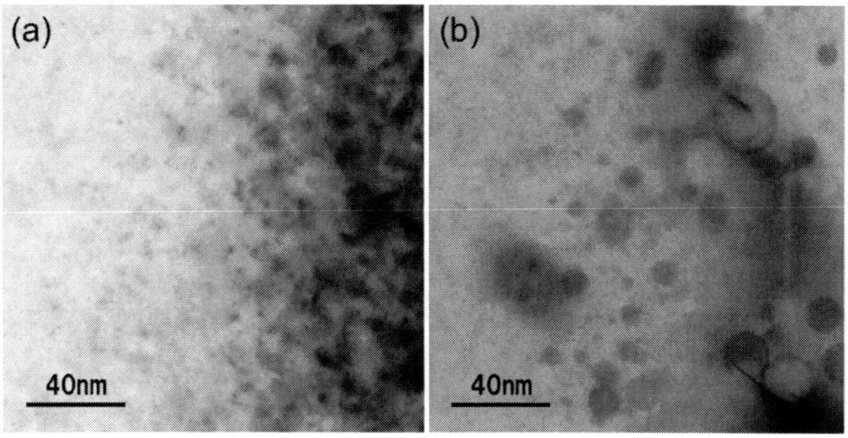

Figure 6. Cross-sectional TEM images obtained after the He implantation and subsequent annealing at (a) 573 K and (b) 1173K.

K probably represent He bubbles where positrons are trapped. During the further annealing process, as a result of coalescence among bubbles and He dissipation, microvoids are developed.

5. Summary

High-dose He-implantated Si has been investigated using positron annihilation spectroscopy. Upon the post-implantation annealing at around 573 K, He bubbles with a few nm size are formed and positrons are trapped there. These He bubbles are probably the source for microvoids formed during the high temperature annealing process up to 1173 K.

References

[1] Griffon C C, Evans J H, De Jong P C, Van Veen A 1987 Nucl. Instr. Methods B **27** 417
[2] Evans J H, Van Veen A and Griffon C C 1987 Nucl. Instr. Methods B **28** 360
[3] Corni F, Nobili C, Ottaviani G, Tonini R, Calzolari G, Cerofolini G F and Queirolo G 1997 Phys. Rev. B **56** 7331
[4] Brusa R S, Karwasz G P, Tiengo N, Zecca A, Corni F, Nobili C, Ottaviani G and Tonini R 1997 Mater. Sci. Forum **255** 665
[5] Brusa R S, Karwasz G P, Tiengo N, Zecca A, Corni F, Tonini R and Ottaviani G 2000 Phys. Rev. B **61** 10154
[6] Corni F, Calzolari G, Frabboni S, Nobili C, Ottaviani G, Tonini R, Cerofolini G F, Leone D and Servidori M 1999 J. Appl. Phys. **85** 1401
[7] Raineri V, Coffa S, Saggio M, Frisina F, Rimini E 1999 Nucl. Instr. Methods B **147** 292
[8] Ziegler J F, Biersack J P and Littmark U, The Stopping and Range of Ions in Solids, 1985, Pergamon, New York
[9] Kirkegaard P, Pederson N, and Eldrup M 1989 Riso-M-**2704**
[10] Puska M J and Nieminen R M 1994 Rev. Mod. Phys. **66** 841
[11] Alatalo M, Barbiellini B, Hakala M, Kauppinen H, Korhonen T, Puska M J, Saarinen K, Hautojärvi P and Nieminen R M 1996 Phys. Rev. B **54** 2397
[12] Gonze X 2002 Comput. Mater. Sci. **25** 478
[13] Clementi E and Roetti C 1974 At. Data Nucl. Data Tables **14** 177
[14] Boroński E and Nieminen R M 1986 Phys. Rev. B **34** 3820
[15] Vishnyakov V M, Donnelly S E, Carter G, Birtcher E C, Haworth L 2002 Solid State Phenomena **82** 267
[16] Krause-Rehberg R and Leipner H S 1998 "Positron annihilation in Semiconductors" Springer
[17] Saito M and Oshiyama A 1996 Phys. Rev. B **53** 7810
[18] Bermond J M, Métois J J, Egéa X and Floret F 1995 Surf. Sci. **330** 48.
[19] Jensen K O, Eldrup M, Singh B N and Victoria M 1988 J. Phys. F **18** 1069
[20] Pivac B, Milat O, Dubek P, Bernstorff S, Corni F, Nobili C, Tonini R 2003 Phys. Status Solidi (a) **198** 29

International Symposium on Advanced Science Research 2009 IOP Publishing
Journal of Physics: Conference Series **225** (2010) 012030 doi:10.1088/1742-6596/225/1/012030

OH Radical in Water Studied by Quantum Beats on Positron Annihilation -The Effect of Water Liquid Structures-

Jeong-Jae Lee[1], Toshitaka Oka[2] and Tetsuya Hirade[1,3,*]
[1] Graduate School of Science and Engineering, Ibaraki University, 4-12-1 Narusawa, Hitachi, Ibaraki, 316-8511 JAPAN
[2] Advanced Science Research Center, Japan Atomic Energy Agency, Tokai, 319-1195 Japan
[3] Nuclear Science and Engineering Directorate, Japan Atomic Energy Agency, Tokai, 319-1195 Japan

E-mail: hirade.tetsuya@jaea.go.jp

Abstract. Singlet radical pairs of ortho-positronium and hydroxyl (OH) radical were generated in positron radiolysis in water. Quantum beats on the energy distribution of positron annihilation γ-rays caused by the reaction of electron-spin-correlated pairs of ortho-positronium and OH radical were newly measured at 32°C by using positron annihilation age-momentum correlation (AMOC) measurement. The quantum beats can provide information of hyperfine coupling constants of the OH radical in water liquid structures, and hence can provide the information of change of the water liquid structures. The quantum beats appeared at 32°C could not be explained easily. The reason was continuous and slow change of water liquid structures at 32°C. The water liquid structures probably can be understood by using the positron annihilation quantum beat measurements.

1. Introduction

Positron that is anti-particle of electron has the same mass and the opposite charge of electron. Positronium (Ps) is a bound state of positron and electron. The Ps formation process in insulators is well-explained by the spur reaction model proposed by Mogensen in the case of positrons with energies of more than 100eV [1]. The model describes the Ps formation process as a combination of a thermalized positron and one of the excess electrons formed by the positron irradiation at the terminal spur of the positron track within about 1ps. A quarter of the Ps forms as para-Ps (p-Ps, singlet Ps), and three quarters form as ortho-Ps (o-Ps, triplet Ps).

Basically, in a vacuum, p-Ps annihilates with a lifetime of 125ps through the intrinsic annihilation of two annihilation γ-rays being emitted while o-Ps annihilate with an intrinsic annihilation lifetime of 142ns through three γ-rays being emitted. In condensed matter a positron in Ps can annihilate with one of the electrons from the surrounding molecules, which is known as "pick-off" annihilation. This mainly occurs with o-Ps because of its long intrinsic annihilation lifetime. Pick-off annihilation mainly provides two annihilation γ-rays. A positron that escapes the Ps formation, i.e. a free positron, can also annihilate with one of the electrons from the surrounding molecules. The lifetime of the free positrons is about 400ps, with free positron annihilation also mainly providing two annihilation γ-rays.

Because of Ps negative work functions in many materials, Ps is repelled by surrounding substance,

* To whom any correspondence should be addressed.

and hence Ps can form a bubble in liquids. The size of the bubble is controlled by the zero point energy of Ps in the bubble and the energy of the bubble by the surface tension and the pressure. Therefore the bubble size should be larger with increasing temperature, for the less surface tension at higher temperatures. For the larger bubble size, the pick-off annihilation probability of Ps in the bubble is smaller because of the less overlapping of the positron in Ps and electrons on the wall of the bubble.

In water, o-Ps gave an opposite tendency, the shorter lifetimes at higher temperatures. [2] Stepanov suggested o-Ps reactions with spur reactants, such as hydroxyl (OH) radicals, H_3O^+ cations, and hydrated electrons to explain the temperature dependence observed in water. [3] In water these species form as indicated below within about 1ps.

$$H_2O \;\text{^^^}\!\!\rightarrow\; H_2O^{\bullet+} + {}^{\bullet}e^- \qquad \text{(ionization)} \qquad (1)$$

$$\begin{aligned} {}^{\bullet}e^- + e^+ &\rightarrow \text{p-Ps}^{\bullet}\ (25\%) \\ &\rightarrow \text{o-Ps}^{\bullet}\ (75\%) \qquad \text{(Ps formation)} \qquad (2) \end{aligned}$$

$$H_2O^{\bullet+} + H_2O \rightarrow H_3O^+ + {}^{\bullet}OH \qquad \text{(hydroxyl radical formation)} \qquad (3)$$

$${}^{\bullet}e^- + nH_2O \rightarrow {}^{\bullet}e^-_{aq} \qquad \text{(hydration)} \qquad (4)$$

The expected o-Ps reactions with OH radicals, H_3O^+ cations, and hydrated electrons are as follows:

$$\text{o-Ps}^{\bullet} + H_3O^+ \rightarrow e^+ + H_2O + {}^{\bullet}H \qquad \text{(oxidation)} \qquad (5)$$

$$\begin{aligned} \text{o-Ps}^{\bullet} + {}^{\bullet}OH &\rightarrow \text{PsOH} \qquad \text{(complex formation)} \qquad (6) \\ &\rightarrow e^+ + OH^- \qquad \text{(oxidation)} \qquad (7) \end{aligned}$$

$$\text{o-Ps}^{\bullet} + {}^{\bullet}OH(\uparrow) \rightarrow \text{p-Ps}^{\bullet} + {}^{\bullet}OH(\downarrow) \qquad \text{(spin conversion)} \qquad (8)$$

$$\text{o-Ps}^{\bullet} + {}^{\bullet}e^-_{aq}(\uparrow) \rightarrow \text{p-Ps}^{\bullet} + {}^{\bullet}e^-_{aq}(\downarrow) \qquad \text{(spin conversion)} \qquad (9)$$

Here $^{\bullet}$ indicates the presence of an unpaired electron.

Spin conversion reactions (8) and (9) can be easily observed using positron annihilation age-momentum correlation (AMOC) measurements [4]. Most of p-Ps annihilates intrinsically with the momentum of the electron and the positron cancelling each other out. Hence the energy of the annihilation γ-rays is very close to 511keV, which is the energy equivalent to the rest mass of an electron (or a positron). Conversely however o-Ps and free positrons annihilate by picking off one of the electrons from the surrounding molecules, which then means that the energy distribution of the annihilation γ-rays is wider because of the Doppler shift. With spin conversion reactions (8) and (9) the p-Ps annihilation increases and the distribution of the annihilation γ-ray energy becomes narrower. A shape parameter, *S*, is often used for easy comparison of the distribution of the annihilation γ-ray energy. The value of *S* is the ratio between the counts appearing in the fixed central area and those in the overall peak area. The narrower the peak the larger the *S* values. AMOC experiments can provide time resolved *S*-parameters and larger *S* values at the older positron age region can indicate the existence of the spin conversion of o-Ps. [4]

The radical reaction depends on the spin state of the radicals. The radical reaction of the o-Ps and the OH radical, reactions (6) and (7), can just occur when the spin state of the radical pair is a singlet. The reactions between the o-Ps and the OH radical, (6), (7) and (8), are diffusion controlled and the spin conversion reaction or the radical reaction occur when they meat each other. Hence quantum beats in the annihilation γ-ray energy distribution were expected to result from the singlet-state probability of the geminate pair of o-Ps and OH radical formed by one ionization process as indicated in (1). [5] Indeed, the hyperfine coupling constant of Ps is very large and then the oscillation between singlet state and triplet state is very fast, probably 15-20ps in water. This oscillation is too fast to detect by AMOC because of the time resolution of 240ps. However, we do not know the spin direction of the proton in OH radical and then there will be two oscillation frequencies, sum and difference of the Larmor frequencies of electrons in Ps and OH radical, making swells which frequency can indicate the hyperfine coupling constant of the OH radical because the Larmor frequency of the electron in the OH radical is much slower than that in the Ps. The probability of the radical reaction of the geminate pair of the o-Ps and the OH radical is more at the swells than between the swells because the spin state can be in a singlet state every 15-20ps. Hence, the probability of the spin conversion reaction is more between the swells and high $S(t)$ values appear as quantum beats. The quantum beats observed here can give the information of the hyperfine coupling constant of OH radicals. [5]

Hirade indicated that there existed reactions of o-Ps and cation radicals in n-hexane. [6] The reactions between geminate pairs should be possible. It means that the quantum beats are also expected on $S(t)$ for n-hexane. In the literatures, there have been many $S(t)$ observed by AMOC even for liquids, and the $S(t)$ for n-hexane was reported by Stoll. [7] The quantum beats were observed for n-hexane and the beat position can be well explained by the mechanism given here.

The quantum beats observed for water at 18°C and 25°C showing that there were two hyperfine coupling constants. [5] Here new results observed at 32°C will be introduced to provide more knowledge on the water liquid structures where OH radical formed.

2. Experimental

AMOC measurements [8] were performed using ^{22}Na (about 400kBq) sealed in two 7.5μm Kapton foils as the positron source. About 10 % of the positrons annihilate within the foils. Ps does not form in Kapton foil and then there is just one lifetime component that gives about 382ps for the lifetime component from the source. Hence it does not affect the o-Ps reactions. A LOAX-51370/20-P (SEIKO EG&G) type semiconductor detector with an energy resolution (FWHM) of 1.28keV at 662keV was used to detect the annihilation γ-ray energy. The time resolution was about 240ps. Total counts appeared near the 511keV peak and in an about 10ns time range of the AMOC measurement reached about 1 million. The count rate was about 1.2 cps. Ultra-pure water supplied by Wako Pure Chemical Industries was used for the water samples. To remove any oxygen molecules nitrogen gas was bubbled through the samples before use in the experiments. The sample temperature was controlled by a water bath and then there is no magnetic field caused by the heater.

3. Results and Discussion

The positron age dependent S-parameters, $S(t)$, obtained using AMOC in water at 32°C are provided in Fig.1(a). Large S values appeared in the earlier positron age region of up to 0.5ns, which was caused by the annihilation of the p-Ps formed in reaction (2). In the older age region than 1.5-2ns larger S values reappeared that must have been caused by the spin conversion of o-Ps. Moreover, the positron annihilation lifetime spectra constructed by the AMOC experiments for water did not show any beats or oscillations as expected by the mechanism explained above. [5]

The magnitude of the isotropic hyperfine coupling constant, i.e. the Fermi contact term, of the OH radical is 73.25MHz while that of the OH-H_2O complex is 155.3MHz [9]. Just one water molecule can enhance the hyperfine coupling constant to be more than twice the magnitude. The OH radical in liquid water can have OH-nH_2O complex states and hence the expected hyperfine coupling constant will be larger. The hyperfine coupling constant of the OH radicals in water was obtained by fitting the beat positions with the mechanism explained above. [5] The hyperfine coupling constants of OH radicals in the liquid structures as OH-nH_2O were observed here. The structure affects the hyperfine coupling constants making it possible to observe the change in the liquid structures of water using the quantum beats observed there.

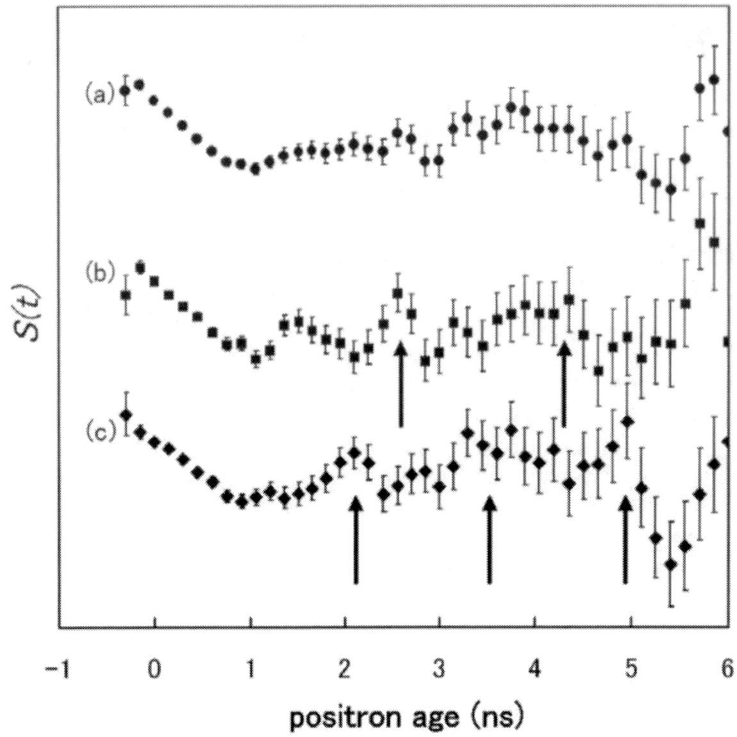

Fig.1 *S(t)* curve observed through AMOC in ultrapure water at 32°C. (a) was obtained through 2weeks measurement. (b) was obtained by the first half and (c) was by the second half of the measurement.

It was reported that the hyperfine coupling constants were about 870MHz and 758MHz for 18°C and 810MHz and 678MHz for 25°C, respectively. [5] However, it might be expected to have more complicated liquid structures at higher temperatures. *S(t)* at 32°C as shown in Fig.1 seems to have complicated structures than those at lower temperatures. It is almost impossible to explain beats in Fig.1 (a) by the mechanism explained above. The effect on the elapsed time of the measurement appeared in (b) and (c) that was obtained by the experiment for the *S(t)* curve (a). (b) is the first half and (c) is the second half of the AMOC measurement shown in (a). (b) and (c) showed some beats and some of them can be explained by the proposed mechanism as indicated arrows in Fig.1. [5] The hyperfine coupling constants were about 709MHz and 580MHz estimated by the arrows indicated in (c) and (b) respectively. These are not all of the beats appeared here. It is indicating that there exist many water liquid structures where OH radicals formed. Moreover water is not just a simple mixture of all of water liquid structures. The fraction of every structure changes continuously and very slowly. It was reported that there were two hyperfine coupling constants observed at 18°C and 25°C and it was indicating that there could be two water liquid structures. [5] However, there seems to be more structures observed at 32°C and any structures are probably not stable. These things say that there are many water liquid structures and only some of them, probably mainly two structures, are more stable than others. That is the reason why two sets of beats were observed at lower temperatures such as 18°C. [5]

4. Conclusion

Quantum beats were detected in the reactions of electron-spin-correlated o-Ps and OH radical pairs in water at 32°C. These newly observed quantum beats in the annihilation γ-ray energy distribution caused by the o-Ps reactions provide the opportunity for new research to be undertaken. The changes in the beats can provide information on the change on the liquid structures of water. The *S(t)* curve measured at 32°C for about 2weeks did not show clear quantum beats expected by the mechanism of appearance of the quantum beats. It was caused by the change of the water liquid structures during measurement. The $S(t)$ curves obtained with the shorter measurement time could indicate the quantum beats. There probably exist many water liquid structures and only some of them are more stable than others. That is the reason why two sets of beats were observed at lower temperatures such as 18°C. Moreover the water liquid structures changed continuously and slowly.

Acknowledgements

This research was partially supported by a Ministry of Education, Culture, Sports, Science and Technology Grant-in-Aid for Scientific Research (C), 18550084, 2006-2008 and also by the Inter-University Program for the Joint Use of JAEA Facilities. The author would also like to acknowledge Prof. Nagai for his assistance with the AMOC measurements, and Mr. Sawahata, Mr. Kawate and Mr. Ishimoto for their assistance with the laboratory equipments.

References

[1] Mogensen O E 1974 *J. Chem. Phys.* **60** 998–1004
[2] Kotera K, Saito T, Yamanaka T 2005 *Phys. Lett.* A **345** 184–190
[3] Stepanov S V, Byakov V M, Hirade T 2007 *Radiat. Phys. Chem.* **76** 90–95
[4] Castellaz P, Major J, Mujica C, Schneider H, Seeger A, Siegle A, Stoll H, Billard I 1996 *J. Radioanal. Nucl. Chem.* **210** 457–467
[5] Hirade T 2009 *Chem. Phys. Lett.* **480** 132–135
[6] Hirade T and Mogensen O E 1994 *Hyperfine Interact.* **84**, 491-498
[7] Stoll H, Koch M, Lauff U, Maier K, Major J, Schneider H, Seeger A, Siegle A 1995 Appl. Surf. Sci. **85** 17–21
[8] Suzuki N, Nagai Y, Hyodo Y 2000 *Radiat. Phys. Chem.* **58** 777–780
[9] Brauer C S, Sedo G, Grumstrup E M, Leopold K R, Marshall M D, Leung H O 2004 *Chem. Phys. Lett.* **401** 420–425

Charge Particle Diffusion and the Effective Interactions in the Liquid Phase : Based on Positron Diffusion Studies

I. Kanazawa, H. Suzuki, and H. Kitahata

Department of physics, Tokyo Gakugei University, Nukuikita-machi 4-1-1, Koganei-shi, Tokyo, 184-8501, Japan

E-mail: kanazawa@u-gakugei.ac.jp

Abstract. Based on the theoretical formula, which explains an anomalous diffusion mechanism of positrons in the liquid phase, we propose an origin of the attractive interactions between like charged ions in the liquid phase. It is suggested that the attractive interactions are attributed to exchange interactions of the massive gauge fields between changed ions.

1. Introduction

The interionic interactions in fluids can be described by pairwise additive solvent-meditated potentials. Further, one can describe the long range ion-ion interparticle correlations at infinite dilution via a dielectric continuum description, emloying bulk solvent properties. It is possible to study the short ranged solvent-ion and ion-ion correlations in the context of a correction to the continuum picture. The calculation [1], by the use of Ornstein-Zernike-like integral equation, has predicted the attractive interaction between an anionic ions, Cl^{-}-Cl^{-}, in water near contact with respect to infinite separation. The origin of the attractive interaction between like changed ions has not been confirmed so far.

Since fluctuation from the equilibrium medium is preferable in liquid phases, localization of charged partilcles like positrons is highly probable [2, 3]. Free energy density functional theories [4, 5] provide self-trapping as a solution of the charged particle in a given host liquid. The sensitivity of positrons to changes induced by melting has already been reported by angular correlation [6, 7, 8] and lifetime [9, 10] studies of liquids. Gramsh et al. [11, 12] have observed very different behavior of the diffusion length L_+ of positrons in liquid and solid metals. That is, on melting, L_+ decreases remarkably, and in the liquid phase, L_+ increases with temperature. Kanazawa and coworkers [13-17] proposed a qualitative explanation for the increase with temperature of the positron diffuison length in the liquid phase, by using the theoretical formula, whichis based on the gauge-invariant effective Lagrangian with spontaneously broken density (the hedgehog-like a fluctuation) and the massive internal gauge fields. In this study, extending that theoretical formula, we shall propose an origin of the attractive interaction between like changed ions in the liquid phase.

2. The spontaneously broken density with internal Yang-Mills fields and the effective interaction

The connection on gauge field describes how the orientation of the fiber changes as one goes along a path in base space, or how the path in the base space is lifted into a path in the fiber bundle. We introduce a field-theoretical model to treat the problem of a charged particle in a fluid host in three spatial dimensions. It has been proposed that the parameter $\rho(t, r, u) = \rho(t, x, y, z, u)$ in three dimensional liquids is specified by the rotation, which is related to the gauge fields A^a_μ of SO(4) symmetry of S^3. The curvature can be represented by using a component, u, in the other-axis direction, if the three spatial dimensional axes are x, y, and z. It is preferable that we think of the anomalous fluctuation around the charged particle in the three dimensional liquid as the hedgehog-like fluction (soliton), taking into account the curvature. We adopt the parameter $\rho(r, u) = \rho^a (a = 1, 2, 3, 4)$, which is similar to that in the Sachdev and Nelson model [18]. The SO(4) quadruplet fields, A^a_μ, are spontaneously broken through the Higgs mechanism similar to the way in which the fluid host is broken around a charge particle. When the hedgehog-like fluctuation (soliton) around a charged particle is created, we set the symmetry breaking of the quadruplet fields, $\langle 0|\rho|0\rangle$, equal to $(0, 0, 0, \nu_4)$. Now, we can introduce the approximate Lagrangian as follows,

$$\begin{aligned} L = \; & \psi^+(i\partial_0 - g_2 T^a A^a_0)\psi - \frac{1}{2m}\psi^+(i\nabla - g_2 T^a A^a_{\mu\neq 0})^2\psi \\ - \; & \frac{1}{4}(\partial_\nu A^a_\mu - \partial_\mu A^a_\nu + g_1 e_{abc} A^b_\mu A^c_\nu)^2 \\ + \; & \frac{1}{2}(\partial_\mu \rho^\beta - g e_{\beta\alpha\gamma} A^\alpha_\mu \rho^\gamma)^2 + c_3^2 \rho^a \rho^a - \lambda_3(\rho^a\rho^a) \end{aligned} \tag{1}$$

Then, we set the symmetry breaking as follows,

$$\rho^a \to (0, 0, 0, \nu_4) + (\rho^1, \rho^2, \rho^3, \rho^4).$$

Thus, we can introduce the effective Lagrange density,

$$\begin{aligned} L = \; & \psi^+(i\partial_0 - g_2 T^a A^a_0)\psi - \frac{1}{2m}\psi^+(i\nabla - g_2 T^a A^a_{\mu\neq 0})^2\psi \\ - \; & \frac{1}{4}(\partial_\nu A^a_\mu - \partial_\mu A^a_\nu + g_1 e_{abc} A^b_\mu A^c_\nu)^2 \\ + \; & \frac{1}{2}(\partial_\mu \rho^\beta - e_{\beta\alpha\gamma} A^\alpha_\mu \rho^\gamma)^2 + \frac{m_1^2}{2}[(A^1_\mu)^2 + (A^2_\mu)^2 + (A^3_\mu)^2] \\ + \; & m_1(A^1_\mu \partial_\mu \rho^2 - A^2_\mu \partial_\mu \rho^1) + m_1(A^2_\mu \partial_\mu \rho^3 - A^3_\mu \partial_\mu \rho^2) + m_1(A^3_\mu \partial_\mu \rho^1 - A^1_\mu \partial_\mu \rho^3) \\ + \; & g_1 m_1\{\rho^4[(A^1_\mu)^2 + (A^2_\mu)^2 + (A^3_\mu)^2] - A^4_\mu(\rho^1 A^1_\mu + \rho^2 A^2_\mu + \rho^3 A^3_\mu)\} \\ - \; & \frac{m_2^2}{2}(\rho^4)^2 - \frac{m_2 g}{2m_1}\rho^4(\rho^a)^2 - \frac{m_2^2 g^2}{8m_1^2}(\rho^a\rho^a)^2. \end{aligned} \tag{2}$$

where m_1 is $\nu_4 \cdot g$ and m_2 is $2\sqrt{2}\lambda_3 \cdot \nu_4$. The effective Lagrange density, L_{eff}, represents three massive vector fields $A^1_\mu, A^2_\mu, A^3_\mu$, and one mass less vector field A^4_μ. Based on the effective Lagrangian in eq.(2), Kanazawa and coworkers [13-17] has proposed the explanation for the increase of the positron diffusion length with temperature in liquid metals. That is, as the temperature increases, the effective mass of the positron decreases due to the restoration of the spontaneously broken density around the positrons in the liquid phase. The generation function $Z[J]$ for Green's functions is shown as follows;

$$\begin{aligned} Z[J] &= \int \mathcal{D}A\mathcal{D}B\mathcal{D}\rho\mathcal{D}c\mathcal{D}\bar{c}\mathcal{D}\psi^+\mathcal{D}\psi \\ &\quad \cdot \exp i \int d^4x \left(\mathcal{L}_{eff} + \mathcal{L}_{\text{GF+FP}} + J \cdot \Phi\right), \end{aligned} \tag{3}$$

$$\mathcal{L}_{\text{DF+FP}} = B^a \partial^\mu A^a_\mu + \frac{1}{2}\alpha B^a B^a + i\bar{c}^a \partial^\mu \mathcal{D}_\mu c^a, \tag{4}$$

where B^a and c^a are the Nakanishi-Lautrup(NL) fields and Faddeev-Popov fictitious fields, respectively.

$$J \cdot \Phi \equiv J^{a\mu} A^a_\mu + J^a_B B^a + J_\rho \cdot \rho^a + \bar{\eta}\psi + \eta\psi^+ + \bar{J}^a_c c^a + J^a_c \bar{c}^a \tag{5}$$

BRS-quartet [19, 20] in the present theoretical formula are$(\phi_1, B^1, C^1, \bar{C^1})$, $(\phi_2, B^2, C^2, \bar{C^2})$, $(\phi_3, B^3, C^3, \bar{C^3})$, $(A^4_{L,\mu}, B^4, C^4, \bar{C^4})$ where $A^4_{L,\mu}$ is the longitudinal component of A^4_μ. So we need these fields are unobservable and fictious ones. Since these masses are created through the Higgs mechanism by introducing a charged particle in the fluid host, the massive gauge fields A^1_μ,A^2_μ,A^3_μ are localized around the charged particle. Now, extending that theoretical formula, we shall discuss an origin of the attractive interaction between like charged ions in the liquid phase. Integral equation methods have been used by Petti and Rossky [1] to study solvent-averaged forces and the effective interactions or the potentials of mean force for the alkali halides in water at infinite dilution. Integral equations revealed that the potential of mean force of unlike charged ions in water display clear minima in the ion-contact and solvent-separated region. The presence of such minima and maxima is not predicated by continuum solvent theoly. Especially it is of significance that the Cl^--Cl^- pair indicates the possibility of a substantial minimum at the ion-contact pair distance [1]. Furthermore, these studies [1, 21] suggest strongly the attractive interaction in the distance between Cl^--Cl^- ions from ~3.5 to ~5.0$\AA$.

Here, if inverse of the mass, m_1, of the massive gauge fields A^1_μ, A^2_μ and A^3_μ in eq.(2) is ~5.0$\AA$, the attractive potential, $-\frac{g_2^2}{4\pi}\frac{e^{-m_1 r}}{r}$, in the distance (below ~5.0$\AA$) between negative two ions is derived from massive gauge fields A^1_μ, A^2_μ and A^3_μ exchange interactions. This attractive potential might correspond to one in the distance between Cl^--Cl^- ions from ~3.5 to ~5.0$\AA$.

3. Conclusion

Based on the gauge-inuariant effective Lagrangian with the spontanously broken density and internal Yang-Mills gauge fields, we have discussed the origin of attractive interactions between like changed ions in the liquid phase. It is shown that the attractive interactions might be attributed to exchange interactions of the massive gauge fields between changed ions.

References

[1] Reffitt B M, Rossky P 1986 *J. Chem. Phys.* **84** 5836.
[2] Iaknbov I T, Khrapak A G 1982 *Rep.Prog. Phys.* **45** 697.
[3] Hernandez J P1991 *Rev. Mod. Phys.* **63** 675.
[4] Miller B N, Resse T L 1989 *Phys. Rev.* A**39** 4735.
[5] Miller B N, Resse T L 1990 *Phys. Rev.* A**42** 6068.
[6] West R N, Borland R E, Cooper J R A, Cusak N E 1967 *Proc. R. Soc. London. Ser.* A**92** 195.
[7] Gustafson D R, Mackintosh A R 1963 *Phys. Lett.* **45** 234.
[8] Gustafson D R, Mackintosh A R, Zaffarano D J 1963 *Phys. Rev.* **130** 1455.
[9] Brandt W, Waung H F 1968 *Phys. Lett.* A**27** 700.
[10] Crisp V H C, Lock D G, West R N 1974 *J. Phys.* F**4** 830.
[11] Gramsch E, Lyun K G, Throwe J, Kanazawa I 1991 *Phys. Rev. Lett.* **67** 1282.
[12] Gramsch E, Lyun K G, Throwe J, Kanazawa I 1999 *Phys. Rev.* B**59** 14282.
[13] Kanazawa I, Radiuanal J 1996 *Nucl. Chem.* **210** 451.
[14] Kanazawa I 2000 *Rad. Phys. Chem.* **58** 457.
[15] Matsushita Y, Kitahata H, Kanazawa I 2007 *Phys. Stat. Sol.* (c)**4** 3546.
[16] Kitahata H, Matsushita Y, Kanazawa I 2008 *Appl. Surf. Sci.* **244** 167.
[17] Matsushita Y, Kitahata H, Kanazawa I 2008 *Trans. Mater. Res. Soc. Japan* **33** 271.
[18] Sachdev S, Nelson D R 1984 *Phys. Rev. Lett.* **53** 1947.
[19] Becchi C, Rouet A, Stora R 1975 *Commun. Math. Phys.* **42** 127.
[20] Kago T, Ojima I 1979 *Prog. Theor. Phys.* Suppl. **66** 1.
[21] Dang L X, Peffitt B M 1987 *J. Am. Chem. Soc.* **109** 5531.

International Symposium on Advanced Science Research 2009 IOP Publishing
Journal of Physics: Conference Series **225** (2010) 012040 doi:10.1088/1742-6596/225/1/012040

Development of elemental analysis by muonic X-ray measurement in J-PARC

K Ninomiya[1], T Nagatomo[2], K M Kubo[2], P Strasser[3], N Kawamura[3], K Shimomura[3], Y Miyake[3], T Saito[4] and W Higemoto[1]

[1] Japan Atomic Energy Agency, Tokai, Naka, Ibaraki, 319-1195, Japan
[2] International Christian University, Mitaka, Tokyo, 181-8585, Japan
[3] High Energy Accelerator Research Organization, Tsukuba, Ibaraki, 305-0801, Japan
[4] National Museum of Japanese History, Sakura, Chiba, 285-8502, Japan

E-mail: ninomiya.kazuhiko@jaea.go.jp

Abstract. Muon irradiation and muonic X-ray detection can be applied to non-destructive elemental analysis. In this study, in order to develop the elemental analysis by muonic X-ray measurement we constructed a new X-ray measuring system in J-PARC muon facility. We performed muon irradiation for Tempo-koban (Japanese old coin) for test experiment of elemental analysis. Muonic X-rays originating from muon transition in muonic silver and gold atoms were identified. The contents of Tempo-koban (Au:61%) was determined by muonic X-ray intensities.

1. Introduction

When a free negative muon stopped in matter, the muon was captured in the Coulomb field of a nucleus and a muonic atom is formed. The captured muon deexcites to the 1s state through Auger electron and muonic characteristic X-ray emissions. The energies of muonic X-rays are higher than these of electronic X-rays because of a large mass of a negative muon (105.7 MeV/c^2, 207 times larger than that of electron). Such high energy muonic X-rays (0.1-6 MeV) are emitted from bulk samples without photon self-absorption, and the source elements of these muonic X-rays are easily identified using a semiconductor detector. The depth of muon implantation in sample can be tuned by adjusting the incident muon momentum. In addition, the samples are hardly activated by muon irradiations. These are very important advantages for elemental analysis compared with other methods (e.g. X-ray fluorescence and prompt gamma-ray analysis by neutron irradiation). It is expected that muonic X-ray spectroscopy can be applied to non-destructive elemental analysis [1, 2].

In spite of such advantages, this technique has not been used widely. There are few fundamental studies for development of this technique because this method is hardly known in the field of analytical sciences. In our present knowledge, we are not able to determine the elemental contents of sample material quantitatively only from muonic X-ray intensity because of the molecular effects on the formation process of muonic atoms. Muonic X-ray intensities and muon atomic capture ratios depends on the muon-capturing molecule. Although some muon capture models were studied [4-6], the molecular effects on the formation process have still been unclear. Therefore, it is essential to know the relation between muonic X-ray intensities and material contents from muonic X-ray spectra for the standard samples that components have already known. In this study, to develop elemental analysis by muonic X-ray measurement and to investigate the molecular effects on the muonic atom formation, we constructed a new X-ray measuring system in J-PARC.

2. Experimental

The system for muonic X-ray measurement was constructed at D2-area in Muon Facility (Muon Science Establishment; MUSE) in Materials and Life Sciences Facility (MLF), J-PARC (see Figure 1). This system consisted of radiation shields, beam collimator with 8.5 cm diameter, two germanium detectors and MCA in CAMAC system. The primary proton beam intensity during the present work was about 20 kW. We selected the momentum of negative muon as 35 MeV/c. For muonic X-ray measurement, photo events that correlated with muon pulse signals were detected by germanium detectors. We optimized experimental conditions such as detector position, arrangement of radiation shields, the beam transport optics and muon stopping region in zinc plate with 0.5 mm thickness and 10 x 10 cm^2. Muon irradiation for Tempo-koban (provided from National Museum of Japanese History, old Japanese coin produced in the middle of the 19th century) was also performed about 4 hours for test experiment of elemental analysis. The sample dimension of the Tempo-koban was about 0.5 mm thickness and 6.5 x 4 cm^2 with an oval shape. The range of incident muon in the Tempo-koban was estimated as ~0.3 mm. The detection efficiencies of germanium detectors were determined from EGS-5 calculations [7] in which parameters were adjusted to reproduce experimental efficiencies of ^{137}Cs and ^{60}Co standard radiation sources.

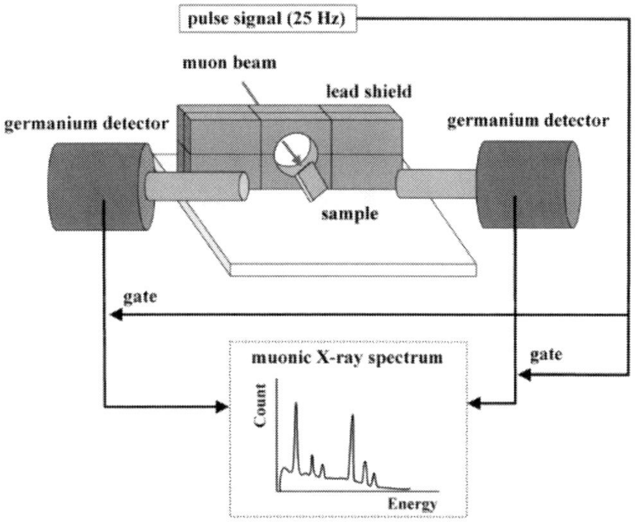

Figure 1. A schematical drawing of the experimental setup.

3. Results and discussion

3.1. Muon irradiation for zinc plate

Figure 2 shows a typical muonic X-ray spectrum by muon irradiation for zinc plate. Muonic K, L, M and N X-ray lines were clearly found. Muonic X-ray intensities per muon capture were estimated from cascade calculation [8]. In this calculation, we gave the well accepted initial parameter as follows; the initial principal quantum number is equal to 14 with statistical initial angular momentum distribution and L and M electrons are never ionized completely by muon-electron and electron-electron Auger processes during muonic cascade. In this way, we estimated that 15 muons per pulse were stopped in the zinc plate from muonic 2-1 X-ray intensity. The consistent values were obtained from analysis of 5-4, 4-3 and 3-2 muonic X-ray intensities.

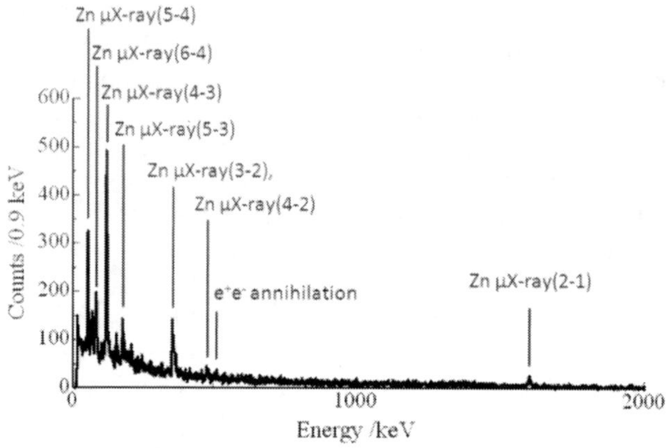

Figure 2. Muonic X-ray spectrum for zinc plate. The muon irradiation time was 5500 s with 20 kW primary proton power (total incident muon: 2 x 10^6). The numbers in each bracket show a change of principal quantum numbers by muonic X-ray transition.

3.2. Muon irradiation for Tempo-koban

Muonic silver and gold X-rays were identified by muon irradiation for the Tempo-koban (see Figure 3). This result shows the Tempo-koban is silver-gold alloy. We analyzed the number of stopped muons per muon pulse by the same method of zinc experiment from muonic silver 4-3 and muonic gold 5-4 X-ray lines. These muonic X-ray lines had good statistics and well separated from other peaks. The muon capture ratio in gold atom is estimated as 0.59 ± 0.06.

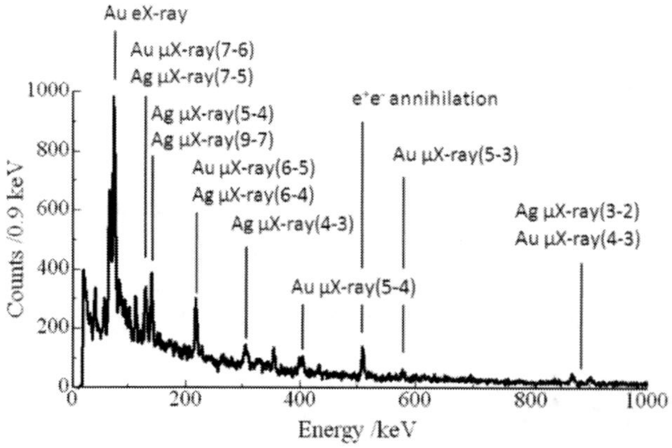

Figure 3. Muonic X-ray spectrum for Tempo-koban (measuring time: 14000 s).

The atomic muon capture ratio ($A(Z_1/Z_2)$) for silver-gold alloy is essential to determine the contents of the Tempo-koban by muonic X-ray intensities. However, the accurate atomic capture ratio has not been reported. We estimated the atomic capture ratio for silver-gold alloy as $A(Au/Ag) = 1.68$ from the Z-law by Fermi and Teller ($A(Z_1/Z_2) = Z_1/Z_2$ [4]). From the muon atomic capture ratio and the number of captured muons in silver and gold atoms, the ratio of atoms in Tempo-koban was determined as $I(Ag/Au) = 1.17$, Au = 61 ± 6 % (Ag = 39 %) in elemental mass ratio. This result does not include the systematic errors originating from muonic X-ray intensity per captured muon (cascade calculation) and atomic capture ratio for silver-gold alloy (muon capture model).

Our result is consistent with the result for well known contents of Tempo-koban (Au: 57%) and for the same Tempo-koban that was analyzed from prompt gamma-ray analysis by neutron irradiation in J-PARC NOBORU (Au: 54 ± 1%) [9].

4. Conclusion

We developed a muonic X-ray measuring system for elemental analysis at MUSE in MLF, J-PARC. Low background muonic X-ray spectra were successfully obtained by extracting correlation photo events with muon beam pulse, and the contents of the Tempo-koban were determined from muonic X-ray intensities without destruction.

In the next step, we are planning to perform muon irradiation for silver-gold alloys that are already known their contents. From these studies, atomic capture ratio and muonic X-ray intensity pattern will be understood. As a result, the systematic errors will be decrease and more accurate elemental analysis will be available. We are also planning to investigate muon capture phenomena for various materials and establish the muon capture model to determine muonic X-ray intensities per captured muon and atomic capture ratios for sample materials without muon irradiation for standard samples.

5. References

[1] Daniel H, 1984 *Nucl. Instr. Meth.* B**3**, 65

[2] Kubo M K, Moriyama H, Tsuruoka Y, Sakamoto S, Koseto E, Saito T and Nishiyama K, 2008 *J. Radioanal. Nucl. Chem.* **278**, 777

[3] Jacot-Guillarmod R, Bienz F, Boschung M, Piller C, Shaller L A, Schellenberg L, Schneuwly H and Siradovic D, 1988 *Phys. Rev. A* **37**, 3795

[4] Fermi E and Teller E, 1947 *Phys. Rev.* **72**, 399

[5] Daniel H, 1975 *Phys. Rev. Lett.* **35**, 1649

[6] Schneuwly H, Pokrovsky V I and Ponomarev L I, 1978 *Nucl. Phys,* A**312**, 419

[7] Hirayama H., Namito Y., Bielajew A. F., Wilderman S. J. and Nelson W. R. 2005 *SLAC-R* 730

[8] Akylas V R and Vogel P, 1978, *Comput. Phys. Comm.* **15**, 291

[9] Matsue H, Kasugai Y, Harada M, Maekawa F, Kubo K and Saito T, 2009 *the 53rd Symposium on Radiochemistry, Tokyo*, 3A01

Simultaneous, coincident 2-D ACAR and DBAR using segmented HPGe detectors incorporating sub-pixel interpolation

Christopher S Williams[1], Larry W Burggraf[1], Paul E Adamson[2], James C Petrosky[1] and Mark E Oxley[1]
[1]Air Force Institute of Technology, AFIT/ENP, 2950 Hobson Way, Wright-Patterson Air Force Base, OH 45433 USA.
[2]High Power Microwave (HPM) Technologies Branch, HPM Division, Directed Energy Directorate, Air Force Research Laboratory, Kirtland AFB, NM 87117 USA.

E-mail: christopher.williams@afit.edu

Abstract. A three-dimensional Positron Annihilation Spectrometry System (3D PASS) for determination of 3D electron-positron (e^--e^+) momentum densities by measuring coincident annihilation photons was designed, constructed and characterized. 3D PASS collects a single data set including correlated photon energies and coincident photon positions which are typically collected separately by two-dimensional angular correlation of annihilation radiation (2D ACAR) and two-detector coincident Doppler broadening of annihilation radiation (CDBAR) spectrometry. 3D PASS is composed of two position-sensitive, high-purity germanium (HPGe) double-sided strip detectors (DSSD(s)) linked together by a 32-channel, 50 MHz digital electronics suite. The DSSDs data were analyzed to determine location of photon detection events using an interpolation method to achieve a spatial resolution less than the 5-mm width of the DSSDs' charge collection strips. The interpolation method relies on measuring a figure-of-merit proportional to the area of the transient charges observed on both strips directly adjacent to the charge collection strip detecting the full charge deposited by the annihilation photon. This sub-pixel resolution, corresponding to the error associated with event location within a sub-pixel was measured for both DSSDs using the approach outlined in Williams et al [1] and was on the order of ± 0.20 mm (± one-standard deviation). As a result of the sub-pixel resolution, the distance between the DSSDs and material sample was reduced by a factor of five compared to what is typically required in 2D ACAR systems was necessary to achieve 0.5-mrad angular resolution. This reduction in the system's footprint decreases attenuation of the annihilation photons in the air between the material sample and the DSSDs and increases the solid angle between the sample and the DSSDs, ultimately resulting in higher system detection efficiency. 3D PASS was characterized in the same manner comparable to state-of-the-art 2D ACAR and CDBAR spectrometers. 3D PASS spectra were collected and analyzed for single-crystal copper (Cu) and silicon carbide (6H SiC) and compared with the results in the literature.

1. Introduction

Several methods are currently used in the scientific community and industry to measure and characterize defects in semiconductor materials including electron paramagnetic resonance photoluminescence and positron annihilation spectroscopy (PAS). PAS relies on the measurement of radiation from the annihilation of an electron and a positron. PAS encompasses several experimental techniques; the most commonly used are positron annihilation lifetime spectroscopy, coincident Doppler broadening of annihilation radiation (CDBAR), and angular correlation of annihilation radiation (ACAR). These non-destructive PAS techniques have been gaining increasing popularity with technological improvements in detector performance and affordability of digital electronics.

The annihilation of an electron and a positron usually produces two oppositely directed 511-keV photons. Due to the conservation of energy and momentum, the two annihilation photons produced from the annihilation event are emitted exactly co-linearly in the center-of-mass frame-of-reference. In the laboratory frame, the emission angle deviates from 180° due to the momentum that the electron-positron (e^--e^+) pair posses prior to their annihilation. CDBAR uses two high-energy resolution detectors to measure Doppler shifts of the coincident annihilation photon energies due to the distribution of e^--e^+ momenta in the direction parallel to the photon emission. The CDBAR technique has been successfully demonstrated on metallic [2-3] and silicon carbide (SiC) [4] samples by several researchers. 2D ACAR, on the other hand, determines the e^--e^+ pair's momentum distribution in the 2D plane perpendicular to the photon emission direction using position-sensitive detection systems to measure the deviation of the emission angle from co-linearity between coincident annihilation photons. The 2D ACAR technique has been effectively demonstrated on metals [5-6] and SiC [7] by several researchers. Previously, ACAR and CDBAR techniques have been applied separately resulting in an incomplete description of the e^--e^+ momentum distribution because momentum in the measurement plane was not correlated with momentum in the photon emission direction. Simultaneous momentum measurements can provide the complete, correlated 3D momentum distribution of the e^--e^+pair. This capability will be useful for future comparisons to theoretical models of e^--e^+ wavefunctions, which can be used to predict PAS, as illustrated in Adamson et al [8].

While experimental PAS techniques can provide some information on the structure of the material interrogated, they have several measurement limitations. First, for CDBAR applications, the detection system should have high-quality energy resolution in order to reveal information about the material's core electron environment, which is only marginally achieved by the resolution of the best HPGe semiconductor detector systems. Next, in order to adequately describe the momentum distribution, 2D ACAR measurements have required high activity sources and large distances between detectors and positron source. Typical 2D ACAR systems have used large footprints to achieve adequate angular resolution (~0.5 mrad), often having about 10 m distance between the sample and detectors depending on the pixelation of the detection system, resulting in inefficient detection. In order to record the large number of coincident events ($>10^6$) necessary to achieve sub-milliradian (~0.5 mrad) angular resolution, sources having 100's of mCi activity are typically used in order to collect a spectrum in a reasonable amount of time. Finally, the copious information produced by PAS measurements have not been effectively used due to the inability to efficiently collect, process and store the data. Although these problems continue to limit PAS effectiveness, we have reduced some impact of these limitations by applying two new engineering techniques to state-of-the-art PAS. On our way to making a single system to measure simultaneous, coincident 2D ACAR and CDBAR spectra in order to determine complete 3D e^--e^+ momentum distributions, we have incorporated sub-pixel resolved DSSDs and multiplexed digital electronics on a common clock.

Pixelation created by double-sided orthogonal strip detectors (DSSDs) using planar HPGe crystals has been used for position-sensing gamma detection applications. By applying sub-pixel resolution, our first engineering improvement, in these high-resolution solid state detectors, we have reduced the 2D ACAR footprint and measurement time for a selected source activity. Many researchers [9-10] have shown that transient charge analysis in DSSDs can be used to significantly improve gamma image resolution for all but the external detector strips. Using the relative areas of transient (or induced) charge pulses on strips adjacent to the full-charge collection strips on DSSDs, location of charge deposition events across the width of strips produces sub-pixel spatial resolution much finer than the intrinsic pixilation of the orthogonal charge collection strips. Calibration of this sub-pixel interpolation method is described in detail by Williams et al [1]. Incorporation of transient charge analysis allows considerable reduction of the distance between the sample and detectors without degrading angular resolution. This increases the solid angle between the sample and detector allowing more un-scattered annihilation photons to reach the DSSDs, increasing absolute efficiency. Our second engineering improvement incorporated multiplexed digital electronics using FPGA-logic on a common clock into the system design. These electronics permit characterization of charge deposition dynamics across many detector elements enabling us to ascertain coincident event locations and energies. Additionally, the electronics can be programmed to perform functions specific to our application, eliminating unnecessary data acquisition and storage.

These engineering techniques were applied to a new PAS technique designed to fully characterize 3D e^{-}-e^{+} momentum densities by measuring coincident photons from e^{-}-e^{+} annihilation events. The detection system is composed of two, position-sensitive, HPGe DSSDs coupled together with a digital electronics suite programmed specifically for measuring interpolated sub-pixel charge deposition event locations and energies. We demonstrated this 3D Positron Annihilation Spectroscopy System (3D PASS) by simultaneous measurement of 2D ACAR and CDBAR spectra for single-crystal Cu and 6H SiC.

2. Equipment

Two position-sensitive DSSDs and a digital electronics suite were the main components of our 3D PASS used for this research.

2.1. Detectors

The Ortec DSSD is the same HPGe detector described in Williams et al [1]. The second detector used for the 3D PASS is also a position-sensitive, planar HPGe DSSD. This HPGe DSSD, manufactured by PHDS Co, has a 9 cm diameter Ge crystal and an active depth of 11 mm, shown in Figure 1. The detector has 16 – 5 mm wide charge collection strips on each of the front and rear faces of the Ge crystal, orthogonal to each other. The charge collection strips on the front and back face of the crystal are metalized, amorphous germanium with a thickness of ~2 μm. Since the crystal is circular, only the 8 center strips on each face have a length of 8 cm. The strips beyond the center strips are shorter the further they are from the center of the crystal. Crystal volumes created by intersections of these 5-mm orthogonal strips create 220 intrinsic pixels. The rear strips, denoted as the AC side, are vertical, and are labelled 0 - 15 with 0 being the left-most strip and 15 the right. The front strips, denoted as the DC side, are horizontal and are labelled 16 - 31 with 16 being the top strip and 31 the bottom.

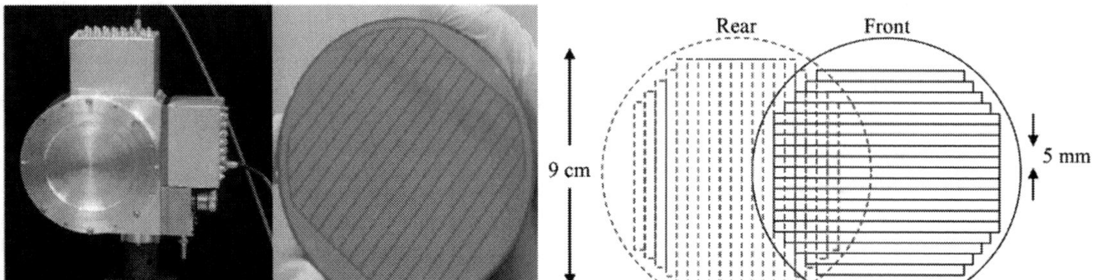

Figure 1. Photo of PHDS DSSD sitting on its LN_2 dewar. Middle: Photo of PHDS DSSD's Ge crystal with electrode masking. Right: PHDS detector's electrode masking layout (not to scale).

2.2. Digital Electronics

The Spec32 digital electronics system is a 32-channel, 50 MHz digitizer. The system contains four independent electronics boards, each with two, 12-bit ADCs, 8 input channels and two Alterra Cyclone FPGAs. Waveforms from the DSSDs' preamplifiers are directly digitized and then processed through trapezoidal filters defined within the Spec32's software, Imager. The FPGA forwards the user-defined output data to the buffer and then over-writes the waveform data, saving valuable buffer memory and reducing the number of buffer spills. For our experiments, the FPGA's were programmed to calculate the energy collected in the full-charge collecting strip and figure(s)-of-merit (FOM)'s proportional to the area of the transient charges in the strips directly adjacent. Because the Spec32 only accepts 32 inputs, all 10 of the Ortec (F1-F5 and R1-R5) and only 16 of the center-most strips of the PHDS (4-11 on the AC side and 20-27 on the DC side) DSSDs outputs were used.

Output data is stored in two data files: raw data and raw event logs. The raw data log stores energy data for each of the 32-channels, initiated by a single trigger of a recordable full-charge event on any channel. This type of file can be extremely large if there are a large number of event triggers. The raw event log stores energy data only for the channels that achieve the trigger threshold, based on that channel's calibration. The FOMs for the transient charges in the strip to the left, denoted the predecessor (Pred) and the strip to the right, denoted the successor (Succ), as well as, the timestamp are also stored. Since the Spec32 is only a 12-bit system, the time stamp is reset every 256, 20 ns clock ticks, meaning multiple events must trigger within a 5.12 μsec window before the timestamp resets to know their relative event timing. No counter is available to list the time between timestamps.

2.3. Source/Sample Configuration
The positron source, 106.3 μCi ^{22}Na, was positioned exactly orthogonal relative to the face of the detector. Two samples were measured: 1-cm^2 x 1.0-mm thick single-crystal Cu, growth orientation (100), procured from MTI Corp and 1-cm^2 x 0.268-mm thick research-grade, on-axis, N-type 6H SiC having growth orientation (0001), purchased from CREE Inc. The crystal face of each sample was turned 45° on a vertical axis orthogonal to the axis between the centers of the faces of the DSSDs in order to maximize positron interaction with the sample near the center point between DSSD faces. In order to reduce positron annihilation with air as positrons travelled from the source to the sample, the ^{22}Na source and material sample were housed in a vacuum chamber, reduced to a pressure of 1 torr.

3. Experiment
Several independent experiments were conducted to determine the performance of the DSSDs in order to finalize the design of 3D PASS. First, in order to determine the distance the detectors must be placed from the sample, the sub-pixel resolution was determined. The interpolation method from Williams et al [1], shown in equation (1), was used to determine event locations.

$$\text{Distance} = \left(\text{Strip \#-1} + \left(\frac{FOM_{\text{Pred}}}{FOM_{\text{Pred}} + FOM_{\text{Succ}}} \right) \right) \times \text{strip width} \tag{1}$$

This interpolation method employs a FOM proportional to the area of the transient charges observed on both strips directly adjacent to the full-charge collecting strip (Strip#), termed FOM_{Pred} for predecessor and FOM_{Succ} for successor strips to estimate sub-pixel location. Sub-pixel resolution, the standard deviation of the error associated with locating charge deposition events within a sub-pixel that met the energy and timing criteria for the interpolation method, was measured for both DSSDs using the same approach outlined in Williams et al [1]. To quantify the spatial resolution, the experimental sub-pixel count distribution from each DSSD was analyzed by a minimum-error least squares fit to a Gaussian point-spread function on a circular collimator aperture. The resulting sub-pixel resolution for the Ortec DSSD was ± 0.22 mm and was ± 0.19 mm for the PHDS DSSD.

Next, the energy resolutions for the DSSDs for full-energy peak (FEP) events, corresponding to complete charge-collection of the full-energy photoelectrons, were quantified. The resolution, of critical importance to CDBAR measurements, depends not only on the individual detector performance, but also on the noise in the Spec32 electronics. ^{22}Na was not used to quantify detector resolution due to the inherent Doppler broadening of the annihilation photons. Photons at 514 keV from a 96.79 μCi ^{85}Sr source were used to estimate the resolution of each DSSD at 511 keV. A 72-hour data set using the ^{85}Sr source was measured using both DSSDs and the Spec32 electronics. The centers of the detector faces were axially aligned and the ^{85}Sr source was located on the axis, centered between the detector faces at a detector/source distance of 12.0 in.

An in-house MATLAB code was used to process the raw event log from the Spec32. The code first discarded all recorded events outside of the energy range 514 ± 30 keV to isolate un-scattered ^{85}Sr gammas. Then, only coincident full-energy events between the front and rear strip in both DSSDs, in time (within 60 ns) and energy, were accepted and examined. The 514-keV FEPs were constructed for each intrinsic pixel in both DSSDs. The average FWHM for the intrinsic pixels in the ORTEC and PHDS DSSDs' was calculated to be 1.76 ± 0.24 keV and 1.49 ± 0.17 keV, respectively. Additionally, these data sets were analyzed for potential correlation of the FOMs with FEP energy with the intent of using the FOMs to evaluate the sub-pixel response and possibly improve energy resolution. Each DSSD's coincident events were plotted against their associated FOMs, both successor and predecessor, and fitted to linear, quadratic, and cubic functions. The resulting best fits, which were linear with slopes on the order of 0.04 keV, indicated no significant correlation between event energy and FOM values.

Based on the measured sub-pixel spatial resolution, the dimensions of sub-pixels for ACAR analysis were established as 1 mm^2 squares, corresponding to two standard deviations of resolution error associated with the interpolation method. To obtain an angular resolution of 0.5 mrad using a 1 mm^2 sub-pixel size, the face of the DSSDs were each placed 2.0 m from the center of the source/sample location, assuming a 1-mm^2 sample. Figure 2 displays the final 3D PASS configuration. The sub-pixel size was chosen 2.5 times larger than the spatial resolution, which compensates for some increase in the sample size used. Once this technique has been demonstrated and our positron beam has been built, our sample/source dimensions will not degrade the 0.5 mrad angular resolution.

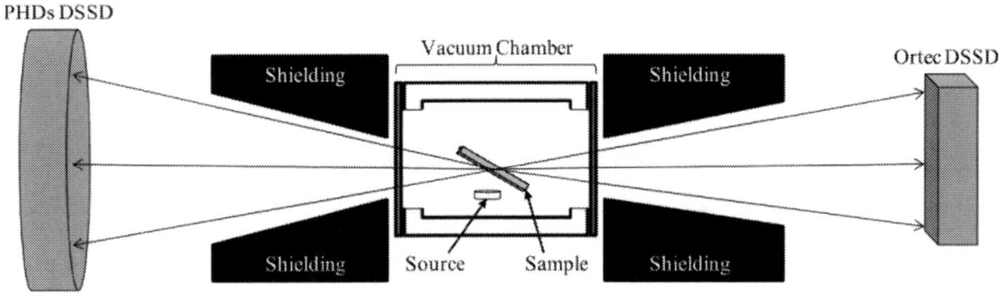

Figure 2. Top view of 3D PASS configuration with two DSSDs, each 2.0 m from the center of sample.

With the footprint established, the 3D PASS was used to collect simultaneous momentum data sets, one for single-crystal Cu and one for single-crystal 6H SiC. Each data set was collected to accumulate 10^6 coincident events that passed the energy and time criteria for the interpolation method. Then, 2D ACAR and CDBAR spectra were reconstructed from each data set of qualified coincident events. The 2D ACAR spectra were binned according to the angular resolution, in 0.5 mrad x 0.5 mrad bins, and the CDBAR spectra were binned into 0.1 keV intervals. Cu spectra were analyzed to evaluate and compare the 3D PASS performance to published results. Cu was chosen for calibration for two reasons: Cu has been extensively studied and because its 2D ACAR distributions perpendicular to the (100) is approximately symmetric.

4. Results

To evaluate the simultaneous 3D momentum capability, the 3D PASS data set for single-crystal Cu was collected and the 2D ACAR was reconstructed as shown in Figure 3. The 2D ACAR spectrum was populated by processing coincident annihilation events using the interpolation method (equation (1)) to determine event location. A benefit of using Cu to calibrate the ACAR spectrum is that the spectrum's inherent symmetry can be used to precisely correct the source/sample misalignment shown in the contour plot of the Cu 2D ACAR spectrum. The system was 3.12 mm off in the X-direction and 2.51 mm off in the Y-direction, which was corrected by re-aligning the sample vacuum chamber, relative to the DSSD faces, also shown in Figure 3. The ACAR response of 3D PASS was evaluated using two alignment-corrected projections at X=0 and Y=0, slightly smoothed using a Savitsky-Golay 2nd order polynomial filter over a frame size of 5 data points. The smoothed projections, shown in Figure 4, compare closely to scaled ACAR data for single-crystal Cu collected by Senicki et al [2], measured using the 1D long-slit ACAR technique. The un-smoothed projections, however, exhibit notches at distinct angles in the raw Cu spectra. By applying the slight smoothing routine, these features are greatly reduced. The appearance of these features on the sides and the peak can be the result of either one or a combination of the following. First, our Cu sample was oriented 45° relative to the face of the DSSDs face, where Senicki et al's was orthogonal. The slight anisotropy of Cu as discussed by Tanigawa et al [6] could contribute to the observed features. Second, Williams et al observed a decrease in efficiency in the charge collection strips the closer events occurred to the edge of the strips which could distort spectra depending on which sub-pixels detect the annihilation photons. Finally, many positrons from the ^{22}Na source probably interacted and annihilated with the vacuum chamber materials, causing momentum data from those annihilations to alter the Cu momentum data.

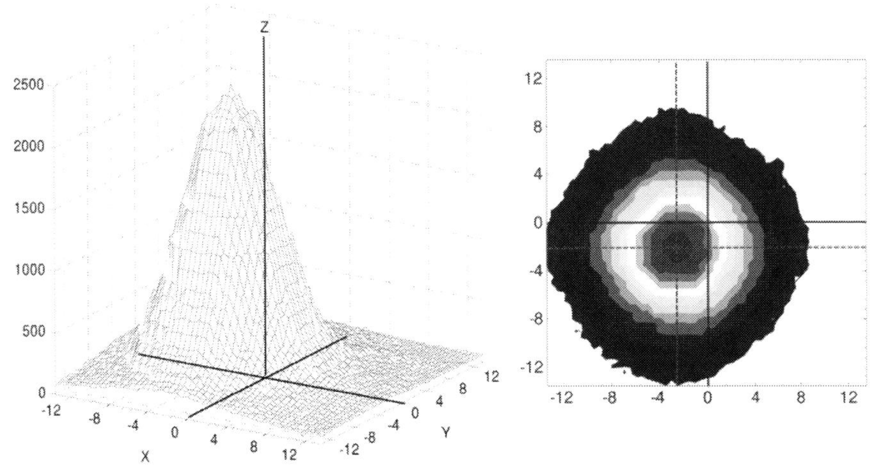

Figure 3. Left: Cu 2D ACAR spectra with X and Y in mrad units and Z in count per bin. Right: Contour plot (at contour intervals of 10% of max counts) for the Cu 2D ACAR spectrum.

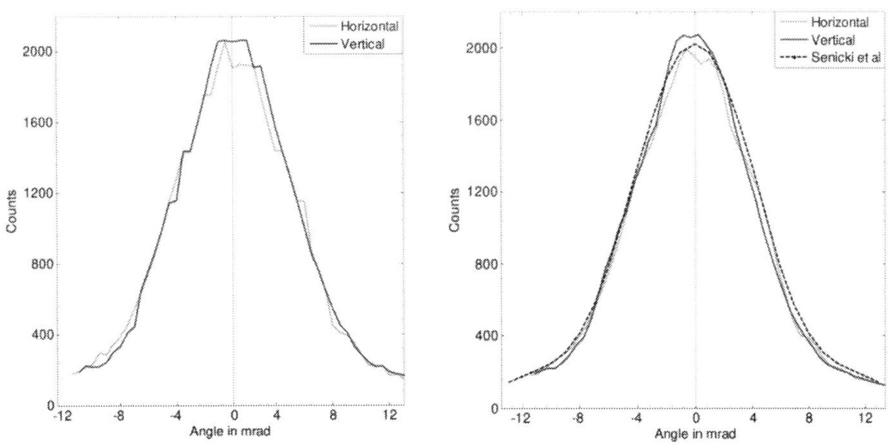

Figure 4. Left: Raw data for horizontal (X) and vertical (Y) projections. Right: Smoothed data compared to Senicki et al's reconstructed data.

To investigate the CDBAR performance of 3D-PASS, the Doppler broadened (DB) lineshape was extracted from the Cu CDBAR spectra for coincident annihilation events. The resulting Cu DB lineshape, shown in Figure 5, compares favorably with Szpala et al's [3] scaled annihilation lineshape in both the high and low momentum regions, except for the slightly larger background contribution on the low-energy side of the DB lineshape and the shoulder features on the sides of the lineshape. These shoulder features most likely are caused rotation of our sample and influenced by the slight anisotropy of Cu. Regardless of these slight features, the performance of 3D PASS for measuring simultaneous

2D ACAR and CDBAR in a one measurement system compares well with published results for single-crystal Cu.

Since performance of 3D PASS compared well with published results for Cu, the 6H SiC 2D ACAR and CDBAR were constructed in the same manner and are shown in Figure 6. Note the differences in the 2D ACAR spectra and CDBAR spectra when compared with the Cu data. It is evident from both spectra that the positrons interacted with higher momentum electrons in the 6H SiC than in the Cu sample. This is supported by the significant amount of interactions at larger angular deviations in the ACAR spectra and the larger shoulders on the DB lineshape slopes typically characteristic of interactions with higher momentum core electrons. This was expected by the anisotropies at higher momentum regions present in the 6H SiC 2D ACAR Kawasuso et al [7] published. Although their spectra have many similarities to ours, we see a distortion particularly about the vertical axis due to our sample rotation. Because the sample orientations differed by 45°, our spectra was significantly transformed due to the high degree of anisotropy in the 6H SiC e^{-}-e^{+} momentum density. The CDBAR, however, did compare well with Rempel et al's [4] CDBAR results on similar 6H SiC samples. Their CDBAR lineshape was reconstructed from the data they presented and included in Figure 6 and was compared with the 3D PASS' DB response for 6H SiC. Even though higher momentum interactions were expected in the SiC, the larger shoulders on the sides could be partly influenced by the source/sample orientation, low count statistics or less likely, the DSSD's efficiencies.

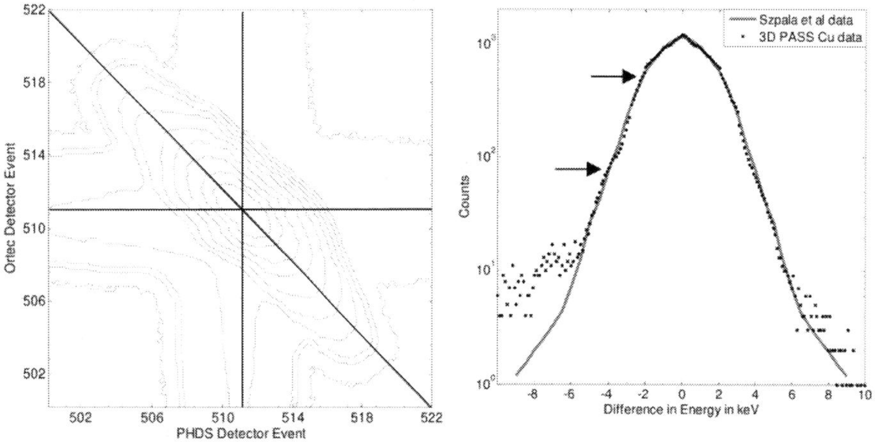

Figure 5. Left: CDBAR spectra for single-crystal Cu. Right: DB lineshape for Cu compared with Szpala et al's lineshape reconstructed from their data. Shoulder features (arrows) suggest discrete momenta.

5. Conclusions and future work

The 3D PASS performance was verified by comparing experimental data with published 2D ACAR and CDBAR results for single-crystal Cu and 6H SiC. Further experimentation will be conducted to determine the origin of the ridge features on the peak and sides of the vertical and horizontal (X and Y = 0) projections for the ACAR spectrum and the slight increased background contribution in the DB lineshape. Currently, 3D PASS measurements of oxygen-atom inclusions in 6H SiC are underway. 3D PASS has been shown to be promising for simultaneously collecting the 3D e^{-}-e^{+} momentum densities using a single measurement. In future work, we will use a pulsed slow positron beam source and moderator for synchronous 3D PASS measurements. This approach will allow positrons to be focused directly into a sample surface, minimizing contribution of extraneous annihilations while greatly increasing the number of coincident annihilation events for improved measurement efficiency. Additionally, the beam will allow surface annihilation measurements to be made using selected positron energies and varying sample orientations.

Acknowledgements

This research was sponsored in part by the Air Force Office of Scientific Research and the Defense Threat Reduction Agency. Views expressed in this article are those of the authors and do not reflect the official policy or position of the Air Force, Department of Defense, or the U. S. Government.

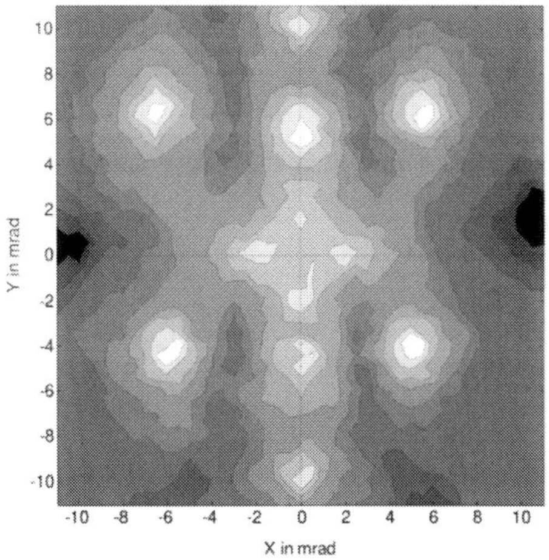

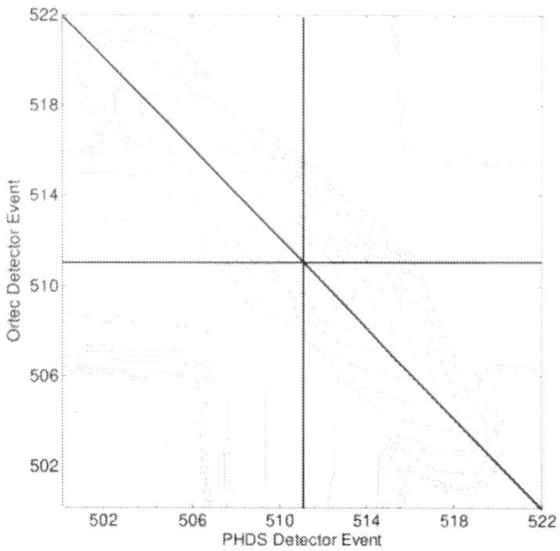

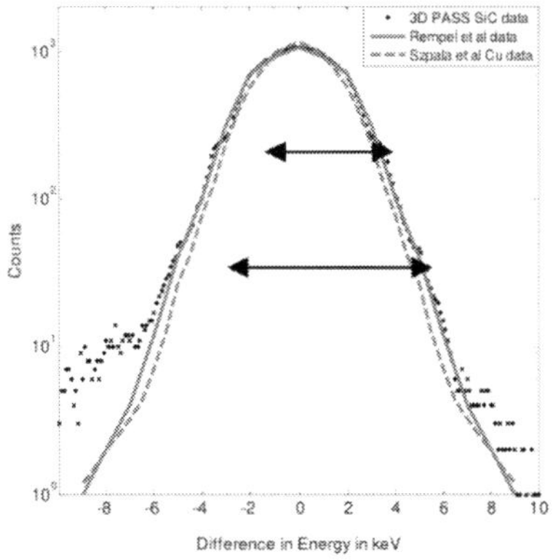

Figure 6. Upper left: 2D ACAR spectra for 6H SiC with contours at 10% of max counts (lighter indicates higher counts). Upper right: CDBAR spectra for 6H SiC. Bottom left: DB lineshape for 6H SiC extracted from CDBAR spectrum compared to Rempel and Szpala et al's. Cu DB spectra reconstructed from their data. Arrows indicate high angular momentum features observed in our spectrum that were not previously observed.

References

[1] C. Williams, W. Baker, L. Burggraf, P. Adamson, and J. Petrosky, accepted by IEEE Trans Nuc. Sci. January 3, 2010.

[2] K. Lynn, J. MacDonald, R. Boie, L. Feldman, J. Gabbe, M. Robbins, E. Bonderup, and J. Golovchenko 1977 *Phy. Rev. Lett.* **38** pp 241-4.

[3] S. Szpala, P. Asoka-Kumar, B. Nielsen, J. Peng, S. Hayakawa, K. Lynn, H. Gossmann, 1996 *Phys. Rev. B* **54** pp 4722-31.

[4] A. Rempel, K. Blaurock, K Reichle, and H. Schaefer 2002 *Mat. Sci. Forums* 389-393 pp 485-8.

[5] S. Berko 1979 *Proc. 5th Int. Conf. on Positron Annihilation (Jpn:Jpn Institute of Metals)* pp 65-87.

[6] S. Taniigawa, K. Ito, S. Terakado, A. Morisue, S. Fujii, Y. Iwase 1985, *Proc. 7th Int. Conf. on Positron Annihilation (New Delhi:World Scientific)* pp 288-90.

[7] A. Kawasuso, T. Chiba, and T. Higuchi 2005 *Phy. Rev. B* **71**, pp 193204.

[8] P. Adamson, X. Duan, L. Burggraf, M. Pak, C. Swalina, S. Hammes-Schiffer, 2008 *J. Phys. Chem. A.* **112** (6) pp. 1346-51.

[9] M. Amman and P. Luke 2000 *Nuc. Inst.Meth. Phys. Res. A* **452** (1) pp. 155-66.

[10] M. Burks, E. Jordan, E. Hull, L. Mihailescu, and K. Vetter K, *2004 IEEE Nuclear Science Symposium Conference Record* Italy pp. 1114-8.

Doppler broadening of annihilation radiation of some single-element materials from the second to the sixth periods

A Kawasuso[1], M Maekawa[1] and K Betsuyaku[2]
[1]ASRC, Japan Atomic Energy Agency, 1233, Watanuki, Takasaki, Gunma 370-1292, Japan
[2] Mizuho Information and Research Institute, 2-3, Kanda, Nishikicho, Chiyoda-ku, Tokyo 101-8843, Japan

E-mail: kawasuso.atsuo@jaea.go.jp

Abstract. The Doppler broadening of annihilation radiation spectra of some single-element materials from the second to the sixth periods measured by the coincidence technique have been compared with the theoretical calculation based on the projector-augmented wave (PAW) method. Having appropriate valence electron configurations the calculation well reproduces the measured spectra.

Introduction

Electron-positron momentum distribution obtained through the Doppler broadening of annihilation radiation (DBAR) measurement is quite useful for identifying defects species in materials since the spectrum shape reflects the elemental difference where positrons annihilate. To interpret the experimental DBAR spectra, detailed theoretical calculation is needed. It is pointed out that the ordinal pseudo-potential manner hardly calculates the realistic electron wavefunctions near core region and this is especially serious problem when calculating the DBAR spectra for transition metals [1]. To avoid this difficulty, one way is to employ the all-electron manner for calculating electronic state. In this study, we measured the DBAR spectra of some single-element materials from the second to the sixth periods and compared them with the theoretical spectra based on the projector-augmented wave (PAW) method developed by Blöhl [2] and the two-component density functional theory.

Experiment

Samples used in this study were Be polycrystals, highly oriented pyrolytic graphite (HOPG) from the second period, Mg and Al polycrystals, Si single crystals from the third period, Ti, V, Cr, Mn, Fe, Co, Ni, Cu, Zn and Ga polycrystals and Ge single crystals from the fourth period, Zr and Mo polycrystals from the fifth period, Gd, Au and Pb polycrystals from the sixth period. After electro-chemical polishing, polycrystals were annealed in appropriate temperatures to reduce defects.

A ^{22}Na positron source deposited onto a titanium film of 5 μm thick was sandwiched by two samples and coincidence Doppler broadening measurements were carried out using two high-purity Ge detectors and the FAST-ComTec MPA-Win system. In each spectrum, more than 10^7 counts were accumulated. The energy resolution was approximately 1 keV in the full width at half maximum (FWHM). Here, the Doppler energy shift of 1 keV corresponds to 3.92 mrad in angle deviation from π of two annihilation radiations and 0.54 a.u. (3.92×10^{-3} m_0c) in momentum shift.

Journal of Physics: Conference Series **225** (2010) 012027 doi:10.1088/1742-6596/225/1/012027

Calculation

The DBAR spectra were calculated within the local density approximation [3]. The valence electron momentum distribution is calculated using the conventional scheme:

$$\rho_V(\mathbf{p}) = \pi r_e^2 c^2 \sum_n \left| \int e^{-i\mathbf{pr}} \Psi_+(\mathbf{r}) \Psi_n(\mathbf{r}) \sqrt{\gamma(\mathbf{r})} d\mathbf{r} \right|^2, \qquad (1)$$

where r_e is the classical electron radius, c is the speed of light, $\Psi_+(\mathbf{r})$ is the positron wave function, $\Psi_n(\mathbf{r})$ is the valence electron wavefunction of the n-th band, and $\gamma(\mathbf{r})$ is the enhancement factor. The summation was performed over all the occupied states. The core electron momentum distribution is calculated by

$$\rho_C(\mathbf{p}) = \pi r_e^2 c^2 \sum_{i,nlm} \left| \int e^{-i\mathbf{pr}} \Psi_+(\mathbf{r}) \Psi_{i,nlm}(\mathbf{r} - \mathbf{R}_i) \sqrt{\gamma(\mathbf{r})} d\mathbf{r} \right|^2, \qquad (2)$$

where $\Psi_{i,nlm}(\mathbf{r}-\mathbf{R}_i)$ represents the core electron wavefunction specified by the principal, azimuthal, and magnetic quantum numbers (nlm) for the i-th atom and $\mathbf{R}_i$ denotes the position vector. Here, we follows the Alatalo's method. That is, the positron wavefunction is assumed to be spherically symmetric and the core wavefunction is given by $\Psi_{i,nlm}(\mathbf{r})=R_{i,nl}(r)Y_{lm}(\theta,\phi)$, where $R_{nl}(r)$ is the radial part of the wavefunction and $Y_{lm}(\theta,\phi)$ is the spherical function. Consequently, Eq. (2) is rewritten as

$$\rho_C(p) = \pi r_e^2 c^2 \sum_{i,nl} (2l+1) \left| \int j_l(pr) R_{i,nl}(r) \Psi_+(|\mathbf{r} + \mathbf{R}_i|) \sqrt{\gamma(\mathbf{r})} r^2 dr \right|^2, \qquad (3)$$

after integrating on the θ and ϕ axes, where $j_l(pr)$ is the spherical Bessel function. Thus, $\rho_C(\mathbf{p})$ becomes isotropic. The total momentum distribution is given by $\rho(\mathbf{p})=\rho_V(\mathbf{p})+\rho_C(\mathbf{p})$. One-dimensional angular correlation of annihilation radiation (1D-ACAR) spectrum is obtained by integrating the total momentum density in two momentum directions. Subsequently, DBAR spectrum is obtained by convoluting the 1D-ACAR spectrum with the resolution function having a FWHM of 3.92×10^{-3} m_0c.
Supercells with a 1x1x1 conventional unit cell were constructed for individual crystal systems considered in this study. The valence electron wavefunctions were calculated based on the projector augmented-wave (PAW) method [2] using the ABINIT4.6.4 code [4]. The potentials and projectors were generated using the ATOMPAW code [5]. The k-point mesh was 2x2x2 or 4x4x4. The cut-off energy of the plane wave basis set was 60 Ryd. The radial part of core electron wavefunction was represented by the Slater function parameterized by Clementi and Roetti [6]. A self-consistent positron wavefunction was calculated based on the two-component density functional theory in order to minimize the energy functional. The Borónski–Nieminen enhancement factor was adopted [7].

Results and Discussion

Figure 1 shows the DBAR spectra for the Be and HOPG samples as examples of the second period elements. The electron configuration of a Be atom is given as $1s^22s^2$. We first calculated the DBAR spectrum for Be assuming 1s core electrons. However, the calculation hardly reproduced the experimental spectrum. We therefore treated 1s electrons as valence electrons. As shown by the solid curve, the experimental spectrum is well reproduced. In the case of graphite, due to the sp^2 bonding among C atoms, a layered structure is formed. Therefore, the assumption of 1s core electrons seems to be plausible. However, again the calculation with 1s core electrons failed to reproduce the experimental spectrum. The reason is that the positron wavefunction was approximated as a spherically symmetric function following the Alatalo's method despite that real positron wavefunction is anisotropically localized at the interlayer spacing [8]. Thus, by assuming 1s valence electrons, and hence considering an anisotropic positron wavefunction, the calculation reproduces the experimental spectrum. Probably, for diamond, the 1s electrons can be approximated as isotropic core electrons.
Figure 2 shows the DBAR spectrum for the Al samples as an example of the third period elements. In this case, the calculation assuming $1s^22s^22p^6$ core and $3s^23p^1$ valence electrons reproduces the experimental spectrum. A similar result was obtained for the Si samples. For the Mg samples, both 2p 3s electrons should be treated as valence electrons.

Figure 3 shows the DBAR spectra for the Ti, Fe, Cu and Ge samples as examples of the fourth period elements. From Ti to Ni, 3s and 3p electrons should be treated as valence electrons in addition to 3d and 4s electrons. From Cu to Ge, 3s and 3p electrons may be treated as core electrons. Probably, from Ti to Ni, 3d and 4s orbitals are only partially filled and hence underlying 3s and 3p wavefunctions are modulated, while, from Cu to Ge, the nature of valence electrons may be mostly determined by 3d, 4s and 4p electrons.

Figures 4 shows the experimental DBAR spectra for the V, Cr, Mn, Fe, Co, Ni, Cu, Zn, Ga and Ge samples differentiated by the spectrum for the Ti samples. It is seen that the DBAR spectra change systematically with increasing atomic number from Ti. For the V, Cr and Mn samples, the change in the spectrum shape is relatively small, but the broad bump at around 20×10^{-3} m_0c increases dramatically from Fe to Cu. This change corresponds to the filling of 3d bands. From Zn to Ge, the bump at around 20×10^{-3} m_0c decreases and the intensity at low momentum region increases. This reflects the enhancement of annihilation rate between positrons and the 4sp band electrons. Figure 5 shows the calculated differential DBAR spectra corresponding to the elements in Fig. 4. The calculation reproduces the experimental differential spectra. Although the effect of crystal structure on the DBAR spectrum shape seems to be small for bcc, fcc and hcp, the diamond structure (Ge) gives rise to a dramatic change.

Figures 6 shows the DBAR spectra for the Zr and Mo samples as examples of the fifth period elements. Similarly to the Ti and Cr in the third period, by treating 4s, 4p, 4d and 5s electrons as valence electrons, the calculation reproduces the experimental spectra. Figure 7 shows the DBAR spectra for the Gd and Pb samples as examples of the sixth period elements. In their calculations, the valence electron configurations are assumed to be $4f^7 5s^2 5p^6 5d^1 6s^2$ and $5d^{10} 6s^2 6p^2$, respectively. The calculations reproduce the experimental spectra suggesting that the assumed valence electron configurations are valid.

Conclusions

In the present work, we compared the measured and calculated DBAR spectra for a series of single element materials from the second to sixth periods. Assuming appropriate valence electron configurations, the calculation based on the PAW method for electronic state and the two-component density functional theory reproduced the measured spectra. Having this theoretical calculation, the DBAR measurement will be a useful manner for analysing electronic state of materials as well as defect identification and elemental analysis.

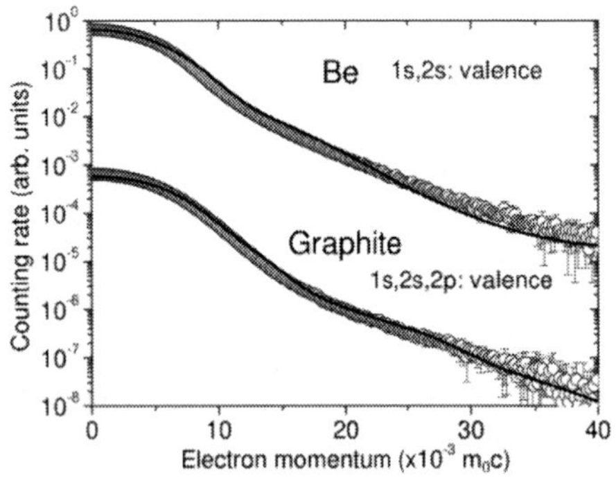

Figure 1. DBAR spectra for the Be and HOPG samples (the second period). Open circles and solid lines denote measured and calculated spectra, respectively.

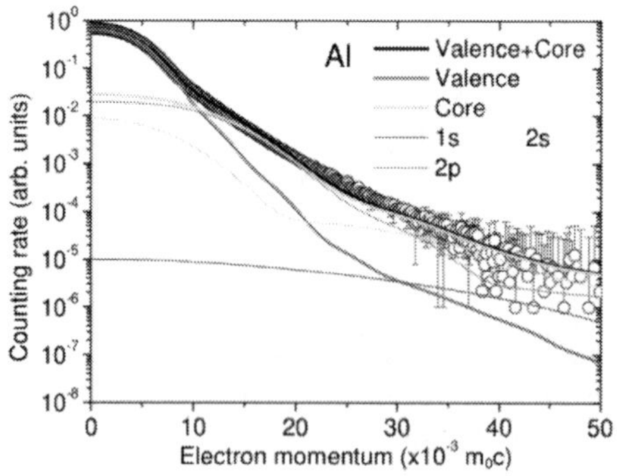

Figure 2. DBAR spectrum for the Al samples (the third period). Open circles and solid lines denote measured and calculated spectra, respectively.

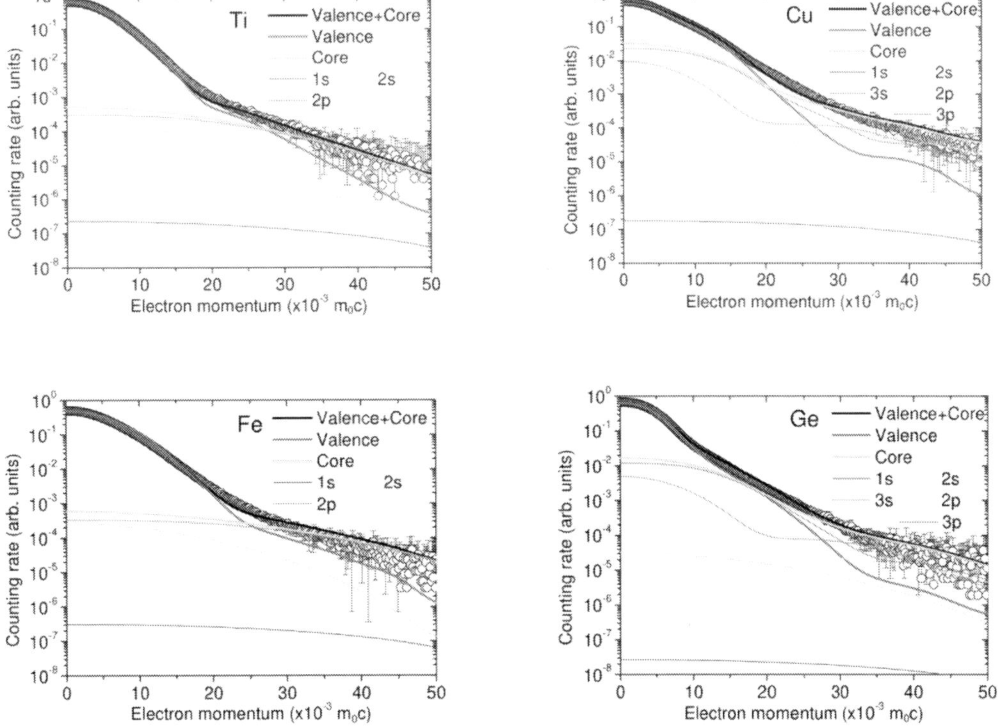

Figure 3. DBAR spectra for the Ti, Fe, Cu and Ge samples (the fourth period). Open circles and solid lines denote measured and calculated spectra, respectively.

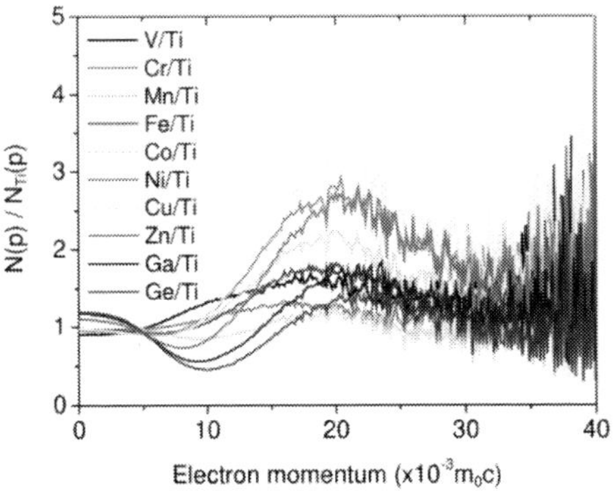

Figure 4. Experimental DBAR spectra for the V, Cr, Mn, Fe, Co, Ni, Cu, Zn, Ga and Ge samples differentiated by the spectrum for the Ti samples.

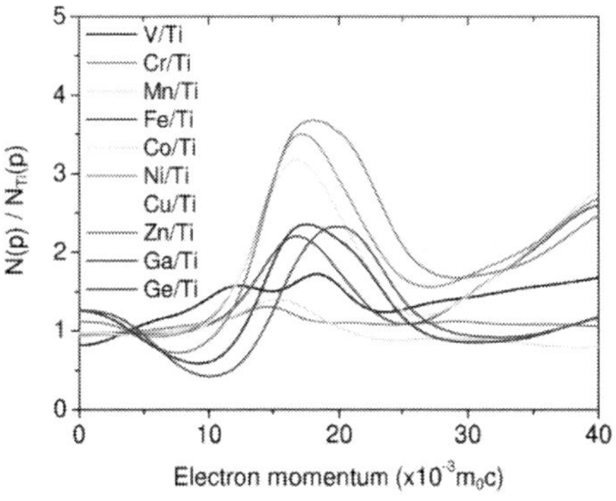

Figure 5. Calculated DBAR spectra for V, Cr, Mn, Fe, Co, Ni, Cu, Zn, Ga and differentiated by the spectrum for Ti.

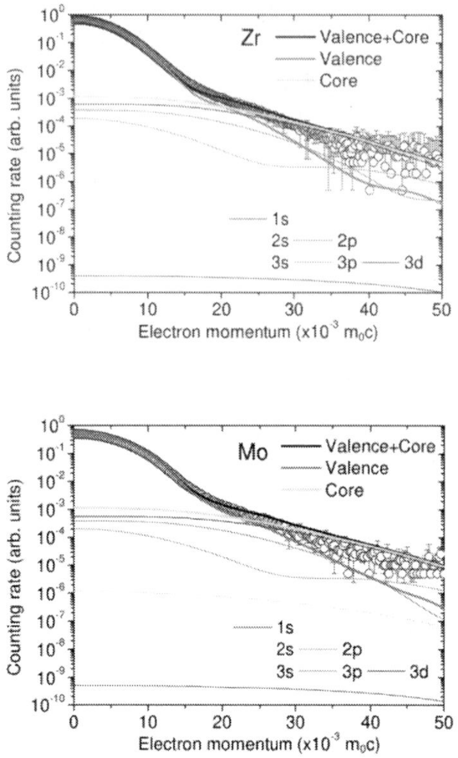

Figure 6. DBAR spectra for the Zr and Mo samples (the fifth period). Open circles and solid lines denote measured and calculated spectra, respectively.

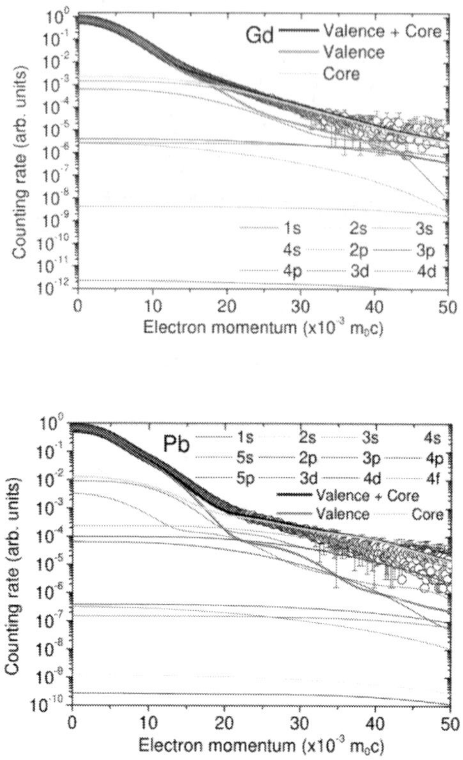

Figure 7. DBAR spectrum for the Gd and Pb samples (the sizth period). Open circles and solid lines denote measured and calculated spectra, respectively.

References

[1] Ishibashi S 2004 *Mater. Sci. Forum* **445-446** 401-403.
[2] Blöhl P E 1994 *Phys. Rev.* B**50** 17953-17979.
[3] Puska M J and Nieminen R M 1994 *Rev. Mod. Phys.* **66** 841–897.
[4] Gonze X, Beuken J -M, Caracas R, Detraux F, Fuchs M, Rignanese G -M, Sindic L, Verstraete M, Zerah G, Jollet F, Torrent M, Roy A, Mikami M, Ghosez Ph, Raty J -Y, and Allan D C 2002 *Comput. Mater. Sci.* **25** 478–492.
[5] Holzwarth N A W, Tackett A R and Matthews G E 2001 *Comput. Phys. Commun.* **135** 329–347.
[6] Clementi E and Roetti C 1974 *At. Data Nucl. Data Tables* **14** 177-478.
[7] Borónski E and Nieminen R M 1986 *Phys. Rev.* B **34** 3820–3831.
[8] Alatalo M, Asoka-Kumar P, Ghosh V J, Nielsen B, Lynn K G, Lruseman A C, Van Veen A, Korhonen T, and Puska M 1998, *J. Phys. Chem.* **59** 55-59.

Development of a spin-polarized positron source

A Kawasuso, Y Fukaya, M Maekawa and A Yabuuchi
ASRC, Japan Atomic Energy Agency, 1233, Watanuki, Takasaki, Gunma 370-1292, Japan

E-mail: kawasuso.atsuo@jaea.go.jp

Abstract. We have started the development of a spin-polarized positron beam. In order to obtain a highly spin polarized positron beam, a ^{68}Ge-^{68}Ga source will be produced using a nuclear reaction of ^{69}Ga(p,2n)^{68}Ge. The optimum proton bombardment energy for ^{69}Ga target is simulated to be 25 MeV using the Induced Radioactivity Analysis Code (IRAC). A newly designed ^{69}Ga target holder will also be used as a ^{68}Ge-^{68}Ga source capsule.

Introduction

Possibility of detecting the excess spins of magnetic substances by angular correlation of annihilation radiation with longitudinally spin-polarized positrons was demonstrated [1]. This will be used in the studies of spin-electronics materials such as half-metals, diluted magnetic semiconductors, magnetic surfaces etc., and of spin-dependent phenomena such as the spin Hall effect, the giant Rashba effect etc. The magnetic quenching of surface positronium [2] and spin-polarized positron diffraction will be promising methods to determine the magnetic susceptibilities of the first surface layers. To realize the above applications of spin-polarized positrons, a highly spin-polarized positron source is needed. Positrons generated through β^+-decay are longitudinally polarized because of the parity non-conservation law. Spin-polarized positrons can also be obtained using the spin-orbit interaction between depolarized positrons and some heavy metal targets [3]. However, this method is less efficient. Therefore, in order to obtain highly spin-polarized positron beam, we are developing a newly designed β^+-emitter source.

The longitudinal spin-polarization of positrons from β^+-emitters is given by

$$P = \frac{N_+ - N_-}{N_+ + N_-} = \frac{v}{c}\frac{1+\cos\alpha}{2}, \tag{1}$$

where, $N_{+,-}$ denote the numbers of up-spin and down-spin positrons, v is the positron speed and α is the open angle from the beam direction (assumed to be the same as the source surface normal direction) [4]. The velocity factor is further given by

$$\frac{v}{c} = \sqrt{1 - \frac{1}{\{1 + E/(mc^2)\}^2}}, \tag{2}$$

where E is the positron kinetic energy. Therefore, to obtain highly spin-polarized positrons, it is important to select β^+-emitters with higher endpoint energies. There are two ways to further enhance the spin-polarization. One is to filter-off low energy positrons and the other is to regulate the solid angle of positron emission. For long-term and stable experiments, it is also important to select a β^+-emitter with long half-life.

Among a number of β^+-emitters, ^{22}Na and ^{68}Ge-^{68}Ga have relatively long half-lives i.e., 2.6 years and 288 days, respectively. Although ^{44}Ti-^{44}Sc has an extremely long half-life (48 years) and hence it is very attractive, the cross section of nuclear reaction (^{45}Sc(p,2n)^{44}Ti) is very small. We thus compare ^{22}Na and ^{68}Ge-^{68}Ga. Figure 1 shows the velocity factor of the spin-polarization for ^{22}Na and ^{68}Ge-^{68}Ga as a function of filtered-off low energy positron fraction calculated from their energy spectra. Here, we replace the energy in eq. (1) with its mean value. The original spin-polarisation of ^{68}Ge-^{68}Ga is 0.92. Whereas, the spin-polarisation of ^{22}Na is less than that of ^{68}Ge-^{68}Ga even after filtering-off low energy positrons. The dump of low energy positrons causes the decrease of intensity. Therefore, to obtain highly spin-polarized positrons with less intensity loss, obviously, ^{68}Ge-^{68}Ga is better than ^{22}Na.

By regulating the solid angle of positron emission, the geometrical factor of spin polarization ($(1+\cos\alpha)/2$) in eq. (1) increases, but it causes the loss of positrons. For example, $(1+\cos\alpha)/2=0.93$ for $\alpha=\pi/3$, but the available positron fraction is only 13% compared to the case of $\alpha=\pi/2$, where all positrons in the forward solid angle can be used. Hence, ^{68}Ge-^{68}Ga is again better than ^{22}Na, because of less intensity loss by filtering-off low energy positrons.

In this study, we carried out a simulation to determine the optimum condition of proton bombardment to produce ^{68}Ge-^{68}Ga source and a design of its capsule.

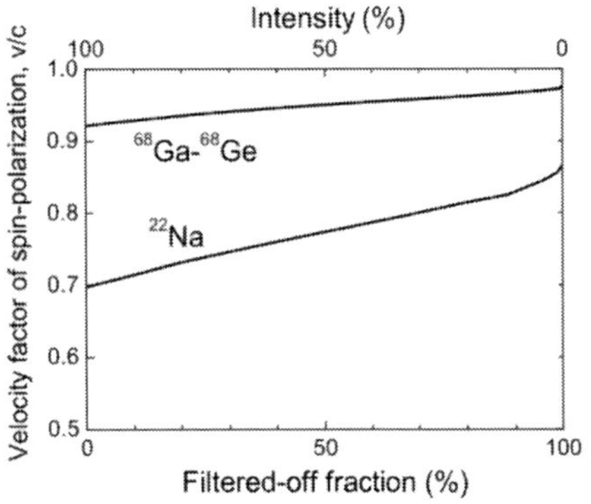

Figure 1. Velocity factor of the spin-polarization (v/c) for ^{68}Ge-^{68}Ga and ^{22}Na as a function of filtered-off low energy positron fraction.

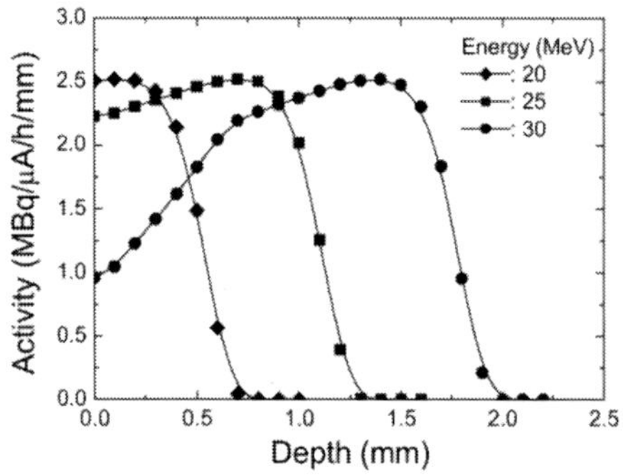

Figure 2. IRAC simulation of depth profiles of ^{68}Ge-^{68}Ga produced by proton bombardment with energies of 20, 25 and 30 MeV.

Simulation

To produce ^{68}Ge-^{68}Ga by proton nuclear reaction, a plausible target would be Ga since this material is placed just before Ge in the periodic table. The stable isotopes of Ga are ^{69}Ga(60.1%) and ^{71}Ga(39.9%). The total amount of ^{68}Ge-^{68}Ga produced by ^{69}Ga(p,2n)^{68}Ge and ^{71}Ga(p,4n)^{68}Ge continuously increases when proton energy (E_p) from 15 to 150 MeV. However, net positron number emitted from the target is reduced for higher proton energies since ^{68}Ge-^{68}Ga is produced in deeper region of the target. As shown later, the optimised proton energy to obtain a high positron emission efficiency from the target surface is ranging from 20 to 30 MeV. The cross section of ^{69}Ga(p,2n)^{68}Ge is three orders of magnitude greater than that of ^{71}Ga(p,4n)^{68}Ge for E_p=30 MeV. Therefore, it is better to use high purity

of ^{69}Ga as a target material. Figure 2 shows the depth profile of ^{68}Ge-^{68}Ga simulated by the Induced Radioactivity Analysis Code (IRAC) [5] at various proton energies. Assuming the proton current of 10 μA and the irradiation duration of 10 hours, the total activities of ^{68}Ge-^{68}Ga are estimated to be 140, 280 and 370 MBq for E_p=20, 25 and 30 MeV, respectively.
Net positron intensity emitted outside from the target is roughly estimated by

$$I = \frac{1}{2}\int_0^d N(z)T(z)dz \,, \tag{3}$$

where N(z) is the depth profile of ^{68}Ge-^{68}Ga, T(z) is the transmission probability of positrons from the depth z to the vacuum and d is the source thickness. The transmission probability may be given by $T(z)=\exp(-\alpha z)$, where $\alpha=17\rho/E_{max}^{1.43}$, ρ is the target density and E_{max} is the energy endpoint. Figure 3 shows the net positron number emitted from the source as a function of the thickness considering the depth profile shown in Fig. 2. Figure 4 shows the case assuming uniform depth profile. Thus, in any cases, more positrons are obtained with E_p=25 MeV and source thickness of 1 mm. Whereas, a higher emission efficiency to the total activity is obtained for E_p=20 MeV and target thickness of 0.5 mm.
The positron emission rate obtained in a feasible proton beam condition (10 μA, 10 hours) is of the order of 10^7 e^+/sec. By repeating proton bombardment, this value could be enhanced owing to the long half-life of ^{68}Ge-^{68}Ga.

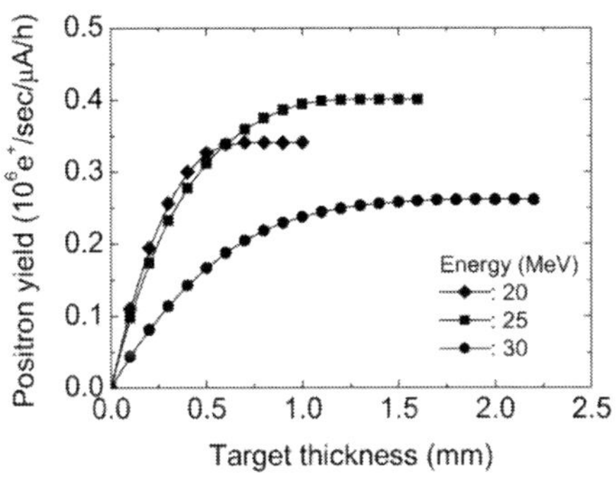

Figure 3. Net positron emission rate from ^{68}Ge-^{68}Ga produced by proton bombardment with energies of 20, 25 and 30 MeV as a function of target thickness assuming non-uniform depth profile.

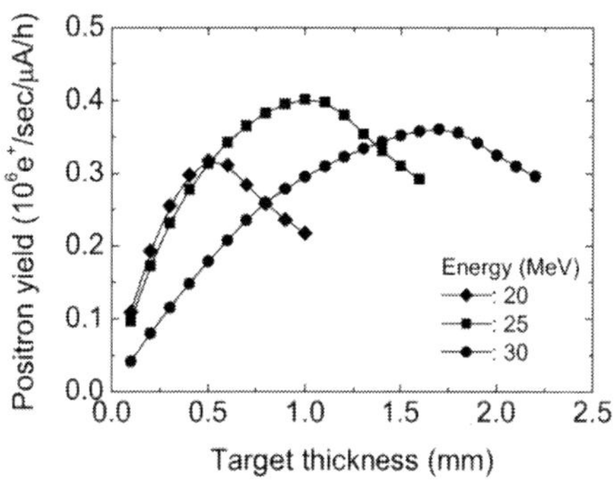

Figure 4. Net positron emission rate from ^{68}Ge-^{68}Ga produced by proton bombardment with energies of 20, 25 and 30 MeV as a function of target thickness assuming uniform depth profile.

Source capsule
We designed the source capsule as shown in Fig. 3. The basic concept is similar to that designed by the Halle group [6] and some points are modified considering proton bombardment to Ga target. A proton absorber (tungsten disc) and a graphite tray filled with source material (^{69}Ga) are put into an aluminium pedestal. These are covered with an aluminium inner cap with a beryllium window (25 μm thick) and by an aluminium outer cap with a titanium window (5 μm thick). For avoiding unintentional activation and for cooling efficiently by running water during proton bombardment, the source capsule components are basically made of aluminium. The source tray and the inner cap can not react with melted Ga. Also the source tray should be made of low-Z materials to reduce the backscattering

probability of positrons. Therefore, the source tray and the inner cap window are made of graphite and beryllium, respectively. During proton bombardment with the inner cap, protons transmit the Ga target and the graphite tray and eventually stop at the surface of the tungsten disk as an absorber. When the source is taken outside from the irradiation chamber, it is covered with the outer cap. The air leak is prevented by rigid contact of the edges of pedestal and outer cap. Thus, the source can be used in a vacuum condition. Figure 4 shows the pictures of fabricated source capsule components.

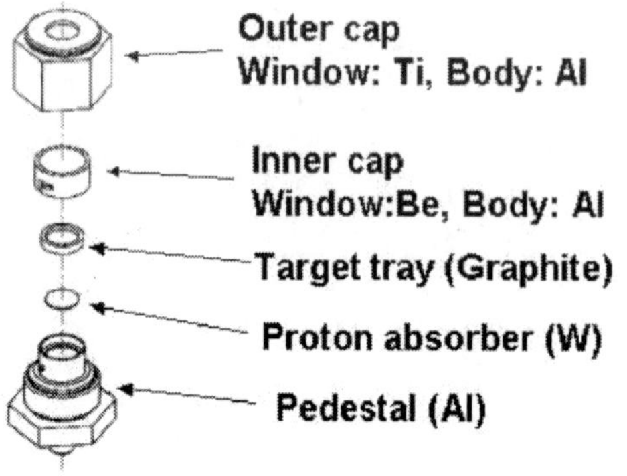

Figure 3. Schematic view of the source capsule components.

Figure 4. Pictures of fabricated source capsule components: (a) Pedestal, (b) Graphite tray where Ga is put into, (c) Inner cap with a Be window and (d) Outer cap with Ti window.

Conclusions

In the present work, we estimated the optimum condition to produce a highly spin-polarized positron source (^{68}Ge-^{68}Ga) by means of ^{69}Ga(p,2n)^{68}Ge nuclear reaction. A practical ^{68}Ge-^{68}Ga source could be produced by proton bombardment with energies of 20-25 MeV to high purity of ^{69}Ga targets of 0.5-1 mm thick. We also made a newly designed ^{69}Ga target holder, which is also used a sealed ^{68}Ge-^{68}Ga source capsule. Proton bombardment will be conducted using a Cyclotron accelerator at Japan Atomic Energy Agency.

References

[1] Berko S 1967 *Positron Annihilation in Ferromagnetic Solids*, in: Stewart A T and Roellig L O (Eds.) *Positron Annihilation*, Academic Press pp. 61-80.
[2] Gidley D W, Koymen A R and Weston T 1982 *Phys., Rev. Lett.* **49** 1779-1783.
[3] Mott N F 1932 *Proc. Roy. Soc.* (London) A**135** 429-458.
[4] Major J 1999 *Spin-Polarized Positron Beams in Condensed–Matter Studies*, in: Coleman P (Ed.) *Positron Beams and their applications*, World Scientific pp.259-306.
[5] Tanaka S, Fukuda M and Nishimura K 1997 *JAERI-Data-Code* **97-019** 1-91.
[6] Krause-Rehberg R, van der Walt N, Büttner L, Börner F 2004 *Nucl. Instr. and Meth. in Phys. Res.* **B 221** 165-167.

International Symposium on Advanced Science Research 2009 IOP Publishing
Journal of Physics: Conference Series **225** (2010) 012002 doi:10.1088/1742-6596/225/1/012002

Muon, positron and antiproton interactions with atoms and molecules

Edward A. G. Armour

School of Mathematical Sciences, University of Nottingham, Nottingham NG7 2RD, UK

E-mail: `edward.armour@nottingham.ac.uk`

Abstract. In this paper, a description is given of some interesting processes involving the interaction of a muon, a positron, or an antiproton with atoms and molecules. The process involving a muon is the resonant formation of the muonic molecular ion, $dt\mu$, in the muon catalyzed fusion cycle. In the case of a positron, the process considered is positron annihilation in low-energy positron scattering by the hydrogen molecule. The antiproton is considered as the nucleus of an antihydrogen atom interacting with simple atoms. Attention is given to antiproton annihilation through the strong interaction. An outline is given of proposed tests of fundamental physics to be carried out using antihydrogen.

1. Introduction

In my talk at the Advanced Science Research 2009 Symposium at Tokai, I described particular interactions of muons, positrons and antiprotons with atoms and molecules.

2. Muon catalyzed fusion

The particle the muon that is central to muon catalyzed fusion was discovered by Anderson and his first graduate student, Neddermeyer, in 1936 when studying cosmic rays [1]. The muon has a mass of $207m_e$, where m_e is the mass of the electron. It has a mean lifetime of 2.2 microseconds. See, for example, [2]. The muon may have a positive or a negative charge. In what follows, I will assume that the muon being considered is negatively charged.

If the electron in a hydrogen atom (H) is replaced by a muon, the result is a hydrogen-like atom $p\mu$. The reduced mass of this atom is $186m_e$. Thus its bohr radius is $\frac{a_0}{186}$, where a_0 is the bohr radius of H. The binding energy of each of its bound states is $186\times$ (the corresponding value for the H atom). $p\mu$ is thus very compact and strongly bound.

An electron can bind two protons to form a weakly bound ion, H_2^+, the hydrogen molecular ion. If the electron in this ion is replaced by a muon, the resulting ion in its ground state is very compact and strongly bound like $p\mu$. As is to be expected, these properties remain if either or both protons are replaced by a deuteron (d) or a triton (t).

A natural question to ask is whether any way could be found of bringing about fusion that did not require very high temperatures, such as are necessary for fusion in the core of the sun and in the ITER project. In 1947, Frank [3] suggested that in the presence of protons and deuterons a slow muon might bind to a p to form $p\mu$. This could come close to a d and form $pd\mu$ which we have seen is tightly bound. Frank thought that it was sufficiently tightly bound

that p-d fusion might occur. With luck, this would leave the muon free to catalyze further fusions, i.e. bring about p-d fusion while remaining unchanged at the end of the reaction. Experiments on d-d fusion catalyzed by a muon carried out by Dzhelepov et al. [4] at the Joint Institute for Nuclear Research (JINR) in Dubna, north of Moscow, in 1966 revealed a strong and unexpected temperature dependence in the rate of formation of $dd\mu$. This was a very exciting discovery. It strongly suggested that a resonant process was involved and held out the possibility of a large increase in the fusion rate under suitable conditions.
It was not long before a form was suggested for this resonant process. In 1967, Vesman [5] proposed that the formation of $dd\mu$ could occur by the reaction

$$d\mu + \mathrm{D}_2 \rightarrow [(dd\mu)^* dee] \tag{1}$$

where the species on the right-hand side is a muonic molecular complex in which $dd\mu$, in a weakly bound excited state, forms one of the nuclei. Such a mechanism depends crucially on the existence of a weakly bound state with binding energy less than the 4.5 eV dissociation energy of D_2. Gershtein and Ponomarev and their group [6, 7] showed that $dd\mu$ did indeed have a weakly bound state which had rotational and 'vibrational' quantum numbers $(J, v) = (1, 1)$ and binding energy $\sim$2 eV. They were also able to show that a corresponding weakly bound state of $dt\mu$ exists with binding energy $\sim$1 eV.
The most favourable muonic molecular ion for muon catalyzed fusion is $dt\mu$. For example, the d-t fusion rate for this ion is more than 1000 times larger than for any other ion. Also, it produces nearly the highest energy per fusion (17.6 MeV). See, for example, [8].
In the resonant reaction

$$t\mu + \mathrm{D}_2 \rightarrow [(dt\mu)_{11} dee] \tag{2}$$

$dt\mu$ is formed in its $J = 1$, $v = 1$ state. However, rapid fusion only takes place if $dt\mu$ is in a state with $J = 0$. It is thus important to know the binding energies and wave functions of the various states of $dt\mu$ below the very weakly bound $(1, 1)$ state. These binding energies are given in [9]. Deexcitation to these lower states is brought about by Auger decay.
Reaction (2) is incomplete without an indication of what results from the formation of the $[(dt\mu)_{11} dee]$ complex. With inclusion of the most important decay products, it becomes

$$t\mu + \mathrm{D}_2 \rightleftarrows [(dt\mu)_{11} dee] \longrightarrow [(dt\mu)_{Jv} de]^+ + e, \tag{3}$$

where $(J, v) = (0, 1), (2, 0), (1, 0)$ or $(0, 0)$. If the total angular momentum of the $t\mu + \mathrm{D}_2$ is taken to be zero for simplicity, the cross section $\sigma_r(E)$ for this resonant process is determined by the Breit–Wigner formula [10].

$$\sigma_r(E) = \frac{\pi}{k^2} \frac{\Gamma_e \Gamma_a}{(E - E_r)^2 + \frac{1}{4}(\Gamma_e + \Gamma_a)^2}, \tag{4}$$

where k is the wave number of the relative motion of $t\mu$ and D_2,

$E =$ the energy of the system, $E_r =$ the energy of the resonant state,
$\Gamma_e =$ partial width for back decay into $t\mu + \mathrm{D}_2$, and $\Gamma_a =$ partial width for Auger decay.

Fig 1 shows how $dt\mu$ formation fits into the overall muon catalyzed fusion cycle.
Cohen at Los Alamos, who has made many contributions to the theory of muon catalyzed fusion over the years, encouraged me to apply the methods of quantum reactive scattering to reaction (3). This reaction has some similarities with the chemical reaction

$$\mathrm{H} + \mathrm{D}_2 \rightarrow \mathrm{H} + \mathrm{D}_2 \text{ or } \mathrm{HD} + \mathrm{D}. \tag{5}$$

This and other chemical reactions have been extensively studied by these methods. See, for example, [12].
However, comparison of reactions (3) and (5) shows that (3) has a special feature not present in (5), namely the Auger decay process which leads to the loss of the electron. Reaction (5) corresponds quite closely to the much slower side reaction

$$t\mu + D_2 \leftrightarrows [(dt\mu)_{11}dee] \rightarrow [(dt\mu)_{Jv}e] + D \tag{6}$$

where $(J, v) = (0, 1), (2, 0), (1, 0)$ or $(0, 0)$.
This reaction proceeds through the *same* resonances as reaction (3). In this case, the resonant complex decays forward into the muonic molecule $[(dt\mu)_{Jv}e]$ and a D atom or backward into $t\mu + D_2$.
One of the methods that has been applied to reaction (5) is the method of Pack and Parker [13]. They use adiabatically adjusting, principal axes hyperspherical coordinates (APH) in their calculations. These are elegant coordinates that transform smoothly between different channels such as $H + D_2$ and $HD + D$. Together with Pack, my postdoc Zeman and I applied this method to reaction (6) [14–16].
Unfortunately, there was no easy way of including the Auger decay channel directly in our treatment. However, as pointed out by Men'shikov and Faifman [10], the coupling between the resonant channels is small as the lifetime of the resonant complex is much longer than the time the complex takes to complete a vibration. Thus the various decay processes operate essentially independently. This made it possible for us to obtain what we expect to be accurate values for Γ_e and Γ_n, the partial widths for back decay and for the decay of the complex into $[(dt\mu)_{Jv}e] + D$, by analysing our results for the cross section for the resonant reaction (6) using the Breit–Wigner formula. Details of our calculation are given in [15].
It was found, as expected, that the reaction proceeded only through the resonances. The resonant states were found to be the vibrational states of the complex with vibrational quantum number $v_c = 3$ and 4. The center of the $v_c = 2$ resonance was calculated to be just slightly below threshold.
Somewhat to our surprise, we found that our calculated value of the back decay rate [16] was much larger than the value obtained for it by Lane [17] and comparable with the calculated values of the Auger decay rate in [18]. Further work is necessary to resolve this discrepancy.

3. Low-energy positron-hydrogen-molecule scattering

In this section, I wish to consider low-energy scattering of a positron, the antiparticle corresponding to the electron, by the hydrogen molecule.
The calculations I wish to describe were carried out with the nuclei in a fixed position. The quantity that we have calculated is $Z_{\text{eff}}(k)$, the effective number of electrons in the H_2 target that are available to the incident positron for annihilation. k is the wave number of the positron. The positron annihilation rate, λ, is given by

$$\lambda = \pi r_0^2 c D Z_{\text{eff}}(k), \tag{7}$$

where r_0 is the classical radius of the electron and D is the density of H_2 molecules in the vicinity of the positron.
There is currently great interest in the very large positron annihilation rates, and hence $Z_{\text{eff}}(k)$ values, that have been observed in low-energy positron scattering by some organic molecules. These are observed experimentally to occur at energies just below those of excited vibrational states of the molecule concerned [19]. This has been explained by Gribakin [20] as being due to resonant behaviour involving a positron-molecule quasi-bound state. It is hoped that useful information about this process can be obtained from our study of e^+H_2 scattering.

The protons are fixed at A and B, which have coordinates $(0, 0, -\frac{R}{2})$ and $(0, 0, \frac{R}{2})$, respectively, w.r.t. Cartesian axes. R is the internuclear distance. The prolate spheroidal coordinates (λ, μ, ϕ) used in the calculations are of the form

$$\lambda = \frac{r_A + r_B}{R} \quad \text{and} \quad \mu = \frac{r_A - r_B}{R},$$

where r_A and r_B are the distances of the point with coordinates (λ, μ, ϕ) from A and B, respectively. ϕ is the usual azimuthal angle of spherical polar coordinates. Unless stated otherwise, the internuclear distance R was fixed at $1.4a_0$, the equilibrium value for the H_2 molecule.
The Kohn trial function was taken to be of the form

$$\Psi_{\text{trial}} = \psi_p \phi_{\text{target}} + \sum_{i=0}^{n} g_i \chi_i. \tag{8}$$

The open channel function ψ_p was taken to have the form

$$S + a_t T$$

where

$$T = S + iC \tag{9}$$

$$S = N \frac{\sin[c(\lambda_1 - 1)]}{\lambda_1 - 1} \tag{10}$$

$$C = N \frac{\cos[c(\lambda_1 - 1)]}{\lambda_1 - 1} \quad (1 - \exp[-\gamma(\lambda_1 - 1)]) \tag{11}$$

where λ_1 is the ellipsoidal prolate spheroidal coordinate for the positron,

$$c = \tfrac{1}{2} kR,$$

N is a normalisation constant, γ is a non-linear parameter and ϕ_{target} is the H_2 target wave function. A linear combination of S and C, as above, is the simplest form for the function, ψ_p. It represents the s-wave component of the part of the overall scattering wave function that is of $\sum_g^+$ symmetry. $\{\chi_i\}_{i=0}^n$ are short-range correlation functions and a_t and $\{g_i\}$ are linear parameters to be determined by the complex Kohn method.
The short-range correlation functions are of the form

$$\chi_0 = C \exp[-\gamma(\lambda_1 - 1)] \phi_{\text{target}} \tag{12}$$

$$\chi_i = S_{23} \bar{N} \lambda_1^{a_i} \lambda_2^{b_i} \lambda_3^{c_i} \mu_1^{d_i} \mu_2^{e_i} \mu_3^{f_i} C(s_i, t_i) \exp[-\alpha\lambda_1 - \beta(\lambda_2 + \lambda_3)] \Omega, \quad (i \neq 0), \tag{13}$$

where particle 1 is the positron and particles 2 and 3 the electrons, S_{23} is the electron symmetriser, α and β are positive, non-linear parameters and $\bar{N}$ is a normalisation constant. $a_i, b_i, c_i, d_i, e_i, f_i$ are non-negative integers. The forms used for $C(s_i, t_i)$ are given in Table 1. If the method of models [21] is used

$$\Omega = \phi_{\text{target}}.$$

Otherwise Ω is taken to be equal to one.
Details of the first calculations carried out without using the method of models are given in [22]. Earlier calculations by the generalised Kohn method used the method of models [23]. In this method, the target Hamiltonian, $\hat{H}_t$, is taken to be

$$\hat{H}_t = -\tfrac{1}{2}\nabla_2^2 - \tfrac{1}{2}\nabla_3^2 + V_{\text{mod}} \tag{14}$$

where V_{mod} is such that

$$\hat{H}_t \phi_{\mathrm{target}} = E_t \phi_{\mathrm{target}} \tag{15}$$

where E_t is the expectation value of the energy of ϕ_{target}. No knowledge of V_{mod} is required when carrying out Kohn calculations. Also we do not have to consider the $\frac{1}{r_{23}}$ electron repulsion term in the exact form of the target potential.

This method had the very considerable advantage that it was possible to include the key correlation functions in Table 1 containing $C(1,1) = \rho_{12} = \frac{2}{R} r_{12}$ without calculating integrals involving $\frac{\rho_{12}\rho_{13}}{\rho_{23}}$. These functions are Hylleraas-type functions that correlate the positron and the electrons. They play a very important role in improving the accuracy of the wave function at points at which the positron and an electron coincide that are the key to calculating $Z_{\mathrm{eff}}(k)$ accurately. However, the complexity of the calculation prevented us from using a very accurate target wave function.

Recently, we have taken account of the motion of the nuclei in our calculations as R has been fixed at its equilibrium value, 1.4 a_0. As a first step we have determined how $Z_{\mathrm{eff}}(k)$ at $k = 0.04$ varies with the internuclear distance, R. The results obtained both with and without the use of the method of models are given in Table 2. It can be seen that the results obtained using the method of modules are larger than those obtained without using this method. The vibrationally averaged value of $Z_{\mathrm{eff}}(k)$, obtained using a Morse function, is 13.5 when obtained using the method of models and 9.7 without using this method [24, 25].

Thus the value obtained with the method of models is closer to the experimental value of 14.6 [26] than that obtained without using it. In a very recent, very accurate calculation of $Z_{\mathrm{eff}}(k)$ at very low energies, Zhang et al. [27] find similar behaviour to the behaviour we find using the method of models as R increases beyond $1.6a_0$. The more accurate value obtained by the method of models, despite using a less accurate target wave function, may be due to the better convergence behaviour that is obtained using this method [28, 22].

4. Antihydrogen

There is great current interest in the preparation of cold antihydrogen ($\bar{\mathrm{H}}$), the simplest antiatom made up of an antiproton and a positron. This work is being carried out by the ALPHA and ATRAP collaborations at CERN [29, 30]. It is proposed to carry out tests of fundamental principles of physics using trapped antihydrogen. There is a very good article on the underlying theory by Shore [31]. For my summary with comments, see [32].

To this end, it is proposed to carry out the following experiments at CERN.

(1) Examination of the spectrum of $\bar{\mathrm{H}}$ to test the CPT and Lorentz invariance of relativistic quantum field theory.

(2) The weak form of Einstein's principle of equivalence predicts that $\bar{\mathrm{H}}$ should fall at the same rate as H in a gravitational field. This is to be tested by an experiment to be carried out by Kellerbauer et al. [33].

The proton and the antiproton are hadrons. Thus they can interact through the strong interaction that brings about proton-antiproton annihilation. This interaction is much stronger than the electromagnetic interaction that brings about electron-positron annihilation. The strong interaction is described by Quantum Chromodynamics (QCD). Current applications treat QCD as a perturbation. However, the coupling constant for QCD at very low energies is too large for perturbation theory to be applicable. This situation is not very satisfactory. However, the underlying problem 'is a very difficult, long standing problem in QCD' [34].

The most accurate practical treatment is to use a complex phenomenological potential taking into account the following: the meson exchange part of the strong interaction, proton-antiproton annihilation, the dipole interaction between the nuclear spins and spin-orbit coupling between nucleon spins and partial waves. The form of the terms in the potential is determined from experiment. A detailed treatment of this type is possible for $\mathrm{H} + \bar{\mathrm{H}}$ [35], but, as far as I am aware, not for more complicated systems.

Calculations that we have carried for $\mathrm{H}\bar{\mathrm{H}}$ scattering [36] show that cross sections for $\bar{\mathrm{H}}$ loss through the rearrangement reaction with product $p\bar{p} + Ps$ are larger than for $\bar{\mathrm{H}}$ loss by antiproton annihilation [35]. However, this is not the case for $\mathrm{He}\bar{\mathrm{H}}$ scattering. Cross sections for $\bar{\mathrm{H}}$ loss through the three rearrangement reactions with products $\mathrm{He}^{+}\bar{p} + Ps$, $\mathrm{He}\bar{p} + e^{+}$ and $\alpha\bar{p} + Ps^{-}$, respectively, are much smaller than for $\bar{\mathrm{H}}$ loss by antiproton annihilation [37, 25].

5. Conclusion

I have described interesting ways in which a muon, a positron and antihydrogen, which contains an antiproton, can interact with simple atoms and molecules. It is to be expected that new facilities such as J-PARC and further work at CERN will make possible many more studies of interactions of this type.

I have also outlined proposed tests of fundamental physics to be carried out using antihydrogen. These should either reinforce our present understanding of physics or show that modifications to this understanding are necessary. A need for modifications would be a very exciting outcome.

Acknowledgments

I am very grateful to the organizers for the invitation to speak at the Advanced Science Research 2009 Symposium at Tokai. I am also very grateful to Professor Humberston for valuable discussions and to Professor Charlton for encouraging me to work on positron and antimatter physics. I wish to thank EPSRC (UK) for support for this research through an advanced fellowship grant (EP/D069785/1), two postdoctoral research grants (GR/L29170) and (GR/R26672/01), two doctoral training grants, (GR/P01595/01) and (EP/P500389/1), and a project studentship (EP/C548019/1). I also wish to thank the University of Nottingham for support for this research.

Table 1. Forms used for the factor $C(s_i, t_i)$ in the short-range correlation functions.

s_i	t_i	$C(s_i, t_i)$	Type of basis function
0	0	1	Product of σ functions
1	1	ρ_{12}	Positron-electron Hylleraas-type
1	2	ρ_{23}	Electron-electron Hylleraas-type
2	1	$\frac{2}{R}(x_1x_2 + y_1y_2)$	Positron-electron configuration interaction (CI) product of π functions
2	2	$\frac{2}{R}(x_2x_3 + y_2y_3)$	Electron-electron CI, product of π functions

R is the internuclear distance.
The distance between particles i and j,

$$r_{ij} = \frac{R}{2}\rho_{ij}.$$

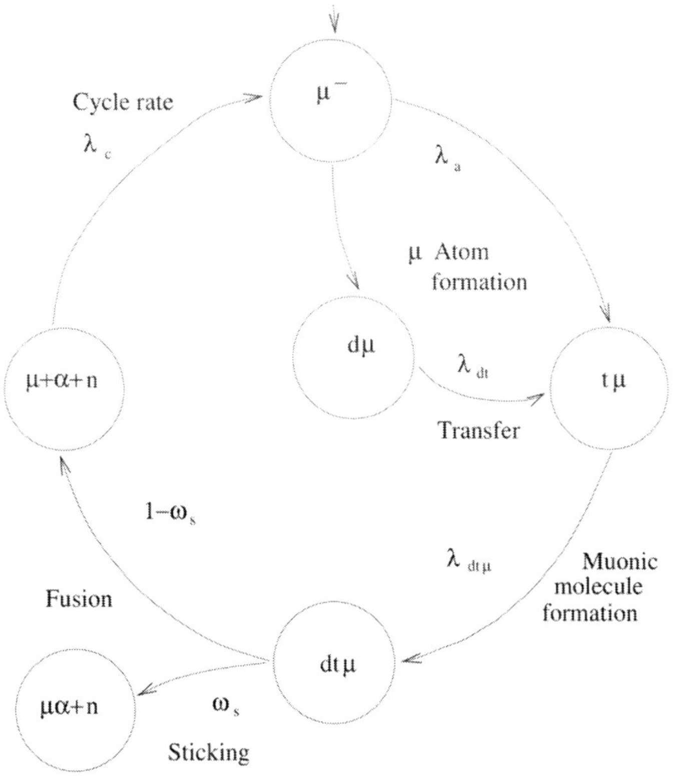

Figure 1. The principal muon catalysis fusion cycle in a deuterium and tritium mixture [8]. Side chains involving *d-d* and *t-t* fusion are not shown. λ_a = muonic atom formation rate $\approx 4 \times 10^{12}\, s^{-1}$; λ_{dt} = muon transfer rate from $(d\mu)_{1s}$ to $(t\mu)_{1s} \approx 3 \times 10^8\, s^{-1}$; $\lambda_{dt\mu}$ = resonant formation rate of $dt\mu \approx 4 \times 10^8\,\mathrm{sec}^{-1}$; ω_s = effective sticking probability $\approx 0.43\%$; λ_c = cycle rate. Values [11] are for $T = 300$K and liquid hydrogen density (4.25×10^{28} atoms m^{-3}).

Table 2. The dependence of $Z_{\rm eff}(k)$ for $k = 0.04$ on the internuclear distance R.

Internuclear distance, R (a_0)	$Z_{\rm eff}(k)$	
	Calculated using the method of models †	Calculated without using the method of models *
0.8	6.4	5.0
0.9	7.2	5.8
1.0	8.0	6.5
1.2	10.0	8.1
1.4	12.6	9.6
1.6	15.8	10.6
1.8	19.8	11.1
2.0	24.3	10.8
2.1	26.7	10.5

† Trial function containing 279 short-range correlation functions, 18 of which correlate the positron and the electrons, and a target wave function that takes into account 57.1% of the correlation of energy of H_2 at $R = 1.4a_0$.

∗ Trial function as above but with a target wave function that takes into account 96.8% of the correlation energy of H_2 at $R = 1.4a_0$.

The percentage correlation energy is given by the percentage of the energy, beyond the SCF energy, taken into account by the wave function under consideration.

References

[1] Neddermeyer S H and Anderson C D, 1937 Phys. Rev. **51**, 884.
[2] Froelich P, 1992 Advances in Physics **41**, 405.
[3] Frank F C, 1947 Nature **160**, 525.
[4] Dzhelepov V P, Ermolov P, Moscalev V I and Filchenkov V V, 1966 Sov. Phys. JETP **23**, 820.
[5] Vesman E A, 1967 Sov. Phys. JETP Lett. **5**, 91.
[6] Gershtein S S and Ponomarev L I, 1977 Phys. Lett. **72 B**, 80.
[7] Vinitsky S I, Ponomarev L I, Puzynin I V, Puzynina T P, Somov L N and Faifman M P, 1978 Sov. Phys. JETP **47**, 444.
[8] Ponomarev L I, 1990 Contemporary Physics **31**, 219.
[9] Bhatia A K and Drachman R J, 1989 Comments At. Mol. Phys. **22**, 281.
[10] Men'shikov L I and Faifman M P, 1986 Sov. J. Nucl. Phys. **43**, 414.
[11] Breunlich W H, Kammel P, Cohen J S and Leon M, 1989 Annu. Rev. Nucl. Part. Sci. **39**, 311.
[12] Manolopoulos D E and Clary D C, 1989 Ann. Rep. Prog. Chem. Sec. C: Phys. Chem. **86**, 95.
[13] Pack R T and Parker G A, 1989 J. Chem. Phys. **87**, 3888 (1987); **90**, 3511.
[14] Armour E A G, 1997 J. Chem. Soc., Faraday Trans. **93**, 1011.
[15] Zeman V, Armour E A G and Pack R T, 2000 Phys. Rev. A **61**, 052713.
[16] Zeman V and Armour E A G, 2001 Hyperfine Interactions **138**, 255.
[17] Lane A M, 1983 Phys. Lett. **98**A, 337.
[18] Bhatia A K, Drachman R J and Chatterjee L, 1988 Phys. Rev. A **38**, 3400.
[19] Gilbert S J, Barnes L D, Sullivan J P and Surko C M, 2002 Phys. Rev. Lett. **88**, 043201
[20] Gribakin G F, 2000 Phys. Rev. A **61**, 022720.
[21] Drachman R J, 1972 J. Phys. B: At. Mol. Opt. Phys. **5** L30.
[22] Cooper J N, Armour E A G and Plummer M, 2008 J. Phys. B: At. Mol. Opt. Phys. **41**, 245201.
[23] Armour E A G, Baker D J and Plummer M, 1990 J. Phys. B: At. Mol. Phys. **23**, 3057.
[24] Cooper J N, 2009 PhD thesis, University of Nottingham, UK.
[25] Armour E A G, Cooper J N, Gregory M R, Jonsell S, Plummer M and Todd A C, 2010 Proceedings of POSMOL 2009, Toronto, J. Phys. Conf. Ser., in press.
[26] Laricchia G, Charlton M, Beling C D and Griffith T C, 1987 J. Phys. B: At. Mol. Phys. **20**, 1865.
[27] Zhang J-Y, Mitroy J and Varga K, 2009 Phys. Rev. Lett. **103**, 223202.
[28] Van Reeth P and Humberston J W, 1995 J. Phys. B: At. Mol. Opt. Phys. **28**, L23.
[29] Fujiwara M et al., 2008 Phys. Rev. Lett. **101**, 053401.
[30] Gabrielse G et al., 2008 Phys. Rev. Lett. **100**, 113001.
[31] Shore G M, 2005 Nuc. Phys. B **717**, 86.
[32] http://www.maths.nott.ac.uk/personal/eaga/
[33] Kellerbauer A et al., 2008 Nuc. Math. and Inst. B **266**, 351.
[34] Signer A, 2006 Private communication.
[35] Armour E A G, Liu Y and Vigier A, 2005 J. Phys. B: At. Mol. Opt. Phys. **38**, L47.
[36] Armour E A G and Chamberlain C W, 2002 J. Phys. B: At. Mol. Opt. Phys. **35** L489.
[37] Armour E A G, Todd A C, Jonsell S, Liu Y, Gregory M R and Plummer M, 2008 Nucl. Instr. and Meth. B **266**, 363; in 'Cold Antimatter Plasmas and Application to Fundamental Physics', 2008 (eds. Kanai Y and Yamazaki Y) AIP CP1037, 333.

Journal of Physics: Conference Series **225** (2010) 012007 doi:10.1088/1742-6596/225/1/012007

Positronium Formation in Porous Materials for Antihydrogen Production

R Ferragut[1,2], A Calloni[1], A Dupasquier[1], G Consolati[2,3], F Quasso[3], MG Giammarchi[2], D Trezzi[4], W Egger[5], L Ravelli[5], MP Petkov[6], SM Jones[6], B Wang[7], OM Yaghi[7], B Jasinska[8], N Chiodini[9] and A Paleari[9]

[1] Dipartimento di Fisica, Politecnico di Milano, via Anzani 42, 22100 Como, Italy
[2] Istituto Nazionale di Fisica Nucleare, via Celoria 16, 20133 Milano, Italy
[3] Dipartimento di Fisica, Politecnico di Milano, piazza L. da Vinci 32, 20133 Milano, Italy
[4] Dipartimento di Fisica, Università degli Studi di Milano, via Celoria 16, 20133 Milano, Italy
[5] Institut für Angewandte Physik und Messtechnik, Universität der Bundeswehr München, Werner Heisenberg-Weg 39, D-85577, Neubiberg, Germany
[6] Jet Propulsion Laboratory - NASA, California Institute of Technology, 4800 Oak Grove Drive, Pasadena, CA 91109, U.S.A.
[7] Center for Reticular Chemistry, Department of Chemistry and Biochemistry, University of California-Los Angeles, 607 Charles E. Young Drive East, Los Angeles, CA 90095, U.S.A.
[8] Institute of Physics, Maria Curie-Sklodowska University, pl. M. Curie-Sklodowskiej 1, 20-031 Lublin, Poland
[9] Dipartimento di Scienza dei Materiali, Università di Milano-Bicocca, via Cozzi 53, 20125 Milano, Italy

E-mail: rafael.ferragut@polimi.it

Abstract. Positronium (*Ps*) formation measurements in several porous materials as: Vycor, germanate Xerogel, Metal-Organic Frameworks MOF-177 and Aerogel with two densities (20 and 150 mg/cm^3), were performed by means of a variable energy positron beam provided with a Ge detector and a positron lifetime spectrometer. An efficient formation of cooled *Ps* atoms is a requisite for the production of antihydrogen, with the aim of a direct measurement of the Earth gravitational acceleration *g* of antimatter, which is a primary scientific goal of AEGIS (Antimatter Experiment: Gravity, Interferometry, Spectroscopy). Porous materials are necessary to form a high yield of *Ps* atoms as well as to cool *Ps* through collisions with the inner walls of the pores. The different materials were characterized and produce *Ps* into the pores. Lifetime measurements give an estimation of the typical pores dimension of the substances. A comparative study of the positron lifetime and the *Ps* fraction values in the above mentioned materials indicates that silica Aerogel, with the appropriate density, is an excellent candidate for an efficient formation of cold *Ps* atoms for the AEGIS project.

1. Introduction

The AEGIS experiment has been approved at the CERN Antiproton Decelerator to directly measure the gravitational acceleration *g* on a beam of cold antihydrogen [1]. Gravitational interaction between matter and antimatter has never been tested experimentally. Some fundamental questions of modern physics relevant to unification of gravity with the other fundamental interactions, models involving vector and scalar gravitons, matter–antimatter symmetry can be enlightened via experiments with antimatter [2].

Figure 1 shows a schematic drawing of the basic experimental setup that should reach an accuracy of 1% in the measurement of the matter-antimatter gravitational acceleration. The experiment is designed to allow higher precision measurements through cooling of the antiproton ($\overline{p}$) beam [3]. The essential steps leading to the production of antihydrogen ($\overline{H}$) and the measurement of its gravitational interaction in AEGIS with the use of the CERN cold $\overline{p}$ are the following: i) production of *Ps*; ii) laser excitation of *Ps* to form Rydberg positronium (Ps^*) and to increase the cross section of the following reaction [4,5]; iii) production of cold Rydberg antihydrogen $\overline{H}^*$ at 100 mK based on the reaction between cold $\overline{p}$ and Ps^*; iv) acceleration of $\overline{H}^*$ beam using inhomogeneous electric fields [3] and v) determination of *g* in two-grating Moiré deflectometer coupled with a position-sensitive detector [6].

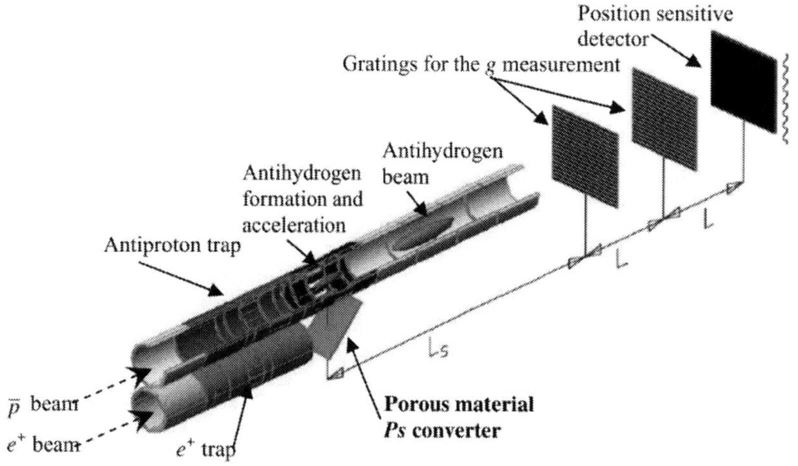

Figure 1. Sketch of the experimental setup region where antiprotons and positrons are manipulated to form and to accelerate antihydrogen. This region is at low temperature (~100 mK) in an axial magnetic field of 1 T (after Ref. 7).

In the present work results regarding *Ps* formation are reported. The *Ps* ionization energy of the bound state between a positron and an electron is 6.8 eV. There are several models of *Ps* formation, which were reviewed in a monograph by Mogensen [8]. The first general theory was published by Ore and Powell [9]. Ground state *Ps* is formed for 75% as ortho-*Ps* (spin 1) and for 25% as para-*Ps* (spin 0). Para-*Ps* annihilates in two γ rays with 511 keV each and ortho-*Ps* in vacuum is required to annihilate into three γ rays at least, with a maximum energy of 511 keV each and with a total energy of 1022 keV ($2m_0c^2$). We are interested only in the fraction of *Ps* emitted by the converter as ortho-*Ps* with a characteristic lifetime in vacuum of 142 ns, since the lifetime of para-*Ps* is too short (125 ps) to allow the necessary laser excitation before its decay. The ortho-*Ps* lifetime can be shortened by collisional pick-off annihilation and ortho-para conversion (both processes lead to annihilation in two γ rays) in ranges typically from a fraction of 1 ns to tens of nanoseconds.

An efficient formation of cooled *Ps* atoms is a requisite for the production of antihydrogen. Porous materials are necessary to form a high yield of *Ps* atoms as well as to cool *Ps* through collisions with the inner walls of the pores. The collisions between *Ps* and the internal surface of the pores involve a weak coupling to phonons or other surface modes. The velocity distribution of the *Ps* atoms coming out of the target should be the order of 10^4 m/s to allow *Ps* laser excitation to a Rydberg state (Ps^*). The formation of cold $\overline{H}^*$ (Fig. 1) will be possible by means of the charge exchange reaction between Ps^* and cold $\overline{p}$

$$Ps^* + \overline{p} \rightarrow \overline{H}^* + e^-, \qquad (1)$$

with a residual electron.

The different porous materials tested in this work (Vycor, germanate Xerogel, Metal-Organic Frameworks MOF-177 and Aerogel with two densities) were characterized by means of a variable energy positron beam with a high purity germanium detector and positron lifetime spectroscopy. The results indicate that these materials produce *Ps* into the pores. Lifetime measurements give an estimation of the typical pores dimension of the substances. A comparative study of the positron lifetime and the *Ps* fraction values in the above mentioned materials indicates that silica Aerogel, with a selected density, is an appropriate candidate for an efficient formation of cold *Ps* atoms for the AEGIS project.

2. Experimental procedure and methodology

The composition of the porous materials used in the present work is shown in Table 1.

Table 1. Composition of the materials.

Porous sample	Composition
Vycor	$(SiO_2)_{0.55}(B_2O_3)_{0.35}(Na_2O)_{0.1}$
Germanate Xerogel	$(Ge)_{0.3}(SiO_2)_{0.7}$
MOF-177	$Zn_4O(O_2C\text{-})_6$
AerogelC-20	$(SiO_2)_{1-X}(C)_X$, X=0.0067
Aerogel-150	SiO_2

The procedure used to produce a mesopore film in the Vycor glass of about 2 μm in depth [10] was the following: i) the glass sample was annealed at 530°C for 16 h; ii) the sample was next converted into a porous sorbent by leaching in hot (80°C) 3N H_2SO_4 for 15 minutes and then in 0.5N NaOH solution at room temperature for 5 min and iii) the sample was washed out to neutrality with distilled water and dried at 80°C for 24 h. The germanate Xerogel was prepared following the procedure indicated in Ref. 11. After drying, the sample was sintered with the following conditions: the Xerogel was heated from room temperature to 450°C under oxygen flux with a ramp of 6°C/h, and then it was kept at such temperature for 1 day in argon atmosphere to replace oxygen with inert gas. Further heating to the final temperature (680°C) was then carried out in 2% vol. H_2 in Argon. Nitrogen sorption measurements showed the presence of a bimodal distribution of pores, comprised of nanopores (below 2 nm) and mesopores (between 20 nm and 50 nm). The number density of the small pores was strongly dominant (150:1). MOF-177 was synthesized according to Ref. 12 in the form of quasi-cubic grains with a side of about 50 μm. MOF-177 is a porous materials with low density (4.27×10^{-1} g cm^{-3}) and large, regular, accessible cages and tunnels. This material is specially used for gas storage [13]. Before the positron measurements granular MOF-177 was compacted with a charge of 8×10^3 N cm^{-2} into a cylindrical pellet of 1.5 cm in diameter and 7 mm in thickness. Silica Aerogel samples were fabricated using a procedure developed for the particle capture collectors of the NASA Stardust project [14]. The sample density was controlled by mixing different amounts of acetonitrile into the sol. The solution was then dried by supercritical solvent extraction, which allows the solvent to be removed while the network gel remains in a highly expanded state in the form of a three dimensional array of filaments. For this comparative study, two samples were selected with densities 20 mg cm^{-3} (AerogelC-20) and 150 mg cm^{-3} (Aerogel-150) and high porosity (99% and 93%, respectively). AerogelC-20 was doped with 0.4 wt.% (0.67 at.%) of graphite.

The measurements of *Ps* formation yield were performed by means of a monoenergetic positron beam using the well-known "3γ method" [15-17]. The positron implantation energy was variable from 0.05 keV up to 18 keV. The measurements were performed at room temperature with a vacuum level between 10^{-7}-10^{-9} mbar. A high purity germanium (HPGe) detector was used to obtain the annihilation γ spectrum. From the annihilation spectrum it was possible to obtain the $R(E)$ parameter

$$R(E) = \frac{V}{P}, \tag{2}$$

where P and V are the integrated counts (after background subtraction) in the peak (511 ± 4.25 keV) and valley area (after the Compton edge, from 350 keV up to 500 keV) of the energy γ spectrum. When there is no pick-off annihilation, i.e. when the *Ps* is emitted in vacuum, the positronium fraction $f(E)$ can be obtained from $R(E)$ using the relationship:

$$f(E) = \left[1 + \frac{P_1(R_1 - R(E))}{P_0(R(E) - R_0)}\right]^{-1}, \tag{3}$$

where P_0 and R_0 are the parameters when the positronium fraction is nil (F=0) and P_1 and R_1 are the parameters when 100% of the positrons becomes *Ps* (F=1). When the *Ps* is formed inside the pores, the pick-off effect reduces the probability of 3γ annihilations by a factor

$$\varepsilon = \frac{\lambda_{3\gamma}}{\lambda_{3\gamma} + \lambda_{p.o.}}, \tag{4}$$

where $\lambda_{3\gamma} = (142\ \text{ns})^{-1}$ is the 3γ annihilation rate in vacuum and $\lambda_{p.o.}$ the pick-off annihilation rate. Thus the proper expression for the *Ps* fraction becomes:

$$F(E) = \varepsilon^{-1} f(E). \tag{5}$$

In the present work the factor ε^{-1} can be evaluated for high positron implantation energies from the available lifetime measurements (see below). In all other cases we report the result obtained from Eq. (3) as "3γ fraction" ($F_{3\gamma} = f$). We warn however that this nomenclature, which is adopted by other authors [18,19], is somewhat misleading, since the true fraction of positrons annihilated in 3γ is smaller by a factor ¾. Nevertheless $F_{3\gamma}$ is the useful parameter to assess the potential of a material to emit o-*Ps* that has lost kinetic energy by collision while meandering through open pores.

The calibration for 0% and ~100% of *Ps* in vacuum was performed using a Ge single crystal (100) at 1000 K [16]. However the R_0 and R_1 parameters depend on the scattering and absorption of the annihilation gamma rays photons in the sample and on the experimental geometry. Corrections of R_0 and R_1 measured with the Ge sample are thus necessary [20]. In thick samples of porous materials a direct measurement of R_0 is impossible because *Ps* is formed even at high implantation depth. These corrections were thus performed following the procedure indicated in Ref. 20. The contribution of these corrections affects $F_{3\gamma}$ from 1% up to 3%.

Lifetime measurements were performed with the pulsed low-energy positron-lifetime spectrometer PLEPS at the high intensity positron beam facility NEPOMUC at the research reactor Heinz Maier–Leibnitz FRM II of the Technical University Munich [21,22]. The resolution of the spectrometer was about 240 ps. The lifetime measurements were performed with a time-windows of 20 ns periodically repeated. In the presence of pores with a size distribution, the lifetime spectra are expected to be the superposition of infinitely many exponential components. In the present case, an approximate analysis in four components was adopted to obtain four mean lives: ($\tau_1 \ldots \tau_4$) and the corresponding intensities ($I_1 \ldots I_4$). The intensities I_j of each lifetime component were corrected by multiplying by [1-exp(-20ns/τ_j)] after the subtraction of the random background and subsequently were re-normalized.

3. Results and discussion

Figure 2 shows the 3γ fraction $F_{3\gamma}$ as a function of the positron implantation energy for the different materials under test. At low implantation energy (≤200 eV) the 3γ fraction for all samples is high, between 31% and 50%. In MOF-177, germanate Xerogel and Vycor an increase of the 3γ fraction vs. the implantation energies was found, followed by an asymptotic decrease. This trend, which is observed also in other materials [18,19,23], can be explained by a model that takes into account the energy dependence of the ortho-*Ps* formation and escape [24]. The increase in MOF-177, germanate Xerogel and Vycor, if it occurs, is documented by only one experimental point, which is not enough to obtain the energy dependence of the *Ps* yield using the model of Ref. 24. On the contrary, the

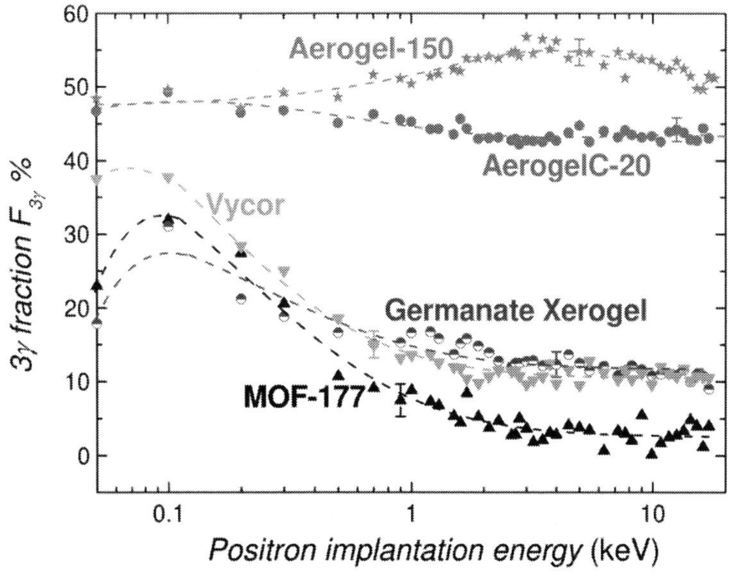

Figure 2. Positronium 3γ fraction $F_{3\gamma}$ as a function of the positron implantation energy in different porous samples: MOF-177, Vycor, germanate Xerogel, and two densities of Aerogel (20 mg cm^{-3} and 150 mg cm^{-3}). The dashed lines upper 0.1 keV in MOF-177, germanate Xerogel and Vycor are fits of the model proposed in Ref. 24. The others dashed lines are only a visual guide. Error bars are shown for one point only in each evolution.

asymptotic decrease is clearly observed and can easily be fitted with the mentioned model (dashed line above 100 eV); the corresponding best-fit ortho-*Ps* diffusion lengths are 17 nm, 5 nm and 3 nm for MOF-177, Germanate Xerogel and Vycor respectively. The model cannot be applied to silica Aerogel, where the low pick-off rate combined with a diffusion length of some microns [25,26] cancels the characteristic decrease of the 3γ yield at high energy. The 3γ fraction in Aerogel samples is very high at all implantation energies, as expected for a material with very large free space between silica particles. The values between 43% and 57% observed above 1 keV are similar or higher than to the maximum values reported in literature [19].

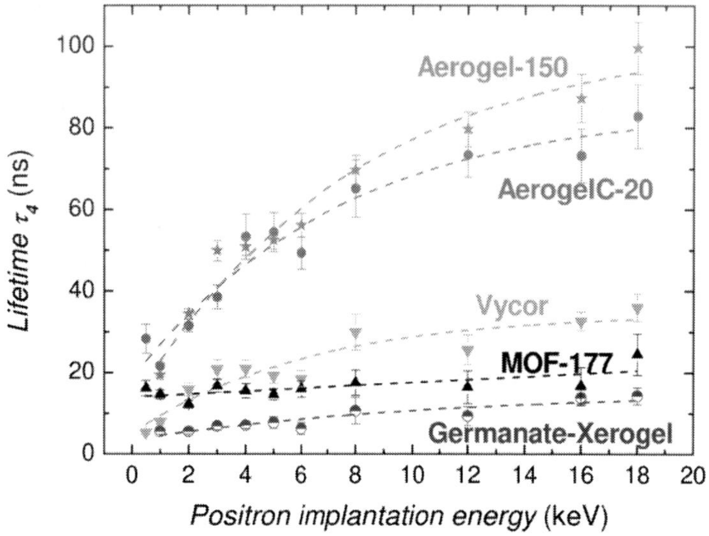

Figure 3. Positron lifetime of the long component τ_4. The dashed lines are only a visual guide.

The relevant results of the lifetime measurements are reported in Figure 3 and Table 2. Figure 3 depicts the implantation energy dependence of the longest lifetime τ_4 isolated in the four-component analysis of the lifetime spectrum. The decrease of τ_4 observed for Aerogel at low implantation energies is symptomatic of *Ps* escape in the vacuum, and is an instrumental effect related to a decreased efficiency of detection of ortho-*Ps* when the pick-off annihilation channel is suppressed.

Table 2. Lifetime and intensity associated to the long component, equivalent pore size for spherical and cylindrical geometries and *Ps* fraction *F* (Eq. 5).

Porous sample	Lifetime τ_4 (ns)	Intensity I_4 (%)	**Diameter** (nm)		*F* (%)
			Spherical	Cylindrical	
Vycor	35	2.7	2.5	2.2	40
Germanate Xerogel	14	3.7	1.6	1.4	~100
MOF-177	20	3.7	1.8	1.7	20
AerogelC-20	82	27	6.0	5.2	74
Aerogel-150	100	23	9.4	8.4	71

Table 2 includes the lifetime τ_4 measured at the highest implantation energy, which can be taken as the upper limit of the ortho-*Ps* lifetime in the pores of the samples. The data reported in Table 2 for Aerogel are consistent with the results obtained by Mincov et al. [27] and by Chang et al. [28]. The lifetime values can be used to estimate the size of the pores with the extended Tao-Eldrup model [29] for spherical pores and its extension to cylindrical geometry [30]. The diameters obtained for spherical and cylindrical (infinitely long) pores are reported in Table 2. This table includes the intensity I_4, which is to be considered the lower limit of the ortho-*Ps* fraction because ortho-*Ps* also contributes to the shorter components and because the energy window of the lifetime spectrometer is set for 2γ annihilations. A different estimate of the ortho-*Ps* fraction may be obtained as ¾ of F, where F is the *Ps* fraction, evaluated from Eq. 5 using the experimental lifetime data to calculate the factor ε (Eq. 4).

4. Concluding remarks

In our quest for a promising positron-*Ps* converter to be chosen in the AEGIS experiment, we have tested a series of materials with known open porosity. We have found that the highest formation of *Ps* occurs in germanate Xerogel and Aerogel. However, high formation is not the only important factor. It must be reminded that *Ps* is emitted into the pores with an initial kinetic energy of the order of a few eV. Since our goal is to obtain ortho-*Ps* in the free space with energies of the order of 10^{-2} eV, it is necessary that the *Ps* suffers a large number of collisions with the walls of the pores before emerging in the free space outside the target. This means that the useful ortho-*Ps* fraction must come from the deep regions of the target. This is certainly not the case of germanate Xerogel, where ortho-*Ps* appears to be strongly quenched in the pores. The reason is certainly the presence in this material of a dense distribution of very small pores, probably with a low degree of interconnections, which act as dominant ortho-*Ps* cages. Most probably, this situation also occurs in Vycor. MOF-177 is a new material; there is no previous information of its ability of producing *Ps*. Our experiment shows that, in spite of its very large free volume, this material is not really interesting as a converter. On the contrary, low-density Aerogel is a promising target candidate under both aspects: high formation and long survival time of ortho-*Ps* inside the pores. In the present work it was not possible to measure the kinetic energy of ortho-*Ps* outside the target. The literature tells however that ortho-*Ps* cooling in porous silica is feasible. Nagashima et al. [31] found an average *Ps* energy of 93 meV in Aerogel of 100 mg cm^{-3} at room temperature in vacuum. Mills et al. [25] observed ortho-*Ps* thermalization also at low temperature in SiO_2 powders compressed up to a similar density to the more dense Aerogel that was studied in the present work. The results of Ref. 25 indicate: 2% at 4.2 K, 8% at 77 K and 12% at 300 K of thermal ortho-*Ps* with positron implantation energies of 19 keV, 10 keV and 7 keV, respectively. These high implantation energies require rather thick samples (several microns, for typical Aerogel densities), which is a condition easily met with homogenous Aerogel. However a compromise must be reached between efficient cooling, which requires deep positron implantation, and high ortho-*Ps* yield outside the target, which is favoured by implantation depths not large in comparison with the ortho-*Ps* diffusion length. In the case of Aerogel, the ortho-*Ps* diffusion length is expected to be about 2 μm, as measured for similar material [26] and to implant positrons at 2 μm in a target with mass density of 150 mg cm^{-3} the positron beam energy should be about 3.5 keV. Time-of-flight measurements of the kinetic energy of the emitted ortho-*Ps* are highly desirable for finding the best compromise.

Acknowledgments

We gratefully acknowledge Prof. Gianpaolo Rosati for the compaction of the MOF-177 samples. Work partially supported by Istituto Nazionale di Fisica Nucleare (Italy).

References

[1] AEGIS web site: http://aegis.web.cern.ch/aegis/
[2] Hughes R J 1993 *Nucl. Phys.* A **558** 605
[3] Testera G et al 2008 *Proc. of Cold Antimatter Plasmas and Application to Fundamental Physics Conference (Okinawa)* vol 1037 (AIP Conference Proceedings) p 5
[4] Giammarchi M G et al 2009 *Hyperfine Interact.* doi: 10.1007/s10751-009-0018-5
[5] Castelli F, Boscolo I, Cialdi S, Giammarchi M G and Comparat D 2008 *Phys. Rev.* A **78** 052512
[6] Bonomi G et al 2009 *Hyperfine Interact.* doi: 10.1007/s10751-009-0015-8
[7] Kellerbauer A et al 2008 *Nucl. Instr. & Methods* B **266** 351
[8] Mogensen O E 1995 *Positron annihilation in chemistry* (Berlin, Springer-Verlag)
[9] Ore A and Powell J L 1949 *Phys. Rev.* **75** 1696
[10] Jasińska B and Dawidowicz A L 2003 *Radiat. Phys. Chem.* **68** 531
[11] Chiodini N et al 2007 *Sol. Stat. Comm.* **144** 429
[12] Chae H K et al 2004 *Nature* **427** 523
[13] Sanderson K 2007 *Nature* **448** 746
[14] McDonnell J A M et al 2000 *Adv. Space Res.* **25** 315
[15] Marder S, Hughes V, Wu C S and Bennet W 1956 *Phys. Rev.* **103** 1258
[16] Mills A P Jr 1978 *Phys. Rev. Lett.* **41** 1828
[17] Lynn K G and Welch D O 1980 *Phys. Rev. B* **22** 99
[18] Liszkay L et al 2008 *Appl. Phys. Lett.* **92** 063114
[19] Petkov M P, Weber M H, Lynn K G and Rodbell K P 2001 *Appl. Phys. Lett.* **79** 3884
[20] Ferragut R, Trezzi D, Calloni A and Dupasquier A (submitted to *Rev. Sci. Instrum.*)
[21] Hugenschmidt C et al. 2008 *Appl. Surf. Sci.* **255** 29
[22] Egger W, Sperr P, Kögel G and Dollinger G 2007 *phys. stat. sol. (c)* **4** 3969
[23] Mariazzi S, Toniutti L, Patel N and Brusa R S 2008 *Appl. Surf. Sci.* **255** 191
[24] Eldrup M, Vehanen A, Schultz P J and Lynn K G 1985 *Phys. Rev.* B **32** 7048
[25] Mills A P Jr, Shaw E D, Chichester R J and Zuckerman D M 1989 *Phys. Rev.* B **40** 2045
[26] Ferragut R et al (to be published)
[27] Mincov I, Petkov M P, Tsou P and Troev T 2004 *J. Non-Cryst. Solids* **350** 253
[28] Chang T B, Wang Y Y, Chang C C and Wang S 1982 in *Positron Annihilation*, ed P G Coleman, S C Sharma and L M Diana (Amsterdam: North-Holland) p 696
[29] Dull T L, Frieze W E, Gidley D W, Sun J N and Yee A F 2001 *J. Phys. Chem.* B **105** 4657
[30] Goworek T, Ciesielski K, Jasińska B and Wawryszczuk J 2000 *Radiat. Phys. Chem.* **58** 719
[31] Nagashima Y et al 1995 *Phys. Rev.* A **52** 258

Development of mini linac-based positron source and an efficient positronium convertor for positively charged antihydrogen production

T Muranaka[1], P Debu[1], P Dupré[1], L Liszkay[1], B Mansoulie[1], P Pérez[1], J M Rey[1], N Ruiz[1], Y Sacquin[1], P Crivelli[2], U Gendotti[2] and A Rubbia[2]

[1]Irfu, CEA-Saclay, F-91191 Gif-sur-Yvette Cedex, France
[2]Institut für TelichenPhysik, ETHZ, CH-8093 Zürich, Switzerland

tomoko.muranaka@cea.fr

Abstract. We have installed in Saclay a facility for an intense positron source in November 2008. It is based on a compact 5.5 MeV electron linac connected to a reaction chamber with a tungsten target inside to produce positrons via pair production. The expected production rate for fast positrons is $5{\cdot}10^{11}$ per second. The study of moderation of fast positrons and the construction of a slow positron trap are underway. In parallel, we have investigated an efficient positron-positronium convertor using porous silica materials. These studies are parts of a project to produce positively charged antihydrogen ions aiming to demonstrate the feasibility of a free fall antigravity measurement of neutral antihydrogen.

1. Introduction

The possibility that antimatter may behave differently from matter in the earth's gravity field is a long-standing issue. However, no direct measurement has ever been successfully done yet [1]. Our physics motivation is to perform the feasibility of the first direct measurement of the Earth's gravitational acceleration on antihydrogen atoms ($\overline{\mathrm{H}}$). The production of $\overline{\mathrm{H}}$ in CERN is currently based on a three-body interaction of an antiproton ($\bar{\mathrm{p}}$) in a dense positron (e^+) cloud [2]. With this method, the produced atoms are highly excited and emitted isotropically at high velocity. It is thus not adequate for free-fall measurement in that the main systematic error on the gravitational acceleration comes from a temperature of $\overline{\mathrm{H}}$. In order to obtain ultracold neutral $\overline{\mathrm{H}}$, Walz and Hänsch suggested to produce positively charged antihydrogen ions ($\overline{\mathrm{H}}^+$). Then once the ions are cooled down to μK, excess positrons are detached by using a laser pulse [3]. We proposed a way to prepare such ions using interactions of the antiprotons with a dense target of positronium atoms (Ps) [4, 5]. The charge exchange process $\bar{\mathrm{p}} + \mathrm{Ps} \rightarrow \overline{\mathrm{H}} + e^-$ is followed by $\overline{\mathrm{H}} + \mathrm{Ps} \rightarrow \overline{\mathrm{H}}^+ + e^-$. In this scheme, it is necessary to prepare firstly a large quantity of e^+ then an efficient e^+-Ps covertor in order to produce the required density of Ps atoms. According to calculated and measured cross sections, one can produce one $\overline{\mathrm{H}}^+$ ion [6] to ten ions [7] when a bunch of 10^7 antiprotons passes through Ps cloud of the order of 10^{13} cm^{-3}.

2. Positron production

The required Ps density may be obtained by sending of the order of 10^{11} slow e^+ onto the convertor that we have developed [8], with the duration less than a lifetime of the ortho-positronium (142 ns).

Taking into account the future installation to an antiproton facility, the intense, low energy and transportable positron source is required for our purpose. We have purchased a small industrial electron linear accelerator from Getinge Linac Technologies [9]. Its nominal kinetic energy and average current are 5.5 MeV and 0.14 mA, respectively. The energy is below the neutron activation threshold, which may be an important issue when considering the heavy regulations associated with radioactive installations. The repetition rate is tunable up to 200 Hz with a fixed bunch length of 4 μs. The magnetron peak power is 1.9 MW. The total power consumption is 35 kVA. The acceleration length is 21 cm after which the beam diameter is 1 mm. The overall dimensions are roughly 1 m x 1 m x 0.8 m. Simulations, performed with the GEANT3 software [10], predict production rates of $5\cdot10^{11}$ e^+ s^{-1} when the beam hits a 200 μm thick tungsten target at 5 degrees incidence angle. The beam diameter at the target position is about 1 cm. At such incident energy of electrons, produced positrons are emitted almost isotropically from the target, with 2/5 of backward emissions from the incident beam direction. The average exit angle with respect to the incident beam is of the order of 50 degrees. The positron energy is widely spread from 0 to 3 MeV with the averaged energy at 800 keV [11]. The setup comprising the linac, the coils with their support and the vacuum chamber, is installed inside an 8 m x 5 m x 3 m room with concrete shielding against X rays. The water-cooling system used for the linac is scaled also to cool the magnet coils. A ventilation system is installed to evacuate the ozone formed inside the room.

2.1. Fast positron selection

Behind the tungsten target, a set of transport coils and a dipole magnet have been mounted to separate the electrons from the fast positrons. In order to measure the fast positron flux, a charge deposited type detector has been installed at 600 mm horizontally and 550 mm vertically from the target position. Once the fast positron optimisation is done, the beam line will be prolonged from here for moderated slow positrons leaving the setup and be guided to an electromagnetic trap. The detector consists simply of 35 aluminum pads of 35 mm x 35 mm square with 5 mm of thickness on a stainless-steel holder covering almost 300 mm of diameter of vacuum chamber. Currents of arrived particles on the pads are sampled by an analog to digital convertor (ADC). The trigger signal to the ADC created by pulse generator is sent also to the linac for defining its repetition frequency. Cancellation of radio frequency noise from linac is essential for pC current measurement. The magnetic field in the vacuum chamber was calculated by TOSCA. Then the positron production and particle transfer simulations have been studied by using GEANT4 simulation code [12]. The simulation predicts that the positrons cannot be completely separated from electrons only by the dipole magnet: 1) certain low energy electrons (less than 10 keV) directly from tungsten target, and 2) backscattered high energy (around 1 MeV) electrons arrive at the positron detector. The former electrons are expected to be localised in one Al pad of the detector where positrons will not be distributed. The latter electrons have to be eliminated for proper positron detection. We have recently installed two Al collimators of 2 cm of thickness, 10 cm of inner diameter after tungsten target. While all the positrons will pass through the centre of collimator halls, more than 90 percent of high energy electrons will be blocked as its wider diameter of the circular motion in the magnetic field. The measurement with this setup is foreseen in the end of 2009.

2.2. Perspective

At the exit of the current setup, the positrons have to be moderated. A tungsten moderator of 10^{-4} efficiency can be placed after the primary target, so that a flux of about 10^7 slow e^+s^{-1} is expected. Another option is to build a moderator based on solid Ar, Kr or Ne at cryogenic temperatures, to be placed away from the diffracted electron flux [11]. We then estimate efficiency at least 10 times higher after including transport losses due to the large beam size.

3. Positronium production

In parallel with positron production, we have investigated efficient positron-positronium convertors. Experiments have been conducted with ETH Zurich's Na-22 based beamline at CERN. As the first step, the re-emission yield of ortho-positronium (o-Ps) into vacuum outside mesoporous silica films on glass plates have been measured in reflection mode with a specially designed lifetime (LT) spectrometer. Values as high as 40% were found. Time-of-flight (TOF) measurements showed that the energy of the orthopositronium emitted into vacuum is below 100 meV in the film with approximately 2 nm pores at 3 keV positron energy [8]. We have recently produced samples with same mesoporous materials put onto the inner wall of glass tubes by spin coating methode. In this configuration, the actual o-Ps density felt by injected $\bar{p}$ is estimated to increase more than ten times from the plate-type samples. Both LT and TOF Measurements are scheduled for the beginning of 2010.

References

[1] Bellucci S and Faraoni V 1996 *Phys. Lett.* B **377** 55-59
[2] Andresen G et al. 2007 *Phys. Rev. Lett.* **98**, 023402; G. Gabrielse et al. 2007 *Phys. Rev. Lett.* **98** 13002
[3] Walz J and Hänsch T 2004 *General Relativity and Gravitation* **36** 561
[4] Pérez P and Rosowsky A 2004 *Nucl. Inst. Meth.* A **532** 523
[5] Pérez P and Rosowsky A 2005 *Nucl. Inst. Meth.* A **545** 20
[6] Biswas P K 2001 *J. Phys.* B **34** 4831
[7] Walters H R J and Starett C 2007 *physica status solidi* (c) **4** 3429
[8] Liszkay L et al. 2008 *Appl. Phys. Lett.* **92** 063114
[9] http://www.linactechnologies.com/
[10] GEANT3.21, CERN Program Library 1994 W5013
[11] Pérez P, Liszkay L, Rey J M, Delferrierre O, Blideanu V, Carty M, Curtoni A, Ruiz N and Sauce Y 2008 *Appl. Surf. Sci.* **255** 33
[12] GEANT4 2003 *Nucl. Inst. Meth.* A **506** 3 250; 2006 *IEEE Transactions on Nuclear Science* **53** 1 Part 2 270

International Symposium on Advanced Science Research 2009 IOP Publishing
Journal of Physics: Conference Series **225** (2010) 012018 doi:10.1088/1742-6596/225/1/012018

Positron accumulation and manipulation for antihydrogen synthesis

H Imao[1,2], K Michishio[3], Y Kanai[1], N Kuroda[2], Y Enomoto[1,2], H Higaki[4], K Kira[4], A Mohri[1], H A Torii[2], Y Nagata[1], C Kim[2], Y Matsuda[2], Y Nagashima[3] and Y Yamazaki[1,2]

[1] Atomic Physics Laboratory, Riken Advanced Science Institute, Wako, Saitama, Japan
[2] Graduate School of Arts and Sciences, The University of Tokyo, Meguro, Tokyo, Japan
[3] Department of Physics, Tokyo University of Science, Shinjuku, Tokyo, Japan
[4] Graduate School of Advanced Sciences of Matter, Hiroshima University, Hiroshima, Japan

E-mail: `imao@riken.jp`

Abstract. Our group ASACUSA-MUSASHI has established an efficient way for accumulating antiprotons in the cusp trap, a combination of an anti-Helmholz superconducting coil and a multi-ring electrode trap. The last piece for synthesizing antihydrogens in the cusp trap is positron. We have developed a compact system to effectively accumulate positrons based on N_2 gas-buffer scheme with a specially designed high precision cylindrical multi-ring electrode trap. Millions of positrons were accumulated in the pre-accumulator just using polycrystalline tungsten moderators. The accumulated positrons were transported as a pulsed beam via three guiding coils and caught in the cusp trap under cryogenic and ultra high vacuum conditions without serious loss. Confinement of two kinds of numerous antiparticles, e.g., 10^8 positrons and 10^7 antiprotons, in the cusp trap becomes feasible.

1. Introduction

Antihydrogen is one of the most attractive systems involving positron because it is an ideal system for testing CPT invariance. Ever since cold antihydrogens were successfully produced [1, 2], a priority was "confinement of ground-state antihydrogen" for its high-precision spectroscopy. Several research groups are pursuing that with various approaches [2, 3]. We ASACUSA-MUSASHI group has developed a unique electro-magnetic trap for synthesizing and confining antihydrogen, the "cusp traph, which is a combination of a multi-ring electrode trap (MRT) and an "axially-symmetrich anti-Helmholtz magnetic field [5, 6, 7]. We established an efficient way [8, 9] to accumulate antiprotons in this trap; millions of antiprotons, about 10^2 times larger than those in other groups, were stably confined. A remaining major difficulty for producing antihydrogen in the cusp trap was the preparation of positrons. The expected recombination mechanisms of antihydrogen are radiative and three-body one. The two mechanisms lead to different quantum state populations of antihydrogen, and the reaction rate have different dependencies on the positron plasma density and temperature, n and $T^{-0.63}$ for the radiative [10], n^2 and $T^{-9/2}$ [11] for the three-body, respectively. In order to produce a large amount of antihydrogens via recombination in the cusp trap it is required to have cold and high-density positrons available. For the purpose of this, we have developed an exceptionally compact

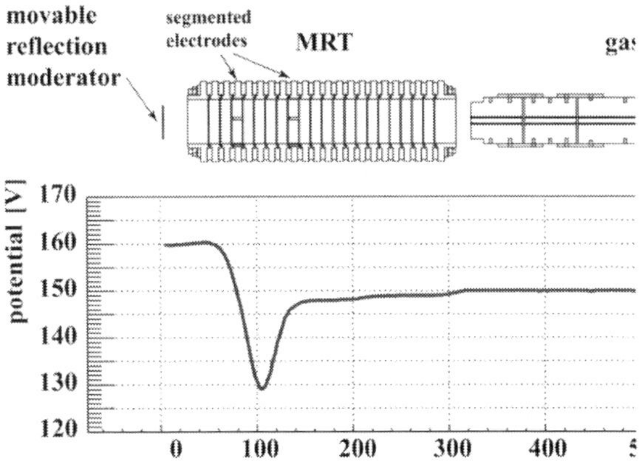

Figure 1. A Cross-sectional view of the trapping region of the positron pre-accumulator and the electrical potential along the axis of the trap and the moderators.

and high-efficient positron pre-accumulator based on N_2 gas-buffer scheme [12] applicable to the cusp trap.

2. Positron pre-accumulator and transport line to cusp trap

The positron pre-accumulator is an instrument for trapping and cooling a continuous beam of slow positrons generated by moderating β^+ particles from a 50-mCi ^{22}Na radioactive source. In our new apparatus, all of the required components, i.e., the radioactive source enclosed a tungsten shield, polycrystalline tungsten moderators, a N_2 gas-buffer cell and a high-precision cylindrical MRT, were situated in a superconducting solenoid (2.5 T). Such all-in-one system in the strong magnetic field has several advantages. A stable trapping by suppressing the diffusion ($\propto$ $(L/B)^2$) and a rapid cooling of positrons by synchrotron radiation ($\propto$ $1/B^2$ [13]) can be expected in the strong magnetic field. A beamline complex between the source and the trapping region was avoided because half of positrons emitted from the source are automatically guided to the main trapping region along magnetic field lines. Also, pre-accumulated positrons in the strong magnetic field can be easily transported to the cusp trap without the reflection due to the magnetic mirror.

A cross-sectional view of the trapping region is shown in Fig. 1. We used two annealed polycrystalline tungsten foil moderators (4 μm for transmission moderator and 25 μm for reflection moderator in thickness) to make slow positron beams. A fraction of positrons into the foil are thermalized within a few picoseconds and isotropically diffuse. Some of the positrons reach the surface and are emitted with the energy around 3 eV depending on the work function for positrons ϕ_+. In this way about ~0.1% of the irradiated positrons directly from the radioactive source emerge from the tungsten surface. The physical dimensions of the N_2 buffer cell and the MRT are designed to allow an efficient pressure gradient to be formed in the small trapping region. The inner pressure of the entrance side of the N_2 buffer cell is maintained to 10^{-3} Torr while the exit side of the trapping region is around 10^{-6} Torr. An efficient differential pumping with the capillary structure (ϕ=6 mm, 400 mm in length) is accomplished. The emitted slow positrons moderated in the tungsten foil lose energy dominantly via the electronic excitation of a N_2 molecule and then trapped. The trapped positrons further collide with N_2 molecules in a MRT and simultaneously accumulated. The MRT has 22 channels of the gold-coated aluminum-ring electrodes which can form flexible potentials for trapping, confining and extracting positrons. We used PEEK (poly-ether-ether-ketone) rings for the

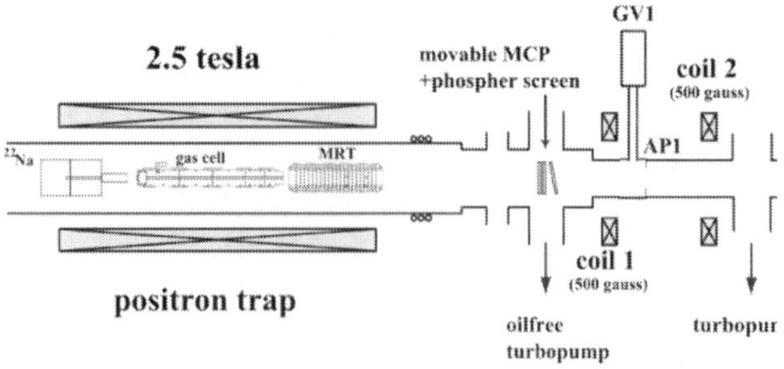

Figure 2. Schematic view of the transport line between the positron pre-accumulator and the cusp trap.

electrical insulation of the electrodes. Two channels of them are segmented in quarter, making it possible to compress or decompress the positrons in the trap by applying rotating electric fields. An outstanding feature of the MRT is the excellent cylindricity, which is important for stable trapping of positrons. The cylindricity of 15 μm involving the segmented electrodes was achieved by utilizing several advanced processing techniques, wire-electrical discharge machining and horning.

We need to transport the positrons after trapping in the pre-accumulator to the cusp trap ($\sim$3 T) under cryogenic and ultra high vacuum conditions where the positrons are mixed with the antiprotons. The specially designed transport line with three guiding coils was developed (Fig. 2). After the accumulation in the pre-accumulator, the vacuum separation valves (GV1 and GV2) are opened and the transfer magnet is energized (500 Gauss) for 3 second. The positrons accumulated in the pre-accumulator are extracted as a low-energy ($\sim$100 eV) pulsed beam and transported to the cusp trap via two aperture (AP1 and AP2) for a differential pumping. The transported pulsed positrons are caught in the MRT of the cusp trap by opening and closing a gate electrode. We found in the particle trajectory calculations that such low-energy positrons can be transported along the magnetic field lines and re-trapped in the cusp trap without serious loss. Two kind of detectors were placed between the positron pre-accumulator and the cusp trap for monitoring the extracted positrons:@(1) A combination of a micro-channel plate and a phosphor screen to get information about the number and the x-y distribution of positrons. (2) NaI and plastic scintillation detectors to detect the positron annihilation signal providing information about the time-of-flight and the z distribution of the pulsed positrons. The whole experimental operations including source opening and closing, moderator movement, potential manipulation of MRTs, detector movement, the data acquisition from the detectors and gating the gate valves are automated to make it possible to run the experiment remotely.

3. Results and discussion

We performed the adjustment and the optimization of the accumulation and transportation process of positrons. The parameter space for them was vast, e.g., trapping potential, buffer gas pressure, moderator treatments, axis adjustment between the MRTs and the magnetic field lines, variety of positron manipulation in both the pre-accumulator and the cusp trap. The optimization of the trapping potential and the gas pressure were especially essential for the trapping efficiency. One of the optimized trapping potential is shown in Fig. 1. Because the length of the trapping region was quite limited in our all-in-one system compared with the conventional positron accumulators [12], the pressure gradient of the buffer gas in the MRT are sharper. The pseudo-harmonic potential for confining positrons is made at the exit side of the MRT where positrons have the lifetime of around 100 s when a proper amount of the buffer gas present. The optimized path energy around ~9 eV, which is the energy of the moderated positrons passing through the gas buffer cell, is consistent with that reported elsewhere [12, 14]. During the measurements we have been able to trap about 2×10^6 e^+ within 30 s accumulation time. Although direct measurements to know our moderator efficiency still not be performed, the expected trapping efficiency referring typical efficiency of the polycrystalline tungsten moderator [15] is more than 25%, which is favorably comparable with the values for conventional "largeh accumulators. About 90% of the positrons in the pre-accumulator are transported and successfully re-trapped in the cusp trap. We are continuing the development works for the manipulation techniques of positrons required for the antihydrogen synthesis in the cusp trap. In the preliminary tests, we have shown that the effectiveness of the stacking of positrons and the positron compression technique for making high-dense positron plasma in the cusp trap.

4. Summary and future prospect

In summary, we have successfully developed the compact positron pre-accumulator applicable for the cusp trap and prepared a significant amount of positrons for $\bar{H}$ synthesis in the cusp trap. The experiment for synthesizing antihydrogen is undergoing. Fine adjustment and further improvement of the positron pre-accumulator are still under developing. The moderator, especially, leave much to be improved. A mesh tungsten moderator [16, 17] or field assisted moderator [18] are possible candidate for the high-efficiency moderators.

5. Acknowledgment

We would like to acknowledge Prof. F. Saito and Prof. T. Hyodo for supplying tungsten mesh moderators. We are indebted to CERN staff for their helpful support with the experiment. This research was partially supported by the Japan Society for the Promotion of Science Fellow, by a Grant-in-Aid for Creative Scientific Research (10P0101) from the Japanese Ministry of Education, Science, Sports and Culture, and by the Special Research Projects for Basic Science of RIKEN.

References

[1] Amoretti A *et. al.* 2002 *Nature (London)* **419** 456.
[2] Gablielse G *et. al.* 2002 *Phys. Rev. Lett.* **89** 233401.
[3] Bertsche W *et. al.* 2006 *Nucl. Instr. Meth. Phys. Res.* A**566** 746.
[4] Gablielse G *et. al.* 2008 *Phys. Rev. Lett.* **100** 113001.
[5] Mohri A *et. al.* 2003 *Europhys. Lett.* **63** 207.
[6] Shibata M *et. al.* 2008 *Rev. Sci. Instr.* **79** 015112.
[7] Saitoh H *et. al.* 2008 *Phys. Rev.* A**77** 051403.
[8] Kuroda N *et. al.* 2005 *Phys. Rev. Lett.* **94** 023401.
[9] Kuroda N *et. al.* 2008 *Phys. Rev. Lett.* **100** 203402.
[10] Stevefelt J, Boulmer J and Delpech J-F 1975 *Phys. Rev.* A**12** 1246.
[11] Glinsky M E and OfNeil T M 1991 *Phys. Fluids* B**3** 1279.
[12] Murphy T J and Surko C M 1992 *Phys. Rev.* A**46** 5696.
[13] O'Neil T M O 1980 *Phys. Fluids* **23** 725.
[14] Charleton M and HumberstonJ W 2001 *Positron Physics*, Cambridge Univ. Press, Cambridge.
[15] Schultz P J and Lynn K G, *Rev. Mod. Phys.* **60** 701.
[16] Nagashima Y *et. al.* 2000 *Jpn. J. Appl. Phys.* **39** 5356.
[17] Saito F *et. al.* 2002 *Appl. J. Surf. Sci.* **194** 13.
[18] Lynn K G and McKee B T 1979 *Appl. Phys.* **19** 247.

Precise Measurement of the HFS of Positronium using the Zeeman Effect I: Experimental Set-up and RF System

G Akimoto[1], A Ishida[1], Y Sasaki[1], A Miyazaki[1], K Kato[1], T Suehara[2], T Namba[2], S Asai[1], T Kobayashi[2], H Saito[3], M Yoshida[4], K Tanaka[4], A Yamamoto[4], Y Urushizaki[5], I Ogawa[5], T Idehara[5] and SSabchevski[6]

[1]Department of Physics, Graduate School of Science,The University of Tokyo, 7-3-1 Hongo, Bunkyo-ku, Tokyo, 113-0033, Japan
[2]International Center for Elementary Particle Physics (ICEPP), The University of Tokyo, 7-3-1 Hongo, Bunkyo-ku, Tokyo, 113-0033, Japan
[3]Department of General Systems Studies,
Graduate School of Arts and Sciences,The University of Tokyo, 3-8-1 Komaba, Meguro-ku, Tokyo, 153-8902, Japan
[4]High Energy Accelerator Research Organization (KEK),1-1 Oho, Tsukuba, Ibaraki, 305-0801, Japan
[5]Research Center for Development of Far-Infrared Region (FIR-FU), University of Fukui, 3-9-1 Bunkyo, Fukui, Fukui, 910-8507, Japan
[6] Bulgarian Academy of Science, 1, 15 Noemvri Str, 1040 Sofia, Bulgaria

E-mail: akimoto@icepp.s.u-tokyo.ac.jp

Abstract. Positronium is a QCD-free system and the measurement of its hyperfine splitting provides a strict test of quantum electrodynamics (QED). Recent research revealed a discrepancy of 3.9σ between the QED prediction and previous experimental results. We report on the prototype run of an improved experimental set-up and the performance of its RF system.

1. Introduction

Positronium (Ps), the bound state of an electron and a positron, is a QCD-free system suitable for strict testing of quantum electrodynamics (QED). The ground state of Ps has two spin states (spin 0 and spin 1), and its so-called "hyperfine splitting" (Ps-HFS) is the energy difference between these states. Ps-HFS is sensitive to new physics beyond the standard model via vacuum oscillation of the S = 1 state of Ps (ortho-positronium, o-Ps). If an unknown light particle (such as an axion or a millicharged particle) exists, it contributes to the energy level, causing a discrepancy from the QED prediction.

Recently, a new calculation of the $\mathcal{O}(\alpha^3)$ correction to Ps-HFS has been developed [1, 2]. It revealed a 3.9σ (15 ppm) discrepancy between the theoretical prediction and previous experimental results [3, 4, 5]. The experimental values are consistent with each other (3.3 ppm) and are all systematically lower than the theoretical prediction (Figure 1). This could be a result of new physics, a miscalculation of the QED correction, or common systematic uncertainties that

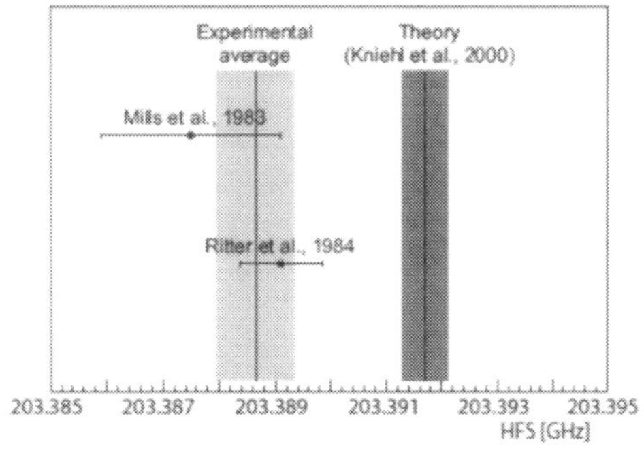

Figure 1. Discrepancy between theoretical prediction and experimental values for Ps-HFS

have not been considered. We investigate some of the possible sources of systematic error using an improved experimental set-up.

2. Measuring procedure

In all previous precision measurements, the Ps-HFS transition was not directly measured, due to technical difficulties with producing the required 203 GHz frequency. Instead, an indirect approach was used. The energy levels of Ps in a static magnetic field are split by the Zeeman effect. The energy splitting of Ps (between perturbed o-Ps and unperturbed o-Ps) is about 3 GHz in the presence of a magnetic field of 0.9 T. This energy difference (ΔE) is a function of Ps-HFS ($\Delta\nu$) and the magnetic flux density (B). Ps-HFS ($\Delta\nu$) can then be derived from ΔE and B.

$$\Delta E = \frac{\Delta\nu}{2}\left(\sqrt{1+x^2}-1\right) \tag{1}$$

$$x \sim 0.2B[T] \tag{2}$$

Figure 2 shows the experimental set-up of the prototype run (July–September 2009). The purpose of this run was to clarify the possible problems and systematic errors in the experiment in preparation for the first run.

Ps is generated by directing positrons from a ^{22}Na positron source (700 kBq) into an RF cavity filled with nitrogen gas. The cavity also stores high-power 3 GHz microwaves. The cavity is surrounded by γ-ray detectors [six $LaBr_3(Ce)$ scintillators] that have good energy resolution (3% at 511 keV). This good resolution means that 2γ decay can be tagged using energy information from a single crystal which yields a higher signal rate than the traditional geometrical back-to-back tagging using the coincidence of two scintillators. The cavity and the γ-ray detectors are located inside a large superconducting magnet, which applies a uniform magnetic field to the cavity. Microwaves are supplied from an external amplifier.

When the Zeeman energy shift ΔE, which changes with B, is resonant with the RF frequency, the transition from $m_z = \pm 1$ to $m_z = 0$ increases, and the $m_z = 0$ state quickly decays into 2γ. Ps-HFS can be identified by the central value of the resonance peak obtained by scanning the magnetic field.

The performance of the transition RF, the uniformity of the magnetic field, and the estimation of material effects are the main sources to systematic uncertainty in this experiment.

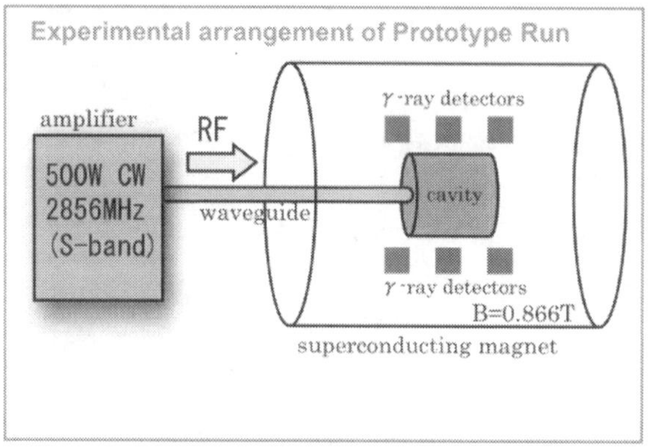

Figure 2. Experimental set-up

The Ps is spread over a large region of the cavity, and the nonuniformity of the static magnetic field over this region was the main source of the uncertainty in past results. By using a large-bore (800 mm) superconducting magnet, this run achieved a uniformity of 10.4 ppm (RMS), which contributes 22 ppm to the systematic uncertainty in Ps-HFS. In future runs we will compensate for this nonuniformity with additional coils (currently under development).

The Ps is produced in the nitrogen gas. Its initial kinetic energy is on the order of 1 eV. It collides with the gas molecules frequently, losing kinetic energy gradually until it reaches 30 meV (thermalization). The electric field of the gas molecules shifts the energy state of the Ps (the Stark effect). In previous experiments, this material effect was estimated by extrapolation to the vacuum from the measured Ps-HFS values at various pressures (about -33 ppm/atm). This procedure assumes that the Ps is well thermalized, and that its mean velocity is the same in the gas at various pressures. In earlier decay rate measurements, we revealed that this thermalization is slow (on the order of the lifetime of o-Ps), and that the assumption of good thermalization is not satisfied, which results in large systematic errors [6]. Nonthermal Ps would also affect the Ps-HFS measurement.

The timing of positron emission, which is negligibly close to the timing of Ps formation, is tagged with a thin (100 μm) plastic scintillator. Using an appropriate timing window cut, we can reduce the amount of the nonthermal Ps and select the well-thermalized Ps. The contribution of the nonthermal Ps as a function of time and pressure can also be measured. (We will measure it precisely in future runs.)

We focus on the RF system used in this experiment and its performance in the following section.

3. The RF system and its performance

The RF system is designed to meet the following requirements.

(i) Unlike the case for transitions of ordinary atoms such as hydrogen, a high-power RF source is necessary in order to obtain a sufficient transition signal. This is because the lifetime of o-Ps is short (142 nsec), which means that relatively little of the produced Ps can transit before decaying. The transition probability is proportional to the RF power, and about 29% of the o-Ps transits at an RF field strength of 2 mT.

(ii) The transition probability of o-Ps depends on the integrated power of the applied RF. We must control this integrated power with an accuracy of less than 0.1%.

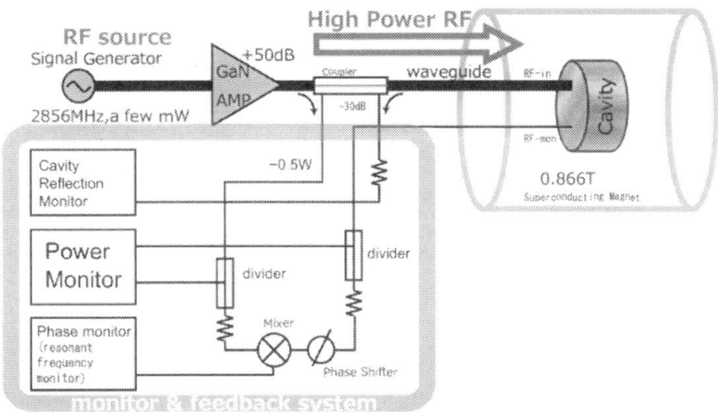

Figure 3. Schematic of RF system

(iii) Ps-HFS is derived from ΔE (the frequency of the RF source) and the strength of the applied magnetic field. The frequency of RF should be stable [less than $\mathcal{O}(1ppm)$].

Figure 3 shows the RF system used in the prototype run, which consists of two parts. The bold black line (upper part of Figure 3) is the system supplying the cavity with high-power RF. The area bounded by the faint red line (lower part of Figure 3) is the monitoring and feedback system, which stabilizes the RF power and frequency.

3.1. Requirement 1: RF power

The stored energy in the cavity is proportional to the product of the Q-value and the applied power. The cavity is cylindrical, and the resonance mode used is TM_{110}. This mode concentrates the RF power in the axial region of the cavity. The B field lines are paralell to the cavity axis. A positron emitted by the ^{22}Na isotope source at the bottom of the cavity spirals around the magnetic field lines and thus form Ps in the axial region.

For this experiment we used a resonance frequency of 2856MHz (S-band). The RF source used in this experiment consists of a signal generator and a gallium nitride amplifier . The maximum output power of the amplifier is about 600 W. The transmission path from the amplifier to the cavity consists of a waveguide which suppresses losses to low about 10%. In this measurement, 409 W of high-power RF is applied to the cavity.

The resonance performance of the cavity is described by its Q-value. The resonance curve is Lorentzian, and the Q-value is defined by the resonant frequency over the full width at half maximum (FWHM), which is equal to the stored energy over the power loss per angular frequency. We monitored the Q-values daily during the prototype run. The measured Q-values ware greater than 14,000 (the resonance curve is shown in Figure 5).

This RF system achieved a sufficient number of Ps transitions to measure Ps-HFS.

3.2. Requirement 2: Stability of the RF power

We monitored cavity reflection, RF power, and resonant frequency drift to get feedback about the power and frequency of the RF system. Figure 4 shows the power fluctuations in the amplifier gain. The green line is the power input to the amplifier, and the red line is the power output from the amplifier. Each line is independently scaled for comparison of the fluctuations. (The amplifier gain is about $+50dB$, so the absolute value of the output power is about 10^5 times larger than the input power.) The ratio of output power to input power fluctuated by a

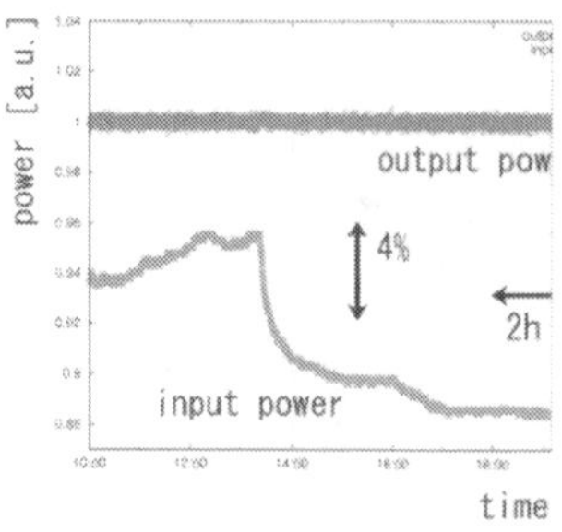

Figure 4. Normalized amplifier power (with power feedback)

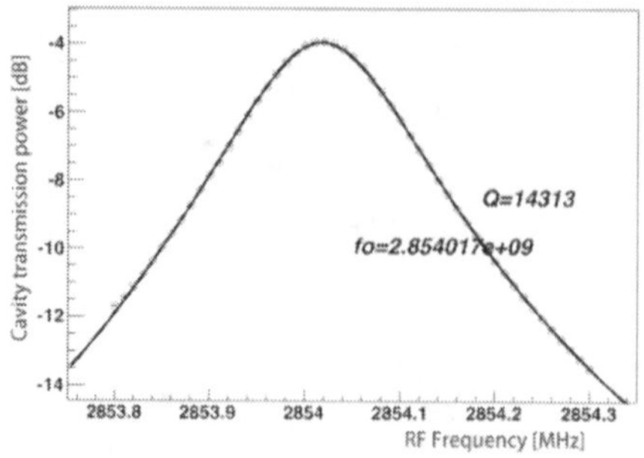

Figure 5. Resonance curve of the cavity (TM_{110})
FWHM is about 200kHz

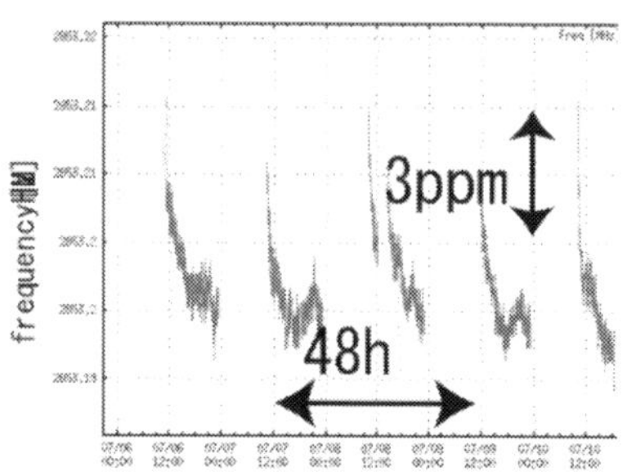

Figure 6. Drift in the resonant frequency

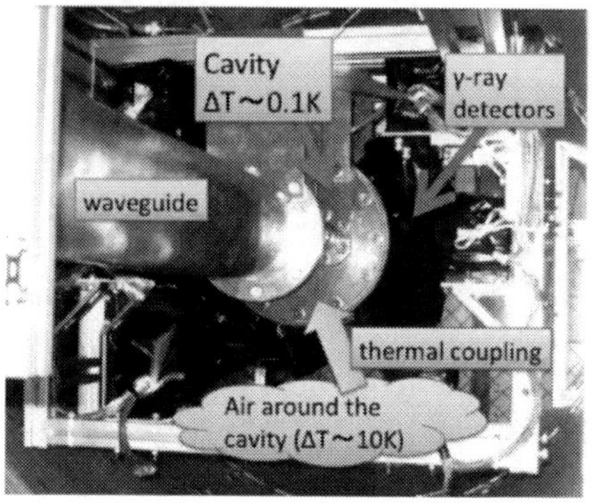

Figure 7. Cavity and thermal environments

few percents. We compensated the input power for the amplifier gain drifts thus stabiliing the output power. As a result the output power is stable enough to measure the Ps-HFS.

3.3. Requirement 3: Stability of RF frequency

The stability of the RF frequency is affected by drifting in the cavity's resonance frequency which arises from thermal expansion/contraction of the cavity. This drift is of order 6 ppm (Figure 6). Though the RF power is stable and the cavity temperature is moderated by a water chiller, the resulting stability is still inadequate. This instability is a result of drifting in the surrounding air temperature. ($\approx 10°C$). (Figure 7). Our goal is to stabilize the temperature around the cavity and reduce the resonant frequency drift to 1 ppm.

4. Conclusion and outlook

This experiment measures Ps-HFS using the Zeeman effect. Its RF system requires stable, high-power output at a stable frequency. The prototype run clarified the problems with the set-up and achieved promising progress toward the first run, as detailed below.

(i) High power: THe RF system successfully supplies the cavity with high-power RF. (409 W of power was supplied, and the Q-values of the RF cavity were greater than 14,000.)

(ii) Stable power: The output power of the amplifier was also stable, fluctuating by less than 0.1%.

(iii) Stable frequency: We will update the RF system to improve the stability of the resonant frequency.

In spring of 2010, we will start the first run, aiming at a precision on the oder of 1 ppm.

References

[1] Hoang A H, Labelle P, and Zebarjad S M 1997 *Phys. Rev. Lett.* **79** 3387-3390.
[2] Kniehl B A and Penin A A 2000 *Phys. Rev. Lett.* **85** 5094-5097.
[3] Carlson E R, Hughes V W, and Lindgern I 1976 *Phys. Rev.* A**15** 1331-1338.
[4] Ritter M W, Egan P O, Hughes V W, and Woodle K A 1984 *Phys. Rev.* A**30** 241-250.
[5] Theriot Jr. E D, Beers R H, Hughes V W, and Ziock K O H 1970 *Phys. Rev.* A**2** 707-721.
[6] Asai S, Orito S and Shinohara N 1995 *Phys. Lett.* B**357** 457-480.

Experiment for the First Direct Measurement of the Hyperfine Splitting of Positronium

A Miyazaki[1], T Suehara[2], A Ishida[1], T Namba[2], S Asai[1], T Kobayashi[2], H Saito[3], M Yoshida[4], T Idehara[5], I Ogawa[5], Y Urushizaki[5] and S Sabchevski[6]

[1] Department of Physics, Graduate School of Science, The University of Tokyo, 7-3-1 Hongo, Bunkyo-ku, Tokyo, 133-0033, Japan
[2] International Center for Elementary Particle Physics (ICEPP), The University of Tokyo, 7-3-1 Hongo, Bunkyo-ku, Tokyo, 113-0033, Japan
[3] Department of General Systems Studies, Graduate School of Arts and Sciences, The University of Tokyo, 3-8-1 Komaba, Meguro-ku, Tokyo, 153-8902 , Japan
[4] Accelerator Laboratory, High Energy Accelerator Research Organization (KEK), 1-1 Oho, Tsukuba, Ibaraki, 305-0801, Japan
[5] Research Center for Development of Far-Infrared Region, University of Fukui (FIR-FU), 3-9-1 Bunkyo, Fukui, Fukui, 910-8507, Japan
[6] Bulgarian Academy of Science, 1, 15 Noemvri Str., 1040 Sofia, Bulgaria

E-mail: miyazaki@icepp.s.u-tokyo.ac.jp

Abstract. Positronium is an ideal system for the research of the bound state QED. The hyperfine splitting of positronium (Ps-HFS: about 203 GHz) is a good tool to test QED and also sensitive to new physics beyond the Standard Model via a quantum oscillation between an ortho-Ps and a virtual photon. Previous experimental results show $3.9\,\sigma$ (15 ppm) discrepancy from the QED calculation. All previous experiments used an indirect method with static magnetic field to cause Zeeman splitting (a few GHz) between triplet states of ortho-Ps, from which the HFS value was derived. One possible systematic error source of the indirect method is non-uniformity of the static magnetic field. We are developing a new direct Ps-HFS measurement system without static magnetic field. In this measurement we use a gyrotron, a novel sub-THz light source, with a high-finesse Fabry-Pérot cavity to obtain enough radiation power at 203 GHz. The present status of the optimization studies and current design of the experiment are described.

1. Introduction

Positronium (Ps), the electron-positron bound state, is a purely leptonic system. The energy difference between ortho-positronium (o-Ps, $^3S_1$ state) and para-positronium (p-Ps, $^1S_0$ state), hyperfine splitting of positronium (Ps-HFS), is a good target to verify bound state QED precisely. The Ps-HFS value is approximately 203 GHz (0.84 meV), which is significantly larger than hydrogen HFS (1.4 GHz). One reason for the large Ps-HFS is the quantum oscillation: o-Ps $\rightarrow \gamma^* \rightarrow$ o-Ps (o-Ps has the same quantum number as a photon). Since some hypothetical particles (such as a millicharged particle) can participate in the quantum oscillation, resulting in a shift of Ps-HFS value, its precise measurement provides a probe for new physics beyond the Standard Model.

Measurements of the Ps-HFS have been performed in 70's and 80's with the combined precision of 3.3 ppm[1, 2]. The results were consistent with 1st and 2nd order calculation of the QED available at that time. However, the 3rd order corrections have been calculated recently, with the new prediction of 203.39169(41) GHz[3]. This differs from the measured value of 203.38865(67) GHz by 3.9σ, 15ppm. This discrepancy may come from new physics or the common systematic errors in the previous measurements.

In all previous measurements, the Ps-HFS transition was not directly measured, since 203 GHz was too high frequency to produce and control. They measured Zeeman splitting of o-Ps instead. A static magnetic field makes Zeeman mixing between $m_z = 0$ state of o-Ps and p-Ps, the resultant energy level of $m_z = 0$ state becomes higher than $m_z = \pm 1$ (Zeeman splitting). This Zeeman splitting, which is approximately proportional to the HFS energy level, is a few GHz frequency if ~ 1 Tesla magnetic field is applied. They applied uniform magnetic field in RF cavities where positronium was produced, causing Zeeman transition and resultant o-Ps to p-Ps transition. However, uncertainty (especially non-uniformity) of the static magnetic field can be a large source of systematic error of HFS.

In contrast to the indirect method, we plan to directly measure the HFS transition, which does not need a static magnetic field, and is thus free from the systematic error from the field. To cause the direct transition, a powerful 203 GHz radiation field is necessary. We are developing sub-THz to THz light source called a gyrotron, and also a high-finesse Fabry-Pérot cavity to accumulate sub-THz photons for the direct HFS measurement.

2. Gyrotron

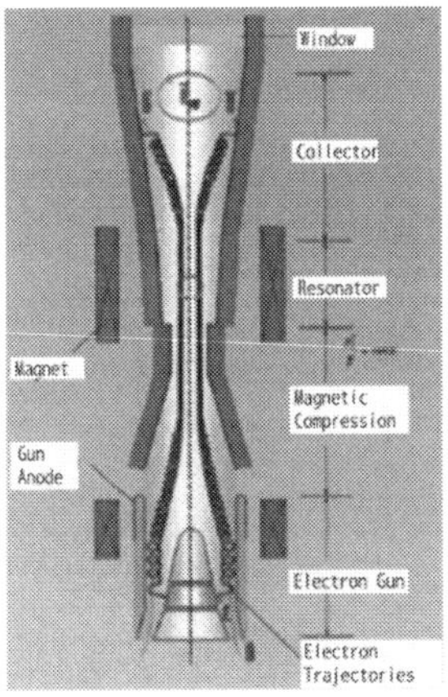

Figure 1. Schematic of gyrotrons.

Figure 2. Picture of the gyrotron for the HFS experiment.

The gyrotron[4] is a novel high power source for sub-THz to THz frequency region. The structure of gyrotrons is shown in Fig. 1. The electrons are produced and accelerated at the DC electron gun, concentrated and rotated as cyclotron motion in the superconducting magnet. The cyclotron frequency f_c is given as

$$f_c = \frac{eB}{2\pi m_0 \gamma}, \tag{1}$$

where B is the magnetic field strength, m_0 is the electron rest mass and γ is the relativistic factor of the electron. A cavity is placed at the maximum magnetic field, whose resonant frequency is tuned just to the cyclotron frequency to enhance the monochromatic light. The electrons stimulate resonance of the cavity and produce coherent photons at the cavity. The photons are guided to the output port through the window, while electrons are dumped at the collector.

We developed a gyrotron operating at $f_c = 203$ GHz with $B = 7.425$ Tesla, $\gamma \sim 1.02$, shown in Fig. 2. The obtained power is 609 W at output of Gyrotron, which is reduced to 440 W during transmission through the waveguide system to the positronium cavity. The frequency width is determined by B uniformity and γ spread by thermal distribution of electrons, and is expected to be less than MHz, which is narrow enough to make resonance at the Fabry-Pérot cavity. Measured result with a similar gyrotron shows the frequency width is less than 10 kHz[5]. The frequency can be tuned by changing the γ factor with different acceleration of electrons, but the tuning range is limited by the resonant width of the cavity to several handreds of megahertz. The power of the gyrotron will be monitored using pyroelectric detectors to account for corrections to the transition probability.

3. Fabry-Pérot cavity

Photons produced at the gyrotron are transported and accumulated in a cavity shown in Fig.3. Since 203 GHz ($\lambda = 1.475$ mm) photons can be treated optically at the centimeter (or larger) size scale, we plan to use a Fabry-Pérot cavity, consisting of two opposing mirrors to confine photons between them. Unlike RF cavities, the confinement in the Fabry-Pérot cavity is 1-dimensional while the other four sides are open. A metal-mesh mirror is used on the input side of the cavity to pass photons from Gyrotron, and a copper concave mirror is used on the other side.

Table 1. Mesh parameters with simulation and measurement result. reflectance and transmittance are from estimations by the simulation, and finesse is from results of the measurements.

mesh material	line width	line separation	reflectance	transmittance	finesse
gold	20 μm	50 μm	99.3%	0.32%	650
gold	10 μm	50 μm	98.6%	0.75%	290
silver	50 μm	130 μm	96.9%	2.70%	180

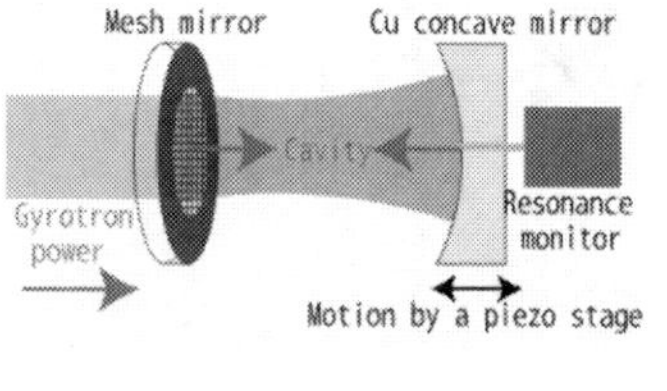

Figure 3. Overview of the cavity.

Figure 4. Picture of setup of the cavity test.

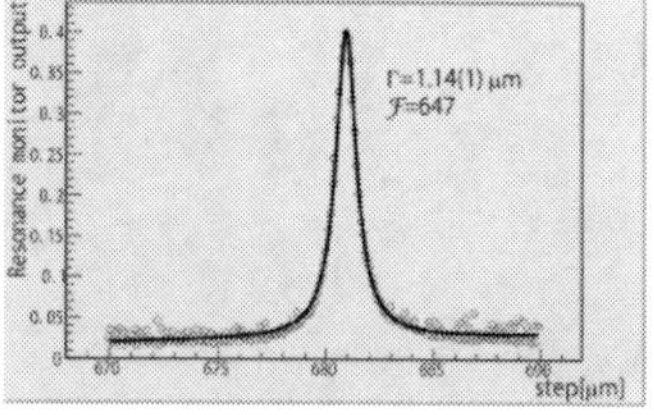

Figure 5. Resonance shape obtained by the test.

The two most important characteristics of cavities are "finesse ($\mathcal{F}$)" and "input coupling". Finesse can be given as $\mathcal{F} = 2\pi/(1-\rho)$, where ρ is the fraction of power left after one round-trip, characterizing the capability of the cavity to store photons inside. To maximize the $\mathcal{F}$, power losses must be minimized, and there are 3 type of loss, diffraction loss, medium loss and ohmic loss. With the confinement of photons by the concave mirror and gas medium, diffraction and medium loss are negligible in our cavity. Ohmic loss occurs at the mirrors, which is around 0.15% at the copper mirror and more at the mesh mirror. The ohmic loss at the mesh mirror depends on mesh parameters, and the calculated results are listed in Table 1. Input coupling is the fraction of input power introduced to the cavity mode. It is an important parameter to introduce photons efficiently into the cavity. In our cavity, the input coupling is mainly determined by transmittance of the input mesh mirror, which is also calculated by the simulation.

The finesse and the input coupling were measured using various mesh and concave mirror parameters. Figure 3 and 4 show the test setup, which is consisted of a mesh mirror on the mirror mount and a concave mirror on a piezo stage. Input, transmitted and reflected powers were monitored by three pyroelectric power monitors. By shifting cavity length precisely by the piezo stage, Breit-Wigner resonance shape was observed in the transmit power monitor as shown in Fig.5. Finesse could be obtained from the width of the resonance, and the input coupling was observed with the reflection monitor. With current best combination of the mesh and the concave mirror, $\mathcal{F} = 650$ was obtained (see Figure 5). For the input coupling, concrete value could not be obtained because of interference of the reflection power and difficulty to determine absolute power of the reflection/transmission because of non-optimal setup of the power measurements. Input coupling of $\sim 20\%$ can be estimated with the reflection data.

4. Detection System

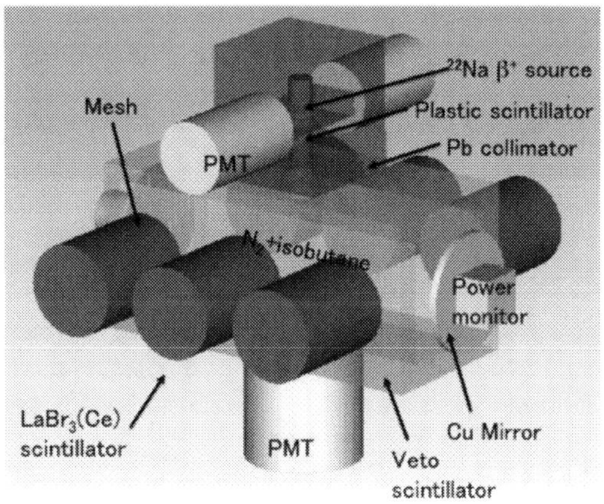

Figure 6. Overview of the detection system.

Figure 6 shows a schematic view of the whole system in preparation now. Gyrotron power is introduced to the cavity via the mesh mirror, and accumulated between the mesh and the Cu mirror. The ^{22}Na β^+-source is located 30 mm above the cavity. The emitted positrons pass through a β-tag scintillator(t$\sim$200mum) to generate start timing. A lead collimator (10 mm length, 15 mm aperture) and a veto scintillator to reach the cavity are introduced to select β^+ stopped in gas. The lead collimator also works as a shield to protect $LaBr_3$(Ce) scintillators from accideltal photons (mainly 1275 keV and 511 keV photons emitted around the source.

The cavity is filled with mixed gas of nitrogen and isobutane to form positronium atoms (20% efficiency). Para-positronium annihilates into two 511 keV photons immediately, while o-Ps remains with $\tau \sim 142$ ns and decays into three photons (< 511 keV). Six $LaBr_3$ scintillators surround the cavity to detect photons with high energy resolution of $\sim$ 4%(FWHM511keV).

Two photon-decay can be separated from three photon-decay easilly using energy information. Two photon-decay will increase when photon frequency from Gytorton is close to HFS. We will obtain Breit-Wigner resonace as a function of frequecy of photon, and HFS will be obtained directly from the center of Breit-Wigner resonace curve. The $LaBr_3$ scintillators have good timing resolution of $\sim$ 300 psec to separate delayed events (signal) from prompt events (positron annihilation) to improve signal ratio significantly.

5. Summary

We are preparing the first experiment probing Ps-HFS with direct method to investigate discrepancy of Ps-HFS value between theory and experiment. We have developed a high power 203 GHz radiation source gyrotron and a Fabry-Pérot cavity with a metal-mesh mirror and a Cu concave mirror. We are now assembling the detection system, and the first data taking is scheduled in about a year.

References

[1] Mills A P Jr and Bearman G H 1975 *Phys. Rev. Lett.* **34** 246, Mills A P Jr 1983 *Phys. Rev.* A **27** 262
[2] Ritter M W, Egan P O, Hughes V W and Woodle K A 1984 *Phys. Rev.* A **30** 1331
[3] Kniehl B A and Penin A A 2000 *Phys. Rev. Lett.* **85** 5094
[4] Idehara T, Ogawa I, Mitsudo S, Pereyaslavets M, Nishida N and Yoshida K 1999 *IEEE Trans. Plasma Sci.* **27** 340
[5] Idehara T, Saito T, Ogawa I, Mitsudo S, Tatematsu Y, La Agusu, Mori H and Kobayashi S 2008 *Appl. Magn. Reson.* **34** 265

Measurement of Positronium Hyperfine Splitting with Quantum Oscillation

Y Sasaki[1], A Miyazaki[1], A Ishida[1], T Suehara[1], T Namba[1], S Asai[1], T Kobayashi[1], H Saito[2], M Ikeno[3], K Tanaka[3], A Yamamoto[3]

[1]Department of Physics and ICEPP, The University of Tokyo, 7-3-1 Hongo, Bunkyo-ku, Tokyo, 133-0033, Japan

[2] Institute of Physics, University of Tokyo, 3-8-1 Komaba, Meguro-ku, Tokyo, 153-8902, Japan

[3] High Energy Accelerator Research Organization (KEK), 1-1 Oho, Tsukuba, Ibaraki, 305-0801, Japan

E-mail: ysasaki@icepp.s.u-tokyo.ac.jp

Abstract. It is well known that a coherent system can bring a quantum interference between different energy eigenstates into an observable oscillation. Since an energy difference and a frequency are related with each other by the equation $\Delta E = \hbar\omega$, we can make use of the oscillation to measure the energy difference of quantum states. In this experiment, the hyperfine splitting of positronium is measured using the quantum oscillation. The result is 203.324±0.044(stat.)±0.028(sys.) GHz. This value is consistent with the theoretical calculations.

1. Introduction

Positronium(Ps) is a bound state of the electron and the positron. It has triplet states and singlet state. The triplet state is called "Ortho-Positronium (o-Ps)" and the singlet state is called "Para-Positronium (p-Ps)". O-Ps has three degenerated eigenstates $|s=1,m_z=\pm1\rangle$,$|s=1,m_z=0\rangle$. Because of the charge conjugation conservation in the electromagnetic interactions, o-Ps mainly decays into three gamma rays, which leads to the relatively longer lifetime of 142 ns in vacuum. On the other hand, p-Ps has only one eigenstate $|s=0,m_z=0\rangle$ and shorter lifetime of 125 ps, mainly decaying into two 511 keV-gamma rays.

The energy difference between o-Ps and p-Ps is called "Positronium Hyperfine Splitting (Ps HFS)", which is about 0.84 meV (203.4 GHz). It has been measured in several experiments, mostly using Zeeman effect as follows; Figure 1 shows the energy level of Ps as function of a static magnetic field. In a static magnetic field, $|s=1,m_z=0\rangle$ and $|s=0,m_z=0\rangle$ get mixed into two new eigenstates $|+\rangle$,$|-\rangle$, while $|s=1,m_z=\pm1\rangle$ states stay in the same energy. The energy difference between $|+\rangle$ and $|s=1,m_z=\pm1\rangle$ is,

$$\Delta_{\text{mix}} = \frac{\Delta_{\text{HFS}}}{2}(\sqrt{1+\chi^2}-1), \tag{1}$$

where $\chi = \frac{g'\mu_B H}{\Delta_{\text{HFS}}}$, H is the magnetic field strength, Δ_{HFS} is Ps HFS, μ_B is the Bore magneton and g' is the electron spin g-factor with correction for bound systems. Since χ is small in a weak magnetic field, we can assume that Δ_{mix} is approximately proportional to the HFS. Δ_{mix} can be measured with a few GHz-RF wave. This Zeeman effect was used in the previous measurements performed by Mills [1] and Ritter [2] and $\Delta_{\mathrm{HFS}} = 203.38865(67)$ was obtained. On the other hand, V.G.Baryshevsky [4] proposed to make use of the quantum oscillation between these states instead of observing the transition, and later, he observed the oscillation experimentally [5]. In 1996, S.Fan [6] measured Ps HFS with this method, obtaining 202.5 ± 3.5 GHz.

Positrons emitted from β^+ source are polarized to their momentum-direction because of Parity-violation in weak interaction. Higher polarization can be obtained with ^{68}Ge-Ga source whose endpoint energy is 1.9 MeV (polarization ratio $P = \beta$). When the positron stops in a material, it forms polarized positronium which is in a superposed state of $|+\rangle$ and $|s=0, m_s = \pm 1\rangle$. Such a superposition oscillates with the frequency corresponding to the energy difference. In this experiment, we produce a coherent superposition of $|+\rangle$ and $|s=0, m_s=\pm 1\rangle$ efficiently using polarized β^+ perpendicularly to the magnetic field.

Oscillation will be observed as follow.

Let us take the momentum direction of the positron as x-axis and that of the magnetic field as z-axis. We define $\theta = \arccos(\frac{z}{\sqrt{x^2+y^2}})$ and $\phi = \arctan(\frac{y}{x})$. The quantum oscillation modulates the angular distribution of three gamma rays emitted from o-Ps decay (Figure 2). As a result, the decay curve of o-Ps beats with the oscillation. The oscillating amplitude is the biggest at $(\theta,\phi) = \left(\frac{\pi}{4},\frac{\pi}{2}\right), \left(\frac{3\pi}{4},\frac{\pi}{2}\right), \left(\frac{3\pi}{4},-\frac{\pi}{2}\right), \left(\frac{\pi}{4},-\frac{\pi}{2}\right)$ (the adjacent pairs have the opposite phases). Unlike the muon precession, in which the decay amplitude of $\mu \to e$ rotates, this oscillation changes its angular distribution as a "vibration" in the yz-plane. This property is unique to Spin-1 systems.

In general, the decay curve of o-Ps which is observed by a detector with finite acceptance can be written as,

$$\begin{aligned} f(t) &= C_1 e^{-\gamma_1 t} + C_2 e^{-\gamma_2 t} \\ &+ C_3 e^{-\frac{\gamma_1+\gamma_2}{2}t} \times \sin(\Omega t + \theta_0) + C_4 , \end{aligned} \tag{2}$$

where γ_1 is the decay rate of $|+\rangle$, γ_2 is that of $|s=1, m_z = \pm 1\rangle$ and Ω is the angular frequency of the oscillation. C_3 is the oscillation depth, which is proportional to the initial polarization ratio of positrons. Since $|+\rangle$ state can decay into two gamma rays, γ_1 is larger than γ_2.

We can calculate Ps HFS from Ω and the magnetic field strength.

2. Experimental Setup

Figure 3 shows a schematic diagram of the setup with a coordinate axis and size, a photograph of vacuum chamber is shown in Figure 4. We use ^{68}Ge-Ga positron source (30 kBq) which is sealed with t=13 μm Titanium. A positron emitted to the x-direction is tagged by the plastic scintillator (t=500 μm, NE102), producing a light pulse which is directed to two photomultipliers through the light guide. The positron stops in the silica aerogel (0.11 g/cm^3, ϕ=10 mm, d=10 mm) and forms a positronium. Ps-formation assembly is put in a vacuum chamber in order to reduce the pick-off effect.

Gamma rays emitted from o-Ps decay are detected by six $LaBr_3$(Ce) scintillators (ϕ=1.5 inch, d=2 inch) as shown in Figure 5, 6. Four of them are placed at $(\theta,\phi) = \left(\frac{\pi}{4},\frac{\pi}{2}\right), \left(\frac{3\pi}{4},\frac{\pi}{2}\right), \left(\frac{3\pi}{4},-\frac{\pi}{2}\right), \left(\frac{\pi}{4},-\frac{\pi}{2}\right)$ to observe the oscillating curves. The rest of the detectors are placed at $(\theta,\phi) = \left(\frac{\pi}{2},\frac{\pi}{2}\right), \left(\frac{\pi}{2},-\frac{\pi}{2}\right)$ to observe the decay curve without oscillation. The average energy resolution of the $LaBr_3$ detectors is 4% (FWHM) at 511 keV in the magnetic field strength of 100mT. The timing resolution of 511keV-511keV coincidence between two $LaBr_3$ detectors is 200 ps (FWHM). However, because timing resolution of plastic scintillator is limited in our setup, β-γ timing resolution is 3.8 ns (FWHM) at 511 keV in the magnetic field strength of 100 mT.

The magnetic field is provided with a superconducting magnet in KEK, which has an excellent uniformity of 20 ppm over the volume of the silica aerogel. The magnetic field strength is

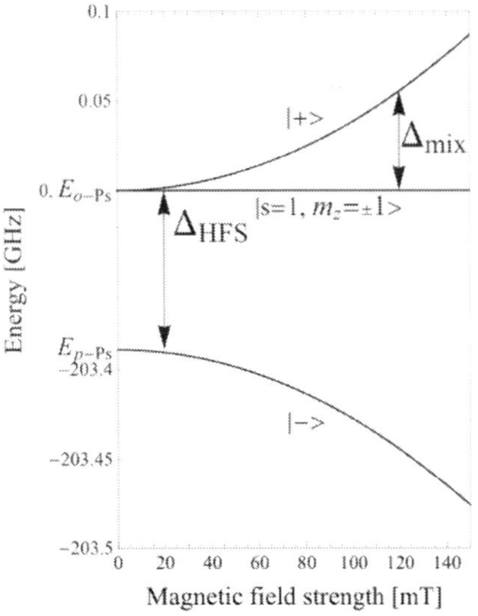

Figure 1. Energy level of positronium in its ground state as function of the static magnetic field strength. Δ_{HFS} represents the HFS and Δ_{mix} shows the Zeeman shift between $|+\rangle$ and $|s = 1, m_s = \pm 1\rangle$.

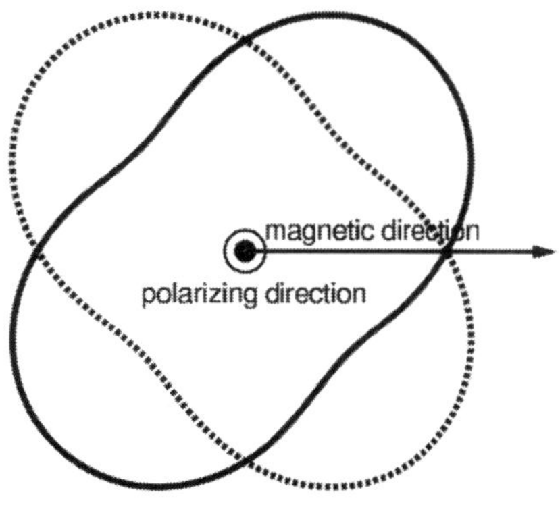

Figure 2. Angular distribution of the gamma rays emitted from o-Ps decay. The radius stands for the amplitude of emitting direction. It changes with time as is shown in the solid and dotted lines. The magnetic direction goes from left to right, the polarizing direction is perpendicular to the sheet.

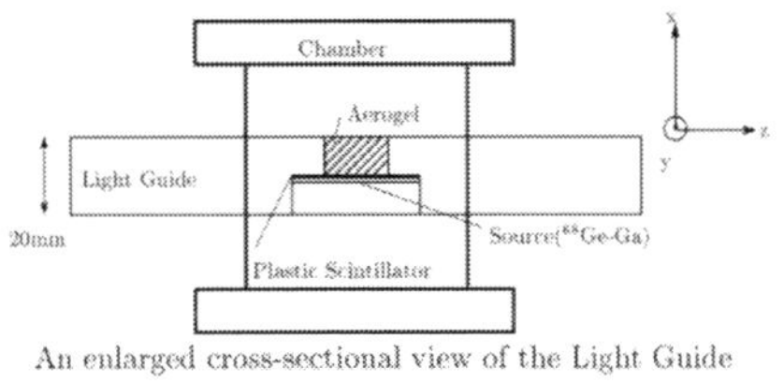

Figure 3. Schematic diagram of the vacuum chamber. β^+ emitted from ^{68}Ge-Ga passes through the plastic scintillator, producing a start signal. β^+ stops in the aerogel and forms a positronium.

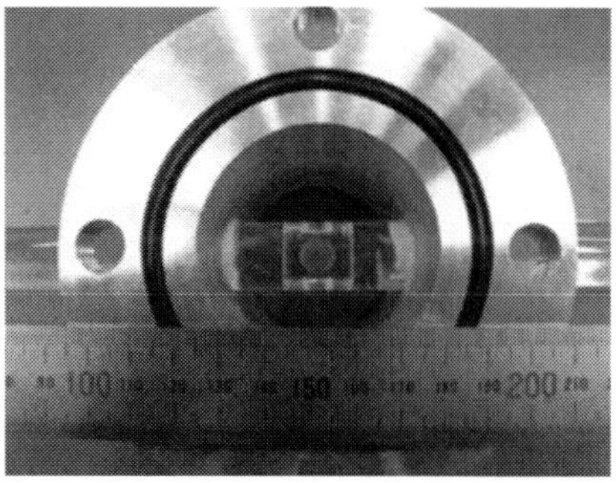

Figure 4. Photograph of the vacuum chamber.

measured with a NMR magnetometer. Uncertainty of the absolute calibration of the magnetic field strength is 70 ppm.

Data acquisition is started when at least one signal of the $LaBr_3$ scintillator and that of the plastic scintillator coincide within -50 ns to 1650 ns. The timing information of the plastic and $LaBr_3$ scintillators are measured by direct clock TDCs, which have good integral and differential linearities. Start signal is produced with β-timing and stop signal is with γ-timing. These TDCs can work on as fast as 8 GHz clock, but most of data-acquisition are performed on 5 GHz clock to obtain good performance.

The experiments are conducted in the magnetic field strength of 0 mT, 100 mT, 118 mT, 135 mT, 138 mT with both $+z$ polarization and $-z$ polarization by changing the emitting

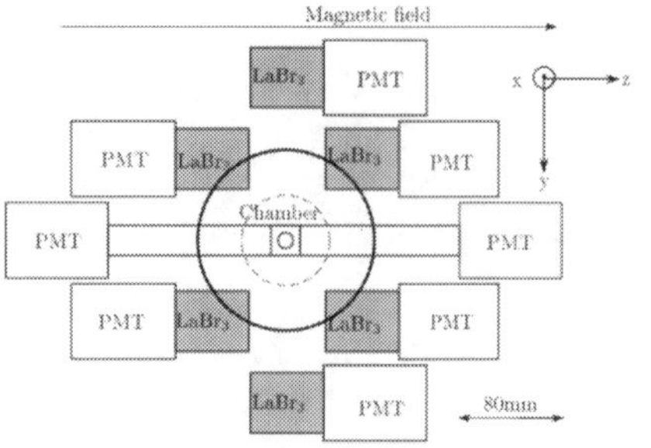

Figure 5. Schematic diagram of the gamma detectors. There are three diagonal pairs for observing the normal phase oscillation, the opposite phase oscillation and the decay curve without oscillation. The vacuum chamber is placed at the center.

Figure 6. Photograph of the gamma detectors.

direction of β^+. The period of each run is about 3 days and total data acquiring period is about 22 days with 740 Hz of DAQ rate. The trigger rate is about 3.0 kHz.

3. Analysis

Detectors at $(\theta, \phi) = \left(\frac{\pi}{4}, \frac{\pi}{2}\right), \left(\frac{3\pi}{4}, \frac{\pi}{2}\right), \left(\frac{3\pi}{4}, -\frac{\pi}{2}\right), \left(\frac{\pi}{4}, -\frac{\pi}{2}\right), \left(\frac{\pi}{2}, \frac{\pi}{2}\right), \left(\frac{\pi}{2}, -\frac{\pi}{2}\right)$ are indexed $1 \sim 6$ respectively. The pairs of 1 st and 3 rd detectors, 2 nd and 4 th detectors observe the oscillation with the same phases. 5 th and 6 th detectors observe the exponential decay curve without oscillation. χ^2 function is defined to fit the histograms of all the 6 detectors at once. The fitting function for n-th detector is,

$$\begin{aligned} f(t,n) &= A_n e^{-\gamma_1 t} + B_n e^{-\gamma_2 t} + C_n e^{-\frac{\gamma_1+\gamma_2}{2}t} \times \sin(\Omega t + \theta_n) + D_n \qquad (3) \\ &\quad \text{for } n = 1,2,3,4\,, \\ f(t,n) &= A_n e^{-\gamma_1 t} + B_n e^{-\gamma_2 t} + D_n \qquad (4) \\ &\quad \text{for } n = 5,6\,. \end{aligned}$$

The two decay rates γ_1, γ_2 and the angular frequency of the oscillation Ω are defined as common variables (see equation (2)), and the others are defined independently. The fitting range is 30 ns to 1430 ns. Figure 7 shows an example of the timing histograms with the best fit functions.

Figure 8 shows the oscillating period($\frac{2\pi}{\Omega}$) as function of magnetic field strength. The black line is the best fit result. The combined result is

$$\begin{aligned} \text{HFS} = 203.324 \;&\pm\; 0.044(\text{stat.})(215\,\text{ppm}) \\ &\pm\; 0.028(\text{sys.})(140\,\text{ppm})\,\text{GHz} \qquad (5) \end{aligned}$$

Main source of the systematic error is absolute calibration of the NMR magnetometer (140 ppm). Accuracy of this measurement is improved by the factor of 70 than the previous measurement conducted by S.Fan [6].

4. Conclusion

Ps HFS is measured with its quantum oscillation. We have used the uniform magnetic field, β^+ source which has high endpoint energy and experimental setup which has good timing resolution

Journal of Physics: Conference Series **225** (2010) 012045 doi:10.1088/1742-6596/225/1/012045

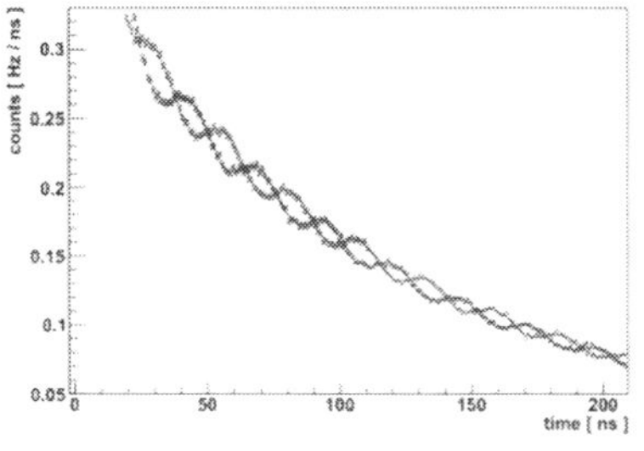

Figure 7. Timing histogram obtained in the run of 100 mT, $+z$ polarization. Red line shows the decay curve with normal phase and blue line shows the one with the opposite phase. The black lines are the best fit results.

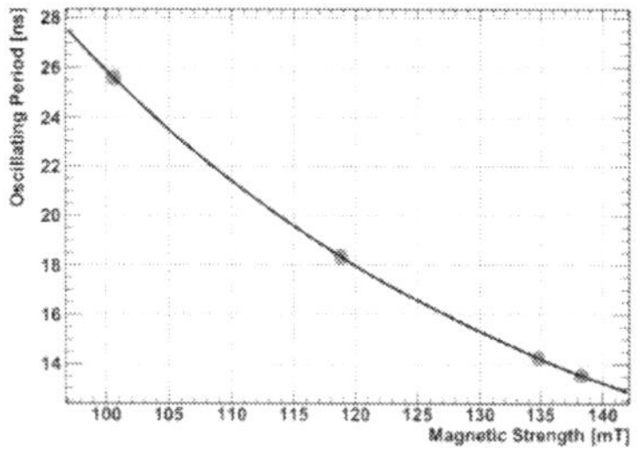

Figure 8. Measured oscillating periods in the experiment. Red circles show the data points. Errorbars are smaller than the circles. The black line is the best fit result.

to obtain big oscillating amplitude. The result is HFS $= 203.324 \pm 0.044(\text{stat.})(215\,\text{ppm}) \pm 0.028(\text{sys.})(140\,\text{ppm})$ GHz, which improves the result of the previous experiment conducted with the same method by the order of two. This result agrees with the theoretical calculations.

There have reported several applications using quantum oscillation of positronium as a probe in material science. See [7], for example.

References

[1] Mills A P Jr and Bearman G H 1975 *Phys. Rev. Lett.* **34** 246, Mills A P Jr 1983 *Phys. Rev.* A **27** 262.
[2] Ritter W M 1984 *Phys. Rev.* A **30** 1331.
[3] Melnikov K 2001 *Phys. Rev. Lett.* **86** 1498.
[4] Baryshevsky G V 1989 *J. Phys.* B **22** 2835.
[5] Baryshevsky G V 1989 *J. Phys.* A **136** 428.
[6] Fan S 1996 *Phys. Rev.* A **216** 129.
[7] Ivanov E 2008 *Appl. Surf. Sci.* **225** 179.

Precise measurement of HFS of positronium using Zeeman effect

A Ishida[1], G Akimoto[1], Y Sasaki[1], A Miyazaki[1], K Kato[1], T Suehara[1], T Namba[1], S Asai[1], T Kobayashi[1], H Saito[2], M Yoshida[3], K Tanaka[3], A Yamamoto[3], Y Urushizaki[4], I Ogawa[4], T Idehara[4] and S Sabchevski[5]

[1] Department of Physics and ICEPP, the University of Tokyo, 7-3-1 Hongo, Bunkyo-ku, Tokyo, 113-0033, Japan

[2] Institute of Physics, the University of Tokyo, 3-8-1 Komaba, Meguro-ku, Tokyo, 153-8902, Japan

[3] High Energy Accelerator Research Organization (KEK), 1-1 Oho, Tsukuba, Ibaraki, 305-0801, Japan

[4] Research Center for Development of Far Infrared Region, University of Fukui, 3-9-1 Bunkyo, 910-8507 Fukui, Japan

[5] Institute of Electronics of the Bulgarian Academy of Sciences, 1784 Sofia, Bulgaria

E-mail: ishida@icepp.s.u-tokyo.ac.jp

Abstract. The ground state hyperfine splitting of positronium, $\Delta_{\rm HFS}$, is sensitive to high order corrections of QED. A new calculation up to $O(\alpha^3)$ has revealed a $3.9\,\sigma$ discrepancy between the QED prediction and the experimental results. This discrepancy might either be due to systematic problems in the previous experiments or to contributions beyond the Standard Model. We propose an experiment to measure $\Delta_{\rm HFS}$ employing new methods designed to remedy the systematic errors which may have affected the previous experiments. Our experiment will provide an independent check of the discrepancy. The measurement is in progress and a result of $\Delta_{\rm HFS} = 203.385 \pm 0.011$ GHz (58 ppm) has been obtained from the prototype run. A measurement with a precision of O(ppm) is expected within a few years.

1. Introduction

Positronium (Ps), a bound state of an electron and a positron, is a purely leptonic system which allows for very sensitive tests of QED. The precise measurement of the hyperfine splitting between orthopositronium (o-Ps, 1^3S_1) and parapositronium (p-Ps, 1^1S_0) (Ps-HFS) provides a good test of bound state QED. Ps-HFS is expected to be relatively large (for example compared to hydrogen HFS) due to a relatively large spin-spin interaction, and also due to the contribution from vacuum oscillation (o-Ps $\to \gamma^* \to$ o-Ps). The contribution from vacuum oscillation is sensitive to new physics beyond the Standard Model.

Figure 1 shows the measured and theoretical values of Ps-HFS. The combined value from the results of the previous 2 experiments is $\Delta_{\rm HFS,exp} = 203.388\,65(67)$ GHz (3.3 ppm) [1, 2]. Recent developments in nonrelativistic QED (NRQED) have added $O(\alpha^3)$ corrections to the theoretical prediction which now stands at $\Delta_{\rm HFS,th} = 203.391\,69(41)$ GHz (2.0 ppm) [3]. The discrepancy of 3.04(79) MHz (15 ppm, $3.9\,\sigma$) between $\Delta_{\rm HFS,exp}$ and $\Delta_{\rm HFS,th}$ might either be due to the common

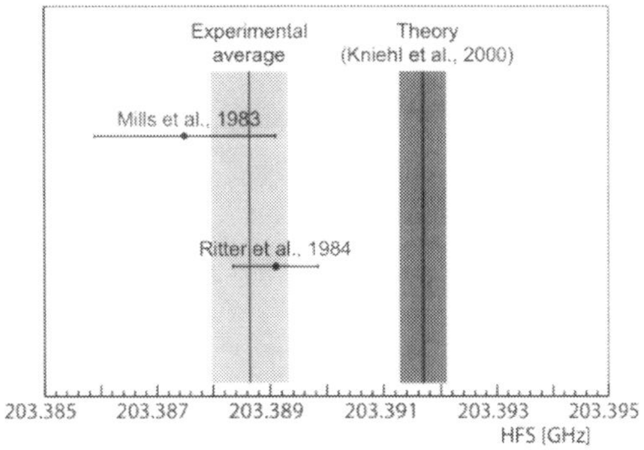

Figure 1. Measured and theoretical values of Ps-HFS.

systematic uncertainties in the previous experiments or to new physics beyond the Standard Model.

There are two possible common systematic uncertainties in the previous experiments. One is the unthermalized o-Ps contribution which results in an underestimation of the material effect. This effect has already been shown to be significant [4]. The other is the uncertainty in the magnetic field uniformity which was cited as the most significant systematic error by previous experimenters.

2. Experimental setup

The energy levels of the ground state of Ps are shown as a function of static magnetic field in Figure 2. Due to technical difficulties in directly stimulating Δ_{HFS}, we make an indirect measurement by stimulating the transition Δ_{mix}. This is the same approach as previous experiments. The relationship between Δ_{HFS} and Δ_{mix} is given by the Breit-Rabi equation

$$\Delta_{\mathrm{mix}} = \frac{1}{2}\Delta_{\mathrm{HFS}}\left(\sqrt{1+x^2}-1\right), \quad (1)$$

in which $x = 2g'\mu_B H/h\Delta_{\mathrm{HFS}}$. $g' = g\left(1-\frac{5}{24}\alpha^2\right)$ is the g factor for a positron (electron) in Ps [5], μ_B is the Bohr magneton, H is the static magnetic field, and h is the Plank constant.

In a static magnetic field, the p-Ps state mixes with the $m_z = 0$ substate of o-Ps hence the $|+>$ state annihilates into 2 γ-rays with a lifetime of about 8 ns (with our experimental conditions). The $m_z = \pm 1$ substates of o-Ps annihilate into 3 γ-rays with a lifetime of about 140 ns. When a microwave field with a frequency of Δ_{mix} is applied, transitions between the $|+>$ state and $m_z = \pm 1$ substates of o-Ps are induced so that the 2 γ-ray annihilation rate increases. This increase is our experimental signal.

The prototype run of the measurement has been performed. A schematic diagram of the experimental setup of the prototype run is shown in Figure 3. Following are the main improvements in our experiment which we expect will significantly reduce systematic errors present in previous experiments.

2.1. Large bore superconducting magnet (Figure 4)

A large bore superconducting magnet is used to produce the magnetic field (~ 0.866 T) which induces the Zeeman splitting. The bore diameter of the magnet is 800 mm, and its length is 2 m. The large bore diameter means that there is good uniformity in the magnetic field in the region where Ps is formed (11 ppm without utilization of any compensation).

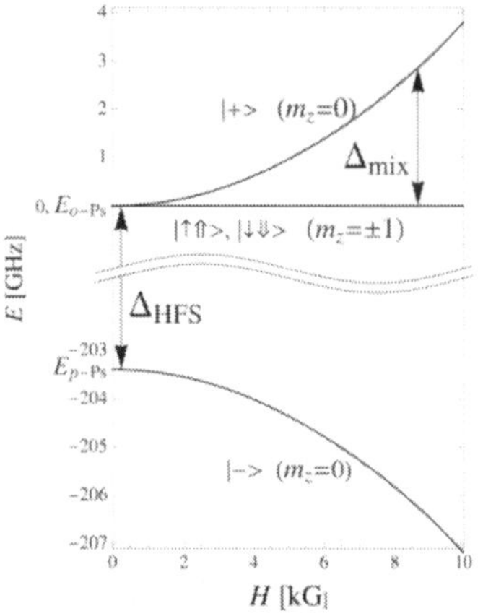

Figure 2. Zeeman energy levels of Ps in its ground state.

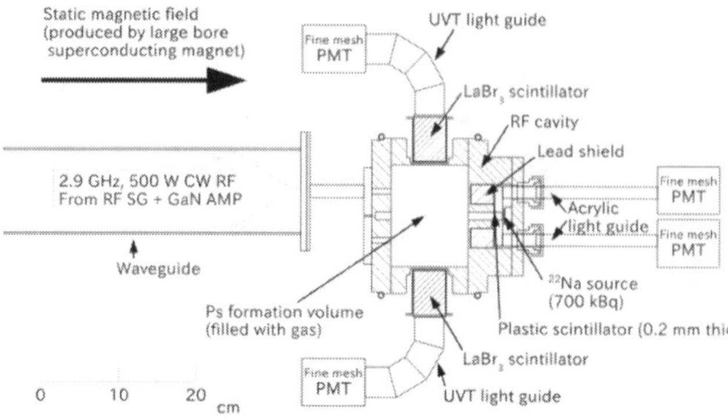

Figure 3. Schematic diagram of the experimental setup of the prototype run (top view in magnet).

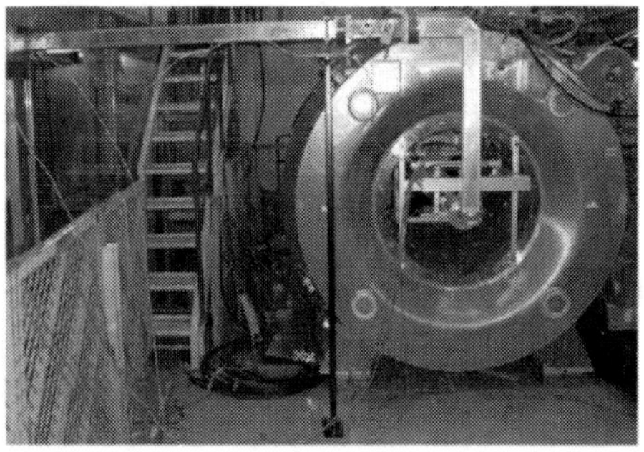

(a) The large bore superconducting magnet and the microwave waveguide. Microwaves are guided through the waveguide into the cavity.

(b) Inside of the magnet. The β-tagging system, the γ-ray detectors and the microwave cavity are located at the center of the magnet.

Figure 4. Photographs of the experimental setup of the prototype run.

Furthermore, the magnet is operated in persistent current mode, making the stability of the magnetic field better than 1 ppm.

2.2. β-tagging system and timing information

The positron source is 19 μCi (700 kBq) of ^{22}Na. A plastic scintillator 10 mm in diameter and 0.2 mm thick is used to tag positrons emitted from the ^{22}Na. The scintillation light is detected by photomultipliers and provides a start signal which corresponds to the time of Ps formation.

The positron then enters the microwave cavity, forming Ps in the N_2 gas contained therein.

Ps decays into photons that are detected with $LaBr_3$ (Ce) scintillators. Accumulating measurements of the times of positron emission and γ-detection results in decay curves of Ps as shown in Figure 5. The timing information is used to improve the accuracy of the measurement of Δ_{HFS} as follows:

(i) Imposing a time cut means that we can select well thermalized Ps, reducing the unthermalized o-Ps contribution. It should also be possible to precisely measure the contributions of unthermalized o-Ps, and of material effects (we plan to make such measurements in future runs).

(ii) A time cut also allows us to avoid the prompt peak (contributions of simple annihilation and of fast p-Ps decay), which greatly increases the S/N of the measurement.

2.3. High performance γ-ray detectors

Six γ-ray detectors are located around the microwave cavity to detect the 511 keV annihilation γ-rays. $LaBr_3$ (Ce) scintillators, 1.5 inches in diameter and 2 inches long are used. $LaBr_3$ (Ce) scintillators have good energy resolution (4 % FWHM at 511 keV) and timing resolution (0.2 ns FWHM at 511 keV), and have a short decay constant (25.6 ns). The good energy resolution and the high counting rate of $LaBr_3$ results in very good overall performance for measuring 2γ decays. In particular the good energy resolution allows us to efficiently separate 2γ events from 3γ events, negating the need to use a back-to-back geometry to select 2γ events, thus greatly increasing the acceptance of our setup.

This γ-ray detector system greatly reduces the statistical error in the measurement.

2.4. RF system

Microwaves are produced by a local oscillator signal generator and amplified to 500 W with a GaN amplifier (R&K A2856BW200-5057-R).

The microwave cavity is made with oxygen-free copper; the inside of the cavity is a cylinder 128 mm in diameter and 100 mm long. The side wall of the cavity is only 2 mm thick in order to allow the γ-rays to efficiently escape. The cavity is operated in the TM_{110} mode. The resonant frequency is 2.9 GHz and $Q_L = 14700 \pm 50$. The cavity is filled with gas (90 % N_2 and 10 % iso-C_4H_{10}) with a gas-handling system. Iso-C_4H_{10} is used as the quenching gas to remove background 2 γ-ray annihilations.

3. Analysis and current status

The prototype run was performed from 29 June 2009 to 18 September 2009 using the large bore magnet with no compensation (compensation magnets to reduce the uniformity to O(ppm) are planned but are not yet installed). In the overall period, the trigger rate was about 3.5 kHz and the DAQ rate was about 0.7 kHz. The data acquisition was performed using CAMAC. The 2 γ-ray annihilation rate has been measured at various magnetic field strengths with a fixed RF frequency.

3.1. Data analysis

Figure 5 shows examples of measured timing spectra. The peak coming from prompt annihilation and p-Ps decay is followed by the decay curve of o-Ps and then the constant accidental spectrum. A timing window of 30–200 ns is applied to select o-Ps events. Figure 6 shows the 2γ transition spectrum, which is obtained as follows:

(i) The accidental contribution is subtracted using the timing window t =700–900 ns.

(ii) The ordinary o-Ps decay spectrum is also subtracted. This spectrum is obtained by running without RF power, and is normalized to the 2γ transition spectrum in the region 380–460 keV.

The 2γ transition spectrum is measured at different magnetic field strengths. The resonance lines obtained are shown in Figure 7.

Fitting the measured points with the Breit-Wigner function results in center values of the magnetic field of $B_0 = 0.865\,5665(71)$ T (8.2 ppm) at 1.5 atm and $B_0 = 0.865\,662(10)$ T (12 ppm) at 1.0 atm.

Systematic errors of the prototype run are summarized in Table 1.

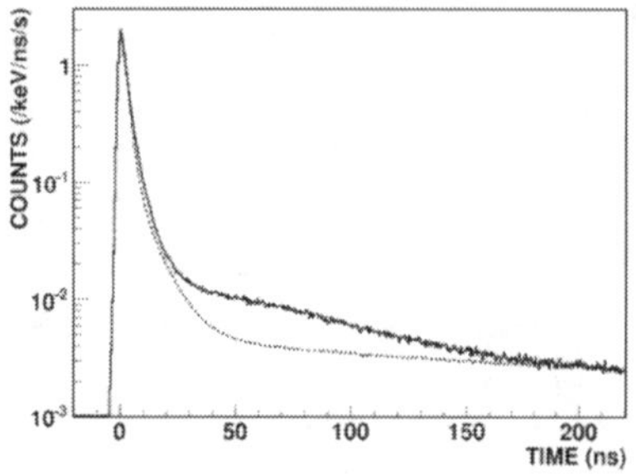

Figure 5. Decay curves of Ps. The solid line is on-resonance (0.8659 T) RF ON, and the dotted line is RF OFF. An energy window of 492–530 keV is applied.

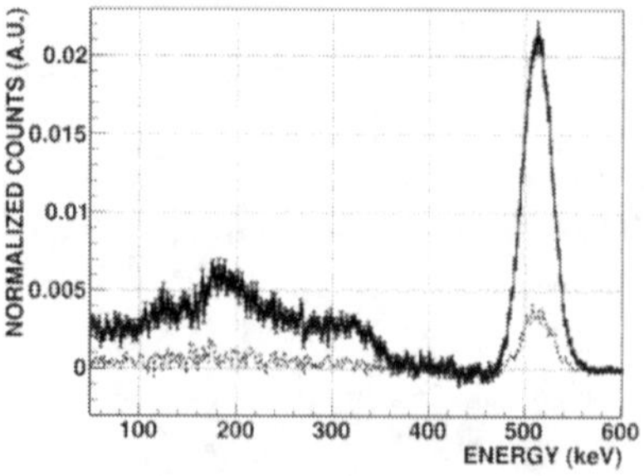

Figure 6. Energy spectra at different magnetic field strengths. The solid line is on-resonance (0.8659 T) RF ON and the dashed line is off-resonance (0.8614 T) RF ON. A timing window of 30–200 ns is applied, and the accidental spectra from 700–900 ns and RF OFF have been subtracted.

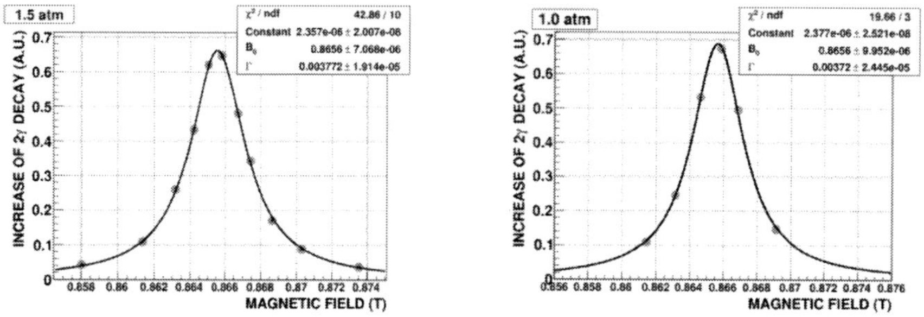

(a) Resonance line at 1.5 atm gas pressure.

(b) Resonance line at 1.0 atm gas pressure.

Figure 7. Resonance lines. Error bars are smaller than the marker size.

Table 1. Systematic errors of the prototype run.

Element	Errors in Δ_{HFS} (ppm)
Non-uniformity of the magnetic field	22
Analysis method[1]	< 40
Line-shape correction[1]	< 20
Gas pressure dependence	8
Thermalization of Ps[2]	< 20
RF frequency	6
Q-value of the cavity	10
Magnetic field correction	4
Stability of the magnetic field	2
NMR measurement	2
Quadrature sum	56

[1] Further analysis will reduce these uncertainties.
[2] Direct measurement of Ps thermalization function will reduce this uncertainty.

The preliminary value of Δ_{HFS} calculated from equation (1) is

$$\Delta_{\mathrm{HFS}} = 203.385 \pm 0.003(\mathrm{stat.}) \pm 0.011(\mathrm{sys.})\,\mathrm{GHz}\,(58\,\mathrm{ppm})\,, \tag{2}$$

which is consistent with both of the previous experimental values and with the theoretical value.

3.2. Next steps

The following improvements are planned for future measurements:

(i) Compensation magnets will be installed and $O(\mathrm{ppm})$ magnetic field uniformity is expected to be achieved.

(ii) Measurements at various pressures of gas will be performed to estimate the material effect (the Stark effect). The accumulation of these measurements will result in an $O(\mathrm{ppm})$ statistical error within a few years.

(iii) The timing information allows for a measurement of Ps thermalization as a function of time [4]. We can thus precisely measure the material effect including the thermalization effect.

4. Conclusion

A new experiment to measure the Ps-HFS which reduces possible common uncertainties in previous experiments has been constructed and the prototype run has been finished. A preliminary value of $\Delta_{\mathrm{HFS}} = 203.385 \pm 0.003(\mathrm{stat.}) \pm 0.011(\mathrm{sys.})\,\mathrm{GHz}\,(58\,\mathrm{ppm})$ has been obtained, which is consistent with both of the previous experimental values and with the theoretical calculation. Development of compensation magnets is underway with a view to obtaining $O(\mathrm{ppm})$ magnetic field homogeneity for the final run. The final run will start in about a year. A new result with an accuracy of $O(\mathrm{ppm})$ will be obtained within a few years which will be an independent check of the discrepancy between the present experimental values and the QED prediction.

References

[1] Ritter M W, Egan P O, Hughes V W and Woodle K A 1984 *Phys. Rev.* B **30** 1331
[2] Mills A P Jr 1983 *Phys. Rev.* A **27** 262
[3] Kniehl B A and Penin A A 2000 *Phys. Rev. Lett.* **85** 5094
[4] Kataoka Y, Asai S and Kobayashi T 2009 *Phys. Lett.* B **671** 219
[5] Grotch H and Hegstorm R A 1971 *Phys. Rev.* A **4** 59

International Symposium on Advanced Science Research 2009 IOP Publishing
Journal of Physics: Conference Series **225** (2010) 012049 doi:10.1088/1742-6596/225/1/012049

Application of GAMESS/NEO to quantum calculations of muonic molecules

Eugene V Sheely[1], Larry W Burggraf[1], Paul E Adamson[2] , Xiaofeng F Duan[3] and Mike W Schmidt[4]
[1]Air Force Institute of Technology, AFIT/ENP, 2950 Hobson Way, Wright-Patterson Air Force Base, OH 45433 USA
[2]High Power Microwave (HPM) Technologies Branch, HPM Division, Directed Energy Directorate, Air Force Research Laboratory, Kirtland AFB, NM 87117 USA
[3]DoD Supercomputing Resource Center (DSRC), Air Force Research Laboratory, 2435 Fifth Street, Bldg 676, Wright-Patterson Air Force Base, OH 45433 USA
[4]Department of Chemistry, Iowa State University, 201 Spedding Hall, Ames, IA 50011 USA

E-mail: Eugene.Sheely@afit.edu

Abstract. The General Atomic and Molecular Electronic Structure System (GAMESS) has been modified to perform studies involving negative muons. This system, coupled with the Nuclear-Electronic Orbital (NEO) method enables the ab-initio study of muonic atoms where both the negative muon and the positive nuclei are modeled as quantum particles. This is of particular usefulness in the study light nuclei, muonic atoms, such as is encountered in muon-catalyzed fusion. NEO was also modified to allow the inclusion of positive exotic-particles to be studied using open and closed shell Hartree-Fock and Configuration Interaction. Capitalizing on these modified methods, the muon density and vibrational dynamics of some light muonic molecules have been analyzed.

1. Introduction

Many quantum mechanical software packages for predicting molecular properties have been written and modified over a number of years. Multiple general-purpose programs exist that do a good job of calculating Hartree-Fock energies, electron correlation energy by various methods, equilibrium geometries, harmonic and anharmonic vibrational energies, relativistic effects, and many other properties of molecular systems. Most of these software packages are written to exclusively calculate properties of molecular systems that contain only neutrons, protons and electrons. As part of research being performed at the Air Force Institute of Technology (AFIT) and the US Air Force Research Laboratory (AFRL) it has become desirable to develop and apply software to model the physical and chemical properties not only of traditional molecular systems, but of systems that contain exotic-particles as well (positrons, as well as negative and positive muons). At this point, two options existed, develop new code from scratch or modify existing code to calculate the desired properties of exotic molecules. While there are pros and cons to both of these approaches, a decision was made to modify existing software, thereby directly capitalizing on the extensive work in quantum chemistry performed by others.

The software package selected for modification was the General Atomic and Molecular Electronic Structure System (GAMESS). GAMESS is a general purpose electronic structure code whose primary focus is on *ab-initio* quantum chemistry calculations. It is a highly capable, integrated system of computational modules designed to predict physical and chemical properties of molecular systems. Like most quantum chemistry codes, GAMESS treated only electrons as fully quantum particles. GAMESS was developed in a non-commercial enterprise under leadership of Prof. Mark Gordon from North Dakota State University beginning in the 1980's[1,2]. Since its inception, Prof. Gordon, who later moved to Iowa State University, his students and contributing colleagues have continued to develop the code and expand its capabilities. The selection of GAMESS has several benefits over other software packages considered. The GAMESS code has always been made available free of charge for other scientists to assist with its development and to employ for application. Prof. Gordon and his group have demonstrated a willingness to assist in further code development to include exotic particles. The greatest single advantage was the recent development of Nuclear-Electronic Orbital (NEO) code to model quantum nuclei in molecules. This was developed by Prof. Sharon Hammes-Schiffer and associates, from Pennsylvania State University, for integration into GAMESS[3]. NEO is designed to incorporate nuclear quantum effects into electronic structure calculations by treating specified nuclei quantum mechanically on the same level as the electrons. The basic objective of the method is to solve a mixed nuclear-electronic time-independent Schrödinger equation with molecular orbital methods. Both electronic and nuclear molecular orbitals are expressed as linear combinations of Gaussian basis functions. The variational method is used to minimize the energy with respect to all molecular orbitals, as well as the centers of the nuclear basis functions. Correlation effects among electrons and nuclei are included with multiconfigurational (MCF) perturbation theory (MP2)[4] and Configuration interaction (CI)[5] methods. The advantages to this approach include: (1) nuclear quantum effects are incorporated during electronic structure calculations, (2) Born-Oppenheimer separation of electrons and nuclei is avoided, (3) excited vibrational-electronic states may be calculated, (4) its accuracy may be improved systematically, (5) NEO has been incorporated into GAMESS general quantum chemistry code, which is well-tested (6) quantum tunneling systems can be handled with multiple basis function centers for a single quantum nucleus, (7) positions of nuclear basis function centers may be fixed or optimized variationally, and (8) symmetry-adapted linear combinations of nuclear molecular orbitals are available[6,7,8,9,10,11,12,13,14]. One of the biggest advantages of applying the NEO approach to calculate exotic particle molecule properties is that the method takes maximum advantage of the *ab-initio* HF, MCSCF and CI codes developed for the GAMESS general quantum chemical software package. The NEO code provided the potential to extend the capability of GAMESS to model molecular systems having both positive and negative quantum particles.

Between 2005 and 2007 Prof. Larry Burggraf's group from AFIT, in collaboration with Prof. Hammes-Schiffer, modified and extended the GAMESS/NEO software to allow the study of positronic molecular systems using mixed positronic-electronic wavefunctions. Computational methods to model positronic-electronic wavefunctions include: Hartree-Fock (HF), second-order Möller-Plesset perturbation theory (MP2), configuration interaction (CI), and complete active space self-consistent field (CASSCF). The ability to calculate positron-electron annihilation rates based on NEO-HF and NEO-MP2 wavefunctions was also implemented[15,16]. Within the NEO framework, the NEO-HF (Hartree-Fock) method provides the energy corresponding to the single-configuration mixed positronic-electronic wavefunction, minimized with respect to the molecular orbitals expressed as linear combinations of Gaussian basis functions. The mixed wavefunction is a product of two Slater determinants, one representing electronic coordinates and one representing positronic coordinates. These are optimized iteratively in the NEO HF formalism. The electron-electron and electron-positron correlation can be treated in the NEO framework with second-order perturbation theory (NEO-MP2) or multiconfigurational methods such as the full configuration interaction (NEO-FCI) and complete active space self-consistent-field (NEO-CASSCF) methods[15,16].

2. Expansion of the GAMESS/NEO Code

The next step in the development of GAMESS/NEO exotic particle code was to add the ability to model muonic systems. The code was modified to allow the study of any mass for the positive particles. This allowed modelling of molecular properties as a function of mass in ways that could not be done if only specific masses were allowed. Being that the NEO code, which was originally designed for positive nuclei, had previously been modified to allow the study of positronic systems, additional alteration of the code to allow for the study of any mass positive quantum particle was rather straight forward. The NEO code, as it currently exists, does have one significant weakness that should not be overlooked, however. Only one mass of positive quantum particles can be studied at a time without modifying the code to include additional Slater determinants. A system containing multiple positrons, or multiple positive muons can be studied, but a system containing both positrons and positive muons cannot. This weakness in the code could be limiting when studying systems containing positive muons and light nuclei. In this case, the nuclei must be considered classically.

While modifying the NEO-positron code to allow for calculations involving any mass positive quantum nuclear particle was rather straightforward, the same cannot be said for modifying GAMESS to handle any mass negative particle. GAMESS, like most molecular *ab-initio* code assumes the reduced mass of an electron and nuclei to be equal to the mass of the electron. For most applications, this approximation works well. In the case of muonic molecules; however, assuming the reduced mass to be equal to the mass of the muon can result in greater than 10% error in the calculation of bond energy and bond length. To account for this, the reduced mass ratios $\left(\frac{\mu_e}{\mu_{ex}}\right)$

$$\left(\frac{\mu_e}{\mu_{ex}}\right)=\frac{\dfrac{m_e m_{nuc}}{m_e+m_{nuc}}}{\dfrac{m_{ex}m_{nuc}}{m_{ex}+m_{nuc}}}=\frac{m_e\left(m_{ex}+m_{nuc}\right)}{m_{ex}\left(m_e+m_{nuc}\right)} \tag{1}$$

m_e = mass of an electron

m_{nuc} = mass of the nucleus or positive particle of interest (e.g. positron)

m_{ex} = mass of the negative exotic particle of interest (e.g. negative-muon)

can be multiplied times each kinetic energy term in the electronic Hamiltonion

$$-\frac{\hbar^2}{2\mu_e}\nabla_i^2 \tag{2}$$

Currently GAMESS requires that the same reduced mass ratio be multiplied times each kinetic energy term. In other words, if the negative particle mass is large enough that its reduced mass cannot be considered equal to its mass, then all positive classical particles included in the calculation must have the same mass. Correction to this weakness in the code is currently underway. In addition to the benefits of being able to study exotic particles, eliminating the error associated with assuming the reduced mass to be equal to the negative particle mass could have other benefits. The study of energy and spectroscopic differences between isotopes of the same element becomes possible (i.e. the study of laser and chemical isotope separation).

While for Hartree-Fock (HF) and MP2 calculations multiplying the kinetic energy terms of the electronic Hamiltonium times the reduced mass ratio is the principle correction that needs to be made to account for differences in the mass of an electron and exotic-negative particles, such is not the case for CI calculations. In this case the CI coefficients must also be scaled.

Just as NEO can only model one type (mass) of positive quantum particle at a time, GAMESS can only model one type of negative quantum particle at a time. It cannot consider electrons and negative-muons in the same system. To get around this, in the future the NEO code may be modified so that it can be used for negative or positive quantum particles.

Most *ab-initio* molecular code assumes the volume of the nuclei to be negligible. The GAMESS code is not an exception. While this approximation makes it impossible to calculate K x-ray isotope shifts and some other spectroscopic properties for ordinary molecules[17], it is a legitimate approximation for most systems involving only electrons and nuclei[18,19]. In the case of an electron and proton the error generated by this approximation is only about $6\text{x}10^{-9}$ eV. The average muonic molecular radius is about 186 times smaller than that of an average electronic radius and therefore considering the volume of the nucleus to be zero for muonic molecules can result in significant errors. As the size and charge of the nuclei increase, the accuracy of assuming the nucleus to be a point volume decreases. For relatively small nuclei (e.g. the first few rows of the periodic table) a first order approximation to the energy correction can be calculated by assuming the potential energy to be

$$V(r) = -\frac{Ze^2}{4\pi \in_0 r} \tag{3}$$

when the negative particle radius (r) is greater than the mean nuclear radius (R) ($R \approx 1.25A^{1/3}$ fm where A is the mass number, Z is the atomic number), and when $r \leq R$

$$V(r) = -\frac{Ze^2}{4\pi \in_0 R}\left[\frac{3}{2} - \frac{1}{2}\left(\frac{r}{R}\right)^2\right] \tag{4}$$

If $\langle V \rangle$ is the expectation value of the potential energy assuming a point mass and $\langle V' \rangle$ the expectation value calculated using the above two equations, then the potential energy correction (ΔE) for nuclear volume is $\Delta E = \langle V' \rangle - \langle V \rangle$.

$$\Delta E = \frac{e^2 Z^4}{\pi \in_0 a_0^3}\int_0^R \left(r - \frac{3r^2}{2R} + \frac{r^4}{2R^3}\right)\exp\left(-\frac{2Zr}{a_0}\right)dr \tag{5}$$

where

$$a_0 \equiv \frac{4\pi \in_0 \hbar^2}{\mu e^2} \tag{6}$$

Equation (5) reduces to

$$\Delta E = \frac{e^2}{8\pi \in_0 a_0 R^3 Z}\left[3a_0^3 - 3a_0 R^2 Z^2 + 2R^3 Z^3 - 3a_0\left(a_0 + RZ\right)^2 \exp\left(-\frac{2RZ}{a_0}\right)\right] \tag{7}$$

When $\frac{R}{a_0}$ is small[18,19], this can be approximated as

$$\Delta E = \frac{e^2 Z^4 R^2}{10\pi \in_0 a_0^3} \tag{8}$$

As Z and A get large, so does ΔE. When ΔE is more than a small fraction of the total energy, as it is for muo-molecules with large nuclei, the above perturbative correction is not an accurate method of accounting for nuclear volume effects. In these cases it is necessary to more accurately approximate the true potential $V(r)$ and solve the corresponding Schrödinger equation. The authors have no current plans to modify GAMESS to study muonic molecules involving muonic binding to large nuclei.

As with all *ab-initio* codes, there are limits to what can be studied. Resources and computational time are limited. In order to expand what can be calculated using GAMESS the code has been modified to work in parallel computing environments. Parallelization of NEO would enhance its usefulness, but as of the writing of this paper, has not been performed.

3. GAMESS/NEO and negative muons

As with most computational efforts, the goal is not to develop computer code, rather it is to obtain the results that can be generated using the code. As part of code development, it is important to test the code and compare to known solutions and previously calculated results. Following are a couple applications of GAMESS/NEO to muonic molecular systems which help to demonstrate the viability of the code and some of what can be studied through its use. Beginning with a problem for which the exact solution is known seemed to be a good starting point. The non-relativistic *μ-p* binding energy can be determined if the nucleus is considered to be a classical particle (Born-Openheimer approximation)[20]

$$E = -\frac{Z^2 e^4}{n^2 8\pi \in_0 a_0} \tag{9}$$

The exact ground state solution of -2.5286 keV can be compared to that calculated using GAMESS. 3s, 4s, 5s, 4s3p and 5s4p3d non-contracted basis sets were optimized using the method of Petersson and Zhong[21]. The results are summarized in table 1. Being that there is very little contribution from *p* and *d* orbitals, the differences in the values presented in table 1 are due almost entirely to differences in the number of *s* basis sets used.

Table 1. Energy as a function of Basis Set used.

Basis Set	Energy (keV)
3s	-2.5062
4s	-2.5262
4s3p	-2.5262
5s	-2.5288
5s4p3d	-2.5289
Exact	-2.5285

A Hartree-Fock 5s4p3d potential energy surface was generated for $p\text{-}\mu\text{-}p^+$. The basis set, which is very sensitive to the fitting coefficients, was again optimized using the method of Petersson and

Zhong[21]. The potential energy as a function of nuclear bond length at 27 points ranging in length from 0.003 Å to 0.03 Å was calculated. The points were interpolated between using cubic spline interpolation with fixed endpoints[22,23].

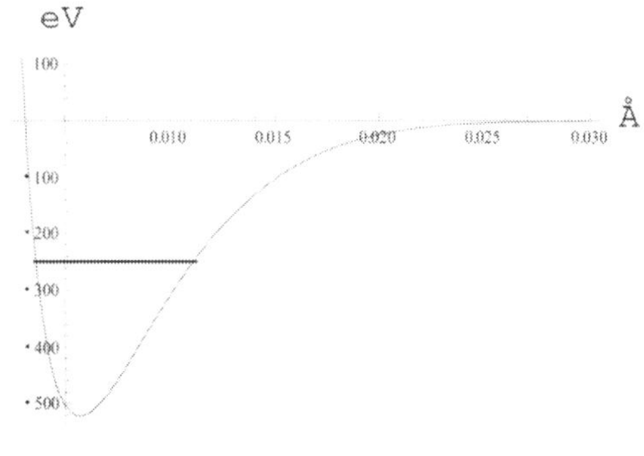

Figure 1. p-μ-p$^+$ Potential energy surface (PES). The horizontal line represents the ground state vibrational energy.

Setting the first derivative of the energy with respect to bond length equal to zero, the potential energy minimum (equilibrium geometry) was found to be .0056839 Å. This corresponds to a classical Equilibrium Dissociation Energy (D_e) of 522.099 eV. The harmonic vibrational frequency of the nuclei

$$\nu = \frac{1}{2\pi}\sqrt{\frac{V''\left(r_{eq}\right)}{\mu}} \tag{10}$$

was calculated to be 1.77197×10^{17} Hz where μ is the reduced mass of the nuclei and $V''\left(r_{eq}\right)$ is the second derivative of the potential energy with respect to bond-length, evaluated at the equilibrium geometry.

In order to observe the anharmonic effects of internal kinetic energy on the vibrational frequency, Hamilton's equation of motion

$$\frac{\partial H}{\partial q} = -\dot{p} \tag{11}$$

was solved for position (q, which in this case is internuclear separation) and momentum *(p)*[24]. H represents the classical hamiltonion and is equal to the sum of the potential and kinetic energy. The vibrational spectra (power spectral density) of the system can be generated by taking the Fourier transform of the position, momentum, kinetic or potential energy as a function of time[25,26] (see figures 2 and 3).

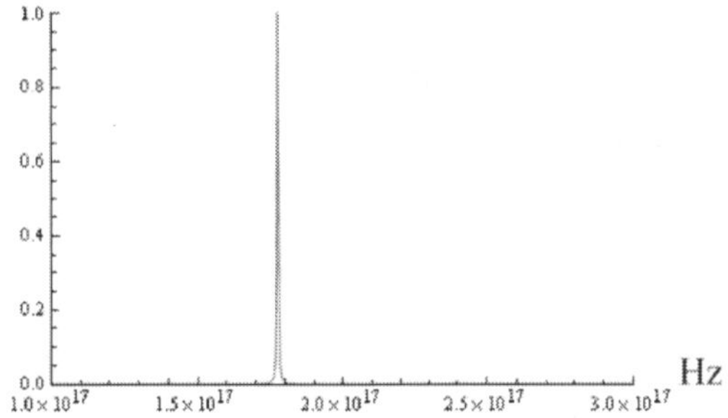

Figure 2. Classical internal vibrational energy of 0.23 eV. The result is nearly Harmonic, as is expected near the bottom of the potential energy well. This minimal vibrational energy cannot be obtained in a quantum system, but does demonstrate a harmonic approximation.

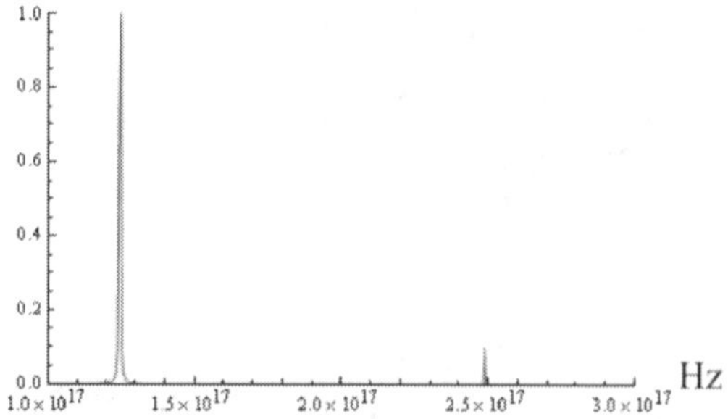

Figure 3. Internal vibrational energy is equal to the ground-vibrational-state dissociation energy $D_0 = 263.75\ eV$. The internal vibrational energy is 258.35 eV. A peak shift and additional peaks which result from anharmonic effects can be seen.

By comparing figures 2 and 3, the effects of anharmanicity on the fundamental vibrational frequency can be seen. As the internal vibrational energy increases, the anharmanicity also increases and additional peaks appear. The energy used in the calculation of figure 3 corresponds to the ground state quantum vibrational level[27]. The ground-vibrational-state dissociation energy D_0 (equilibrium dissociation energy D_e minus the ground state vibrational energy) was calculated for the p-μ-p$^+$ potential energy surface, $D_0 = 263.275\ eV$. This compares to a value of 253.15 eV presented elsewhere[28]. It should be noted that the calculations presented in this paper were presented to demonstrate methods and capablilties of the software. The level of calculations was relativily low (small basis sets, nuclei assumed to be classical and correlation energy ignored). Nuclear volume effects were also neglected in these calculations.

4. Conclusions

The GAMESS/NEO code has been modified to allow the *ab-initio* study of exotic-molecules. GAMESS can model any mass quantum negative particle and NEO any mass quantum positive particle. While molecular systems with large numbers of negative and/or positive quantum particles can be modeled using GAMESS/NEO, all negative quantum particles must have the same mass and all positive quantum particles must have the same mass. This excludes the possibility of modeling molecules that contain both electrons and negative muons. Future modifications of the code are planned which will eliminate this shortcoming by allowing NEO to model negative particles.

Currently if the negative quantum particles are large enough that the reduced mass between them and all positive particles cannot be considered equal to the mass of the negative particle, then only one mass possitive particle (nuclei or positron) can be considered. This is true whether the positive particle is considered classically or quantum mechanically. Modifications to the code that will eleminate this weakness are planned for the next couple of months.

Currently muo-molecular calcuations using GAMESS can be performed at the HF and MP2 level. NEO calculations for closed shell systems can be performed at the HF, MP2, CASSCF, and CI level. Opened shell HF and CI calculations are also possible. The capability to perform CI muo-molecular calcuations in GAMESS should be added to the code shortly.

GAMESS can run in serial or parallel environments. NEO currently only runs on serial systems. The authors are not aware of any effort currently underway to modify NEO to work in a parallel computing environment. Parallelization of NEO would enhance its usefulness and will hopefully be undertaken soon.

Acknowledgements
The authors would like to recognize assistance provided by Mark Gordon and Gary Kedziora. The views expressed in this article are those of the authors and do not necessarily reflect the official policy or position of the Air Force, the Department of Defence, or the United States Government.

References

[1] Schmide M W, Baldridge K K, Boatz J A, Elbert S T, Gordon M S, Jensen J H, Koseki S, Matsunaga N, Nguyen K A, Su S J, Windus T L, Dupuis M and Montgomery J A 1993 *J. Comput. Chem.* **14**, 1347-63
[2] Gordon M S and Schmidt M W 2005 Advances in electronic structure theory: GAMESS a decade later, in Dykstra C E, Frenking G, Kim K S and Scuseria G E Theory and Applications of Computational Chemistry, the first forty years (Amsterdam: Elsevier) 1167-89
[3] Webb S P, Iordanov T and Hammes-Schiffer S 2002 *J. Chem. Phys.* **117**, 4106-18
[4] Pak M V, Swalina C and Hammes-Schffier S 2005 *Chem. Phys. Lett.* **404**, 394-99
[5] Skone J H, Pak M V and Hammes-Schiffer S 2005 *J. Chem. Phys.* **123**, 134108
[6] Iordanov T and Hammes-Schiffer S 2003 *J. Chem. Phys.* **118**, 9489-96
[7] Pak M V and Hammes-Schffier S 2004 *Phys. Rev. Lett.* **92**, 103002
[8] Pak M V, Swalina C,Webb S P and Hammes-Schffier S 2004 *Chem. Phys.* **304**, 227-36
[9] Swalina C, Pak M V and Hammes-Schffier S 2005 *J. Chem. Phys.* **123**, 014303
[10] Reyes A, Pak M V and Hammes-Schiffer S 2005 *J. Chem. Phys.* **123**, 064104
[11] Swalina C and Hammes-Schiffer S 2005 *J. Phys. Chem. A* **109**, 10410-7
[12] Swalina C, Pak M V, Chakraborty A and Hammes-Schiffer S 2006 *J. Phys. Chem. A* **110**, 9983-7
[13] Swalina C, Pak M V, Chakraborty A and Hammes-Schiffer S 2007 *J. Phys. Chem. A* **111**, 4522-6
[14] Ludlow M K, Skone J H and Hammes-Schiffer 2008 *J. Phys. Chem. B* **112**, 336-43
[15] Adamson P E 2007 A General Quantum Mechanical Method to Predict Positron Spectroscopy, *Dissertation, AFIT*
[16] Adamson P E, Duan X F, Burggraf L W, Pak M V, Swalina C and Hammes-Schiffer S 2008 *J. Phys. Chem. A* **112(6)**,1346-51.
[17] Krane K S 1988 *Introductory nuclear physics* (New York: Wiley—Interscience) p 51
[18] Cohen-Tannoudji C, Diu B and Laloë 1977 *Quantum mechanics*, vol. 2 (New York: Wiley—Interscience) pp 1141-6
[19] Krane K S 1988 *Introductory nuclear physics*, (New York: Wiley—Interscience) pp 49-50
[20] Levine I N 1995 Physical Chemistry, 4th ed. (New York, NW:McGraw-Hill) p 602
[21] Petersson G A, Zhong S, Montgomery J A and Frisch M J 2003 *J. Chem. Phys.* **118**, 3 1101-9
[22] Press W H, Flannery B P, Teukolsky S A and Vetterling W T 1992 *Numerical Recipes in FORTRAN: The Art of Scientific Computing, 2nd ed.* (Cambridge, MA: Cambridge University Press) pp 107-10
[23] Burden R L and Faires J D 1993 *Numerical Analysis, 5th ed.* (Boston, MA: PWS Pub.), pp 120-1
[24] van Gunstrene W F and Berendsen H J C 1990 *Angewandte Chemie, International Edition in English* **29**, 992-1023
[25] Press W H, Teukolsky B P, Vetterling W T and Flannery B P 2007 *Numerical recipes, the art of scientific computing, 3rd ed.*, (Cambridge, MA: Cambridge University Press) pp 600-717
[26] Marple S L 1987 *Digital Spectral Analysis with Applications*, (Upper Saddle River, NJ: Prentice-Hall)
[27] McQuarrie D A 1976 *Statistical mechanics*, (New York, NY: Harper & Row Pub.) pp 91-8, 137
[28] Alexander S A and Monkhorst H J 1989 *AIP Conf. Proc.* **188** 246-58

Muonic molecule formation in muon-catalyzed fusion

Naritoshi Kawamura[1], Katsuhiko Ishida[2], Teiichiro Matsuzaki[2], Hiroshi Imao[3,4] and Kanetada Nagamine[5]

[1] Muon Science Section, Materials and Life Science Division, J-PARC Center, Tokai-mura, Ibaraki, 319-1195, Japan
[2] Advanced Meson Science Laboratory, RIKEN, Wako, Saitama, 351-0198, Japan
[3] Atomic Physics laboratory, RIKEN, Wako, Saitama, 351-0198, Japan
[4] Graduate School of Arts and Sciences, University of Tokyo, Komaba, Meguro, Tokyo, 153-8902, Japan
[5] University of California, Riverside, CA 92521, USA

E-mail: nari.kawamura@kek.jp

Abstract. A negative muon induces nuclear-fusion reactions between hydrogen isotopes. In this reaction, a muon plays a role of a catalyst, *i.e.* muon injection causes a chain reaction of nuclear fusions in hydrogen system. This phenomenon is called muon catalyzed fusion or μCF for short. A fusion reaction itself is dominated by nuclear force in the energy range of mega-electron-volt, however μCF reaction is affected by much-low-energy phenomena like temperature and density in the range of milli-electron-volt. This is due to the muonic molecule formation through which a fusion reaction occurs. Therefore, μCF has been an attractive subject for atomic physics as well as nuclear physics.

1. Introduction

A negative muon stopped in a deuterium and tritium (D-T) mixture induces dt-fusion through the formation of $dt\mu$ molecule: $d + t + \mu^- \rightarrow dt\mu \rightarrow \alpha + n + \mu^- + Q(17.59 \text{ MeV})$. After this reaction, the muon is relieved and induces a subsequent fusion reaction. As shown in Fig. 1, this chain reaction (μCF cycle) repeats until the muon is lost from the cycle by either muon decay or muon capture to a higher-Z particle, like the fusion-product (α).

It is widely accepted that a $dt\mu$ molecule can be regarded as being resonantly formed in a two-body collision between $t\mu$ and D_2: $t\mu + D_2 \rightarrow [(dt\mu)dee]$ [1]. In this reaction, the collisional energy and the binding energy of $dt\mu$ molecule are absorbed by the rotational and vibrational energy of the $[(dt\mu)dee]$ complex, as shown in Fig. 2. This resonant formation mechanism has succeeded to explain a high muon cycling rate, which is predominantly determined by the $dt\mu$ formation rate. It was natural that a theoretical calculation based on this mechanism [2] predicted the $dt\mu$ formation rate to decrease steeply with decreasing temperature below 100 K, at which the collisional energy did not reach the lowest resonance level. However, this was not consistent with the experimental results [3, 4, 5], and even in liquid and solid D-T mixture, high fusion rate was observed.

There is another discrepancy between the resonant $dt\mu$ formation mechanism and the experimental result. Let us consider the density dependence of the muon cycling rate (λ_c), which is normalized by the target density in order to remove any trivial density dependence. Because the conventional resonant $dt\mu$-formation mechanism has been based on a two-body

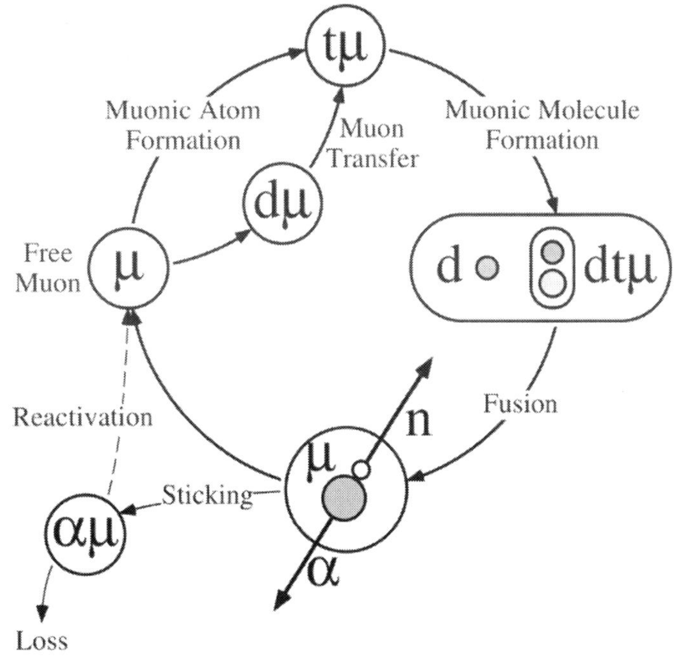

Figure 1. A schematic diagram of μCF cycle.

collision between $t\mu$ and D_2, λ_c is expected not to depend on the density. However, λ_c shows obvious density dependence [4].

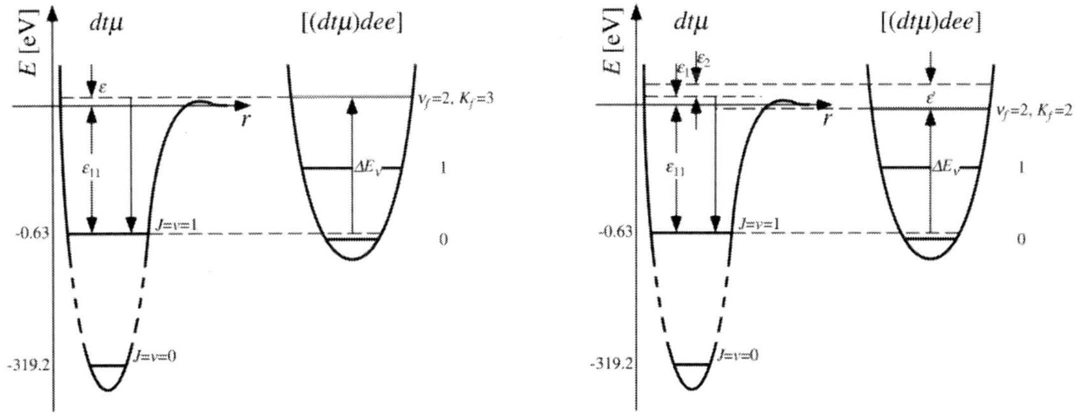

Figure 2. A schematic figure of the resonant $(dt\mu)$ formation in two body collision (left) and in three body (right).

Several theoretical approaches [6, 7, 8] have been provided to solve these problems. The idea of a three-body collision during $dt\mu$ formation was proposed by Menshikov *et al.* [6]. They hypothesized molecular formation in a three-body system, $t\mu + D_2 + {D_2}'$, in which the resonant molecular formation is enhanced compared with that in a two-body system by the third particle, ${D_2}'$, which takes away the excess energy of molecular formation like Feschbach-resonance. This idea allows molecular formation below the above-mentioned threshold energy, and thus can explain the temperature dependence. The contribution of the third particle may also explain the density dependence. Although qualitative explanation was provided, theoretical studies based on the three-body (many-body) effect predict different tendency of molecular formation rate in low temperature from each other.

This study aims to unveil the mystery in muonic molecule formation by performing a comprehensive survey of μCF in solid and liquid.

2. μCF experiment at the RIKEN-RAL Muon Facility

A systematic study of μCF in liquid and solid D-T mixtures was performed at the RIKEN-RAL Muon Facility [9]. An advanced tritium gas-handling system, which had an *in-situ* ^{3}He-removal capability with a palladium filter [10], was applied to prepare a D-T mixture gas with an intended tritium concentration from 20% to 70%. By controlling the cryogenic system, the target cell was adjusted to an intended temperature from 5 to 16 K. By synchronized detection with the intense pulsed muon beam [9], X-rays associated with the muon-to-α-sticking phenomenon were clearly observed [5, 11]. Fusion neutrons were also observed [12].

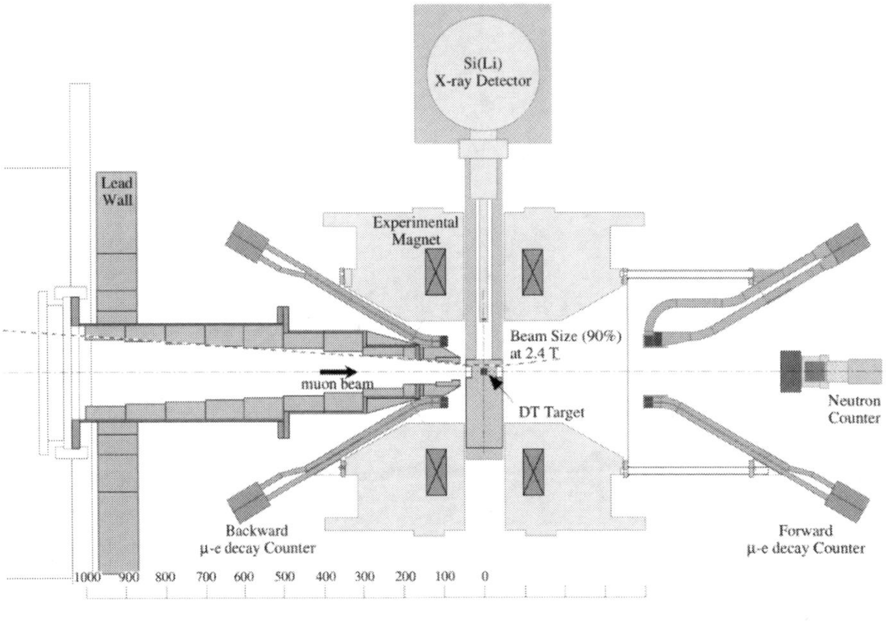

Figure 3. The experimental setup for μCF at the RIKEN-RAL Muon Facility.

The μCF cycle is characterized by the muon cycling rate (λ_c) and the total muon loss probability per cycle (W). These two parameters are associated with quantities determined by neutron detection, *i.e.* the neutron yield (Y_n) and the neutron disappearance rate (λ_n):

$$\begin{aligned} Y_n &= \int \phi\lambda_c \exp(-\lambda_n t)dt = \frac{\phi\lambda_c}{\lambda_n}, \\ \lambda_n &= \lambda_0 + W\phi\lambda_c, \end{aligned} \tag{1}$$

where ϕ is the atomic-number density of the D-T mixture, and λ_0 is the muon decay rate (0.455×10^6 s^{-1}). As shown in the above equations, λ_c has already been normalized by the target density in order to remove any trivial density dependence. Figure 4 shows a typical analysis result. As shown in this figure, with decreasing temperature, unexpected decrease in λ_c and increase in W were observed. The former change is interpreted to be caused by the temperature dependence in the $dt\mu$ formation process. The most probable origin of the latter change is the temperature dependence in the muon reactivation process after muon-to-α sticking. The $dt\mu$ formation rate can be derived as shown in Fig. 5. A theoretical calculation which took into account a three-body effect phenomenologically predicted the enhancement in $dt\mu$ formation [13] and reproduced the present result by the transition energy from $t\mu$ to $dt\mu$ shifting from 0.601 eV to 0.596 eV. The temperature dependence of the $dt\mu$ formation rate is able to be understood by multi-body effect, *i.e.* Feschbach resonance mechanism, and affect λ_c to decrease with decreasing temperature. On the other hand the increase in W, which is predominantly determined by the muon-to-α sticking probability ω_s^{eff}, is still open question. The muon sticking process is divided into two processes, initial sticking and a subsequent reactivation process: $\omega_s^{eff} = (1 - R)\omega_s^0$, where R is the reactivation probability. Because the initial sticking denoted by ω_s^0 is a process immediately after a nuclear reaction, is is unnatural that the surrounding molecules affects this process. Thus, the temperature dependence in ω_s^{eff} is implied that the reactivation process depends on the temperature. Concerning the α-sticking process, we obtained other information by X-ray detection. The K_α X-ray yield from muon-to-α sticking was observed to be constant with the temperature. Although both R and the sticking X-ray yield describe the sticking process, only R shows a temperature dependence. This difference is understandable by taking into account the fact that almost all of the sticking X-rays which have a Doppler width corresponding to the fusion Q-value are emitted before the thermalization of an $(\alpha\mu)^+$ ion [5]. Namely, the temperature dependence in R can be explained by an assumption that the reactivation after thermalization or during thermalization depends on the temperature. Although the reactivation has been interpreted to occur in an high energy collision between an $(\alpha\mu)^+$ ion and surrounding hydrogen molecules, the present study observed the first evidence of the reactivation after thermalization, which is not negligible in comparison with that before thermalization.

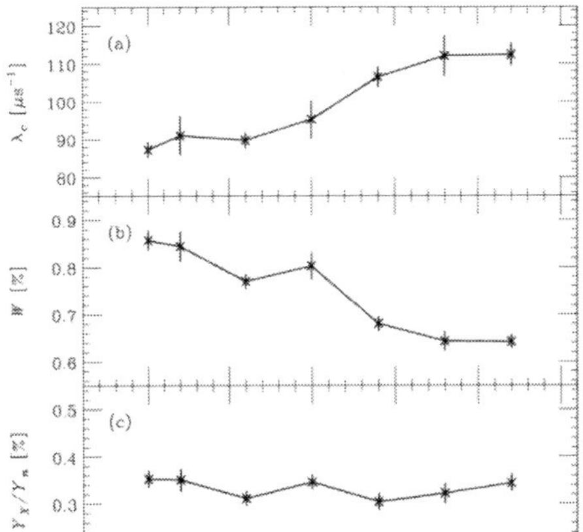

Figure 4. A typical result of the temperature dependence of (a) λ_c, (b) W, and (c) the ratio of the muon-to-α-sticking X-ray yield to the fusion neutron yield at the tritium concentration of 40%.

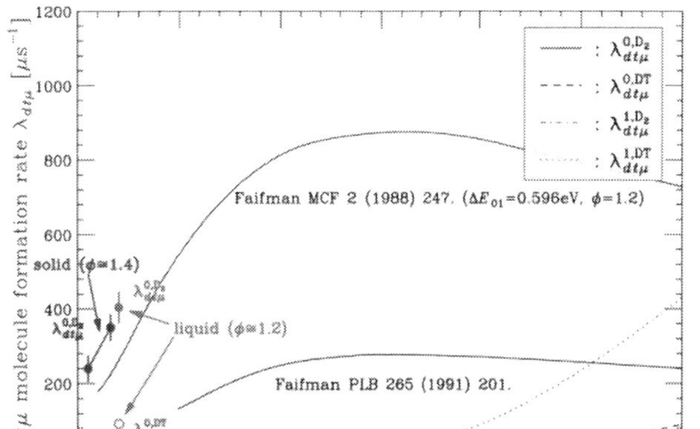

Figure 5. The temperature dependence of the molecular formation rate. The present work and theoretical calculations [2, 13] are shown. In the upper solid curve, the transition energy from $t\mu$ to $dt\mu$ is assumed to be 0.596 eV, while the lower is 0.601 eV.

3. Conclusion

The temperature-dependent phenomena of μCF were clearly observed in solid D-T mixtures. This shows the existence of multi-body effect in $dt\mu$ formation mechanism and also in the muon reactivation process after muon sticking to a fusion product of an α particle. Applying the nuclear-fusion reaction to a research probe, the present work has brought the mystery of μCF concerning the atomic physics in relief. A μCF study of ortho and para enriched deuterium system also shows an evidence of Feschbach resonance in $dd\mu$ formation [14].

In addition to the atomic physics, μCF study produces fruitful results in nuclear physics [10] and condensed matter physics [16]. High intensity pulsed muon beam is a key technique for the present work, and thus J-PARC Muon Facility is expected to promote this interdisciplinary research.

Acknowledgments

The authors acknowledge contributions to the present work by Drs. P. Wright, M. Tanase, H. Sugai, S.N. Nakamura, Y. Matsuda, G.H. Eaton, K. Kudo, N. Takeda, M. Iwasaki and S. Sakamoto and Messrs. P. Demarest, R. Hall, D. Haynes, W.A. Morris, G. Thomas, H.J. Jones, M. Kato and K. Kurosawa.

References

[1] E.A. Vesman, Zh. Eksp. Teor. Fiz. Pisma **5** 50 (1967). [Sov. Phys. JETP Lett. **5** 91 (1967).]; S. S. Gerstein and L. I. Ponomarev, Phys. Lett. B **72** 80 (1977).
[2] M. P. Faifman and L. I. Ponomarev, Phys. Lett. B **265** 201 (1991).
[3] W. H. Breunlich *et al.*, Phys. Rev. Lett. **58** 329 (1987).
[4] P. Ackerbauer *et al.*, Nucl. Phys. A **652** 311 (1999).
[5] K. Ishida *et al.*, Hyperfine Interactions **118** 203 (1999).
[6] L. I. Menshikov and L. I. Ponomarev, Phys. Lett. **167B** 141 (1986).
[7] J. S. Cohen and M. Leon, Phys. Rev. A **39** 946 (1989).
[8] Yu. V. Petrov *et al.*, Phys. Lett. B **331** 266 (1994).
[9] T. Matsuzaki *et al.*, Nucl. Inst. and Meth. A **465** 265 (2001).
[10] T. Matsuzaki *et al.*, Nucl. Inst. and Meth. A **480** 812 (2002).
[11] S.N. Nakamura *et al.*, Phys. Lett. B **473** 226 (2000).
[12] N. Kawamura *et al.*, Nucl. Inst. and Meth. A **465** 525 (2001).
[13] M. P. Faifman, Muon Catalyzed Fusion **2** 247 (1988); M. P. Faifman *et al.*, *ibid.* **2** 285 (1988).
[14] H. Imao *et al.*, Phys. Lett. B **632** 192 (2006).
[15] T. Matsuzaki *et al.*, Phys. Lett. B **557** 176 (2003).
[16] N. Kawamura *et al.*, Phys. Lett. B **465** 74 (1999).

International Symposium on Advanced Science Research 2009 IOP Publishing
Journal of Physics: Conference Series **225** (2010) 012005 doi:10.1088/1742-6596/225/1/012005

Development of new muon source for muon g-2 measurement and muon magnetic microscope

M. Iwasaki and K. Ishida

RIKEN Nishina Center for Accelerator-Based Science, Saitama 351-0198, Japan

E-mail: masa@riken.jp

Abstract. In various fields, from particle physics to condensed matter physics, extremely low emittance muon beam is eagerly awaited. We started a new project to realize the extremely low emittance and high intensity muon beam further beyond previous R&D study performed at RIKEN-RAL. To reduce the muon beam emittance, we plan to utilize the room temperature muonium generator, and furthermore, we apply muonium motion-cancellation mechanism using Doppler shift of $1S$-$2P$ transition energy. By the cancellation, one can expect the beam emittance better than $x\ \theta_x \approx 100$ μm·mrad even at 100 kV extraction field, which can be focused to the size of $\sim 1\mu$m by applying electron microscope technique. We are also developing high intensity and extremely stable VUV laser for the muonium ionization, which can efficiently ionize muonium and produce high muon yield of $\sim$ 10 kHz even at RIKEN-RAL (240kW proton with conventional surface muon channel). This new R&D study will open a totally new application to the muon science, namely a Muon Magnetic Microscope (M^3). This muon beam enable us to observe sub-μmeter magnetic structure of the condensed matter with the depth resolution less than 10 nano-meter.

1. Introduction

At RIKEN-RAL muon facility, we started a new facility oriented project to realize transcendent polarized muon beam, whose characteristics could exceeds previously proposed ultra-slow muon R&D works (RIKEN-KEK collaboration) [1, 2] in many aspects, such as intensity, momentum dispersion, emittance (sharp focusing capability), and time resolution. The limits, we have encountered in the previous ultra-slow muon project, are about 3×10^{-5} in extraction efficiency normalized by the number of incoming muons, focusing of about 3 mmϕ in diameter, and about 8 nsec in time resolution. Compared to the moderator-based slow muon using DC source at PSI [3], the present limits already have several advantages, though the improvement in many aspects are vitally required for the realistic application of the muon based researches. In the new project, we would like to achieve better muon beam characteristics of $\sim$ 100 times in extraction efficiency, $\sim 100 \times 100$ times in focus (horizontal and vertical), and $\sim$ 10 times in time resolution at minimum, starting from room temperature muonium ($\sim$ 300 K) instead of hot tungsten foil ($\sim$ 2100 K). At the same time, we also propose to cool down the muon below 300 K by applying chirped laser and vertical motion compensation field to have better emittance in one direction.

If one can realize $\sim$ micron focusing, it would be feasible to build a muon microscope sensitive to the magnetic properties of condensed matters having spatial resolution of $\sim 1\mu$m, namely a Muon Magnetic Microscope (M^3). Because muon is a self-analyzing particle, the microscopic magnetic property can be studied precisely if one can sharply focus the beam at low momentum.

At about 100 keV re-acceleration, the muon range in matter is few μm (1 μm in copper), so novel three-dimensional magnetic-property mapping become feasible for the first time. Further post-acceleration of the beam up to $\gamma = 3$, one can store the beam in a simple dipole magnet without focusing, which opens a new generation muon g-2 measurement beyond the present systematic error.

2. Muonium in Vacuum

A muonium in vacuum from the muon stopping target (muonium generator, such as hot tungsten and silica powder) is quite dilute system so that the muonium should have Maxwell-Boltzmann distribution $P_{Mu}\Delta p$, which can be written as

$$P_{Mu}\Delta p \propto \exp\left(-\frac{p^2}{2m_\mu kT}\right) p^2 \Delta p = \prod_{j=1}^{3} \exp\left(-\frac{p_j^2}{2m_\mu kT}\right) \Delta p_j, \tag{1}$$

where k is the Boltzmann's constant, T is the temperature of the muonium-generator, p and p_j are the muonium momentum and the projection to each axis, and the m_μ is the mass of the muonium ($\approx$ that of muon). Thus the muonium momentum distributes in a Gaussian function of

$$\sigma_{p_x} = \sigma_{p_y} = \sigma_{p_z} = \frac{1}{\sqrt{3}}\sigma_p = \sqrt{m_\mu kT}, \tag{2}$$

which is 1.66 keV/c for each axis at 300K ($kT = 0.026$ eV). It is clear that the lower temperature target will give us lower momentum dispersion, and the smaller spatial spread resulting higher efficiency for muonium laser re-ionization. On the other hand, muonium cannot come out and be isolated from the target, if the temperature is too low. It takes already about 100 nsec for 1 mm diffusion even at 300 K. Thus the room temperature target would be an ideal target for our purpose.

The resonant laser wave-length depends on the motion proportional to the momentum of the laser direction. Let us define the laser injection direction to be y-axis. The photon wave length of muonium 1S-2P transition λ_0 at rest (Lyman-α) is 122 nm. The Doppler broadening of this photon caused by the muonium thermal motion along y-axis is then given as

$$\sigma_{\nu_y} = \frac{c}{\lambda_0}\sqrt{\frac{kT}{m_\mu}}, \tag{3}$$

which is 38 GHz at 300 K. This means we need to have full band width at least $\sim$ 115 GHz (1.5σ) to ionize the muonium. On the other hand, this also means that one can ionize specific muonium momentum p_y selectively if the laser band width $\Delta\nu$ is much narrower than that caused by the muonium motion. This fact can be utilized for further muon cooling below the target temperature, as descried in later section.

The production of muonium in vacuum from room temperature materials such as silica powder was actively studied in the 1980s [5, 6, 7]. Muonium is formed with about 60% probability inside silica powder particles [8] and is emitted from the particles with 97% probability [9]. For laser re-ionization, silica aerogel would be much more suitable for its self-supporting feature. Actually, silica aerogel is essentially same compound having same micro-structure as silica powder, although less muonium production yield per stopped muon at 0.16 g/cm^3 target is reported [10]. Very lately, hydrophobic aerogel, whose density can be well controlled up to 1.2 g/cm^3, is newly synthesized so that one can search the optimum density in the case of silica aerogel.

To study the best material for the muonium yield at the room temperature, we submitted a proposal S1249 to TRIUMF and the proposal was accepted. In the experiment, we are planning to track a high energy positron from muon decay and detect a left electron at the muonium decay, $(\mu^+e^-)_{atom} \rightarrow (e^+ + \nu_e + \overline{\nu}_\mu) + e^-$, by a MCP at the same time, so that we can substantially reduce the background of decay positrons from the target. We are also aiming to study the muonium chemical potential in the silica grain from the minimum muonium velocity in vacuum, which can be obtained from the time evolution of the muonium spatial distribution.

3. Extraction field

The longitudinal momentum dispersion is defined by the acceleration field and the initial condition of the ionization laser. In our previous work, we applied a static electric field for the ionization muon extraction. This configuration is required to simplify the target design, to operate the hot tungsten target stably. In the case of a room temperature target, we are free from this difficulty.

Let us compare the longitudinal momentum dispersion given by static and pulsed extraction fields. Fig. 1 shows the schematic space (a) and time (b) relation relative to the ionization laser injection, respectively to the static and pulsed extraction fields. We assume that the field gradients are homogeneous in space and flat in time, both given as E_0, in the region of the acceleration. For simplicity, we ignore the initial kinetic motions of muoniums at the laser ionization, because the effect is negligibly small, as described in the previous section.

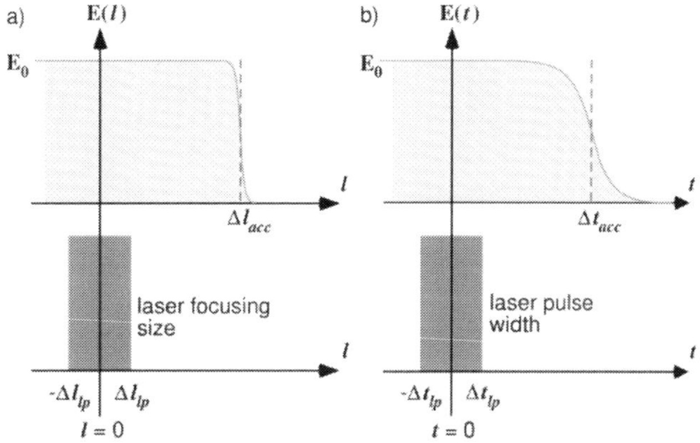

Figure 1. Illustration of the extraction / acceleration field for the laser ionized muon. In the case of a static field (left), the final kinetic momenta are simply defined by the ionization position, and it is difficult to minimize the resulting momentum dispersion, keeping muon yield. In the case of a pulsed field (right), the momentum dispersion is defined by the laser pulse width, which can be as short as a few 100 psec.

In the case of the static field, if the spatial uniformity of the extraction field near the laser ionization position is good enough, then the longitudinal momentum dispersion $\Delta p_l/p_l$ after the acceleration due to Δl_{lp} is given by the spatial distribution of the ionization position as

$$\left.\frac{\Delta p_l}{p_l}\right|_{static} = \frac{1}{2}\frac{\Delta T^{kin}_{static}}{T^{kin}_{static}} = \frac{1}{2}\frac{\Delta \int_l^\infty eE(l)dl}{\int_0^\infty eE(l)dl} = \mp\frac{1}{2}\frac{eE_0\Delta l_{lp}}{eE_0\Delta l_{acc}} = \mp\frac{1}{2}\frac{\Delta l_{lp}}{\Delta l_{acc}}, \tag{4}$$

where Δl_{lp} is the laser pulse width, T^{kin} is the kinetic energy, and Δl_{acc} is the effective acceleration distance, given as

$$\Delta l_{acc} \equiv \frac{1}{E_0} \int_0^\infty E(l) dl. \tag{5}$$

Here we denote the longitudinal axis as l, because the acceleration direction is undecided. It is clear that the dispersion is independent of the ionization timing within a laser pulse.

On the other hand, in the case of the pulsed field, the longitudinal momentum dispersion after the acceleration due to Δt_{lp} is given by the acceleration period, independent of the position. The dispersion is given as

$$\left.\frac{\Delta p_l}{p_l}\right|_{pulse} = \frac{\Delta \int_t^\infty eE(t)dt}{\int_0^\infty eE(t)dt} = \mp \frac{eE_0 \Delta t_{lp}}{eE_0 \Delta t_{acc}} = \mp \frac{\Delta t_{lp}}{\Delta t_{acc}}, \tag{6}$$

where Δt_{lp} is the time width of the laser pulse and Δt_{acc} is that of the effective acceleration, given as

$$\Delta t_{acc} \equiv \frac{1}{E_0} \int_0^\infty E(t) dt, \tag{7}$$

if the field gradient is unchanged during the laser irradiation. Note that these dispersions are insensitive to the field inhomogeneity in space close to Δl_{acc} and to the time dependence around Δt_{acc}.

A pulsed field extraction method is preferable for further muon cooling, and the static method is preferable for a shorter muon bunch width, hence time width, after the re-buncher. The preferred acceleration configuration may be different depending on the application and its purpose. Further design study is needed between initial acceleration and the LINAC for the choice of the extraction method.

4. Additional cooling by angular dispersion compensation

The angular dispersion caused by 300 K muonium thermal motion is already not far from the requirements. In the case of the M^3 source, smaller dispersion is preferable for better focusing in general.

In the case of the g-2 source, the situation is a bit more complicated. The horizontal motion (vertical to the magnetic field) is naturally confined in the ring by the magnetic field, when one put the muon in a storage ring of magnetic dipole field. On the other hand, vertical motion is not confined by the magnetic field. Thus one should minimize the motion parallel to the field, not to let the muon escape from the observation region of the ring.

Even if one can accelerate the beam without heating the muon beam bunch, one cannot be free from the transverse thermal motion. After injecting a parallel muon beam bunch to the storage ring, the beam bunch expands due to the thermal motion, which can be written as

$$\Delta y(t) = \sqrt{\Delta y_{inj}^2 + \left(\frac{\Delta p_y}{m_\mu}\right)^2 \gamma^2 t^2} \quad \text{and} \quad \Delta p_y = \sqrt{m_\mu k T}, \tag{8}$$

where Δy_{inj} is the beam size at the injection and t is time in Lab. frame. For the g-2 measurement, one need to re-accelerate the muon up to $\gamma = 3$ or more [4]. It means that the muon bunch have minimum expansion of about 15 cm in the standard deviation, during the observation time (five times of the muon life (τ_μ) at $\gamma = 3$ and $T = 300$ K). Obviously, this expansion is still not small enough for the precise g-2 measurement, because it is difficult to design the effective region of the decay positron observation much larer than 10 cmϕ. Therefore, further cooling by factor $4 \sim 5$ is required in this direction.

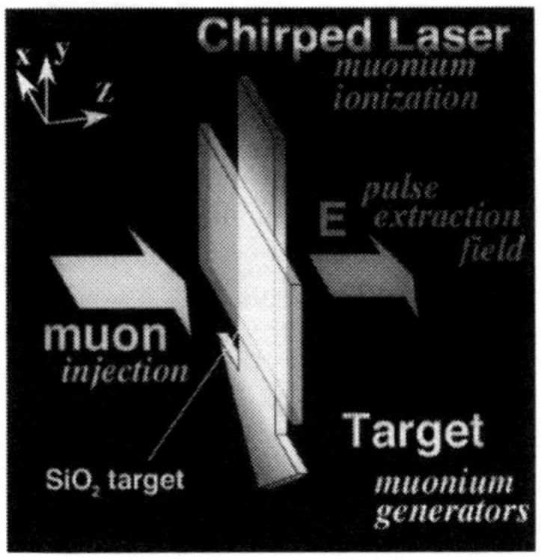

Figure 2. Schematic figure of the proposed configuration of the muon source in the case of g-2 measurement. In this configuration, one can shift the laser central energy in time (chirped laser) to ionize a muonium having specific motion toward the direction of laser axis, as shown in the figure.

Fig. 2 shows the proposed muon extracting configuration, to make it possible to cool down the muon beam in the vertical direction. Parallel pulsed-extraction field along the muon injection direction and applying chirped laser, whose central frequency is shifted by $\Delta\nu$ in a laser pulse, which is described by

$$\nu_0 + \Delta\nu \frac{t - t_0}{\Delta t_{lp}}, \quad \text{where} \quad |t - t_0| < \Delta t_{lp}. \tag{9}$$

The $\Delta\nu$ is defined by the Doppler shift caused by the thermal motion in the y direction. In this way, one can selectively ionize the muonium at p_y by selecting the laser central energy and its band width. Using a narrow band width and sweeping the laser energy during a laser pulse (chirped laser), then the angular dispersion of the extracted muon beam after the acceleration will have a linear correlation with the extracted time. If one wishes to ionize in the range of $n\ \sigma_{p_y}(=\sqrt{kT/m_\mu})$ by a laser pulse, then the formula can be written as

$$= \nu_0 + \left(n \frac{c}{\lambda_0} \sqrt{\frac{kT}{m_\mu}}\right) \frac{t - t_0}{\Delta t_{lp}}. \tag{10}$$

With this laser, one can select upward moving muonium (positive in y direction) in the earlier timing of the laser, and downward one in the later timing. If one makes the band width narrow enough, then one can linearly map the y motion into the momentum dispersion in the acceleration direction as described in section 3.

In this condition, the bunch size in space expands to

$$\Delta z_{bunch} = \sqrt{\Delta z_{lp}^2 + \left(v_0 \frac{\Delta t_{lp}}{\Delta t_{acc}} \frac{L}{v_0}\right)^2} = \sqrt{\Delta z_{lp}^2 + \left(\frac{\Delta t_{lp}}{\Delta t_{acc}} L\right)^2}, \tag{11}$$

after the free space motion of distance L, depending on the ionization timing. The ratio of the spatial expansion M, compared to the original spatial size Δz_{lp}, is given as

$$M = \sqrt{1 + \left(\frac{\Delta t_{lp}}{\Delta t_{acc}} \frac{L}{\Delta z_{lp}}\right)^2}. \tag{12}$$

One can utilize its dispersion to compensate the angular dispersion, and the improvement factor one can expect by this compensation can be naturally defined by M.

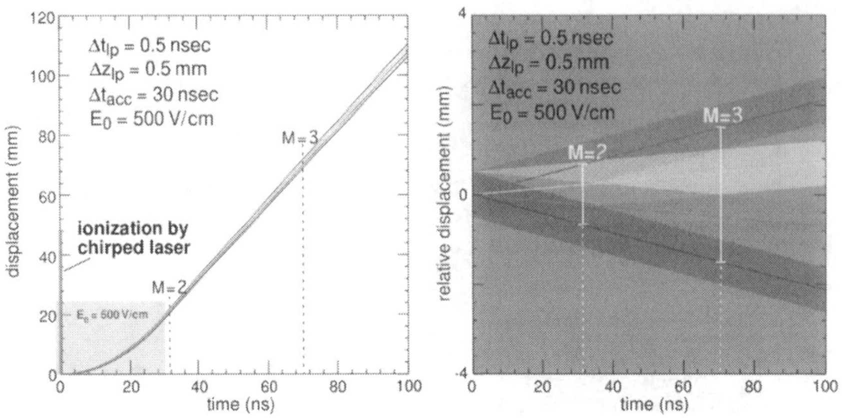

Figure 3. The acceleration of muon bunch (left) and expansion of bunch length depending on ionization timing (right). M is the spatial expansion as dfined in the text.

Fig. 3 shows a typical motion of the muon bunch by a rather weak acceleration field pulse of $E_0 = 500$ V/cm and for $\Delta t_{acc} = 30$ nsec, assuming the ionization laser size of $\Delta z_{lp} =$ 0.5 mm and the laser pulse width of $\Delta t_{lp} =$ 0.5 nsec. The ionization timings are plotted in different colors in a step of 0.25 nsec. As shown in the figures, the muon bunch moves about 2 cm during the effective acceleration field pulse, and each ionization timing region is separated at about 70 nsec after the laser irradiation around 7 cm away from the muonium target.

One can compensate the angular dispersion in the y-direction, as shown in Fig. 4. For the compensation, one should magnify the initial z distribution Δz_{lp} to apply compensation field according to the central p_y for given timing inside a muon pulse. It would be feasible to switch off the electric field in the order of 10 nsec, well before all the muons move out from the acceleration region, because the initial muon motion is very slow.

5. Cylindrical acceleration stage

The simplest and ideal muon $\sim$ 100 kV acceleration scheme, to realize 1μm resolution M^3 and to inject LINACs for new muon g-2 measurement, can be achieved by parallel and static electric field, namely Cockcroft-Walton type, because there exist absolutely no thermal perturbation source during the acceleration. However, simple parallel and static field is not ideal, because the laser-ionized muon is rather large in size. Before the injection to the static acceleration field, one need to focus the laser-ionized muon to utilize only homogeneous region of the field, otherwise the system becomes unreasonably huge.

The better model is electron microscope instead of the simple Cockcroft-Walton. If all the acceleration components have cylindrical symmetry without magnetic field, the equation of the particle tracks close to the beam axis can be represented as

$$\frac{d^2x}{dz^2} + \frac{1}{2\Phi}\frac{d\Phi}{dz}\frac{dx}{dz} + \frac{1}{4\Phi}\frac{d^2\Phi}{dz^2}x = 0 \qquad (13)$$

$$\frac{d^2y}{dz^2} + \frac{1}{2\Phi}\frac{d\Phi}{dz}\frac{dy}{dz} + \frac{1}{4\Phi}\frac{d^2\Phi}{dz^2}y = 0, \qquad (14)$$

where Φ is the electric potential on axis, $\Phi = \phi(0,0,z)$. From the equations, it is clear that the tracks are independent of the charge mass ratio e/m, thus the trajectories should be same

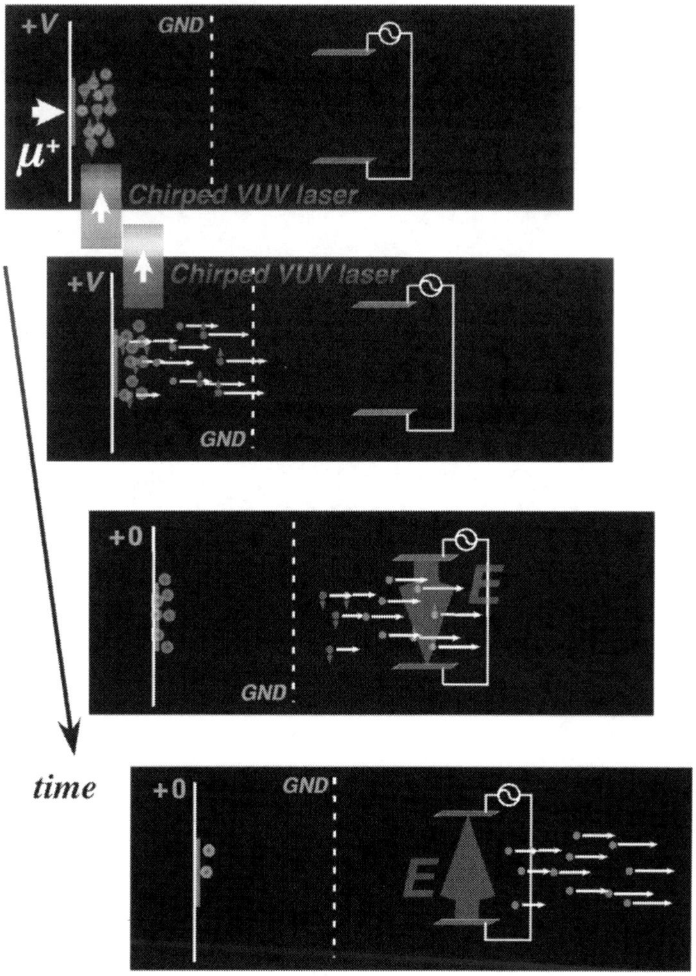

Figure 4. The compensation field to cancel the muonium motion using Doppler effect.

for electrons and muons. If there is higher order aberration in the acceleration stage, the muon bunch can easily heat-up because uncontrollable relative motions are introduced due to the aberration. The equations have cylindrical symmetry, so the second-order aberration term cannot exist, which simplifies the aberration control drastically. Because there exists an MV class electron microscope, it is feasible to apply that technology for the muon acceleration.

6. Primary design and the goal of the R&D

Fig. 5 shows the schematic setup for the acceleration scheme. The surface muon from the production target is transferred to the muonium generation target by a large solid angle solenoid channel, and accumulated to the central axis by a capillary method [11] using capillary shaped heavy material. The ionization laser will be injected just behind the muonium target to have maximum yield.

The extraction field will be generated by applying slightly higher voltage to the muonium target holder than the bias of the initial focusing element. Double einzel lenses will be used to form a weakly-focused parallel beam to confine the muon beam diameter. After the double einzel lenses, one can place an electrical element to cancel the vertical dispersion in the case where a

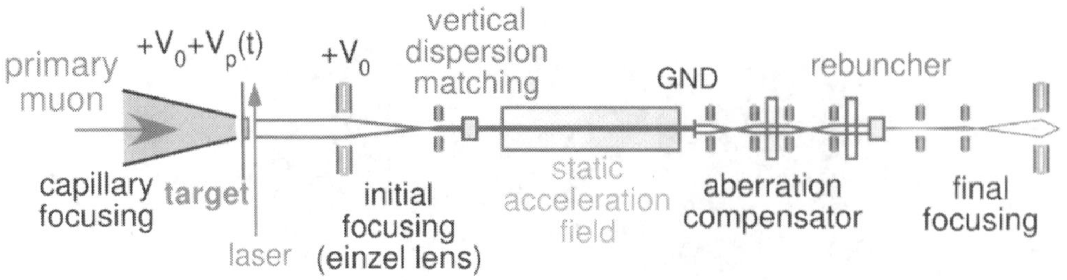

Figure 5. Schematic setup for the initial acceleration.

pulsed extraction field is applied. The weakly-focused parallel beam is further accelerated by the static parallel field, and the higher order aberration is canceled by an aberration compensator composed by four einzel lenses and two sextupole lenses located behind the static parallel field. In the present design, the biggest contribution of higher order aberration is θ^3-term, which come from the spherical aberration of the electro-static lenses.

The present conceptual design is suitable for the the case of muon magnetic microscope application. At Cockcroft-Walton type acceleration of about 100 kV, $\sim$ mm·mrad emittance can be feasible without further cooling method using Doppler shift cancellation mechanism. The cylindrical optics for this initial acceleration ensures good focus better than $\sim$ 10 μm in diameter, when one compensate the higher order aberration. If Doppler shift cancellation mechanism works as expected, one can improve the focusing size one order of magnitude in one direction.

In the case of g-2 application, one needs extremely short-pulse beam to fulfill injection requirement to the post-acceleration LINACs, so it requires a re-buncher to rotate the phase of the muon bunch to compress the muon beam bunch in time, and then apply final focusing elements before injection into the LINAC for the further acceleration up to $\gamma \sim 3$ or more. One needs more detailed design study to finalize the configuration of the final focusing elements to fulfill the matching condition with LINAC injection.

We are presently developing new muon source both for muon g-2 measurement and for muon magnetic microscope application, aiming at the beam emittance better than $\Delta x \Delta \theta_x \approx$ 100 μm·mrad even at 100kV extraction field. With the new high intensity and extremely stable VUV laser, we are expecting the muon yield of $\sim$ 10 kHz even at RIKEN-RAL (240kW proton with conventional surface muon channel). We wish to accomplish this goal earlier than five years.

References

[1] Y. Matsuda et al., Nucl. Phys. Proc. Suppl. 155 (2006) 346-348.
[2] P. Bakule et al., Nucl. Instr. Meth. B 266 (2008) 335.
[3] E. Morenzoni et al., Phys. Rev. Lett. 72 (1994) 2793.
[4] J-PARC proposal found in http://meson.riken.jp/g-2/g-2.pdf
[5] G. A. Beer et al., Phys. Rev. Lett. 57 (1986) 671.
[6] K. A. Woodle et al., Z. Phys. D 9 (1988) 59.
[7] A. C. Janissen et al., Phys. Rev. A 42 (1990) 161.
[8] R. F. Kiefl et al., Hyp. Int. 6 (1979) 185.
[9] G. M. Marshall et al., Phys. lett. 65A (1978) 351.
[10] W. Schwarz et al., J. Non-Crystaline Solids 145 (1992) 244.
[11] T. M. Kojima et al., J. Phys. Soc. Japan 76 (2007) 093501.

Journal of Physics: Conference Series **225** (2010) 012036 doi:10.1088/1742-6596/225/1/012036

J-PARC Muon Facility, MUSE

Y. Miyake[1,2], K. Shimomura[1,2], N. Kawamura[1,2], P. Strasser[1,2], S. Makimura[1,2], A. Koda[1,2], H. Fujimori[1,2], K. Nakahara[1,2,5], S. Takeshita[1,2,6], Y. Kobayashi[1,2], K. Nishiyama[1,2], W. Higemoto[2,3], T. U. Ito[2,3], K. Ninomiya[2,3], M. Kato[1], R. Kadono[1,2], N. Sato[4], and K.Nagamine[1,7]

[1] Muon Science Laboratory, High Energy Accelerator Research Organization, Tsukuba, Ibaraki 305-0801, Japan

[2] Muon section, Materials and Life Science division, J-PARC Center,Tokai, Ibaraki 319-1195, Japan

[3]Advanced Science Research Center, Japan Atomic Energy Agency, Tokai, Ibaraki 319-1195, Japan

[4] Mechanical Engineering Center, High Energy Accelerator Research Organization, Tsukuba, Ibaraki 305-0801, Japan

E-mail: yasuhiro.miyake@kek.jp

Abstract. The muon science facility (MUSE, abbreviation of MUon Science Establishment), along with the neutron, hadron, and neutrino facilities, is one of the experimental areas of the J-PARC project, which was approved for construction in a period from 2001 to 2008. The MUSE facility is located in the Materials and Life Science Facility (MLF), which is a building integrated to include both neutron and muon science programs. Construction of the MLF building was started in the beginning of 2004, and was completed at the end of the 2006 fiscal year. For Phase 1, we managed to install one super-conducting decay/surface muon channel with a modest-acceptance (about 45 mSr) pion injector in the summer of 2008. Finally, on September 19th, 2008, the 20 mm thick edge-cooled, non-rotating graphite target, which is surrounded by a copper frame, was, for the first time, placed into the 3GeV proton beam obtained from the rapid cycling synchrotron (RCS). The nuclear reactions between the 3 GeV proton beam and the nucleus of carbon produce both positively (π^+) and negatively (π^-) charged pions. On September 26th, 2008, we finally succeeded to extract "surface muons (μ^+)", which are obtained from the decay of π^+ near the surface of the pion production target in the proton beam line. First, we commissioned the secondary muon beam line optics by tuning the superconducting magnet, the quadrupole and bending magnets, and the DC separator in order to optimize the transport of the surface muon beam and to eliminate the e^+ contamination. Then, on December 25th, 2008, we also succeeded in the extraction of the "decay muons (μ^+/μ^-)", which are obtained through the in-flight decay of π^+/π^-.

1. Introduction

The J-PARC (Japan Proton Accelerator Research Complex) has been constructed in the south part of the Tokai-JAEA (Japan Atomic Energy Agency) site, and consists of a 181 MeV LINAC

[5] Present address: Physics Department, University of Maryland, College Park, MD 20742, USA

[6] Present address: Quantum Order Research Group, Harima Institute, SPring-8 Center, RIKEN, Sayo-cho , Hyogo 679-5148, Japan

[7] Department of Physics & Astronomy, University of California, Riverside, California 92521, USA

(400 MeV in future), as well as a 3 GeV and a 50 GeV proton synchrotron rings. About 90% of the 3 GeV, 333 μA (1 MW) beam is sent to the MLF for the production of intense pulsed neutron and muon beams, while the remaining 10% will be sent to the 50 GeV ring for further acceleration for the kaon and neutrino physics programs [1].

2. Structure of the MLF Building and the NM tunnel

The MLF consists of the proton beamline tunnel (so called NM tunnel), and two wings for the experimental halls (east wing; Experimental hall No.1, west wing; Experimental hall No.2). The tunnel structure is designed to keep radioactive materials inside the tunnel, considering the safety of operations during maintenance work on the neutron target or muon target. The height of the building is 31 m for the tunnel and 21 m for the experimental halls. The width of the tunnel is 13.5 m, and the east and west wings are 24.5 m and 32 m wide, respectively. The proton beam height is 1.6 m from the floor level. The muon science facility is 30 m long along the proton beamline, and is located upstream of the neutron facility. Figure 1 shows a schematic drawing of the first floor of the MLF building. The proton beamline in the MLF building consists of the M1 line and M2 line regions. The M1 line is located upstream of the muon target, where no significant beam loss occurs. On the other hand, the M2 line is in the vicinity of the muon target where severe beam loss occurs due to the surrounding beamline components. Since a certain fraction of the primary 3 GeV proton beam is scattered preferentially downstream toward the neutron target, two sets of scrapers are installed to prevent severe damage to the beamline components such as quadrupole magnets, beam ducts etc. Although some of the scattered beam is deposited on the scrapers, the various beamline components such as the quadrupole magnets, target chamber, scraper chamber, pillow-seal, vacuum ducts etc. located along both the primary and secondary beamline will suffer not only from tremendously high radiation, but also from corrosion induced by NO_x in irradiated air. Therefore, in the M2 beamline, all of the maintenance work, including power and water connections, is intended to be done remotely from the top of the maintenance area, which is at a level of 4 m from the floor, following the lessons learned from PSI which has been dealing with a 1 MW-class proton beam [4]. Construction of the MLF building was started in the beginning of 2004, and was completed in the end of the 2006 fiscal year[2, 3].

3. Installation of the M2 line beam line components

In the M2 beamline, all beam line components must be installed via remote handling from the maintenance area above (FL 4 m). For that purpose, we installed baseplates with a precision of XY 0.5 mm on the floor of 0.5 m FL, and then we placed the alignment plates matching to the individual beam line components with a precision of XY 0.1 mm, Z 0.1 mm. Iron guide shields equipped with a guiding rail structure were aligned by placing knock pins on the alignment plates. The target chamber, the various beamline magnets (M2 primary line; six quadrupole magnets QM1, QM2, QN1, QN2, QN3, QN4 and four steering magnets, X22, Y22, X23, and Y23 and secondary line; DQ-1-3 , SQ 1-3 triplet magnets, DB1 and SB1 bending magnets), two sets of profile monitor assemblies, 20 sets of pillowseal assemblies, a gate valve assembly and seven sets of duct assemblies were also installed on the alignment plates equipped with the knock pins, which allow precise positioning, guided by the guide shields. The power and control cables are then connected between the magnets on the M2 line and the corresponding power supplies. Finally, a successful test was performed on the magnets on the M2 line by running them continuously for 12 hours. In addition, after completion of the vacuum connection, the ultimate vacuum achieved was as low as 3×10^{-5} Pa. All the beam line components in the M2 tunnel were successfully demonstrated to function properly, delivering the 3 GeV proton beam from the muon target to the neutron source as expected during the period from first proton beam injected into MLF on May 30th, 2008, to the following commissioning period.

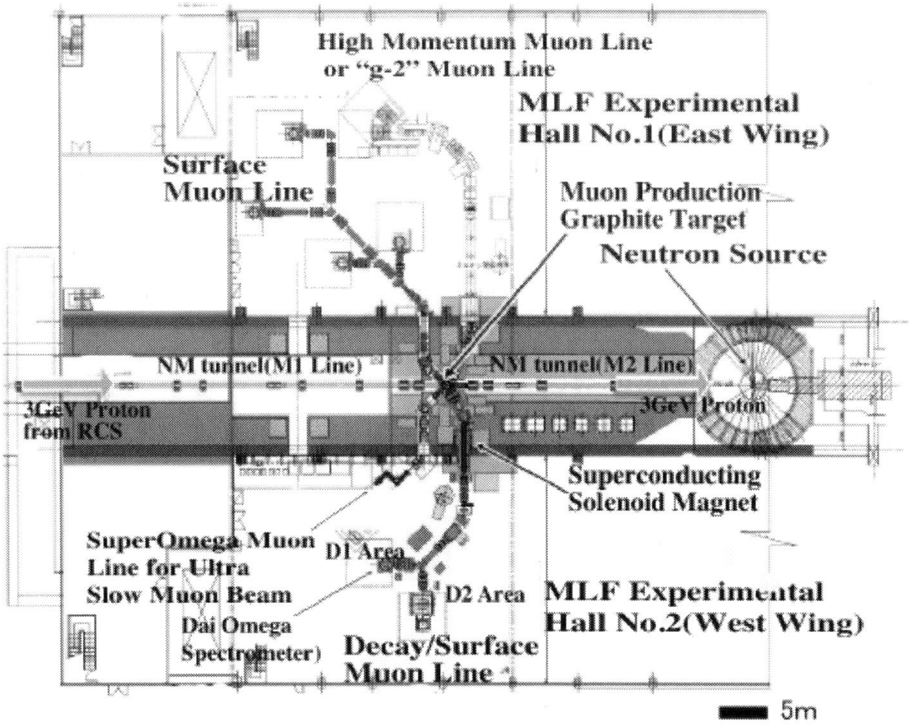

Figure 1. A schematic drawing of the MUSE facility located on the first floor of the MLF building.

4. The Muon Science facility with tandem-type production graphite target

The 3 GeV, 1 MW, 25 Hz proton beam from the RCS synchrotron is transported through the beam transport line (so called 3NBT line) over a distance of about 300 m and focused onto the 24 mm diameter muon target with a spot size that is as small as possible.

First of all, we had to make an important decision whether we should adopt having either a tandem-type target like PSI or RAL, or having a dedicated facility with our own beam dump like KEK-MSL. Finally, in order to stay within the total budget of the project utilizing the common use of the utilities and to avoid the severe beam dump construction costs associated with a tritium water handling facility with high concentration, and also from the viewpoint of the beam sharing with the neutron facility, we decided to construct the muon facility with the tandem-target configuration. The next important issue was how thick a graphite target could be installed in the proton beamline located upstream of the neutron source, causing a beam loss to some extent. After a long discussion and negotiation with the neutron science group, we reached an agreement that the total beam loss induced by the muon production targets should be no more than 10 %. Consequently, we made a basic design of MUSE, consisting of 10 mm and 20 mm thick graphite targets, corresponding to a beam loss of 3.5 % and 6.5 %, respectively, for the pion and muon production upstream of the neutron target, rather than constructing a separate building with our own proton (1 MW) beam dump.

For phase 1, we installed one graphite target with a thickness of 20 mm, from which four sets of the secondary lines can be expected to be built in the experimental hall. It is possible that one line will be used for the decay muon channel, one for high momentum muon channel and two for surface muon channels. In order to satisfy the requirements for the inlet and outlet of the primary 3 GeV line and 4 sets of the secondary lines, we fabricated a specialized target and scraper chamber. In our design concept, taking into account the ease of remote handling during maintenance work, the two sets of scrapers and muon production target are contained in one large vacuum chamber. All the water and cable connections are done at the top in the maintenance area. Consideration was given to the means of mounting the target, how to move and insert the target with the required precision, cooling, monitoring, and changing the target in the hot-cell.

5. Phase 1 of the MUSE

For Phase 1, we installed one super-conducting decay/surface muon channel with a modest-acceptance (about 45 mSr) pion injector. Details of the super-conducting decay/surface muon channel are reported by Strasser et al.[5] in this proceedings.

On September 19th, the 20 mm thick edge-cooled non-rotating graphite target, which is surrounded by a copper frame, was, for the first time, placed into the 3 GeV proton beam from the rapid cycling synchrotron (RCS). The nuclear reactions between the 3 GeV proton beam and the nucleus of carbon produce both positively (π^+) and negatively (π^-) charged pions.

5.1. Beam tuning

At first, we tried to extract a surface muon beam (positive muons) into the D1 area. In order to detect muons, we used two kinds of muon counters. The thinner counter (0.5 mm thick) was placed upstream to detect not only muons, but also positrons having the same momentum as the muons. The thicker (6 mm) counter was placed downstream, where surface muons did not pass through the first counter. Fig.2 shows the first observation of the timing signals in the oscilloscope. The earlier peak corresponds to positrons having the same momentum as the surface muons (having almost the light speed), and the delayed peak corresponds to the surface muons (having 1/4 of the light speed). By optimizing the DC separator, which is a static E $\times$ B filter, the positron contamination was removed. Practically, beam tuning has been done with use of a beam profiler, which is a two dimensional scintillator array (15 $\times$ 15). With use of the beam profiler, optimization of the beam size, timing, slits, and backgrounds, as well as the settings of the magnets along the secondary beam line, has been performed. These magnets include the pion injection magnets, the superconducting magnet, and the muon extraction magnets.

After the beam tuning, we are able to extract, at present, a surface muon (μ^+) rate of 1.1 $\times$ 10^7/s and a decay μ^+ rate of 2 $\times$ 10^6/s at 40 MeV/c and up to 10^7/s at 90 MeV/c, with a beam size of 50 mm in diameter. These rates are normalized to the intensity of the future 1 MW proton beam intensity, although the current intensity is only about 20 kW. These intensities, at the future 1 MW operation, will correspond to more than ten times those at the RIKEN/RAL Muon facility [6].

5.2. μSR experiments at the D1 area

A μSR spectrometer called DAI-OMEGA was placed at the D1 area. It consists of 64 $\times$ 2 forward and 64 $\times$ 2 backward scintillation counters, and is equipped with TF coils (up to 200 G), LF coils (up to 1.5 kG), and correction coils. Programming and adjustment of the DAQ (Data AcQuisition) system were done, as well as adjustment of the coincident timing between telescopes.

In front of a live audience, we demonstrated a μSR asymmetry measurement under a weak transverse magnetic field, adopting an aluminum plate as a sample, with use of the DAI-OMEGA spectrometer. Fig. 3 shows the first μSR asymmetry spectrum witnessed by the audience of about one hundred people. Afterwards, together with the audience, we celebrated the extraction

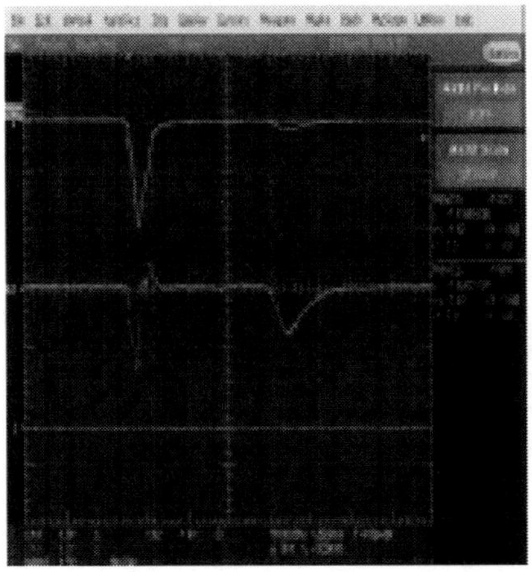

Figure 2. The first muon timing signals obtained in the oscilloscope.

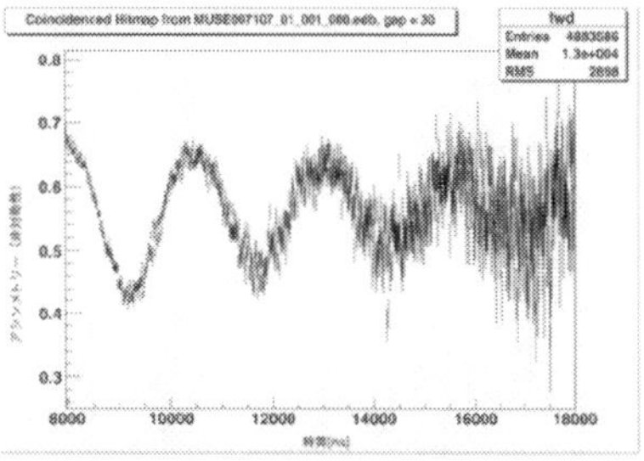

Figure 3. The first μ SR asymmetry spectrum of the aluminum target.

of the first muon beam. Fig. 4 shows a picture celebrating the first muon beam production at J-PARC MUSE.

We have already had more than ten users' runs. Among them, it is our great honor that Takeshita et al. already published a paper demonstrating the presence of a macroscopic phase separation between the superconducting and magnetic phases in Co-doped Iron Pnictide $CaFe_{1_x}Co_xAsF$, utilizing the MUSE facility[7].

5.3. Decay Muon Beam Extraction

On December 25th, 2008, we also succeeded in the extraction of the decay muons which are obtained through the in-flight decay of π^+/π^-, confined by a strong longitudinal field, which can be obtained by the superconducting magnet excited with a full specification of 600 A, corresponding to 6 tesla. As a demonstration of extracting a negatively charged muon beam, we did a trial of the non-destructive analysis of the Tempo-Koban (gold coin in Edo period), by measuring the characteristic muonic X-rays from Au and Ag. Details are shown by Ninomiya et al. in this proceedings [8].

6. Phase 2 of the MUSE

In addition to Phase 1, we are planning to install one surface muon channel with a modest-acceptance (about 50 mSr) and either a high momentum muon beam channel or a dedicated muon channel for the "g-2" experiment in the MLF experimental hall No. 1, and one super omega muon channel with a large acceptance of 400 mSr for the study of thin film magnetism or negative muon physics in the MLF experimental hall No. 2. In the case of the surface muon channel, we can expect a total of 1.6 × 10^7 surface μ^+/s with the 1 MW proton beam. By installing a kicker and beam slicer, we are planning to place four experimental ports. In the case of the super omega beam channel, we are going to install a large acceptance solenoid made of mineral insulation cables (MIC) and a superconducting curved transport solenoid. We can collect either surface or cloud muons with a large acceptance of 400 mSr. Finally, we are expecting 4 × 10^8 surface μ^+/s and 10^7 cloud μ^-/s in the MLF experimental hall No. 2 [9]. Although many of these studies can be performed using either surface or decay muons, at the super omega channel we are aiming to create a new type of muon source: an intense ultra-slow muon source. Slow muons are generated through resonant ionization of muonium (Mu). Mu

Figure 4. A picture celebrating the first muon beam production at J-PARC MUSE.

is formed by stopping an intense surface muon beam on the rear surface of a hot W foil. At the RIKEN/RAL muon facility, 20 slow μ^+/s are obtained out of 1.2 $\times$ 10^6 surface muons/s [10]. Taking into account the repetition rate of the pulsed laser system and the proton beam, as well as the surface muon ratio between RIKEN-RAL and J-PARC MUSE, we can expect 1.3 $\times$ 10^4 slow μ^+/s without any additional laser development. A rate of 1.3 $\times$ 10^6 slow μ^+/s can be achieved with additional laser development, such as the tripling of 366 nm photons with pico second pulse width to match the Doppler broadening of the Mu at 2000 K. When the production of this intense ultra-slow muon beam is realized, the use of its short-range penetration depth will allow muon science to be expanded towards a variety of new scientific fields, such as:

- (i) Surface/boundary magnetism utilizing its spin polarization and unique time-window,
- (ii) Surface chemistry, utilizing a feature of a light isotope of hydrogen; such as catalysis reactions,
- (iii) Precise atomic physics such as QED (Quantum Electro Dynamics) and
- (iv) Low emittance ion sources and further acceleration towards μ^+ μ^- collider experiments in high-energy physics.

7. Summary

Construction of the MLF building, as well as the fabrication and installation of various proton beam line components such as magnets, the muon target, scrapers, shields, monitors, and vacuum components, were completed in 2007, as scheduled. Finally, we succeeded in the extraction of a "surface muon (μ^+)" beam on September 26th, 2008. We also succeeded in the extraction of a "decay muon (μ^+/μ^-)" beam, which is obtained through the in-flight decay of π^+/π^-, on December 25th, 2008.

We have been accepting user's programs open to the world since January, 2009.

References

[1] S. Nagamiya, T. Nagae, Y. Ooyama, Y. Miyake, H. Takano, J.R. Helliwell, J.C. Peng, Energy Review 19, 12(1999) 4-23

[2] Y. Miyake, K. Nishiyama, K. Fukuchi, N. Kawamura, S. Makimura, K. Shimomura, R. Kadono, W. Higemoto, J.L. Beveridge, K. Ishida, T. Matsuzaki, I. Watanabe, Y. Matsuda, and K. Nagamine, Physica B, 326 (2003)255-259

[3] Y. Miyake, K. Nishiyama, N. Kawamura, P Strasser, S. Makimura, A. Koda, K. Shimomura, H Fujimori, N. Nakahara, R. Kadono, M. Kato, S Takeshita, W. Higemoto, K. Ishida, T. Matsuzaki, Y. Matsuda, and K. Nagamine, Nucl. Instr. and Meth. in Phys. Res. A 600(2009) 22-24

[4] G.Heidenreich, P. Baumann, A. Geissler, A. Strinning and W. Wagner, PSI Scientific Report VI (1998) 16

[5] P. Strasser et al., this proceedings

[6] K. Nagamine, T. Matsuzaki, K. Ishida, I. Watanabe, S.N. Nakamura, R. Kadono, N. Kawamura, S. Sakamoto, M. Iwasaki, M. Tanase, M. Kato, K. Kurosawa, G.H. Eaton, H.J. Jones, G. Thomas, and W.G. Williams, Hyperfine Interactions 101/102(1996)521

[7] S. Takeshita, R. Kadono, M. Hiraishi, M. Miyazaki, A. Koda, S. Matsuishi, and H. Hosono, Phy.Rev.Lett. 103 (2009)027002

[8] K. Ninomiya et al., this proceedings

[9] K. Nakahara, Y. Miyake, K. Shimomura, P. Strasser, K. Nishiyama, N. Kawamura, H. Fujimori, S. Makimura, A. Koda, K. Nagamine, T. Ogitsu,A. Yamamoto, T. Adachi, K Sasaki, K. Tanaka, N. Kimura, Y. Makida, Y. Ajima, K. Ishida, Y. Matsuda, AIP conference Proceedings, 981(2007)312-314

[10] P. Bakule, Y. Matsuda, Y. Miyake, K. Nagamine, M. Iwasaki, Y. Ikedo, K. Shimomura, P. Strasser, S. Makimura, Nucl. Instr. and Meth. in Phys. Res. B266 (2008) 335-346

J-PARC Decay Muon Channel Construction Status

P. Strasser[1,2], K. Shimomura[1,2], A. Koda[1,2], N. Kawamura[1,2], H. Fujimori[1,2], S. Makimura[1,2], Y. Kobayashi[1,2], K. Nakahara[1,2,7], M. Kato[1,2], S. Takeshita[1,2,8], M. Hiraishi[3], M. Miyazaki[3], W. Higemoto[4], T. U. Ito[4], K. Ninomiya[4], K. Ishida[5], M. K. Kubo[6], R. Kadono[1,2], K. Nishiyama[1,2], and Y. Miyake[1,2]

[1] Muon Science Laboratory, Institute of Materials Structure Science, High Energy Accelerator Research Organization (KEK), Tsukuba, Ibaraki 305-0801, Japan
[2] Muon Section, Materials and Life Science Division, J-PARC Center, Tokai, Ibaraki 319-1195, Japan
[3] Department of Materials Structure Science, The Graduate University for Advanced Studies (Sokendai), Tsukuba, Ibaraki 305-0801, Japan
[4] Advanced Science Research Center, Japan Atomic Energy Agency (JAEA), Tokai, Ibaraki 319-1195, Japan
[5] Advanced Meson Science Laboratory, RIKEN Nishina Center, RIKEN Wako, Saitama 351-0198, Japan
[6] Department of Material Science, The Graduate School Division of Natural Sciences, International Christian University, Osawa, Mitaka, Tokyo 181-8585, Japan

Abstract. The new Muon Science Facility (MUSE) that is now under construction at J-PARC in the Materials and Life Science Facility (MLF) building will comprise four types of muon channels. In the first stage, a conventional superconducting decay muon channel (D-Line) was constructed, which can extract surface (positive) muons with an expected muon yield of 10^7/s and decay positive/negative muons up to 120 MeV/c, with an expected muon yield of a few 10^6/s at 60 MeV/c for both positive and negative muons. This channel will be used for various kinds of muon experiments like μSR, muon catalyzed fusion and nondestructive elements analysis.

1. Introduction

The new Muon Science Facility (MUSE) [1] is now under construction at J-PARC (Japan Proton Accelerator Research Complex) at the Tokai campus of JAEA (Japan Atomic Energy Agency) in the Materials and Life Science Facility (MLF) building under the collaboration between KEK and JAEA. In the J-PARC project, the 3-GeV 333-μA-proton beam from the rapid cycling synchrotron (RCS) will be transported from the RCS ring to the neutron source situated in the MLF building. There, in the M2 primary proton beamline (M2 tunnel), just 33 m upstream of the neutron source, the muon target made of a 2-cm-thick graphite disc for the production of intense pulsed pion and muon beams is located. To make use of the full potential of this new facility, four dedicated muon channels will be constructed. In the MLF building, the

[7] Present address: Physics Department, University of Maryland, College Park, MD 20742, USA
[8] Present address: Quantum Order Research Group, Harima Institute, SPring-8 Center, RIKEN, Sayo-cho, Hyogo 679-5148, Japan

experimental area is divided into two parts across the primary proton beamline, namely the experimental hall No. 1 (East) and the experimental hall No. 2 (West). The four secondary muon channels that are originating from the muon target in the M2 tunnel are the surface muon channel (S-line) and the high-momentum muon channel (H-line) in the experimental hall No. 1, and the superconducting decay muon channel (D-line) and the ultra-slow muon channel (U-line) in the experimental hall No. 2. At present, only the D-Line is under construction. The other three secondary muon channels were temporarily closed with iron shields and concrete blocks.

This report gives a brief overview of the D-line design and construction, including the superconducting solenoid, the vacuum system, the interlock system and the experimental areas. A previous report was published in [2]. The new muon profile monitor used in the beam commissioning and tuning is also described.

2. Superconducting Decay Muon Channel

The layout of the D-Line in the experimental hall No. 2 of the MLF building is shown in Fig. 1. It consists of three parts: (1) a pion injection, (2) a superconducting decay solenoid, and (3) a muon extraction. Two experimental areas are planed for simultaneous use. This conventional superconducting decay muon channel can extract surface (positive) muons with an expected muon yield of a few 10^7/s and decay positive/negative muons up to 120 MeV/c, with an expected muon yield of a few 10^6/s at 60 MeV/c for both positive and negative muons. The overall transport efficiency is estimated by TURTLE. This channel will be used for various kinds of muon experiments like μSR, muon catalyzed fusion and nondestructive elements analysis.

(1) Pion injection: A quadrupole triplet (DQ1-2-3) is placed at a position of 65 cm from the graphite target, which can accept pions in a solid angle of 65 msr. The following bending magnet (DB1) transports pions to the solenoid up to 250 MeV/c at maximum. Therefore kaon decay muon from the target can also be extracted by upgrading the muon extraction to 250 MeV/c. The coils of the quadrupole triplet and the bending magnet are made by MIC for the hard radiation environment. The magnets are directly installed in the M2 primary proton beamline.

(2) Superconducting decay solenoid: To obtain a high intensity decay muon beam, a superconducting solenoid magnet (DSOL) is employed for the pion to muon decay section. The basic design is similar to those used in the muon beamlines at KEK-MSL, TRIUMF and RIKEN-RAL. The decay solenoid consists of twelve units of superconducting coils with 6 cm in bore radius and 50 cm in length. The applied magnetic field is 5 T. The pions and muons are confined within a radius of 5 cm and therefore transported without any significant loss. The old solenoid that was used at KEK-MSL has been modified for this purpose. The magnet coil is forced-indirectly cooled by a supercritical helium gas supplied from an on-line helium refrigeration system. To achieve the muon extraction at low momentum, this cold bore magnet is directly connected to the muon beamline using only thin thermal insulating aluminum foils of 12.5 μm thick at the entrance and exit of the superconducting solenoid magnet, respectively. An insertion device has been developed for the installation of the superconducting solenoid, because it is set between the M2 tunnel and the experimental hall, as shown in Fig. 1. Two sets of linear guide were adopted for the horizontal transport motion to keep good reproducibility. Six iron blocks, with a total weight of about 50 tons, are also set on the insertion device for the radiation protection from the M2 primary beamline.

(3) Muon extraction: The extraction can transport muons up to 120 MeV/c. A magnetic kicker will be installed in a near future for single-pulse experiments and simultaneous use by the two experimental areas. The major components of the old decay muon channel at KEK-MSL, such as the bending magnets, the quadrupole magnets and the DC separator are being re-used. Three Q-triplets (DQ4-5-6, DQ7-8-9 and DQ13-14-15) and two bending magnets (DB2 and DB3) are used to transport the muon beam to the experimental area D2. The D1-leg at the downstream of DB3 that leads to the experimental area D1, which is equipped with two

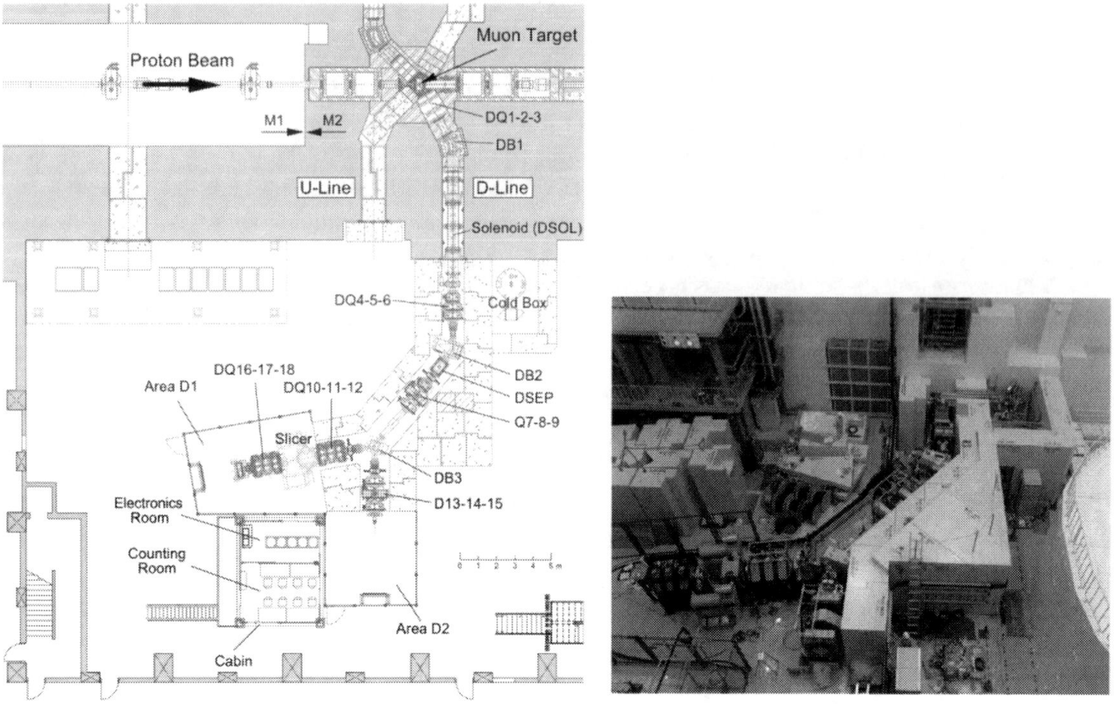

Figure 1. Layout of the decay muon channel in the experimental hall No. 2 of the MLF building (left) and a picture showing the D-Line under construction (right).

Q-triplets (DQ10-11-12 and DQ16-17-18) and a beam-slicing kicker, is constructed by JAEA [3]. All of the magnets in the experimental hall, except those prepared by JAEA, were formerly used at KEK-MSL in Tsukuba. After refurbishment and modification, the magnets are now successfully installed on the D-line. The magnetic poles of both DB2 and DB3, which are made of SS400, were newly fabricated. In particular, the magnetic pole of DB3 was designed so that replacement with a septum magnet in a near future can be easily conducted. The field mapping of both bending magnets was examined after reassembling, resulting that the uniformity of BL integration is less than 2% for DB2 and 5% for DB3, respectively. The assembled Q-triplet, which shares a common star-shaped beam duct, was designed in view of handling by crane operation. The magnets are settled on a common support and placed on a pre-aligned base plate, so that the assemblies can be remotely installed and uninstalled. When placing the triplet magnets, they are guided and precisely positioned using pivots, which are secured on the base plate, in a similar manner to that of the M2 primary beamline magnets.

3. Superconducting Solenoid Refrigeration System

The solenoid magnet coil is forced-indirectly cooled by a supercritical He gas (4.8K at 1.0 MPa) supplied from an on-line helium refrigeration system (TCF 50) for the long-term stable operation. A 80 K copper thermal shield is positioned between the 6 K shield tube and the warm iron cryostat vessel. The 6 K and 80 K thermal shields are supported by insulation rods extending from the cryostat wall. The 80 K shield is cooled by the He gas taken from the intermediate heat exchanger in the cold box.

The helium refrigeration system consists principally of the followings: (1) a He buffer tank (20 m^3, 0.95 MPaG), (2) a cooling tower and a water cooling pump, (3) a helium gas screw compressor (Kaeser), (4) an after cooler, (5) an oil separator, (6) load, unload and bypass valves, (7) a cold box, (8) a VME based digital control device, (9) a PC system for control and data logging, and (10) a cryo panel for safety interlock control.

The screw compressor supplies high-pressure He gas (0.85 MPa) to the cold box. The cold box is designed to supply various types of He requested to the superconducting solenoid. Before the cooling down procedure from the room temperature is commenced, the oxygen concentration of the circulating He gas is reduced to be less than 50 ppm by the operating He gas purifier.

The cooling power is 35 W at 4.5K and 200W at 80K, and it can produce 8 l/h of liquid He. The whole system is monitored and controlled by a VME controller combined with a personal computer with dedicated software based on the LabVIEW system, and cools down automatically. The typical cooling period from the room temperature to the operating temperature (~4 K) is about 4 days. The long-term (3 months) operation is established under quite stable condition.

To protect against any serious damage to the superconducting coils and the He refrigeration system, a VME based interlock system is installed. The system can detect any anomaly in the voltage of the superconducting solenoid, the power lead temperature and the trip signal from the He refrigeration system. Once any emergency status is detected, the electric power supply immediately stops the current supply and the refrigeration system changes to self-operation mode. This system can also record the temperatures and pressures of the superconducting coils, the cryostat and the power leads.

4. Vacuum System

The block diagram of the vacuum system of the decay muon channel is shown in Fig. 2. The vacuum system of the secondary beam channel is isolated from the primary proton beamline by an all-metal gate valve (DGV1; VAT 48146) with a 50 μm-thick Kapton beam window protected

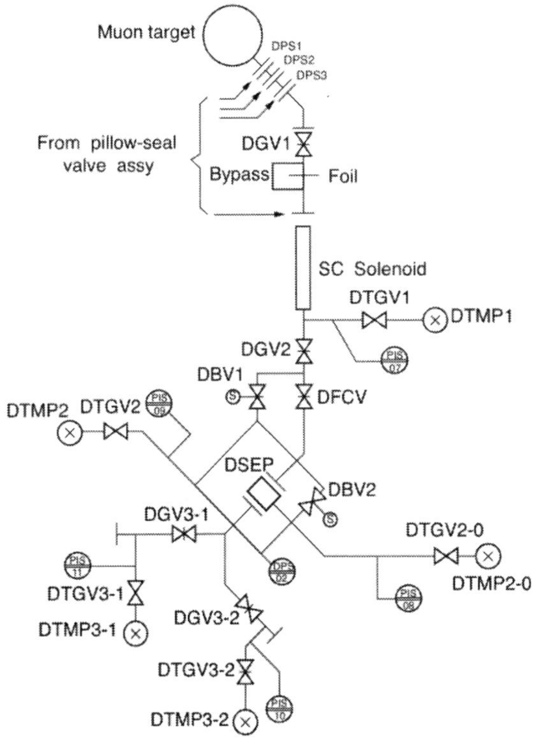

Figure 2. Vacuum system diagram of the decay muon channel.

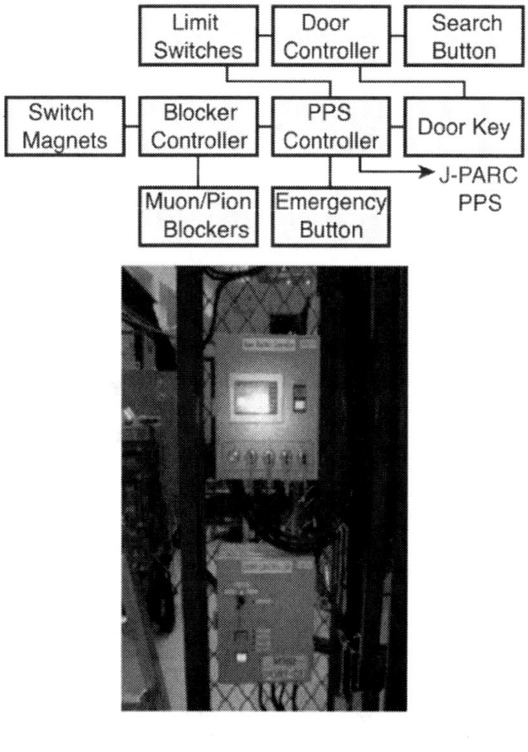

Figure 3. Area entry system diagram (top), and view of the area entry door (bottom).

a bypass. The beam channel is divided into four sections, separated by three gate valves (DGV2, DGV3-1, and DGV3-2; VAT 12150). Each of the sections is separately operated by using turbo molecular pumps with a pumping speed of 0.5 m^3/s. It is noted that the section including the superconducting solenoid should have a separate vacuum system from a viewpoint of operation of the cryogenic system. Particularly, in the event of a quench of superconductivity, tritium gas absorbed through the bypass during the beam time is released due to the temperature raise. Therefore, the exhaust must be led to a vent stack without passing through the other pumping stations. The achieved pressure of the beam channel has been typically $\sim 1\times10^{-4}$ Pa. At the moment, the vacuum chamber of the DC separator (DSEP) is directly connected to the beamline. However, in order to prevent the high-voltage gap from vacuum discharging, it is designed to control pressure independently by inserting beam foils at both sides of the chamber.

A fast closing valve (DFCV; VAT 75046) is installed at the downstream of DGV2 to protect the vacuum of the primary beamline in the event of a sudden vacuum loss. Two FV-type pressure sensors are attached on each side of two beam legs, which respond in 1 ms to close DFCV. If closing of DFCV is detected, DGV2 is also automatically closed. In the case that either of the two experimental areas is in a maintenance condition, the corresponding pressure sensor is turned off depending on the states of DGV3-1 or DGV3-2. At both ends of the beamlines, there are beam foil windows made of Mylar or Kapton with a thickness of 100 μm. The pressure readouts and the open/close status of the gate valves are monitored at the MLF control room through the MELSEC network.

Figure 4. The D-line experimental areas and the cabin in the experimental hall No.2 of the MLF building (top left). Detailed view of the experimental area D1 (top right), the area D2 (bottom left), and the counting room inside the cabin (bottom right).

5. Interlock System

To ensure a safe operation of the facility, a number of safety systems have been incorporated into the interlock system. For example, to prevent users from being exposed to excess dosage of radiation, the entry to experimental areas is strictly controlled. In addition, proper and safe operation of the machines is also necessary. To ensure this, all the muon experimental instruments are controlled by means of the Muon Control System. Its purpose is to control and monitor the safe operation of all the muon instruments. In particular, the personal protection system (PPS) is in place to prevent personnel from being exposed to high levels of radiation, and is the highest priority interlock system at J-PARC. The Muon Control System is related to the operation of the muon target, the D-line and the D1, respectively D2, experimental area entry system. In particular, the experimental area entry system is categorized in the PPS system. All these components worked successfully on Day 1.

The experimental area entry system diagram is shown in Fig. 3 (top). In this system, several status of the PPS related components are being watched. For example, the status of the beam blockers are monitored, which is important to ensure the safety of the personnel. When the muon blocker is opened or the DB3 magnet, which acts as a switching magnet, is turned on, the PPS controller does not allow the area entry door to be opened. This logic guarantees the safety operation for the experimenter. Figure 3 (bottom) shows a view of the area entry door with the muon blocker controller (up) and the door controller (down). The blocker controller has five keys. Four keys are used as personal keys. If all the keys are not back to the muon blocker controller, one can not lock the experimental area door. The operation of the muon blocker can be performed from this muon blocker controller. This system is common with the neutron shutter control system. The following procedure is necessary to enter an experimental area. An experimenter needs first to turn off the switching magnet and close the muon blocker. Then he moves a key from the blocker controller to the door controller and unlock the door. The area entry door can now be opened.

6. Experimental Areas and Cabin

The decay muon channel is divided into two legs after the bending magnet DB3 (see Fig. 1), each leg leading to an experimental area, D1 and D2, respectively. Between the two experimental areas, a cabin was installed to house the electronics room and the counting room. Pictures representing the two experimental areas and the cabin including detailed views of the experimental area D1, the experimental area D2 and the interior of the cabin are shown in Fig. 4. The experimental area D1 is occupied with a beam slicer with slits, a quadruople magnet triplet and a μSR spectrometer for μSR experiments. A more detailed description of this μSR experimental apparatus is reported in these proceedings [3]. The experimental area D2 is presently used for commissioning and beam optics tuning. Each experimental area is equipped with electrical power outlets (AC 100V, 200V and 420V) and cooling water supply. In the future pressurized air, helium recovery line and exhaust line will also be installed. The cabin is divided into two parts, the electronics room and the counting room, respectively. The electronics room is air-conditioned and equipped with five 19-inch rack for the electronics modules of the data taking system. This cabin will be used by users who perform experiments at either one of the experimental areas.

7. Beam Tuning

A new muon beam profile monitor (MBPM) was constructed for the commissioning of the decay muon channel and the beam optics tuning. The objective for the design of this new beam profile monitor was similar to that developed at RIKEN-RAL with Dr. K. Ishida [4], that is to be able to obtain both horizontal and vertical beam profiles simultaneously using surface muons (μ^+), which have a momentum of only 27 MeV/c. At this energy, the range of a muon in a

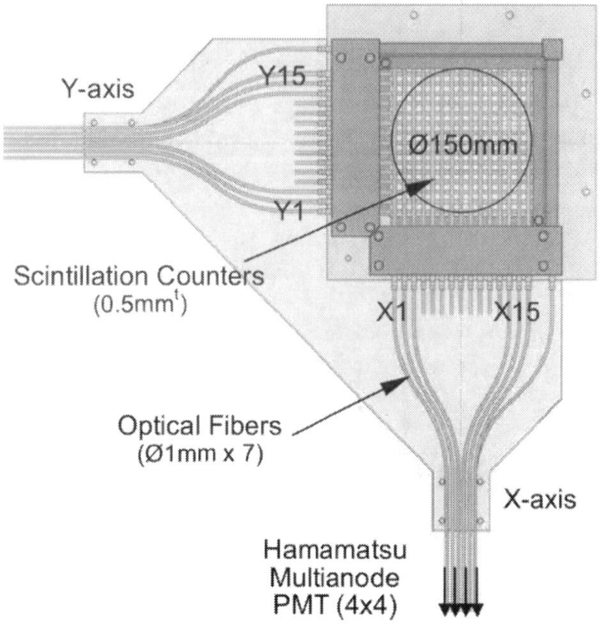

Figure 5. Schematic view of the new muon beam profile monitor.

Figure 6. The new beam profile monitor installed at the area D2.

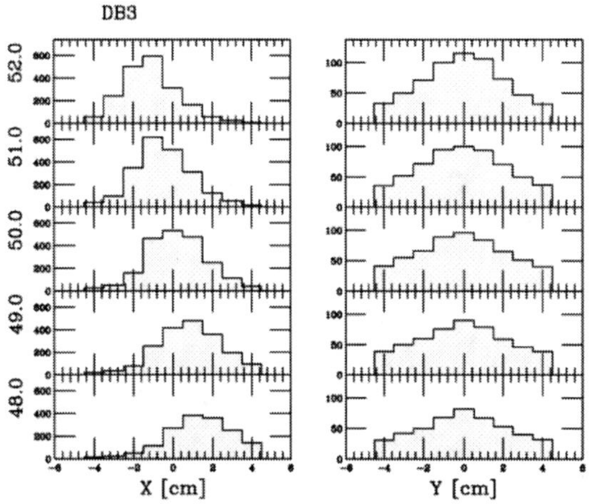

Figure 7. Horizontal (X) and vertical (Y) muon beam profiles as a function of the DB3 bending magnet settings measured with surface muons at the experimental area D1.

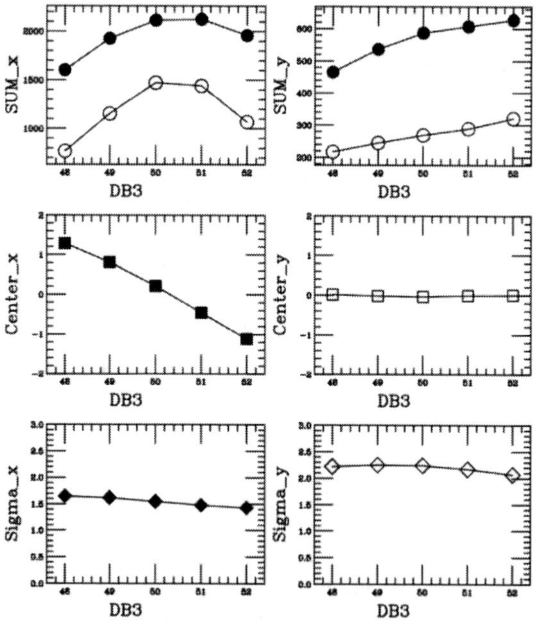

Figure 8. Beam parameters of Fig. 7.

scintillator (NE-102) is only about 1 mm. This profile monitor should also be able to operate with pulsed beams and reject positron or electron contamination in the muon beam. Finally, it should work with high-intensity surface muon beams as well as with low-intensity backward decay muon (μ^+/μ^-) beams.

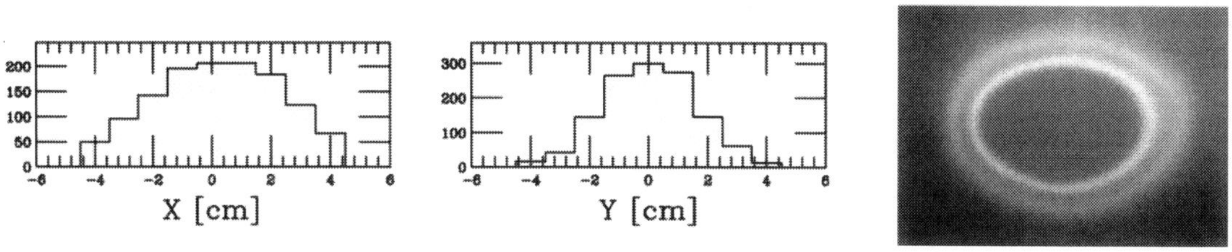

Figure 9. Muon beam profile at the experimental area D1 measured with the new muon beam profile monitor (left) and with an imaging plate (right).

The schematic view of the beam profile monitor is shown in Fig. 5. Each axis comprises fifteen (15) scintillators (4-mm wide and 0.5-mm thick, with 10-mm steps) mounted on a light guide and connected each with seven ϕ1-mm optical fibers to a multianode (4x4) photomultiplier (Hamamatsu type H6568-200-10MOD). Each scintillator is wrapped with two layers of 6-μm-thick Aluminum-Mylar. In addition, the front and the back of the central area are also covered with an additional layer. This proved to be sufficient to reduce light leaks to an acceptable level without significantly slowing down the incoming muons. Each photomultiplier output was sent to a charge ADC (LeCroy model 2249W) with a time gate of 200 ns to be digitized. The same data acquisition program used at RIKEN-RAL that is based on EXP2K and modified by Dr. K. Ishida for the use of the MBPM was used to process the data and fill the histograms. A picture of the new muon beam profile monitor is shown in Fig. 6.

As an example of the performance of this beam profile monitor, the horizontal and vertical muon beam profiles obtained with surface muons at the experimental area D1 for different DB3 bending magnet settings are shown in Fig. 7. Each profile can be obtained with just 1000 spills or about 40 seconds at 25Hz beam operation. Figure 8 shows the different beam profile parameters that are generated automatically after each profile is completed. This graphical interface was found to be very useful in rapid visualization of the response of the beam to each magnet and for optimization of the beamline parameters.

In the future, the three slit box that are used along the D-line to select the muon beam momentum and control the muon beam size will also have a compact muon beam profile monitor just behind the slits for beam optics tuning purposes. Until now, the muon beam tuning has only been performed at the end of a muon channel by looking at the total muon intensity or by measuring the muon beam profile. A vacuum version of this new MBPM using multianode PMT's and optical fibers is now under consideration.

8. Concluding Remarks

At the end of the commissioning beam time of the experimental area D1 in December 2008, the muon beam profile shown in Fig. 9 was obtained. The figure on the left represents the beam profile measured with the new muon beam profile monitor, and the one on the right the profile measured with an imaging plate. The beam tuning is still not completely optimized yet, and the intensity is somewhat lower than expected. Further beam tuning is scheduled to optimize the beam transport efficiency as the beam power of the primary proton beam is gradualy increased.

References

[1] Y. Miyake *et al.*, these proceedings.
[2] K. Shimomura *et al.*, Nucl. Inst. Meth. A 600 (2009) 192.
[3] W. Higemoto *et al.*, these proceedings.
[4] P. Strasser *et al.*, RIKEN Accel. Prog. Rep. 36 (2003) 202.

The EPICS-based remote control system for muon beam line devices at J-PARC MUSE

T. U. Ito[1], K. Nakahara[2*], M. Kawase[3], H. Fujimori[2], Y. Kobayashi[2], W. Higemoto[1], and Y. Miyake[2]

[1] Advanced Science Research Center, Japan Atomic Energy Agency, Tokai, Ibaraki 319-1195, Japan
[2] Muon Science Laboratory, High Energy Accelerator Research Organization (KEK), Tsukuba, Ibaraki 305-0801, Japan
[3] J-PARC Center, Japan Atomic Energy Agency, Tokai, Ibaraki 319-1195, Japan

E-mail: ito.takashi15@jaea.go.jp

Abstract. The remote control system for muon beam line devices of J-PARC MUSE has been developed with the Experimental Physics and Industrial Control System (EPICS). The EPICS input/output controller was installed in standard Linux PCs for slow control of the devices. Power supplies for 21 magnetic elements and four slit controllers for the decay-surface muon beam line in the Materials and Life Science Experimental Facility are now accessible via Ethernet from a graphical user interface which has been composed using the Motif Editor and Display Manger.

1. Introduction

The MUon Science Establishment (MUSE) is located at the Materials and Life Science Experimental Facility (MLF) at J-PARC in Tokai, Japan [1]. A high intensity proton beam is incident on a graphite muon production target [2] in order to produce muons that are captured by four independent secondary beam lines. These include the decay-surface muon beam line, the surface muon beam line, the Super-Omega muon beam line [3], and the high momentum muon beam line. The first beam line to be completed was the decay-surface muon beam line, where the first muon beam has successfully been delivered on September, 2008.

The decay-surface muon beam line captures both surface and decay muons and transports them into experimental area No.2, where the beam line splits into the D1 and D2 channels. The beam line consists of three bending magnets, 18 quadrupole magnets, four slits, a DC separator, a superconducting solenoid, and a beam slicer for the D1 channel [4]. A septum magnet and beam kicker will be installed at the branch point of the D1 and D2 channels in the future for simultaneous operation. Of the 21 magnetic elements currently available, quadrupoles DQ1 through DQ9 and all three bending magnets DB1 through DB3 are common to both channels, while quadrupoles DQ10 through DQ12 and DQ16 through DQ18 provide beam to the D1 channel, and quadrupoles DQ13 through DQ15 transport the beam into the D2 channel.

* Present address: Department of Physics, University of Maryland College Park, MD 20742, USA

Among the beam line elements, the bending magnets, the quadrupole magnets, and the slits are integrated into a common control system. This system can broadly be separated into three categories. They include functions that are performed at the hardware level, processes that are hard-wired into the Programmable Logic Controller (PLC) equipped in each controlled device, and those that are controlled directly by the human machine interface (HMI). In this paper, we focus on the HMI for remote control of the decay-surface muon beam line devices using the Experimental Physics and Industrial Control System (EPICS).

2. EPICS-based remote control system for decay-surface muon beam line devices

2.1. Programmable Logic Controller

The power supplies (PS) for the beam line magnets are located in the power supply yard inside the experimental area No. 2. They can be operated locally using a local control panel available on each power supply, or remotely via Ethernet. Each power supply is equipped with its own PLC with an Ethernet port (FA-M3R, Yokogawa Denki Corp.) and is integrated into a closed control system network available at MLF. The power supply hardware is parameterized through the PLC hardware registry. The variables defined in the registry can then be used to program, via ladder logic, to provide functions such as switches, set points, and interlocks.

The four slits are powered and governed by two controllers with own PLCs. The slit controller for three slits, DSL2, DSL3-1, and DSL3-2, is equipped with a PLC with a serial communication port (KV-3000, Keyence Corp.) while the other one for DSL4 is controlled by a PLC with an Ethernet port (KV-5000, Keyence Corp.). The four slits can be operated locally using local touch-screens, or remotely via RS-232C or Ethernet.

2.2. Experimental Physics and Industrial Control System

The EPICS is a set of Open Source software tools, libraries, and applications to create distributed soft control systems for large scientific experiments [5]. Developed at Argonne National Laboratory, the EPICS is used extensively for real-time and slow control of accelerator devices. At MUSE, the EPICS is used for the slow control of the beam line devices via the PLC. The EPICS software framework was installed in standard Linux PCs, including a Channel Access (CA) client and the Input/Output Controller (IOC) to serve as a server that communicates with the PLC and publish the information to the CA client using the CA network protocol. A graphical user interface (GUI) was composed using the Motif Editor and Display Manager (MEDM) [6], allowing users easily to access the PLCs via EPICS.

2.3. EPICS-based remote control system on MLF control system network

The EPICS-based remote control system for the decay-surface muon beam line devices is modeled after that of the accelerator EPICS control system. A schematic drawing of this system is shown in Fig. 1. Two Linux PCs are currently used as platforms for the EPICS-IOCs which communicate with the PLCs via Ethernet on the closed control system network or RS-232C. The EPICS-IOC for the magnet PSs was installed on Red Hat Enterprise Linux running on the PC1, communicating with the FA-M3R via Ethernet. A device support for the FA-M3R was developed at Japan Atomic Energy Research Institute (JAERI) and Japan Atomic Energy Agency (JAEA). Relations between the PS functions mapped into the PLC registry and EPICS records were defined in EPICS database.

The EPICS-IOCs for controlling the slits were installed on Debian GNU/Linux running on the PC2. A device support for the KV-3000 connected with the PC2 via RS-232C was developed on the top of a serial communication support of asynDriver [7]. On the other hand, netDev [8] including a device support for Keyence PLCs was used for construction of the EPICS-IOC communicating with the KV-5000 via Ethernet.

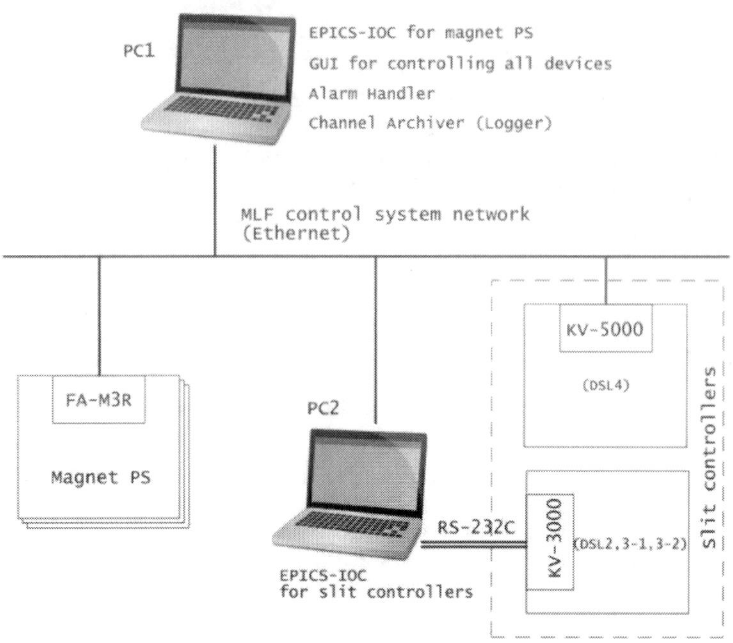

Fig.1: A schematic drawing of the EPICS-based remote control system for the decay-surface muon beam line devices.

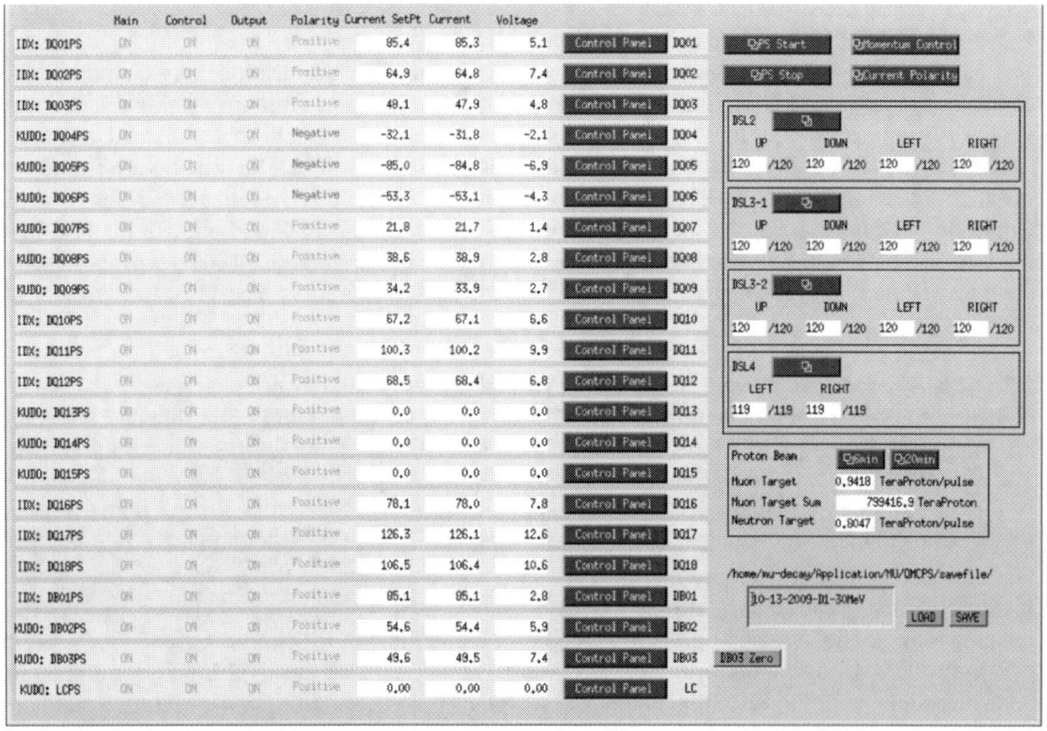

Fig.2: A screenshot of the status monitor. The status of PS for a longitudinal coil installed to a μSR spectrometer at the D1 port is displayed in addition to that for the decay-surface muon beam line magnets.

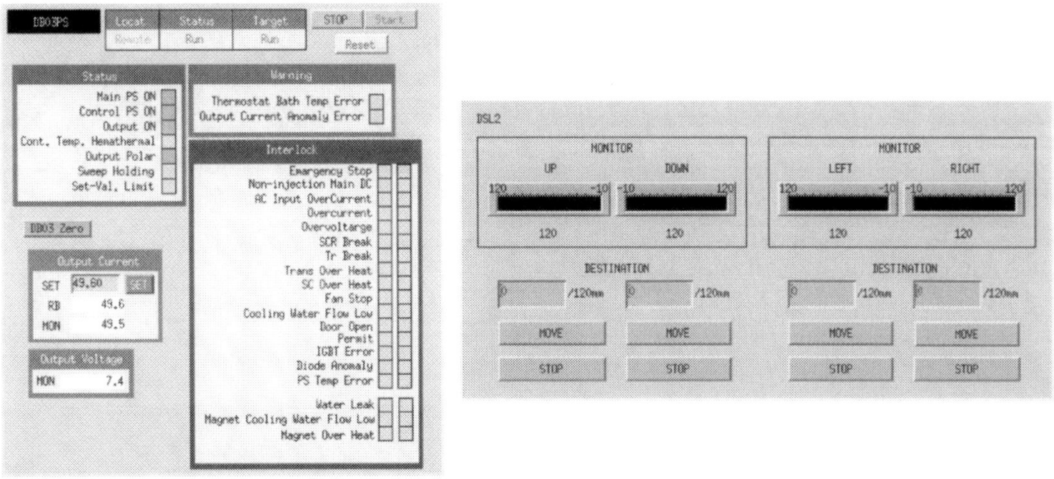

Fig.3: A screenshot of typical examples of the control panel for the magnet PS (left) and the slit (right).

The GUI for monitoring and changing device status via EPICS was also installed on the PC1. A screenshot of a status monitor is shown in Fig. 2. The magnet PS status (main power, control power, current output, polarity, current set point, current monitored, voltage monitored), the positions of the slits, and the proton beam intensity are displayed on the status monitor and updated typically every one second. Since our system and the accelerator EPICS control system are connected to the MLF control system network, the proton beam intensity can be monitored at MUSE.

This panel also works as a launcher of control panels for each beam line component and built-in functions written in perl, such as save/load of particular beam line tunes. Features are added such that it is possible to change the PS currents of group of magnets from a pre-set reference value in order to facilitate beam line tuning. This feature allows the user to scale the currents of all beam line magnets from their present value to match the tune to a different momentum.

Typical examples of the control panel for the magnet PS and the slit are shown in Fig.3. The users can control the on/off switch, current set point and ramp switch, polarity switch, and interlock reset switch of the magnet PS, and the move/stop switch and set point of the slit position from these panels. The control panel for the magnet PS allows for continuous monitoring of interlock and warning status as well.

In addition to local alarms that are available on the magnet PSs, the Alarm Handler is equipped in the remote control system. When the interlock is triggered, the Alarm Handler is designed to issue audible warnings from the PC1. Furthermore, the Alarm Handler will issue warnings when the network connection between the PC1 and any of the PLC is severed for more than a predetermined period of time.

Values of the EPICS records are logged by the Channel Archiver [9] installed in the PC1. The logged variables, such as the output current, the slit positions, and the proton beam intensity, can be displayed as functions of time using a data visualization software written in Java included in a package of the Channel Archiver.

3. Summary

We developed the EPICS-based remote control system for the decay-surface muon beam line devices at J-PARC MUSE. The EPICS-IOCs and the MEDM-based GUI were installed in standard Linux PCs on the MLF control system network for slow control of the magnet PSs with FA-M3R and the slit controllers with KV-3000/5000. The device support for FA-M3R developed at JAERI/JAEA, asynDriver, and netDev were used for construction of the EPICS-IOCs. Major functions of the magnet PSs and the slits are accessible from the GUI via Ethernet.

References

[1] Miyake Y *et al.* 2009 *Nucl. Instr. and Meth.* A **600** 22-24
[2] Makimura S *et al.* 2009 *Nucl. Instr. and Meth.* A **600** 146-149
[3] Nakahara K *et al.* 2009 *Nucl. Instr. and Meth.* A **600** 132-134
[4] Higemoto W *et al.* 2009 *Nucl. Instr. and Meth.* A **600** 182-184
[5] http://www.aps.anl.gov/epics/
[6] http://www.aps.anl.gov/epics/extensions/medm/
[7] http://www.aps.anl.gov/epics/modules/soft/asyn/
[8] http://www-acc.kek.jp/EPICS_Gr/Products/NetDev/netdev-man.html
[9] http://ics-web.sns.ornl.gov/kasemir/archiver/

JAEA-ASRC muon research at J-PARC MUSE

W Higemoto[1,2], T U Ito[1,2], K Ninomiya[1,2], R H Heffner[1,3], K Shimomura[2,4], K Nishiyama[1,2,4], and Y Miyake[2,4]

[1] Advanced Science Research Center, Japan Atomic Energy Agency, Tokai, Naka, Ibaraki, 319-1195, Japan
[2] J-PARC Center, Japan Atomic Energy Agency, Tokai, Ibaraki 319-1195, Japan
[3] Los Alamos National Laboratory, Los Alamos, New Mexico 87545, USA
[4] Muon Science Laboratory, High Energy Accelerator Research Organization , Tsukuba, Ibaraki 305-0801, Japan

E-mail: higemoto.wataru@jaea.go.jp

Abstract. The Japan Atomic Energy Agency (JAEA)-Advanced Science Research Center (ASRC) has developed experimental equipment at the J-PARC MLF muon science facility (MUSE) for muon spin rotation/relaxation experiments. We have extracted part of the muon beam into a muon spectrometer constructed downstream from the Decay/Surface muon beam line. The current status of our project is discussed here.

1. Introduction

The Muon Science Facility (MUSE) at the J-PARC Materials and Life Sciences Facility (MLF) will be utilized for various fields of science, including nuclear and condensed-matter physics and chemistry. Muons are created at MUSE from pions which are generated in a graphite target impacted by a 1MW, 3GeV proton beam; the pions decay into muons with a half-life of 26 ns. These muons are collected by magnets and an intense pulsed muon beam is delivered to the muon experimental area.

The JAEA-Advanced Science Research Center (ASRC) is building the experimental apparatus for muon spin rotation/relaxation experiments at the J-PARC-MLF-MUSE facility. In this article, we present the current status of the ASRC muon research.

2. ASRC muon research

The ASRC muon research program was started in FY2005. The main purpose of our research is the study of f-electron materials using the μSR technique as a microscopic magnetic probe. In some f-electron systems, a tiny and fluctuating magnetic field plays the dominant role in determining the f-electron characteristics, and μSR has great sensitivity to detect these fields. The intense muon beams at J-PARC will allow us to understand the details of a specific magnetic state, and thus J-PARC has great merit in the study of f-electron systems.

For our ASRC μSR experiments a muon beam is extracted downstream of the Decay/Surface muon beam line of MUSE, as a branch of that beam line. A positive muon beam with momentum of 29MeV/c (surface muon) and a positive or negative decay muon beam with a momentum of 20-50MeV/c can be delivered to the experimental area (D1 area). This beam line has a "beam slicer" to obtain higher time resolution, as discussed below. Figure 1 shows a layout of the beamline which we

Journal of Physics: Conference Series **225** (2010) 012012 doi:10.1088/1742-6596/225/1/012012

have installed. The main components are two triplet quadrupole magnets and the beam slicer. This beamline is connected to the bending magnet (DB3) of the Decay/Surface muon line.

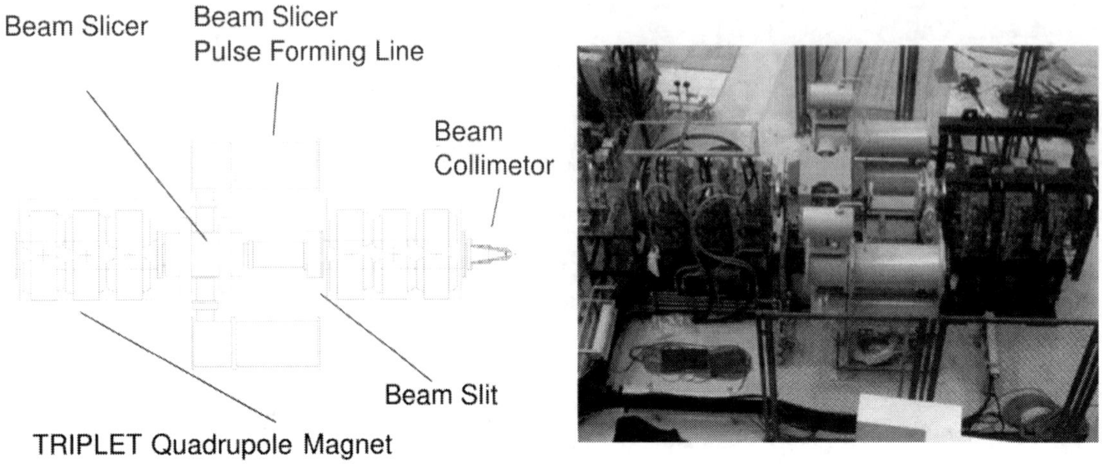

Figure 1. Layout of the ASRC equipment (top view) and photograph of D1 area.

Two sets of triplet quadrupole magnets are used as the focusing device. The maximum magnetic field gradient is 1.675 T/m at 375 A, which was determined from calculations of the beam optics and a desire to produce muons with momenta of up to 50MeV/c. The bore radius of the magnet is 300mm. Star-shape vacuum ducts are installed into each magnet. These magnets are connected to power supplies which are controlled from the Experimental Physics and Industrial Control System (EPICS) [1].

One of the main instruments is the "beam slicer." For pulsed-beam μSR experiments, the observable muon spin precession frequency is restricted by the width of the muon pulse. Since the time structure of the muon pulse is about the same as that of the proton pulse, the width of the muon pulse is longer than 100ns. In addition, the proton beam has two bunch structures, and this fact also suppresses the observable muon spin precession frequency. To overcome the limited time resolution which comes from the time structure of the proton beam, a muon beam slicer was installed. The beam slicer consists of an electric kicker, a pulsed electric power supply, which has a fast rise time, and a correction magnet. A rise time for the electric field less than 20ns (10-90%) and a pulse width of about 300ns have been achieved. This will allow a chopped muon pulse width of less than 30ns.

3. Beam Commissioning

On September 26, 2008, the first muon beam was delivered to the D1 area of J-PARC MUSE. Muon beam commissioning was carried out mainly with a μSR spectrometer (Dai-Omega) used previously at KEK-MSL. An imaging plate [2], or beam profile monitor, was also used to check the beam profile [4]. Recently, negative muons were successfully extracted at the D1 area and a negative muon spin rotation spectrum was observed. The details of the beam commissioning are reported in other articles [3,4].

We also examined the performance of the muon beam slicer. Figure 2 shows oscilloscope traces of the muon beam observed by an in-line scintillator. The beam is bent by the correction magnet located near the muon slicer. When an electric field is applied to the slicer, the muon beam returns to its original trajectory and can reach the scintillator. Here the electric field was synchronized to the timing of the second muon pulse so that only this second muon pulse could reach the muon spectrometer. This "single-pulse operation mode" is used for μSR measurements.

We have also succeeded in slicing the muon pulse itself. When an electric field is applied just before the second muon pulse goes away, only the tail of the pulse can reach the spectrometer and a narrower pulse is obtained. By using this "muon-slice mode," we have obtained a pulse width less than 37ns (FWHM). This muon-slice mode results in a loss of muons, however, so that a high-intensity muon beam is required to obtain sufficient muon intensity. In addition, as shown in Fig. 2, the electric-field pulse shape is not yet rectangular and development of the pulsed power supply is still ongoing. Details of the performance will be reported elsewhere.

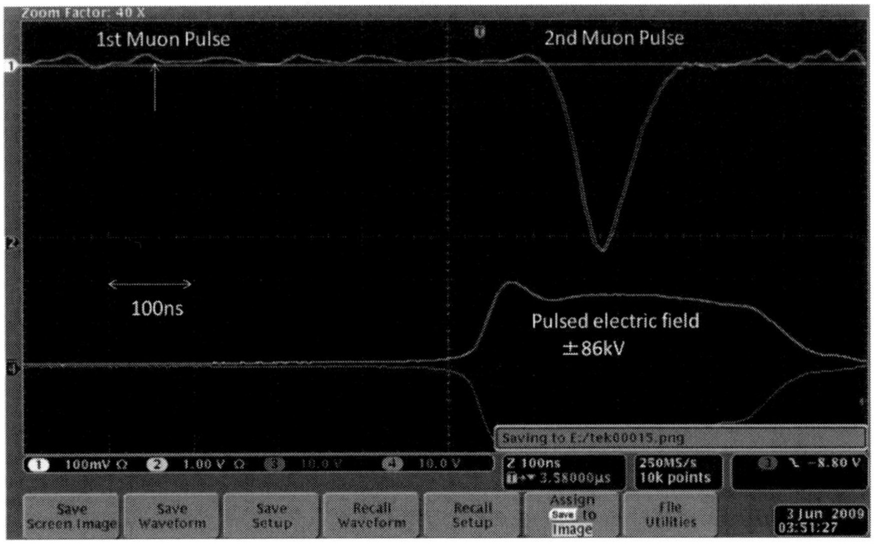

Figure 2. "Single-pulse operation mode" of the muon beam slicer. The first muon pulse is bent by a correction magnet and the second muon pulse is "kicked" by an electric field of ±86kV.

4. Experimental apparatus

We have also developed experimental apparatus for use at the D1 port. For example, a Gifford-McMahon refrigerator is used to cool materials down to 2K without liquid helium. The refrigerator possesses a large sample area and allows use of the "fly-pass method" for background reduction [5]. The development of a uniaxial pressure cell is also underway. The sample can be seen from outside of this cell and surface muons can be used with it, which is different from a hydrostatic pressure cell. Test experiments for this equipment will be performed soon.

5. Scientific Topics

One of the main subjects of the ASRC muon research program is the study of strongly correlated electron systems, especially f-electron systems. Examples of these topics include heavy fermions, anisotropic superconductivity, multipole ordering, physics in the vicinity of a quantum critical point (QCP), and unconventional magnetism (tiny moments, etc.). Studies of these topics are quite suitable for μSR experiments. For example, a small and fluctuating magnetic field is often seen near a QCP, and the high sensitivity of the muon probe is a good tool for such an investigation. The world's highest sensitivity will be obtained by using the intense muon beams of J-PARC, allowing us to obtain new insights into f-electron systems. Up to now we have already measured several Ce-, Sm- and Yb-based systems to clarify their dipole or multipole magnetic properties.

Recently, we have also performed muonic x-ray measurements at the D1 port. Negative muons of 19 MeV/c were implanted into sample gas to clarify the chemical effects of muonic atom formation processes, and several muonic x-rays were clearly observed. These details will be published elsewhere.

6. Summary

We have developed important experimental equipment which is essential for the J-PARC MUSE facility to carry out muon spin rotation/relaxation and muonic x-ray measurements. Our research will add new capabilities to MUSE, highly enhancing our knowledge of muon science.

Acknowledgement

The construction and commissioning was carried out in close collaboration with the KEK-Muon group. Muonic x-ray measurements are carried out collaboration with Professors A.Shinohara, K.M. Kubo, Drs. P. Strasser, N. Kawamura, T. Nagatomo and Ms.Kita.

References

[1] Ito T U, Nakahara K, Kawase M, Fujimori H, Kobayashi Y, Higemoto W, and Miyake Y 2010 *Journal of Physics:Conference Series* these proceedings

[2] Doyama M, Kogure Y, Inoue M, Miyake Y, Nishiyama K, Shimomura K, Nakahara K, Kawamura N, Strasser P and Kurihara T 2010 *Journal of Physics:Conference Series* these proceedings

[3] Miyake Y, Shimomura K, Kawamura N, Strasser P, Makimura S, Koda A, Fujimori H, Nakahara K, Takeshita S, Kobayashi Y,Nishiyama K, Higemoto W, Ito T U, Ninomiya K, Kato M, Kadono R, Sato N and Nagamine K 2010 *Journal of Physics:Conference Series* these proceedings

[4] Strasser P, Shimomura K, Koda A, Kawamura N, Fujimori H, Makimura S, Kobayashi Y,Nakahara K, Kato M,Takeshita S, Hiraishi M, Miyazaki M, Higemoto W, Ito T U, Ninomiya K, Ishida K, Kubo M K, Kadono R, Nishiyama K, and Miyake Y 2010 *Journal of Physics:Conference Series* these proceedings

[5] Lynch M C, Cottrell S P, King P J C and Eaton G H 2003 *Physica B* **326** 270.

Progress in development of new μSR spectrometer at RIKEN-RAL

D Tomono[1], T Kawamata[1], Y Hirayama[2,1], M Iwasaki[1,2], I Watanabe[1], K Ishida[1], T Matsuzaki[1]

[1] RIKEN Nishina Center, RIKEN, 2-1 Hirosawa, Wako, Saitama, 351-0198, Japan
[2] Department of Physics, Tokyo Institute of Technology, 2-12-1 Ookayama, Meguro, Tokyo, 152-8551, Japan

E-mail: tomono@riken.jp (Dai Tomono)

Abstract. A new μSR spectrometer has been developed at the RIKEN-RAL Muon Facility, UK. The spectrometer comprises two significantly advantageous devices of accurate magnets and multi-channel counter system for efficient use of the highly intense muon beam. A commissioning test to commence operation is in progress toward new μSR experiments under various conditions.

1. Introduction

A new research project to explore various matters under multiple extreme conditions, such as high pressure, low temperature, laser irradiation as an additional parameter combined with μSR technique, has recently been launched using the high intensity beam [1] at the RIKEN-RAL muon facility, UK [2]. In order to promote the project more intensively, a high performance and multi-channel spectrometer is newly designed in the Port-4 area. The new spectrometer has two significantly advantageous devices: one is precise Helmholtz magnets which can apply magnetic field to a sample with good homogeneity. They have a wide gap in which many kinds of apparatuses for making various conditions can flexibly be fabricated. The other is a new high-performance counting system for μ-e decay positrons/electrons. Since the intensity of primary proton beam becomes higher owing to upgrade of the ISIS proton synchrotron at the Rutherford Appleton Laboratory (RAL), an instantaneous muon beam rate reaches approximately 3×10^6 muons/sec. However, distortion of the time spectrum becomes severe due to pile-up of signals. The larger number of segments of the counter system than existing spectrometers can reduce the pile-up and then, the systematic error can be suppressed even in the high intense beam. In addition, μ-e decays from the sample can be observed selectively. This "direction-sensitive method" is employed to observe the long-time relaxation by reducing background particles.

This article concentrates on the design of these two devices and a current status of the development using the muon beam.

2. Spectrometer Design

2.1. Main Frame

Figure 1 shows a vertical cross section of the spectrometer comprising five pairs of Helmholtz magnets and electron/positron counters. A pair of large longitudinal field coils coaxially along with the muon beam, transverse field coils on top and bottom, and three pairs of zero

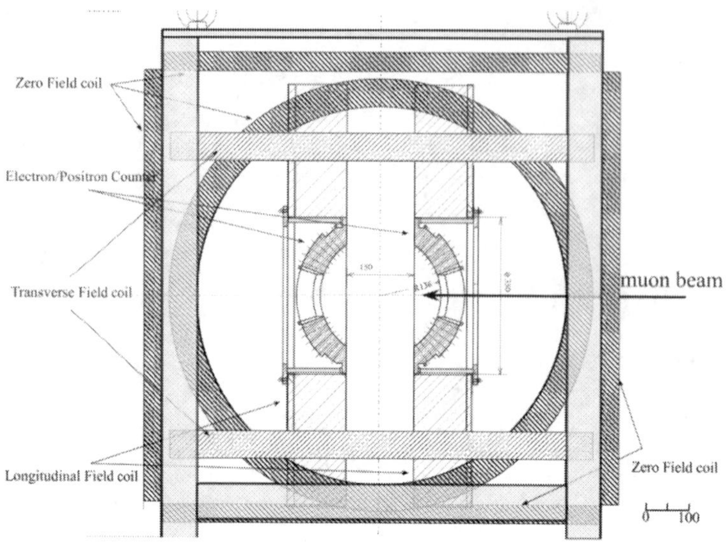

Figure 1. Schematic figure of the magnets and counter system in the spectrometer.

field coils around the sample were mounted in a square aluminum frame. Highly segmented electron/positron counters were mounted coaxially inside of the bore of the longitudinal field magnet. There was an option to rotate the frame of the magnets and counters at an angle of 90 degrees to the muon beam axis, enabling us to cancel the spin asymmetry of both sides of the counters. It was also available for a general purpose of μ-e decay detection.

2.2. Magnets

The longitudinal field coils could apply the external magnetic field up to 0.4 Tesla to the sample space of 150 mm gap in the muon spin direction. The transverse field coils could apply it up to 0.015 Tesla in order to rotate polarized muon spins on a horizontal plane. The other three pair of magnets were installed to cancel the magnetic field around the sample for precise zero field μSR measurements. The spatial homogeneity of the magnetic field is less than 10^{-4} at their maximal field around the sample volume from -15 mm to 15 mm of x, y and z coordinates.

2.3. Counter System

The spin direction of muons which were stopped in the sample was determined by two counter assemblies located at the core of longitudinal coils, which were very limited spaces upstream and downstream to the muon beam direction. A special counter system was newly designed to fit the longitudinal coils. Our requirements were a compact read-out to mount the narrow spaces of the core, a multi-channel counters with less pile-up under highly intense pulsed muon beam and no magnetic field effect for the photon detection from the scintillator, in particular, for photomultiplier tubes (PMT). For this purpose, we employed a spindle scintillator, wavelength shifting fibers, clear fibers and multi-anode photomultiplier tubes (MAPMTs) and then, optimized their size to maximize their performance of μSR spectrometers.

Figure 2 shows the counter system including spindle scintillators and optical fibers. The wavelength shifting fiber of 1.5 mm diameter (Kuraray Y-11(400)MS) was buried at the center of a spindle plastic scintillator ($10\times10\times40$ or 50 mm^3) coated with TiO_2 and tightly connected the clear fiber of 2.5 m length (Kuraray) as a light guide. The 16 clear fibers were bundled and connected to the 16ch MAPMT (H6568-10-200mod). The spindle counter is mounted to point to the center of the sample position. Our recent feasibility test of the prototype counters [3]

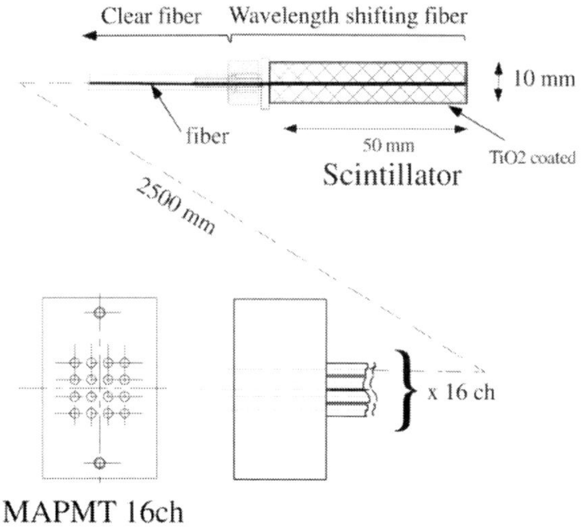

Figure 2. Schematic figure of the counter system composed of spindle scintillator, wavelength shifting fiber, clear optical fiber and MAPMT connection.

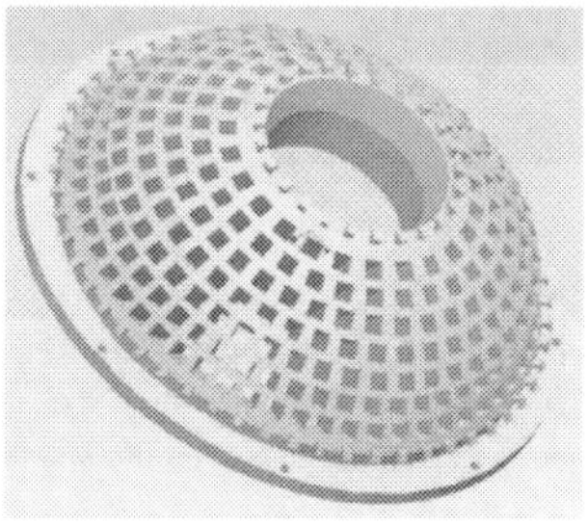

Figure 3. A special mold to house 303 counters which was attached to the longitudinal magnet.

showed that we can selectively observe positrons within 7 degrees from the sample by setting an appropriate threshold level of an analog output. It was determined in the commissioning test using muon beam for each counter.

The seven layers of counter were mounted coaxially along with the beam axis and in total 303 scintillator segmented the detection areas upstream and downstream, respectively. Figure 3 shows a special mold made of ABS to house counters, which mounted at the bore of the longitudinal magnet with aluminum supports. It is also noted that this large number of segments can reduce count-loss due to pile-up of positrons hitting the counter sequentially, resulting in suppressing deviation, at least, in less than 1% of the time spectrum at the time zero just after the muon beam arrival, which is calculated by a present positron hit rate per counter.

Moreover, the wavelength shifting fiber and the optical fiber could guide photons from the scintillator to the MAPMT which was remote from the magnetic field below the magnets. The MAPMT detected approximately 70 photons in average including efficiencies of scintillator variability, photon collection efficiency of WLS fiber, its transmission, fiber connection and MAPMT photocathod. It was covered by double-layered magnetic field shield made of iron and permalloy. It could suppress the magnetic field in less than 0.01 Tesla at the maximal field of the longitudinal magnet of 0.4 Tesla and the remaining magnetic field coming from the magnetized iron layer could be canceled by the permalloy layer in a zero field measurement.

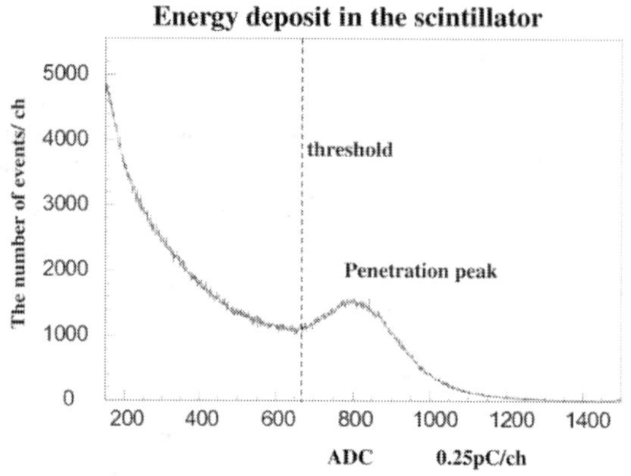

Figure 4. A typical spectrum of energy deposit in a scintillator in a prototype detector.

2.4. Data Acquisition
A new data acquisition to dedicate the multi-channel counter system was built to accumulate large amount of data from the counters. The VME-based 128ch multi-hit TDC (Kaen V1190A) was operated with Linux software. The front end software (MACS/Windows [4]) controlled overall of the data acquisition system and slow control system for temperature and magnet control modules. A detail of the DAQ system is reported in Reference [5].

3. Counter Commissioning
A counter commissioning test using muon beam is being performed with an ADC module built into the new data acquisition system. Firstly, an energy deposit of decay positron for each counter was measured. Since analog pulse outputs from counters had distinctive variations, a threshold level of the discriminator was individually determined by observing an energy deposit spectrum. As shown in the prototype test [3], the counter system could be work as a direction sensitive counter if the threshold level is set at the 70 % of this peak height. Figure 4 shows a spectrum of a prototype counter for the new spectrometer. It shows that a peak on the decreasing component is made by the positrons which penetrates through scintillator in a spindle direction. These positrons are particularly chosen by setting the appropriate threshold level. It enables us to discriminate the signals from a background component in the muon decay time spectrum, called "direction-sensitive method". At present, this study for all counters of new spectrometer is underway.

4. Summary
The new spectrometer was designed and commissioning test is now performed at Port-4, RIKEN-RAL. At present, commissioning of the spectrometer and preparation of apparatus such as cryogenic system are in progress. A various remote control system is now being improved to commence operation toward various new μSR experiments.

References
[1] Watanabe I, *et al.*, *Physica* B **404** 993.
Suzuki T, *et al.*, *Phys. Rev.* B **80** 064407.
Suzuki T, *et al.*, these proceedings.
Ishii Y, *et al.*, these proceedings.
[2] Matsuzaki T, *et al.*, *Nucl. Inst. and Meth.* A **465** 365.
[3] Tomono D, *et al.*, *Nucl. Inst. and Meth.* A **600** 44.

[4] Pratt F.L, *RIKEN-RAL Report,* **4** 26.
[5] Kawamata T, *et al.*, these proceedings.

International Symposium on Advanced Science Research 2009 IOP Publishing
Journal of Physics: Conference Series **225** (2010) 012026 doi:10.1088/1742-6596/225/1/012026

Generation of negative slow muon beam in J-PARC Muon Facility

Naritoshi Kawamura, Tetsuya Masuda, Patrick Strasser, Yasuhiro Miyake

Muon Science Section, Materials and Life Science Division, J-PARC Center, Tokai-mura, Ibaraki, 319-1195, Japan

E-mail: nari.kawamura@kek.jp

Abstract. The world strongest pulsed muon facility has been operated since 2008 in J-PARC (Japan Proton Accelerator Research Complex). This facility utilizes 3-GeV proton to produce muon beam, and thus the negative pion and also negative muon yields are superior to the other meson factories in the world. We try to slow down these negative muons. Negative slow muon beam is desired to check the standard model and search a new physical rule, as well as various applications in material science.

1. Introduction

The intensity of proton accelerator has attained to the order to mega-watt, and several MW-class proton accelerators start to operate in the world. A high intensity proton accelerator (Japan Proton Accelerator Research Complex, J-PARC) project was proposed jointly by Japan Atomic Energy Agency (JAEA) and High Energy Accelerator Research Organization (KEK) [1]. The new facility consists of a 400-MeV linac, 3-GeV and 50-GeV synchrotron rings, which provide an intense proton beam to generate a variety of secondary beams, *i.e.* muon beam, neutron beam, meson beam and neutrino beam for the purpose of particle physics, nuclear physics, material science, life science and nuclear technology. The muon facility established in order to provide the pulsed muon beam for various experimental programs.

The J-PARC Muon Facility succeeded to transport the first muon beam to the experimental area in September 2008 [2]. Although the proton beam power is still one fourth of its design value of 1 MW (3GeV, 333μA), the beam power increases step by step and is almost comparable to ISIS Facility (800MeV, 200μA) in the UK at the present. In comparison with ISIS beam energy of 800 MeV, in J-PARC, 3 GeV is about four times higher. The negative pion production cross section increases steeply above 1GeV, while positive one does not show a big change [7]. Thus, the negative pion yield in J-PARC Muon Facility is higher than the value simply scaled up by the beam current. Therefore, the negative muon characterizes J-PARC Muon Facility.

The other characteristic point in J-PARC Muon Facility is ultra slow muon beam by laser resonant-ionization (LRI) method [3, 4], by which positive muons almost at rest are dissociated from thermal muoniums emitted from hot tungsten surface. LRI method has been developed at KEK-MSL [3, 4] and the RIKEN-RAL Muon Facility. Based upon their developments, in J-PARC Muon Facility, the design work of ultra slow beamline by LRI is in progress for practical use. The key technique of this method is the intense laser synchronized with the muon pulse. In general, the time resolution of signals form pulse muon beam is inferior to that from continuous muon beam due to the finite width of the muon-beam pulse. However, the time width of ultra slow muon beam by LRI is determined by the width and spot size of laser, and thus is able to be quite narrow. Furthermore, the beam emittance by LRI is also able to be quite small. The LRI method provides not only low energy positive muons but also high quality beam with low emittance and low time and momentum dispersion.

This superior LRI method is available only to a positive muon and not to a negative muon, because a positive muon forms a muonium which is dissociated by laser, while a negative muon forms a muonic atom. A negative muon has 200-times higher binding energy than a normal atom, in which a negative muon waits for the decay to an electron and two neutrinos or a competitive process of nuclear capture. In the case of a heavier nucleus, the nuclear-capture process becomes dominant, and the effective lifetime of muon becomes much shorter than the free-decay lifetime of 2.2 μs. Therefore the behavior of negative muon in matter is much different from positive one, and LRI method is not applicable.

An approach to negative slow muon beam was performed by applying Ramsauer-Townsend effect in solid hydrogen [5]. Namely, a muon stopped in solid hydrogen with 1000-ppm deuterium forms a muonic deuterium, and its mean free path reaches 1 mm due to Ramsauer-Townsend effect. From a 1-mm film of solid hydrogen, a muonic deuterium can be extracted. Although a high energy laser like a free electron laser is required to dissolve a muonic deuterium with a a-few-keV binding energy, the muon is able to be released by leading a muonic deuterium to solid deuterium, where a muon catalyzed fusion phenomenon is expected to occur and a 10-keV muon is released. The efficiency of slowing down by this method is quite small, and is not suitable for applying a practical beamline, although it is a novel idea.

In order to generate slow negative muon beam, *i.e.* low energy and low emittance beam, from the world-intense negative muon beam in J-PARC, we start a fundamental study.

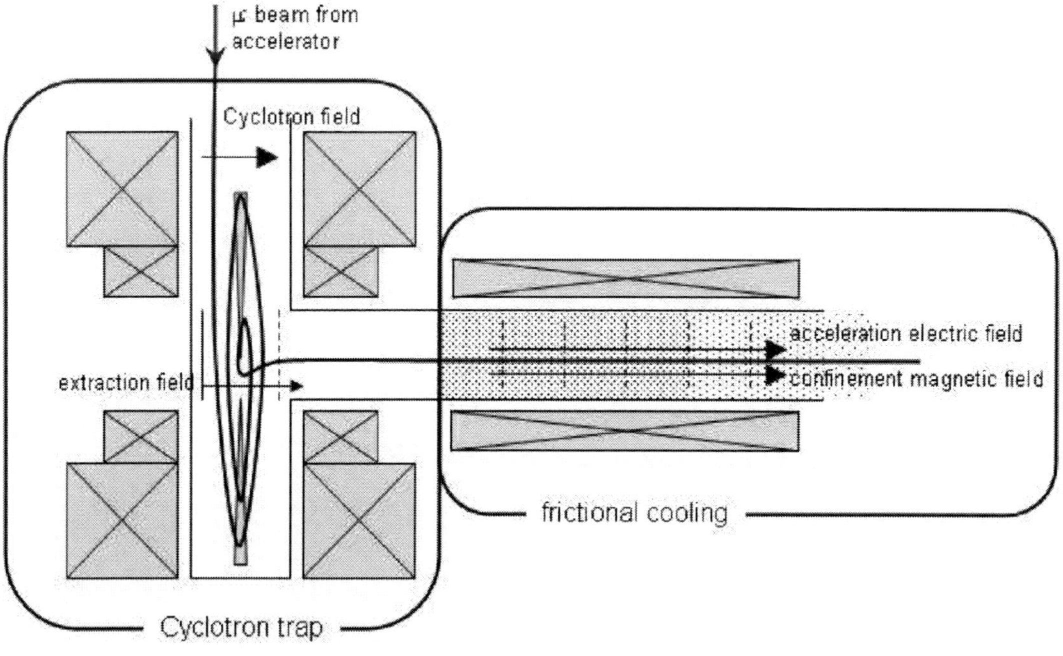

Figure 1. A schematic figure of negative slow muon setup in J-PARC Muon Facility

2. Negative muon slowing down

Technically, slowing down of a negative muon is much difficult in comparison with positive one. We can not use any cooling method applied to positive muons like the degrader method established at PSI [6] nor LRI method. Furthermore, muon has a 2.2-μs life time, which is too short to apply any cooling methods for any other particles with infinite life time. Slowing down have to be complete enough earlier than the muon life time to free-decay loss. About 10% of incident muon is lost by free-decay every 200 ns.

For the first step of negative slow muon production at J-PARC Muon Facility, we will use a cyclotron to trap a few MeV muons from pion decay. Simons *et al.* succeeded to trap cloud muons, 4MeV μ^-, and slowed them down to the range of around 10 keV [8]. Negative muons are injected toward the outer side of the weak-focusing cyclotron field and are bound into the orbit where they lose the energies through moderators placed at the mid-plane of cyclotron. According to the energy loss, the orbit gets smaller and smaller, and the low energy, up to a few tens of keV, muons are accumulated around the center of the cyclotron. By a pulse kicker device, the accumulated muons are extracted to the axis direction of cyclotron and are injected into a frictional cooling chamber. As reported by Mühlbauer *et al.* [9], the frictional cooling method, moderation in matter and acceleration in an electric field, is well match with connecting to the cyclotron trap. The balanced energy is typically a few keV, and could be around 1 keV.

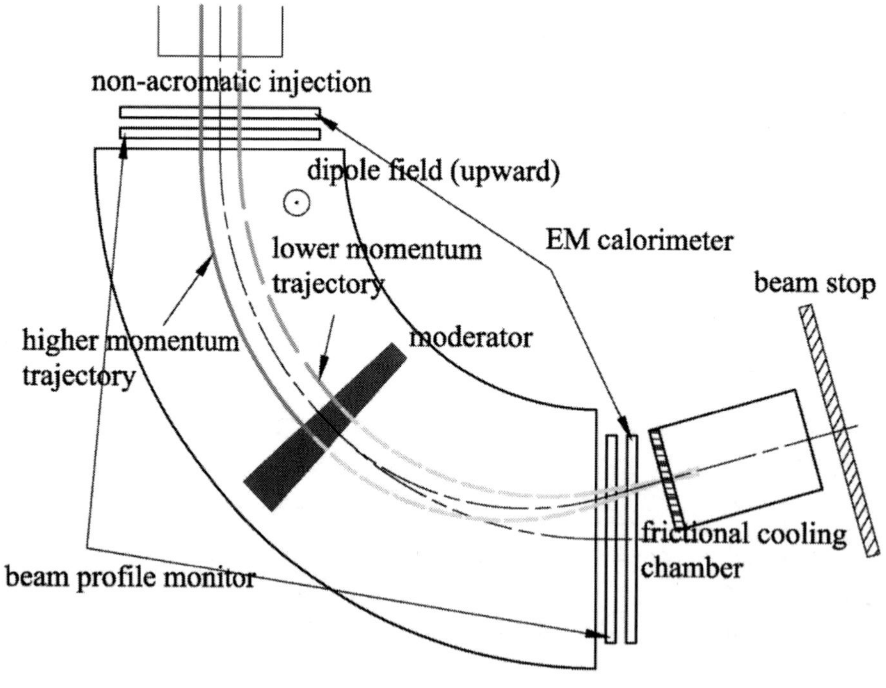

Figure 2. A schematic figure of the setup for test experiment

Simons developed a cyclotron trap [8], by which about 0.3% of 4-MeV muons were converted to 5-25 keV. Frictional cooling method becomes efficient below about 100 keV, and the conversion efficiency to a few keV was estimated to be about 30% [9]. Assuming a negligible transmission loss between a cyclotron trap and a frictional cooling chamber, 0.1% of conversion efficiency are expected. Practically, both cyclotron trap and frictional cooling chamber are axial focusing elements, and thus low-loss extraction and injection require well-studied magnetic field design. A pulse extraction kicker in the cyclotron trap is one of the key devices. Making the transmission efficiency higher, the cyclotron and the frictional cooling chamber is planed to be placed just beside (Fig. 1).

In J-PARC Muon Facility, 30-MeV/c μ^- rate is expected to be above 10^5/s for 1-MW proton beam, and this yields 10^2 keV-muons per second. The yield above 100 MeV/c is two orders of magnitude higher than that of 30 MeV/c, and thus low-loss slowing down from 100-MeV/c μ^- is important to increase the keV-muon rate. By slowing down from higher momentum, the beam becomes more dispersive and more difficult to be trapped in the cyclotron. The optimization of the moderator and cyclotron field is important in the design work. The setup for the test experiment starts to construct in J-PARC, as shown in Fig. 2. The first step is to check the effect of moderators placed in the magnetic field. For this purpose, a dipole magnet will be used, and several moderators to be placed in the cyclotron containing rare-gases solids will be examined.

3. Conclusions and Discussion

The most challenging and attractive usage of the world strongest negative muon in J-PARC is slowing them down. Connecting an intense positron source with the present setup, creation of anti-muoniums, *i.e.* $(\mu^- e^+)$, is one of the goal for checking the CPT theory and search a new physical rule beyond the standard model. Not only particle physics but also material science requires a high quality beam with low emittance and low energy. As the ultra slow positive muon beam pioneers a new application of muon use, slow negative muon also opens new muon science. Tomography by muonic X-ray is one of possible application by using high quality negative slow muon beam of which stopping position is easily controllable with high precision.

References

[1] The Joint Project for High-Intensity Proton Accelerators, KEK Report 99-4; JAERI-Tech 99-056; JHF-99-3 (1999).
[2] Y. Miyake *et al.*, *ibid.*
[3] K. Nagamine *et al.*, Phys. Rev. Lett. **74** 4811 (1995).
[4] Y. Miyake *et al.*, J. Nucl. Sci. Technol. **39** 287 (2002).
[5] P. Strasser *et al.*, Phys. Lett. B **368** 32 (1996).
[6] E. Morenzoni *et al.*, Hyperfine Interactions **106** 229 (1997).
[7] B.J. VerWest *et al.*, Phys. Rev. C **25** 1979 (1982).
[8] L.M. Simons, Hyperfine Interactions **81** 253 (1993).
[9] M. Mühlbauer *et al.*, Hyperfine Interactions **119** 305 (1999).

AUTHOR INDEX

AUTHOR INDEX

AUTHOR INDEX